Brock/Springer Series in Contemporary Bioscience

Drinking
Water
Microbiology

Brock/Springer Series in Contemporary Bioscience

Series Editor: Thomas D. Brock
University of Wisconsin–Madison

Tom Fenchel
ECOLOGY OF PROTOZOA: The Biology of Free-living
Phagotrophic Protists

Johanna Döbereiner and Fábio O. Pedrosa
NITROGEN-FIXING BACTERIA IN NONLEGUMINOUS
CROP PLANTS

Tsutomu Hattori
THE VIABLE COUNT: Quantitative and Environmental Aspects

Roman Saliwanchik
PROTECTING BIOTECHNOLOGY INVENTIONS:
A Guide for Scientists

Hans G. Schlegel and Botho Bowien (Editors)
AUTOTROPHIC BACTERIA

Barbara Javor
HYPERSALINE ENVIRONMENTS: Microbiology and Biogeochemistry

Ulrich Sommer (Editor)
PLANKTON ECOLOGY: Succession in Plankton Communities

Stephen R. Rayburn
FOUNDATIONS OF LABORATORY SAFETY: A Guide for the
Biomedical Laboratory

Gordon A. McFeters (Editor)
DRINKING WATER MICROBIOLOGY: Progress and
Recent Developments

Gordon A. McFeters
Editor

Drinking Water Microbiology

Progress and Recent Developments

With 80 Figures and 99 Tables

Springer-Verlag
New York Berlin Heidelberg
London Paris Tokyo Hong Kong

Gordon A. McFeters
Department of Microbiology
Montana State University
Bozeman, Montana 59717

Library of Congress Cataloging-in-Publication Data
Drinking water microbiology : progress and recent developments /
   [edited by] Gordon A. McFeters.
      p.   cm. — (Brock/Springer series in contemporary bioscience)
   Includes bibliographical references.
   ISBN 0-387-97162-9 (alk. paper)
   1. Water—Microbiology. 2. Drinking water—Analysis.
   I. McFeters, Gordon A. II. Series.
TD384.D75 1990
628.1'6—dc20   89-22041

Production and editorial supervision: Science Tech Publishers
Printed and bound by Edwards Brothers, Ann Arbor, Michigan.
Printed in the United States of America.

9 8 7 6 5 4 3 2 1

ISBN 0-387-97162-9 Springer-Verlag New York Berlin Heidelberg
ISBN 3-540-97162-9 Springer-Verlag Berlin Heidelberg New York

Dedication

This volume is dedicated to Edwin E. Geldreich who continues to be a positive worldwide influence in the area of drinking water microbiology. His efforts have effectively fostered and guided research in the field toward the solution of the most critical concerns relating to the microbiology of drinking water. He has also fostered sharing of research findings among individuals, organizations, agencies, and governments at national and international levels. With more than 40 years of research experience, he has directed investigations into the microbiology of water supplies, treatment processes, methods development, criteria, and standards. He has been a frequent advisor for the World Health Organization on projects worldwide. A member of the American Water Works Association (AWWA), American Society for Microbiology, and the International Association on Water Pollution Research and Control, Geldreich has received the Kimble Methodology Research Award, the United States Environmental Protection Agency's Silver and Bronze medals, AWWA's Research Award, and the 1989 Abel Wolman Award of Excellence. He has published numerous articles in research journals. His interest in the past 15 years has focused on treatment barriers and problems in water supply distribution systems—their cause and control. It is an honor that he has agreed to contribute important chapters to the present book.

v

Preface

The microbiology of drinking water remains an important worldwide concern despite modern progress in science and engineering. Countries that are more technologically advanced have experienced a significant reduction in waterborne morbidity within the last 100 years. This reduction has been achieved through the application of effective technologies for the treatment, disinfection, and distribution of potable water. However, morbidity resulting from the ingestion of contaminated water persists globally, and the available epidemiological evidence (*Waterborne Diseases in the United States*, G.F. Craun, ed., 1986, CRC Press) demonstrates a dramatic increase in the number of waterborne outbreaks and individual cases within the United States since the mid-1960s. In addition, it should also be noted that the incidence of waterborne outbreaks of unknown etiology and those caused by "new" pathogens, such as *Campylobacter* sp., is also increasing in the United States. Although it might be debated whether these increases are real or an artifact resulting from more efficient reporting, it is clear that waterborne morbidity cannot be ignored in the industrialized world. More significantly, it represents one of the most important causes of illness within developing countries. Approximately one-half the world's population experiences diseases that are the direct consequence of drinking polluted water. Such illnesses are the primary cause of infant mortality in many Third World countries. These realities have led to the dedication of the period 1981 to 1990 as the International Drinking Water Supply and Sanitation Decade by the United Nations. It is appropriate that this volume provides a review of recent and significant studies relating to the microbiology of drinking water at the end of this decade.

The purpose of this book is to provide individuals in the numerous disciplines that deal with water quality with a single source of information concerning recent research advances. To that end, authors were asked to provide an overview of their findings in a format useful for the rather wide range of interests represented in the intended audience. The resulting collection of monographs and reviews is of value to microbiologists, engineers, epidemiologists, sanitarians, health officials, and scientists within governmental and international agencies, as well as others who deal with issues relating to the microbiology of drinking water. The diversity of backgrounds and interests represented in these various disciplines suggests that many readers are not likely to be familiar with the primary scientific literature covered in all of the chapters. For that reason, the authors were requested to present their results in some depth, as well as provide context in their chapters. Therefore, readers will find information that will be useful in directing research efforts and explaining field observations from the drinking water industry. Additionally, this book will provide guidance on matters relating to policy decisions and regulations. It will also serve as a reference for students in specialized courses concerning the microbiology of drinking water, although it assumes some knowledge of basic environmental microbiology. Practical considerations have limited the number of participants in this effort, so the scope of this volume is not intended to be exhaustive or complete.

Most of the findings described in this volume have been generated in laboratory and field studies during the past 10 to 15 years, a period of vigorous research activity. More recently, reduced support for research on drinking water in the United States has forced many active laboratories to turn to other areas of investigation. This observation suggests that now is the right time for this volume to summarize and review the status of the field.

This volume has been organized into five parts. Part 1 deals with the microbiology of source water, and Part 2 discusses the microbiology of drinking water treatment. Both of these sections address issues relevant to conditions in and technologies appropriate for both industrialized and developing countries. Part 3 includes chapters concerning the distribution of drinking water, and Part 4 deals with the occurrence and control of some pathogenic microorganisms of importance in drinking water. Finally, Part 5 deals with microbiological and statistical methods used to monitor water quality.

Providing safe drinking water is a complex multidisciplinary task requiring the collective efforts of individuals from a wide variety of disciplines and backgrounds. The intent of this volume is to provide useful overviews of some of the more important recent scientific contributions in the area of drinking water microbiology. Hopefully, it will assist in meeting the critical objective of making safe drinking water available worldwide.

Gordon A. McFeters
Bozeman, Montana

Contents

Contributors

Jean Claude Block Centre des Sciences de l'Environnement, Université de Metz, 1 rue des Recollets, 57000 Metz, France

William G. Characklis Institute for Biological and Chemical Process Analysis, Montana State University, Bozeman, MT 59717, USA

Malay Chaudhuri Department of Civil Engineering, Indian Institute of Civil Engineering, Kanpur 208016, India

James A. Clark Ministry of the Environment, 25 Resources Road, Box 213, Rexdale, Ontario M9W 5L1, Canada

Louis E. Conley City of Pittsburgh Water Department, 226 Delafield Road, Pittsburgh, PA 15215, USA

A.H. El-Shaarawi Canada Centre for Inland Waters, P.O. Box 5050, 867 Lakeshore Road, Burlington, Ontario L7R 4A6, Canada

Edwin E. Geldreich U.S. Environmental Protection Agency, 26 West St. Clair Street, Cincinnati, OH 45268, USA

Charles P. Gerba Department of Microbiology and Immunology, University of Arizona, Tucson, AZ 85721, USA

W.O.K. Grabow Department of Medical Virology, University of Pretoria, P.O. Box 2034, Pretoria 0001, South Africa

Charles N. Haas Pritzker Department of Environmental Engineering, Illinois Institute of Technology, Chicago, IL 60616, USA

Carrie M. Hancock CH Diagnostic and Consulting Service, Inc., 2012 Derby Court, Fort Collins, CO 80526, USA

S. Harakeh Linus Pauling Institute of Science and Medicine, 440 Page Mill Road, Palo Alto, CA 94306, USA

Terry C. Hazen Savannah River Laboratory, Westinghouse Savannah River Company, Aiken, SC 29808, USA

Barbara R. Heller Pritzker Department of Environmental Engineering, Illinois Institute of Technology, Chicago, IL 60616, USA

Charles P. Hibler CH Diagnostic and Consulting Service, Inc., 2012 Derby Court, Fort Collins, CO 80526, USA

John M. Kuchta City of Pittsburgh Water Department, 226 Delafield Road, Pittsburgh, PA 15215, USA

Mark W. LeChevallier American Water Works Service, 1115 S. Illinois Street, Belleville, IL 62220, USA

Richard V. Levy Millipore Corporation, Bedford, MA 01730, USA

Gary S. Logsdon U.S. Environmental Protection Agency, 26 W. St. Clair Street, Cincinnati, OH 45268, USA

Abdul Matin Department of Microbiology and Immunology, Stanford University, Stanford, CA 94305, USA

A. Maul Centre des Sciences de l'Environnement, Université de Metz, 1 rue des Récollets, 57000 Metz, France

Gordon A. McFeters Department of Microbiology, Montana State University, Bozeman, MT 59717, USA

Wesley O. Pipes Department of Civil Engineering, Drexel University, 32nd and Chestnut Streets, Philadelphia, PA 19104, USA

Donald J. Reasoner Drinking Water Research Division, U.S. Environmental Protection Agency, 26 West St. Clair Street, Cincinnati, OH 45268, USA

Joan B. Rose Department of Environmental and Occupational Health, College of Public Health, University of South Florida, Tampa, FL 33612, USA

Syed A. Sattar Department of Microbiology and Immunology, School of Medicine, University of Ottawa, 275 Nicholas Street, Ottawa, Ontario K1H 8M5, Canada

Donald A. Schiemann Department of Microbiology, Montana State University, Bozeman, MT 59717, USA

Ajaib Singh Delaware Department of Public Health, P.O. Box 618, Dover, DE 19903, USA

Stanley J. States City of Pittsburgh Water Department, 226 Delafield Road, Pittsburgh, PA 15215, USA

Gary A. Toranzos Department of Biology, University of Puerto Rico, Rio Piedras, PR 00931, USA

Dirk van der Kooij Biology Division, KIWA Ltd., P.O. Box 1072, 3430 BB Nieuwegein, The Netherlands

Ewout van der Wende Institute for Biological and Chemical Process Analysis, Montana State University, Bozeman, MT 59717, USA

Robert M. Wadowsky School of Medicine, University of Pittsburgh, Pittsburgh, PA 15261, USA

Randy S. Wolford City of Pittsburgh Water Department, 226 Delafield Road, Pittsburgh, PA 15215, USA

Robert B. Yee Department of Infectious Disease and Microbiology, University of Pittsburgh, Pittsburgh, PA 15261, USA

Drinking
Water
Microbiology

Part 1

Microbiology of Source Water

1

Microbiological Quality of Source Waters for Water Supply

Edwin E. Geldreich

Environmental Protection Agency, Cincinnati, Ohio

Protection of drinking water from contamination by human or other animal excrement in sewage, food processing wastes, and stormwater runoff is of paramount importance to everyone. Public health concerns must also include consideration of the availability of a continuous supply since water is essential to sustain life, with a prime consideration given to the production of water that is free of pathogenic agents and significant levels of toxic chemicals and is aesthetically pleasing in taste and appearance. More often than not, these basic requirements involve protection of water sources, conservation of water resources, and treatment to varying degrees to achieve the desired objective—a safe, continuous supply of drinking water.

1.1 Origins of Source Water Pollution

Surface Water Contaminants   Surface water quality is subject to frequent, dramatic changes in microbial quality as a result of a variety of activities on a watershed. These changes are caused by the discharges of municipal raw waters or treated effluents at specified point source locations into receiving waters (river, lake) or by stormwater runoff into the drainage basin at nonpoint locations all over the watershed.

While sewage collection systems have decreased public health risk in urban communities, wastewater collections only serve to divert the massive sewage contributions (Table 1.1) to a single selected focal point, where treatment must be applied to minimize the release of pathogens into a watercourse that is the water resource for other downstream users. Sewage discharges to receiving waters have often been shown to contain a variety of pathogens, with the density and variety being related to the size of population served, seasonal

Table 1.1 Composition of indicator organisms in raw sewages from various communities

Sewer Source	Estimated sewered population	Densities per 100 mL			Percent FC	Ratio FC/FS
		Total coliforms	Fecal coliforms (FC)	Fecal streptococci (FS)		
Esparto, California	—	23,500,000	6,200,000	—	26.4	—
Shastina, California	—	9,600,000	2,300,000	—	24	—
Los Baños, California	6,090	62,000,000	23,000,000	—	37	—
Anoka, Minnesota	9,500	47,400,000	10,200,000	—	21.5	—
Newport, Minnesota	800	13,600,000	3,580,000	—	26.3	—
Red Wing, Minnesota	11,000	17,700,000	4,050,000	—	22.9	—
Mankato, Minnesota	81,490	5,525,000	2,630,000	—	47.6	—
Oakwood Beach, New Jersey	45,000	13,250,000	4,240,000	—	32.0	—
Perth Amboy, New Jersey	38,000	1,600,000	387,000	—	24.2	—
Middlesex, New Jersey	300,000	12,900,000	1,070,000	—	8.3	—
Keyport, New Jersey	5,600	2,210,000	641,000	—	29.0	—
Omaha, Nebraska	180,000	45,800,000	5,360,000	—	11.7	—
Anderson Township, Ohio	11,000	17,200,000	4,600,000	—	26.7	—
Mt. Washington, (Cincinnati), Ohio	20,000	34,800,000	4,900,000	—	14.1	—
Linwood (Cincinnati), Ohio	22,000	—	10,900,000	2,470,000	—	4.4
Preston, Idaho	3,640	—	340,000	64,000	—	5.3
Fargo, North Dakota	50,500	—	1,300,000	290,000	—	4.5
Moorehead, Minnesota	22,934	—	1,600,000	330,000	—	4.9
Lawrence, Massachusetts	67,000	—	17,900,000	4,500,000	—	4.0
Monroe, Michigan	22,968	—	19,200,000	700,000	—	27.9
Denver, Colorado	520,000	—	49,000,000	2,900,000	—	16.9
Average values		21,900,000	8,260,000	1,610,000	37.7	5.1

Data from Geldreich (1978).

Table 1.2 Microbial densities in municipal raw sewage

	Average count per 100 ml	
	Worcester sewage	Pietermaritzburg sewage
Aerobic plate count (37 °C; 48 hr)	1,110,000,000	1,370,000,000
Total coliforms	10,000,000	—
E. coli type 1	930,000	1,470,000
Fecal streptococci	2,080,000	—
C. perfringens	89,000	—
Staphylococci (coagulase positive)	41,000	28,100
Ps. aeruginosa	800,000	400,000
Salmonella	31	32
Acid-fast bacteria	410	530
Ascaris ova	16	12
Taenia ova	2	9
Trichuris ova	2	1
Enteroviruses and reoviruses ($TCID_{50}$)	2,890	9,500

Data from Grabow and Nupen (1972) for two cities in South Africa

patterns for certain diseases, and the extent of community infection. Some indication of the relative occurrences of various pathogens in raw sewage are shown by the representative data in Table 1.2 (Grabow and Nupen, 1972). Since not all pathogens were assessed (due to lack of appropriate methods), these findings in sewage represent only a portion of the potential health threat. If socioeconomic factors and epidemic eruptions are overlayed on these normal occurrences of pathogen discharge, the potential magnitude of the problem is clearly much greater.

It has been suggested that a sewage collection network of 50 to 100 homes is the minimal size required before there is a reasonable chance of successfully detecting salmonellae in the wastewater collection system (Callaghan and Brodie, 1969). *Salmonella* strains were regularly found in the sewage system of a residential area of 4,000 persons (Harvey, 1969). With appropriate methodology, presence of viruses in these small systems should be seen on a seasonal basis.

Food, beverage, meat packing, wood pulp, and paper wastes may contain substantial concentrations of organic nutrients and some pathogens if not properly treated. *Salmonella* is often introduced directly or indirectly with this material through fecal droppings of rodents or birds foraging in the source materials or from addition of sanitary wastes to process wastewaters at the plant (Table 1.3). Segregating plant sanitary wastes is helpful but treatment of both types of discharges to reduce excessive organic and microbial loading to receiving waters should be a major environmental concern to protect downstream uses of these waters for public water supply source. Without significant nutrient reductions in appropriate treatment operations, there will be little improvement in bacterial loading to the receiving waters (Table 1.4)

Table 1.3 Enteric bacterial profiles in raw industrial wastes

	Wood pulp and paper	Beverage	Food processing	Meat processing
Percent of various coliform species				
E. coli	0.4	5.6	35.0	56.9
Klebsiella pneumoniae	92.3	68.0	55.0	21.5
Enterobacter species	6.7	15.0	3.3	13.8
Pectobacterium	0.6	7.0	6.0	0.5
Salmonella strains	0.008	4.4	0.7	7.3
Viable counts/100 ml				
Klebsiella	10^5–10^6	10^4–10^7	10^3–10^5	10^5–10^8
Fecal coliform	10^1–10^4	10^3–10^5	10^2–10^4	10^6–10^9

Data from Herman (1972)

Table 1.4 Enteric bacterial profiles in treated industrial waste effluents

Waste effluent	Klebsiella	E. coli.	Entero-bacter	Pecto-bacterium	Salmonella
	%				
Wood pulp & paper	85.0	4.4	9.5	0.8	0.3
Potato	81.1	0.9	9.4	0.9	1.6
Beverage production	68.9	5.6	15.0	7.0	4.4
Food processing	55.0	24.0	3.3	6.0	0.7
Meat processing	21.5	56.9	13.8	0.5	7.3
Municipal sewage	18.0	62.0	14.3	3.6	2.1

Data adapted from Herman (1972). Numbers are percent of each species, as determined by viable counts.

Stormwater is often a major cause of source water quality deterioration and relates to all land uses over the drainage basin. Each storm event brings elevations in suspended solids, organic demand materials, and organisms to the drainage basin. Rainfall over the watershed introduces very little nutrients in terms of nitrogen and carbon and insignificant densities of bacterial contaminants (Table 1.5). Source for the bacteria in rainfall is dust particles swept up in windstorm events and carried hundreds of miles in the upper atmosphere. Obviously, the high quality of rainfall deteriorates upon passage through layers of air pollutants and in final contact with the earth's surface.

Characterizing stormwater from small urban and rural sites (Table 1.6) reveals wide fluctuations in quality that reflect human activities over the watershed, as well as the magnitude and frequency of storm events (Weibel et al., 1964; Geldreich et al., 1968). Scaling up the magnitude of the contribution

Table 1.5 Selected characteristics for rainwater

Parameter	Season			
	Spring	Summer	Autumn	Winter
Nitrogen (mg/L)	0.93	0.83	0.59	0.49
Carbon (mg/L)	<0.01	0.05	0.01	0.05
pH	4.5	5.6	5.6	5.6
Total coliforms*	<1.0	<1.0	<0.4	<0.8
Fecal coliforms*	<0.3	<0.7	<0.4	<0.5
Fecal streptococci*	<1.0	<1.0	<0.4	<0.5

Data adapted from Geldreich et al. (1968).
* Median densities per 100 ml

Table 1.6 Selected characteristics for stormwater runoff

Parameter	Urban (27 acre—Residential)		Rural (1.45 acre—winter wheat)	
	Range	Mean	Range	Mean
Turbidity	30–1,000	70	—	—
Color	10–380	81	—	—
pH	5.3–8.7	7.5	—	—
Suspended solids (mg/l)	5–1,200	227	5–2,074	313
Volatile suspended solids (mg/l)	1–290	53	—	—
BOD (mg/l)	1–173	17	0.5–23	7
Bacterial indicators (densities per 100 ml)				
Total coliforms	22,000–190,000	110,000	4,400–58,000	31,000
Fecal coliforms	2,500– 40,000	21,000	55–9,000	4,500
Fecal streptococcus	13,000– 56,000	35,000	2,100–790,000	400,000

Data from Weibel et al. (1964, 1965) and Geldreich et al. (1968).

to larger urban areas can be illustrated in a study of data published on the Detroit-Conner Creek combined (sewage plus stormwater) collection system serving approximately 420,000 people and the separate stormwater drain system of Ann Arbor, Michigan, population 67,300 (Table 1.7) (Burm and Vaughan, 1966). There are two important observations that can be made from these data: 1) fecal contamination in a separate urban stormwater system is more often derived from wildlife, dogs, and cats (note the low fecal coliform to fecal streptococcus ratios); and 2) combined sewer systems have a greater magnitude of fecal pollution (higher fecal coliform densities) derived from human populations. Note the higher fecal coliform to fecal streptococcus ratios in samples containing domestic sewage. While wastes and stormwater runoff

Table 1.7 Summary of microbiological data from Detroit and Ann Arbor overflows

Month	Analysis	Separate System—Ann Arbor			Combined System—Detroit		
		Density per 100 ml*	Percent fecal coliform	FC/FS ratio	Density per 100 ml*	Percent fecal coliform	FC/FS ratio
April	Total coliform	340,000	—	—	2,400,000	—	—
	Fecal coliform	10,000	2.9	—	890,000	37.1	—
	Fecal streptococci	20,000	—	0.5	—	—	—
May	Total coliform	510,000	—	—	4,400,000	—	—
	Fecal coliform	51,000	10.0	—	1,500,000	34.1	4.7
	Fecal streptococci	200,000	—	0.26	320,000	—	—
June	Total coliform	4,000,000	—	—	12,000,000	—	—
	Fecal coliform	78,000	2.0	—	2,700,000	22.5	3.7
	Fecal streptococci	120,000	—	0.65	740,000	—	—
July	Total coliform	4,000,000	—	—	37,000,000	—	—
	Fecal coliform	120,000	3.0	—	7,600,000	20.5	21.7
	Fecal streptococci	390,000	—	0.31	350,000	—	—
August	Total coliform	1,700,000	—	—	26,000,000	—	—
	Fecal coliform	350,000	20.6	—	4,400,000	16.9	8.3
	Fecal streptococci	310,000	—	1.1	530,000	—	—

Data of 1964 adapted from Burm and Vaughan, 1966
*Geometric means per 100 ml

are collected in combined systems for transport to a wastewater treatment plant, there are times during prolonged heavy rains when the volume of this mixed waste may be bypassed to receiving waters because treatment capacity of the plant is exceeded and temporary storage is not available.

There is always an opportunity for a chance occurrence of some pathogens in stormwater runoff. For example, 4,500 *Salmonella thompson* per 100 ml were detected from a stormwater sample in the small urban drainage system discussed previously (Geldreich et al., 1968). This particular sample contained 450,000 fecal coliforms and 370,000 fecal streptococci per 100 ml. The most concentrated opportunistic pathogens in urban stormwater around the Baltimore area (Olivieri et al., 1977) were found to be *Pseudomonas aeruginosa* and *Staphylococcus aureus* at levels of 10^3 to 10^5 and from 10^0 to 10^3 per ml, respectively. *Salmonella* and enteroviruses were frequently isolated but at much lower densities ranging from 10^0 to 10^4 per 10 liters of this urban runoff. These pathogens originated in human sewage that mixed with rain water in storm sewer drains and passed through the combined sewer overflow. Heavy loads of fecal pollution are common to stormwater runoff from animal feedlots where thousands of cattle are confined (Geldreich, 1972) and may equate to the discharge of raw sewage from cities with populations that approach 10,000 people.

Surface water quality is also influenced by the occasional recirculation of organisms entrapped in bottom sediments. Based on the examination of a wide variety of bottom sediments from bathing beaches, recreational lakes, creeks and rivers, it became evident that there is an approximate 100 to 1000-fold increase in fecal coliform bacteria in bottom sediments as compared to the overlying waters (Van Donsel and Geldreich, 1971). Bottom sediments will also receive various pathogenic agents passing in the overlying polluted water. *Salmonella*, for example, was found with greater frequency in bottom sediments from grossly polluted waters than from sediments in clean water environments (Table 1.8).

In streams, instability of sludge banks below sewage outfalls can result in recirculating residual pathogens into the overlying water. This condition results from the impact of rapid increases in water velocity generated from major storm events upstream and from spring thaws over the watershed. Human contribution to this instability of sludges and their recirculation is in river channel dredging to maintain navigation. There are two types of dredges used; the mechanical or clamshell dredge and the hydraulic dredge. The bacteriological water quality effects from clamshell dredging are much less because disposal of dredged material is less disruptive and exposure to large volumes of water is minimal (Grimes, 1982). By contrast, hydraulic dredges create tremendous turbulence through the use of high velocity pumps. It is important to note that neither turbidity nor bacteriological effects of dredging extended far downstream in the navigation channel of the Upper Mississippi River (Grimes, 1980). Within less than 2 km below the dredge spoil discharge area, water quality had apparently recovered from the effects of dredging. This was

Table 1.8 Isolation of salmonellae from muds of various quality waters compared with isolation from water

Source	Fecal coliforms per 100 ml water	Total examinations	Isolation of salmonella Positive occurrence	Percentage
Mud	1–200	21	4	19.1
	201–2000	12	6	50.0
	over 2000	15	12	80.0
Fresh water	1–200	29	8	27.6
	201–2000	27	19	70.3
	over 2000	54	53	98.1
Estuarine water	1–200	258	33	12.8
	201–2000	91	40	44.0
	over 2000	75	45	60.0

Data from Geldreich (1970) and Van Donsel and Geldreich (1971).

probably a result of rapid resettling of bottom materials as turbulence subsided. However, aggregates of pathogenic bacteria and virus in fecal cell debris might migrate some distance beyond, so downstream water treatment plants should be on the alert for increased chlorine demand and possible higher densities of pathogens to inactivate in their treatment basins.

Recirculating pathways for pathogens in bottom sediments of lakes and reservoirs are another concern, especially where these water resources are located in temperate zones in which pronounced seasonal changes in water temperature occur. These water bodies may be subject to periods of thermal stratification in the summer and winter, with destratification (water turnover) periods in spring and autumn. During the interval of stratification, water movement in the deeper portions of raw water reservoirs becomes restricted, generally creating a zone of maximum bacterial occurrence in the water layer above the thermocline (Weiss and Oglesby, 1960; Collins, 1963; Niewolak, 1974; Drury and Gearheart, 1975).

As an illustration, vertical sampling of Cruises Creek Reservoir (Walton, KY) revealed (Table 1.9) that during July, bacterial stratification began to occur at mid-depth of about 3 to 3.6 m (10 to 12 ft) and correlated with the formation of a thermocline (Geldreich et al., 1980). Increased bacterial densities in bottom waters during July, August, and September were related to natural siltation in the impoundment. Throughout the summer, there was a slow release of bottom water and a major draw-off of bottom water occurred about October 9 to 12, 1974. This draw-off, done to prevent septic bottom water from entering the water treatment process, decreased the pool of septic water in the bottom of the reservoir and accelerated complete destratification of the reservoir by November.

Table 1.9 Bacterial characterization of Cruises Creek Reservoir*

Month	Surface			Mid-Depth (3–3.6 m)			Bottom (6.1–6.7 m)		
	Fecal coliform per 100 ml	Standard plate count per 100 ml	Temp. °C	Fecal coliform per 100 ml	Standard plate count per 100 ml	Temp. °C	Fecal coliform per 100 ml	Standard plate count per 100 ml	Temp. °C
May	26	390	21.4	55	1,000	13.3	80	1,200	10.3
June	5.5	1,700	26.3	92	2,700	16.4	230	2,100	11.6
July	3.0	3,200	26.6	45	16,000	16.7	190	7,400	11.1
August	3.4	11,000	25.1	58	4,900	17.0	190	4,000	11.2
September	26	21,000	19.2	80	3,700	18.2	450	8,400	13.8
October	18	530	14.8	21	650	14.8	110	1,900	12.7
November	6	180	9.1	4	160	11.8	11	340	11.0

*Reported as geometric mean values. Data of 1974 from Gelreich et al., (1980).

With destratification, water from the bottom and top layers mix and a more uniform dispersion of water quality develops throughout. This mixing process causes decaying vegetation and entrapped organisms in settled particulates to reenter the water column (Niewolak, 1987a, 1987b). As a result, coliform bacteria, turbidity, and humic acids from partially decomposed vegetation will temporarily increase in the water. Upon entering the intake, this water quality change may create a significant but temporary increase in chlorine demand.

The impact of lake destratification on changes in total coliform densities can be seen in the analysis of monitoring data from Lake Gaillard, New Haven, Connecticut (Geldreich et al., 1989). In this example, autumn destratification of the lake (Figure 1.1) produced numerous total coliform densities above 100 organisms per 100 ml for several weeks. A smaller rise in total coliform densities occurred in early spring of that year. Precipitation over the watershed was minimal during these periods and was not considered to be a factor modifying the impact of destratification.

Contamination from Subsurface Sources   Concentrated reservoirs of pathogens may also be associated with nontraditional methods of waste disposal. Land application of minimally treated wastewaters can become a pathway to groundwater pollution in areas where there is rapid infiltration of soils (high percolation rates) and to surface waters through excessive overland spraying in the drainage basin (Kowal, 1982). Under these conditions, the natural self-purification factors (desiccation, acid soil contact, sunlight exposure, soil

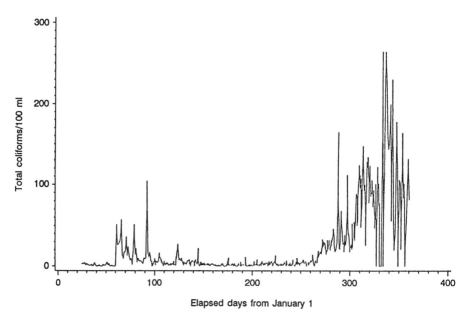

Figure 1.1   Impact of lake destratification at Lake Gaillard (Geldreich et al., 1988).

organism competition, antagonism, and predation) become either inoperable or ineffective. Sanitary landfills of animal wastes from feedlot operations, water plant sludges, or garbage wastes which are improperly located or poorly constructed can also contribute significant pathogen releases to ground and surface waters (Geldreich, 1978). Urban solid waste (garbage) contains not only food discards, garden rubbish, disposable products, soil, rock, and ash residues but also fecal material (Table 1.10). This fecal material is derived from disposable diapers, household animal pet discharges in litter material, and the droppings of rodents and birds foraging for food in exposed waste collections.

Sanitary landfills, if properly designed for the burial of solid wastes, avoid many of the problems associated with open dumping. However, if the impermeable liner at the bottom of the landfill is penetrated, there is a risk that solid waste leachates will leak into the surrounding groundwater or emerge

Table 1.10 Microbial characterization of solid wastes from various cities

Waste collection site[b]	Number of samples	Organisms per g wet wt[a]			
		Total viable bacteria[c]	Spores	Total coliforms	Coliforms
A	4	110,000,000	270,000	3,000,000	260,000
B	3	450,000,000	110,000	6,700,000	510,000
C	6	78,000,000	38,000	1,600,000	1,200,000
D	3	480,000,000	31,000	1,100,000	630,000
E	4	680,000,000	1,900,000	51,000,000	8,100,000
F	2	54,000,000	35,000	13,000,000	5,600,000
G	2	4,000,00	25,000	340,000	15,000
H	3	300,000,000	160,000	8,600,000	3,000,000

[a] Average moisture content was 41%.
[b] Waste collection sites in Cincinnati, Chicago, Memphis, Atlanta, and New Orleans.
[c] Spread plates (trypticase soy agar+blood) at 35°C for 48 hours.
Data from Peterson, (1971).

Table 1.11 Fecal coliform and fecal streptococcus survivals in leachates from sanitary landfills

Days after placement	Densities per 100 mL	
	Fecal coliforms	Fecal streptococci
42	2,600,000	240,000,000
43	4,900,000	790,000
56	2,000	79,000
63	9,000	33,000
70	33,000	170,000

Data from Blannon and Peterson, (1974).

Table 1.12 Microorganisms identified in commercial landfill leachates

Organism	Coliform	Coliform antagonist	Pigmented organism	Opportunistic pathogen*	Primary pathogen
Bacteria					
Acinetobacter sp.				X	
Alcaligenes sp.		X		X	
Bacillus sp.		X		X	
Clostridium perfringens				X	
Corynebacterium sp.				X	
Enterobacter agglomerous	X			X	
Enterobacter cloacae	X			X	
Listeria monocytogenes				X	
Micrococcus sp.			X	X	
Moraxella sp.		X		X	
Mycobacterium				X	
Proteus sp.		X		X	
Providencia alcalifaciens				X	
Pseudomonas sp.		X	±	X	
P. fluorescens		X	X	X	
Salmonella sp.					X
Staphyloccoccus sp.			±	X	X
Streptococcus faecalis				X	
S. durans				X	
Fungi					
Aspergillus niger					X
Cephalosporium sp.					X
Fusarium sp.					X
Neurospora					
Penicillium sp.				X	
Pseudallescheria boydii				X	

Data adapted from Donnelly and Scarpino (1984).
* Species is an opportunistic pathogen. Information source, Lennette et al. (1985).

from the site and contaminate surface waters. Stormwater penetration of landfill sites may provide the mechanism for migration of the entrapped fecal organisms which survive in substantial numbers even after two months of containment (Table 1.11). Many factors influence the persistence of indicator bacteria, including the chemistry of the solid wastes, heat of waste decomposition, moisture content, and the antagonistic action of numerous organisms in the microflora. This latter factor can be seen in the numerous bacteria and fungi isolated from commercial landfill leachates (Table 1.12) that also include a number of primary pathogens and opportunistic pathogens found on occasion in some water supplies.

1.2   Raw Source Water Quality

A number of factors enter into the choice of a best available raw source water for drinking water supply. These include adequate quantity throughout the year, water quality that is amenable to treatment, and some measure of watershed protection from domestic, industrial, and agricultural pollution (Office of Drinking Water, 1978). These sources of pollution introduce a wide range of potentially infectious agents to waters that may ultimately be used in water supply (Smith & Twedt, 1971; U.S. Environmental Protection Agency, 1971; Olivieri et al., 1977). The more usual infectious agents may be classified within four broad groups: bacteria, viruses, protozoa, and helminths (Table 1.13). That these infectious agents, derived principally from infected persons and other warm-blooded animals, are found in polluted water is unquestionable.

Upon discharge into a warm source, pathogens will have a variable persistence which is predicated on many factors. For example, 32 *Salmonella* serotypes were found to be common to sewage effluent samples and the downstream sections of the Oker River in Germany (Popp, 1974). By contrast, *Salmonella* strains were detected in surface water only 250 meters downstream from a wastewater treatment plant with regularity, but never at sample sites 1.5 to 4 kilometers downstream (Kampelmacher and van Noorle Jansen, 1976). *Salmonella* transported by stormwater through a wastewater drain at the University of Wisconsin experimental farm was isolated regularly at a swimming beach 800 meters downstream (Claudon et al., 1971). Excessive BOD or TOC, low stream temperature, and a source of *Salmonella* discharge (wastewater

Table 1.13 Major infectious agents found in contaminated drinking waters worldwide

Bacteria	Viruses	Protozoa
Campylobacter jejuni	Adenovirus (31 types)	*Balantidium coli*
Enteropathogenic *E. coli*	Enteroviruses (71 types)	*Entamoeba histolytica*
Salmonella (1700 spp.)	Hepatitis A	*Giardia lamblia*
Shigella (4 spp.)	Norwalk agent	*Cryptosporidium*
Vibrio cholerae	Reovirus	
Yersinia enterocolitica	Rotavirus	
	Coxsackie virus	

Helminths
Ancylostoma duodenale
Ascaris lumbricoides
Echinococcus granulosis
Enterobius vermicularis
Necator americanus
Strongyloides stercoralis
Taenia (spp.)
Trichuris trichiura

effluent) can produce an impact that will depress stream self-purification processes. As an illustration, salmonellae were isolated in the Red River of the North, 22 miles downstream of sewage discharge from Fargo, North Dakota and Moorhead, Minnesota during September and prior to the sugar beet processing season (U.S. Department of Health, Education, and Welfare, 1965). By November, salmonellae were found 62 miles downstream of these two sites. With increased sugar beet processing, wastes reaching the stream in January, under cover of ice, brought high levels of bacterial nutrients and *Salmonella* strains were then detected 73 miles downstream or four days flow time from the nearest point source discharges of warm-blooded animal pollution.

Receiving waters between the point discharge of sewage effluents and water intakes may not only serve as a buffer to accidental spills and treatment bypasses, but can contribute to water quality improvement through natural self-purification processes. Every stream, lake, estuary, and groundwater aquifer has some limited capacity to assimilate waste effluents and stormwater runoff entering the drainage basin. This self-purification process is a complex and ill-defined mechanism that involves bacterial adsorption with sedimentation, predation, dilution, hydrologic tributary effects, water temperature, and solar radiation (Geldreich, 1986). For example, sedimentation can rapidly move pathogens from the overlying water into the bottom deposits. Over a year, approximately 90 percent of *Salmonella* species isolated from the North Oconee River were recovered in bottom sediments downstream. Adsorption of the organisms to sand and clay and the action of sedimentation resulted in their concentration in the river bottom sediments (Hendricks, 1971).

The rate at which these natural processes progress is dependent upon time, generally referred to as flow time downstream to achieve 99 percent reduction in the microbial population. Mitigating circumstances that may counterbalance these beneficial effects include discharges of industrial wastes that alter the water pH, food processing and paper mill wastes that provide excessive bacterial nutrients, stormwater runoff characteristics of the surrounding watershed, magnitude of pollutional loading from all discharges and their degree of treatment. While natural self-purification may be effective in a given water course during periods of warm temperatures and limited conditions of precipitation, results will be less impressive following wet seasonal periods and be particularly poor during winter environmental conditions.

All things considered, the total impact of various environmental factors results in a microbial barrier of varying effectiveness to many organisms of public health significance, given sufficient time for interaction before the water is used downstream. A study of coliform reductions observed in various rivers, reveals wide differences in the number of hours needed in flow time downstream to achieve a 90 percent kill by natural self-purification reactions. These variations from river to river are in response to the rate of water flow downstream and the amount of pollution load to be assimilated (Table 1.14).

Recently, several states have proposed suspension of the requirement for sewage effluent disinfection at waste treatment plants that are greater than 20 to 40 miles upstream of a water supply intake (Huff, 1981). The argument is that chlorination is not cost-effective for controlling fecal coliform densities in

Table 1.14 Effectiveness of natural self-purification on coliform reduction in surface waters

Ambient waters	Season	Hours*	Reference
Ohio River	Summer	47	Streeter (1929)
	Winter	51	
Missouri River	Winter	115	Kittrell et al. (1963)
Tennessee River	Summer	53	Kittrell et al. (1963)
Sacramento River	Summer	32	Kittrell et al. (1963)
Cumberland River	Summer	10	Kittrell et al. (1963)
Glatt River	Summer	2.1	Waser et al. (1934)

Data revised from Wuhrmann (1972)
*Flow time to achieve 90% reduction.

sewage effluents. Furthermore, based on bacteriological studies, it is argued that the water quality at downstream water supply intakes has not improved substantially as a direct result of sewage effluent disinfection practices. The crux of the problem is that chlorination of sewage effluents would be more effective if disinfection demand was reduced through effective nutrient removal (TOC or BOD), total suspended solid reductions, and minimization of ammonia content. Since many plants have difficulty in consistently achieving limits below 20 mg/l BOD and 30 mg/l suspended solids, sewage treatment plant operators have generally applied excessive chlorine dosages to meet a specified bacterial limit. Reduction to organics should be given equal consideration not only to reduce chlorination applied in sewage treatment but also to reduce organics in raw source waters that must be processed into water supply. The concern here is that these organic residuals combine with disinfection in water supply treatment to create byproducts that may be carcinogens.

Rivers   Surface waters surrounding metropolitan areas are generally polluted with the wastes from these concentrated centers of human activities. To provide maximum protection to water quality, intakes for metropolitan water supplies were originally located upstream from the points of domestic waste discharge and from sources of industrial effluents. In recent years, there has been both a rapid growth of cities and a development of satellite residential areas. These population expansions and growth of numerous recreational activities have frequently adversely affected the water supply intake zone of the stream or lake, bringing new sources of sewage and stormwater runoff to further degrade raw source water quality. For example, there are 56 water supply intakes along the 977.8 miles of the Ohio River and 43 of these intakes are within 5 miles downstream of the effluent discharge of a waste treatment plant. While low flow conditions are minimized by navigational dams on this river, other surface water sources from numerous tributary streams are not so protected. In a study of surface water supplies used by 20 cities serving a total population of seven million people, it was estimated that the minimum wastewater component of the raw source water ranged from 2.3 to 16 percent and increased to predominately wastewater for several municipal intakes during low flow periods (Swayne et al., 1980).

Bacteriological examination of raw water quality at the water treatment plant intakes of Omaha, Nebraska, St. Joseph, Missouri, and Kansas City, Missouri on the Missouri River (Table 1.15) frequently revealed fecal coliform densities in excess of 2,000 organisms per 100 ml (U.S. Environmental Protection Agency, 1971). This fecal pollution load resulted from inputs of raw sewage, effluents from primary and secondary treatment plants of differing efficiencies, cattle feedlot runoff, and discharges from meat and poultry processing plants. The concern is that pathogens are also present, their numbers and kinds being related to the diseases prevalent in upstream human and farm animal populations. To demonstrate this point, field investigation revealed various serotypes of *Salmonella* at each water plant intake. Virus examinations performed on raw water from the intake at St. Joseph resulted in the reported recovery of poliovirus type 2 and 3 and ECHO virus types 7 and 33.

In another study of Missouri River water quality, total coliforms, fecal coliforms virus, and turbidity levels were quantified at the Lexington, Missouri intake during 1976 to 1978 (O'Connor et al., 1982). Data in Table 1.16 were grouped by results obtained during water temperature periods that closely corresponded to seasonal periods. During the winter period, the resulting fecal contamination was derived largely from municipal wastewater effluent discharges and meat-processing plants upstream of the water supply intake. Spring thaws brought increasing flood conditions and associated increases in turbidity and movement of fecal contamination in runoff to the main-stem portions of the river. Occasional high turbidities in the summer were related to major storm periods that increased river flow and turbidities periodically to maximum values. During wet weather periods, stormwater runoff included bypasses of raw sewage in treatment facilities and movements of fecal wastes from numerous feedlot operations. Autumn periods showed a decline in turbidities, river flows, and fecal coliforms as a reflection of reduced rainfall events. The occurrences of enterovirus in these samples suggest that virus levels correlated inversely with coliform levels, temperature, and turbidity. This unexpected finding could be due to more rapid virus die-off at higher water temperatures or, more likely, the decreased recovery efficiency for virus because of interference with virus detection methodology from the higher stream turbidities created by stormwater runoff silts. An average of 2.5 viruses per sample were recovered in 216 samples. Polioviruses were the most abundant (69%), followed by Coxsackie (11%) and ECHO viruses (5%). Only 10% of the total virus particles detected were recovered on 3 to 5 μm pore-size prefilters, suggesting that most of the recovered virus particles were not associated with the suspended sediment in the Missouri River, being more likely a part of fecal cell debris or viral aggregates in the water. The evidence presented in Tables 1.15 and 1.16 illustrates some of the potential health hazards that become

Table 1.15 Fecal coliform densities and pathogen occurrences at the Missouri River public water supply intakes

Raw water intake	River mile	Date	Fecal coliforms per 100 ml*	Pathogen occurrence
Omaha, NB	626.2	Oct. 7–18, 1968	8,300	N.T.**
		Jan. 20–Feb. 2, 1969	4,900	N.T.
		Sep. 8–12, 1969	2,000	*Salmonella enteritidis*
		Oct. 9–14, 1969	3,500	*Salmonella anatum*
		Nov. 3–7, 1969	1,950	N.T.
St. Joseph, MO	452.3	Oct. 7–18, 1968	6,500	N.T.
		Jan. 20–Feb. 2, 1969	2,800	N.T.
		Sep. 18–22, 1969	4,300	N.T.
		Oct. 9–14, 1969	N.T.	*Salmonella montevideo*
		Nov. 3–7, 1969	6,500	N.T.
		Jan. 22, 1970	N.T.	19 virus PFU;** Polio types 2, 3: Echo types 7, 33
		Apr. 23, 1970	N.T.	3 virus PFU, not yet typed
Kansas City, MO	370.5	Oct. 28–Nov. 8, 1968	6,500	N.T.
		Jan. 20–Feb. 2, 1981	8,300	N.T.
		Sep. 18–22, 1969	3,800	*Salmonella newport*
				Salmonella give
				Salmonella infantis
				Salmonella poona
		Sep. 25–29, 1969	3,800	N.T.

Data from (U.S. Environmental Protection Agency, 1971)
*Geometric means
**N.T., no test for pathogens done. PFU, plaque-forming units.

Table 1.16 Raw water quality at Lexington, MO water treatment plant

Temp. Range[a]	No. Samples	Turbidity Range	Total coliforms[b] per 100 ml	Fecal coliforms[b] per 100 ml	Virus (PFU) per 100 gal (378.5 l)	
					Range	Occurrence
0.10 °C	68	0.05–200	93,000	4,600	0–38	79.4%
10.5–20 °C[c]	32	32–600	114,000	7,900	0–9.1	53.1
20.5–28 °C	42	40–1100	162,000	8,600	0–10	42.9

Data from (O'Connor et al., 1982).
[a] Sampling period December 3, 1976 to December 27, 1978.
[b] Geometric mean values.
[c] Spring and autumn periods combined.

more prevalent and more challenging to a water treatment system processing the widely fluctuating water qualities in a major river.

Lakes and Impoundments   Surface water retention in lake and impoundments of raw water supply in reservoirs is an important contributor to water quality enhancement. Storage and sedimentation of surface water often improves the chemical quality (Symons et al., 1970). Many organisms are inactivated by natural self-purification processes while others are removed from the water column through natural siltation (Romaninko, 1971; Dzyuban, 1975; Geldreich et al., 1980). These water quality improvements are variable, being directly proportional to holding time and the magnitude of pollutional inputs. Large impoundments and lakes generally have a relatively uniform retention time and greater dilution potential for pollution inputs. While these conditions will be a positive influence on self-purification, the rate of water quality improvement will be retarded to a varying degree by cold water temperatures typical of large deep lakes.

Giardia outbreaks over the past 10 years have changed forever the time-honored concept that restricted-use watersheds surrounding lakes or impoundments insure freedom from pathogen risk (Mackenthun and Ingram, 1967; Hibler et al., 1975; Allard et al., 1977; Kirner et al., 1978; and Pluntze, 1984). A total of 22,897 people became ill with giardiasis acquired from 84 waterborne outbreaks in the United States during 1961 to 1983. The source of the pathogen was quickly revealed to be in the wildlife population living in the "protected" watershed environment. Beavers were most often considered to be the primary animal reservoir because of the high incidence of infection in beaver colonies and their association with near-shore water environments. However, coyotes, muskrats, voles, cattle, and animal pets may also be involved in the perpetuation of the pathogen in the natural environment (U.S. Environmental Protection Agency, 1979; McCall et al., 1985). Many of these animals are either part of the natural fauna or are introduced onto the watershed as domestic farm animals.

Birds and waterfowl can also be sources of fecal pollution. Sea gulls are scavengers that frequent open garbage dumps, eat contaminated food wastes, and contribute *Salmonella* in their fecal droppings to overnight roosting sites on lakes and open reservoirs in coastal areas (Fennel et al., 1974). Therefore, it is not surprising that seagulls were the contributors of *Salmonella* in an untreated water supply reservoir located in an Alaskan community and the cause of several cases of salmonellosis (Anonymous, 1954). Bird sanctuaries may also create a water pollution problem of increased fecal contamination during the nesting season and at periods of seasonal bird migrations.

Consumption of untreated or poorly treated surface waters of low turbidity that was contaminated with *Campylobacter jejuni* has resulted in several large waterborne outbreaks (Blaser et al., 1983; Blaser and Cody, 1986). Shortly before an outbreak in Bennington, Vermont that involved 3,000 residents, heavy rains increased runoff which led to surface contamination of the municipal water supply (Vogt et al., 1982). Unchlorinated surface water from a reservoir and a creek caused an outbreak of *Campylobacter* enteritis involving 700 individuals in a small community in northern British Columbia (Health and Welfare, Canada, 1981). Illness among vacationers to national parks in Wyoming who drank water from mountain streams and lakes were the cause of a 25 percent increase state-wide in *Campylobacter* enteritis (Taylor et al., 1983). The source of this pathogen was often wildlife, farm animals, and animal pets living on the watershed. Pathogens, suspended solids, humics, fertilizers, and pesticides are the major adverse influences on lake or reservoir water quality that can impact adversely on minimal water treatment systems used by small water-treatment operations. Often these treatment operations are inadequate to cope with sudden increased disinfection demand and elevated turbidities in the raw water as a result of major storm events and spring thaws on the watershed. As a consequence, protozoan cyst penetration of a poorly operated filtration process or passage of pathogenic bacteria or virus beyond an inadequate disinfection application may occur.

Ground Waters   Groundwater has long been considered to be of unquestionable excellence because the soil barrier is often effective in providing isolation of this high quality source water from surfaced pollutants. As a consequence, treatment is frequently nonexistent or limited to water hardness reduction, taste and odor removal, or limited to disinfection. Thus, it is not surprising to note that seventy-five percent of approximately 60,000 public water supplies in the United States use groundwater as the sole raw water source, and seven percent of the public systems use a mixture of ground and surface water (McCabe et al., 1970). Other compelling factors for groundwater selection include the fact that this is the sole source of water in arid regions, the available surface waters may be grossly polluted and require extensive treatment to make a safe supply, or there is insufficient volume of surface water available throughout the year to meet water supply demand.

Groundwaters derived from deep aquifers are generally of good bacteriological quality because vertical percolation of water through soil results in the removal of much of the microbial and organic pollution. By contrast, the waters from shallow wells are frequently grossly polluted with a variety of wastes from surface runoff. As greater land areas are used to accommodate increasing populations and industrial growth, there is an increased risk of polluting high quality groundwater resources through loss of soil barrier protection.

Contaminants enter groundwater supplies by direct injection through wells, by percolation of liquids sprayed over the land or leached from soluble solids at the surface, by leaking or broken sewer lines, seepage from waste lagoons, infiltration of polluted surface streams, irrigation return waters, interaquifer leakage, leachates from landfills, septic tank effluents, and upwelling of salt water into fresh water aquifers. As a result of improper source protection and inadequate treatment, it is not surprising that many small public water systems have the poorest records for compliance with the drinking water standards.

A variety of waterborne disease outbreaks have been attributed to untreated or poorly treated groundwater containing pathogenic forms of bacteria, viruses or eucaryotic organisms (i.e. amoebas, *Giardia*, worms, etc.). Specific bacterial pathogens that have been isolated from well waters include enteropathogenic *E. koli*, *Vibrio cholera*, *Shigella flexneri*, *S. sonnei*, *Salmonella typhimurium*, *Yersinia enterocolitica*, and *Campylobacter jejuni* (Greenberg and Jongerth, 1966; Schroeder et al., 1968; Lassen, 1972; Evison and James, 1973; Lindel and Quinn, 1973; Center for Disease Control, 1973; Woodward et al., 1974; Center for Disease Control, 1974; Dragas and Tradnik, 1975; Highsmith et al., 1977; Schiemann, 1978; Center for Disease Control, 1980; Mentzing, 1981). Poliovirus and enterovirus have also been isolated from a well water used to provide water to restaurant patrons (Vander Velde and Mack, 1973).

A brief review of completed groundwater studies which include a national survey of community water supply systems (McCabe et al., 1970), national statistical assessment of rural water supplies (Francis et al., 1982), Tennessee-Georgia rural water supplies, Interstate Highway drinking water systems (Kansas, Oregon, Virginia), and the Umatilla Indian Reservation groundwater survey (Allen and Geldreich, 1975), illustrate some bacteriological problems in groundwater (Table 1.17). In these five surveys, 9 to 51 percent of the groundwaters examined contained total coliforms. Those households served by cisterns, springs, and surface waters in a national rural water survey (Francis et al., 1982) had the poorest water quality with 77 percent of these supplies containing coliforms. Of all the waters examined in the five different surveys, 2 to 27 percent were positive for fecal coliforms, indicating recent contamination of fecal origin that could also have contained some intestinal pathogens. Compared to surface water systems, the lower percentage of contamination for groundwater systems in the community water supply study (McCabe et al., 1970) reflect better protection of the source and, in many cases, application of disinfection. The high percentages of rural supplies contaminated with coli-

Table 1.17 Microbiological summary of small water systems

Survey	Number of supplies	Percent total coliforms**	Percent fecal coliforms**
Community Water Supply Study			
Small water systems	621	9.0	2.0
Interstate Highway Study*			
Rest-stop water supplies	241	15.4	2.9
Umatilla Indian Reservation*			
Tribal water supplies	498	35.9	9.0
National Rural Water Study*			
Individual wells, cisterns	1,086	42.1	12.2
Small public supplies[a]	241	43.3	7.7
Small community supplies[b]	1,327	15.5	4.5
Tennessee–Georgia Rural Survey*			
Individual wells, cisterns, springs	1,257	51.4	27.0

Data from Allen and Geldreich (1975).
* Data from Francis et al., (1982).
[a] Supplies with >15 service connections
[b] Supplies with <15 service connections
** Percent of samples containing one or more coliforms per 100 ml.

forms were attributed to faulty construction practices and inadequate sanitary safeguards for source water of marginal quality (Francis et al., 1982).

1.3   Raw Source Water Limits for Water Supply Treatment

Water-treatment plants built at the turn of the century are frequently hard-pressed to adequately treat surface waters of increasingly poor bacterial or chemical quality. Even newer treatment plants are being operated beyond their designed capacity. These conditions decrease protection from waterborne pathogen exposures, increase the chance occurrence of momentary breaks in the water treatment barrier, and elevate the risk for a waterborne disease outbreak in the community.

The quality of raw water at public intakes is quite variable. Among the various water systems evaluated in the Community Water Supply Study (McCabe et al., 1970), groundwater was by far the major raw water source, with 820 well or spring sources supplying 570 water plants. During this study, 774 (94.4 percent) of these groundwater supplies contained no detectable fecal coliforms per 100 ml (Table 1.18). Other groundwater sources were of a lesser quality. One supply in Vermont processed a contaminated groundwater containing a fecal coliform density in excess of 2,000 organisms per 100 ml.

Surface water was the source for 160 water plants in this nationwide survey. The results (Table 1.19) showed that 99.4 percent of these surface water sources had fecal coliform densities of less than 2,000 organisms per 100 ml. The one supply treating a surface with a fecal coliform density in excess of 4,000 per 100 ml during the study period was a small military depot in the Kansas City metropolitan area. Other studies on the Missouri River indicate that Omaha, St. Joseph, and Kansas City could have been added to those utilities which frequently process water with similarly high fecal coliform densities. These data suggest very few, possibly less than 0.1 percent, public supplies must treat water with fecal coliform levels above 2,000 organisms per 100 ml.

In order to determine how effective conventional treatment is in providing

Table 1.18 Groundwater quality for 820 public water intakes

Fecal coliforms/ 100 ml	Public supplies			
	Total	Percent	Cumulative total	Cumulative percent
<1	774	94.4	774	94.4
1–20	29	3.5	803	97.9
21–200	14	1.7	817	99.6
201–2,000	2	0.2	819	99.9
2,001–4,000	1	0.1	820	100.0

These intakes include wells and springs serving as source water for 570 public water systems. Data from McCabe et al., (1970).

Table 1.19 Surface water quality for 160 public water intakes

Fecal coliforms/ 100 ml	Public supplies			
	Total	Percent	Cumulative total	Cumulative percent
<1	70	43.8	70	43.8
1–20	30	18.8	100	62.5
21–200	37	23.1	137	85.6
201–2,000	22	13.8	159	99.4
2,001–4,000	0	0.0	159	99.4
4,001–8,000	1	0.6	160	100.0

These intakes included streams, lakes and reservoirs serving as source water for 160 public water systems.
Data from a survey of community water systems in 9 metropolitan areas (McCabe et al., 1970).

a microbiologically safe water supply when the raw source water quality at times exceeded 2,000 fecal coliforms per 100 ml, Akin et al. (1975) examined 6 different plants in three states. Although the primary objective of this study was to determine whether human enteric viruses could be detected in finished waters, parallel multiple-tube tests involving large-volume sampling (5.5 liters) were performed to establish coliform densities in finished water. Test results with this special MPN procedure indicated a value of less than 0.02 coliforms per 100 ml. Total coliforms were detected (Table 1.20) in 11 samples, most of which were samples of finished water processed from the East Fork of the White River in Indiana. The raw water source in this case had fecal coliform concentrations ranging up to 14,000 organisms per 100 ml. However, coliform values in the treated water ranged from 0.02 to 0.23 coliform per 100 ml; all were below the maximum contaminant level of one coliform per 100 ml set by the Federal Drinking Water Standards (U.S. Environmental Protection Agency, 1976). From the operational data reviewed, a 6 log coliform reduction was readily attained by these water treatment plants. Based on analysis of large sample concentrates (1,100 liter samples), no viruses were isolated in 56 finished water samples collected from these plants. These results would suggest the adequacy of conventional treatment processes to remove bacterial and viral pathogens from poor quality waters that exceed the 2,000 fecal coliform limit, the recommended maximum contaminant level for raw source waters (National Academy of Science, 1973), but at an incalculable increased risk of reducing the barrier to microbial penetration. Fortunately, relatively few water-treatment plants are forced to treat waters with fecal coliform densities greater than 2,000 organisms per 100 ml.

Table 1.20 Bacteriological quality of source water and finished potable water at six water treatment plants

Location	Source water(s)	Water plant intake Fecal coliforms per 100 ml	Finished potable water* Total coliform occurrences per 5.55 l	Finished potable water* Fecal coliform occurrences per 5.55 l
Columbus, OH	Scioto River	38– 3,000	2/4	0/4
Sidney, OH	Tawana Creek, wells	40– 2,800	2/3	1/3
Muncie, IN	White River	50– 2,000	0/4	0/4
Seymour, IN	E. Fork, White River	11–14,000	5/11	0/11
St. Joseph, MO	Missouri River	280– 7,000	0/1	0/1
Kansas City, MO	Missouri River	1,700– 4,300	2/2	0/0

Data from Akin et al., (1975)
*Negative results indicate a coliform value less than 0.02 per 100 ml; largest positive value obtained was 0.23 coliforms per 100 ml.

Achieving a safe water supply quality from these poor-quality raw source waters was largely due to careful treatment control involving pre- and post-chlorination combined with other conventional processes (flocculation, sedimentation and filtration) in a treatment train. Now, with attention focused on reducing carcinogenic agents in finished water, prechlorination is considered less desirable due to greater potential for disinfectant byproduct formation as a reaction to organic contaminants not reduced in early stages of treatment. This change in the water treatment chain places a greater, more exacting, emphasis on final disinfection as the major barrier to raw water viral and bacterial penetration into the public water supply.

For disinfection of raw surface water alone to be an effective treatment technique, the source water must be of relative high quality and contain no interferences that either provide organism protection or chlorine demand (Geldreich et al., 1988). When using coliforms to measure the probable magnitude of fecal contamination, a relatively clean raw water from large lakes and impoundments has generally been considered to have an average density of less than 100 total coliforms per 100 ml (Water Supply Program Division, 1971). At this level of contamination, fecal coliforms (a more precise measure of fecal contamination) can be expected to approximate 10 to 30 percent of the less definitive total coliform population (Table 1.21) in surface water impoundments or lakes. In addition to meeting the bacterial limit of 100 total coliforms or 20 fecal coliforms per 100 ml, there is also the necessity to restrict raw water turbidity to 1 nephelometric turbidity units (NTU), color to 15 apparent color units (ACU) and chlorine demand to below 2 mg/l for optimum disinfection effectiveness. If either *Giardia* or *Cryptosporidium* is detected in the source water, minimal treatment must be redefined. Treatment options to consider include extended contact time for increased disinfectant concentrations applied, use of ozone or other alternative disinfectants to inactivate cysts, and filtration to physically remove these pathogenic agents from the water supply.

Table 1.21 Relation between total coliform and fecal coliform number in reservoirs

Reservoir	Number of samples	Total coliforms	Fecal coliforms	Percent fecal coliforms
		——— per 100 ml* ———		
Lake Seymour	49	7.0	0.2	2.9
Hemlock Lake	363	64	4.1	6.4
Lake Coquintan	50	3.9	0.5	12.8
Lake Gaillard	58	16.8	2.2	13.1
Casitas	113	8.7	2.3	26.4
Bull Run	78	29	8.5	29.3

Data from Geldreich et al., (1988).
*Mean values

Groundwater from deep aquifers is often characterized as being of excellent quality requiring no treatment. This condition may have been true in the past but much environmental stress from human activities has been placed on the soil barrier. Contamination of groundwater with volatile organic compounds (VOCs) has become common throughout the United States (Coyle et al., 1988). Various waterborne outbreaks have been reported from use of contaminated well supplies. As a result, the revised Federal Regulations for groundwater will focus on treatment options, one of which will be disinfection requirement for all groundwater sources used in public supplies except for those groundwater sources for which there is data demonstrating no pathogen hazard, no detectable coliforms per 100 ml, and heterotrophic bacterial densities (standard plant count) that do not exceed 500 organisms per ml.

Summary

The greatest impact of water pollution on human health comes through drinking water, the source of which may be degraded by municipal sewage, stormwater runoff, cattle feedlot drainage, and discharges from meat and poultry processing plants. Groundwater used (by an estimated 60 million people) as untreated private to marginally treated public supplies is not always free of these surface water contaminants. In fact, contaminated groundwater supplies were responsible for over 50 percent of the waterborne outbreaks during the period 1940–1970.

While water treatment technology can successfully process poor quality source waters containing in excess of 2,000 fecal coliforms per 100 ml, producing potable water that meets the accepted drinking water standards, there is serious concern that the barrier to microbial breakthrough in the finished water may be dangerously small. Any momentary break in the treatment chain could allow substantial levels of microorganisms to enter the final product—potable water.

For this reason, multiple barriers to pollution are necessary as an additional safeguard to prevent a breakthrough of pathogens into potable water supplies. These barriers include collection and treatment of all wastes, limiting treated effluent loading to receiving waters so that accumulative discharges do not exceed the self-purification capacity of the individual body of water, and appropriate water treatment to insure adequate consumer protection from any public health risk.

References

Akin, E.W., Brashear, D.A., Lippy, E.C., and Clarke, N.A. 1975. A virus-in-water study of finished water from six communities. U.S. Environmental Protection Agency, *Project Summary*, EPA–600/1–75–003. Cincinnati, Ohio.

Allard, J., Champaign, D.A., Delisle, R., Mires, H., and Lippy, E. 1977. Waterborne giardiasis outbreaks—Washington, New Hampshire. *Morbidity and Mortality Weekly Report* 26:169–170 and 175.

Allen, M.J. and Geldreich, E.E. 1975. Bacteriological criteria for groundwater quality. *Ground Water* 13:45–52.

Anonymous. 1954. Ketchikan laboratory studies disclose gulls are implicated in disease spread. *Alaska's Health* 11:1–2.

Blannon, J.C. and Peterson, M.L. 1974. Survival of fecal coliforms and fecal streptococci in a sanitary landfill. U.S. Environmental Protection Agency, News of Environmental Research in Cincinnati, April 12, 1974.

Blaser, M.J., Taylor, D.N., and Feldman, R.A. 1983. Epidemiology of *Campylobacter jejuni* infections. *Epidemiologic Reviews* 5:157–176.

Blaser, M.J. and Cody, H.J. 1986. Methods for isolating *Campylobacter jejuni* from low turbidity water. *Applied and Environmental Microbiology* 51:312–315.

Burm, R.J. and Vaughan, R.D. 1966. Bacteriological comparison between combined and separate sewer discharges in southeastern Michigan. *Journal of Water Pollution Control Federation* 38:400–409.

Callaghan, P. and Brodie, J. 1969. Laboratory investigation of sewer swabs following the Aberdeen typhoid outbreak of 1964. *Journal of Hygiene* 66:489–497.

Center for Disease Control. 1973. Typhoid fever—Florida. *Morbidity and Mortality Weekly Report* 22:77–78 and 85.

Center for Disease Control. 1974. Acute gastrointestinal illness—Florida. *Morbidity and Mortality Weekly Report* 23:134.

Center for Disease Control. 1980. Waterborne disease outbreak in the United States—1978. *Morbidity and Mortality Weekly Report* 29:46–48.

Claudon, D.G., Thompson, D.I., Christenson, E.H., Lawton, G.W., and Dick, E.C. 1971. Prolonged *Salmonella* contamination of a recreational lake by runoff waters. *Applied Microbiology* 21:875–877.

Collins, V.G. 1963. The distribution and ecology of bacteria. *Proceedings of the Society for Water Treatment and Examination* (England). 12:40–73.

Coyle, J.A., Borchers, H.J. Jr., and Miltner, R.J. 1988. Control of volatile organic contaminants in groundwater by in-well aeration. U.S. Environmental Protection Agency. *Project Summary*, EPA–600/S2–88/020, Cincinnati, Ohio.

Donnelly, J.A. and Scarpino, P.V. 1984. Isolation, characterization, and identification of microorganisms from laboratory and full-scale landfills. *Project Summary*, EPA–600/S2–84–119, Cincinnati, Ohio.

Dragas, A-Z and Tradnik, M. 1975. Is the examination of drinkable water and swimming pools for presence of entero-pathogenic *E. coli* necessary? *Zentralblatt für Bakteriologie und Hygiene*, I. Abteilung, Original B 160:60–64.

Drury, D.D. and Gearheart, R.A. 1975. Bacterial population dynamics and dissolved oxygen minimum. *Journal American Water Works Association* 67:3–154.

Dzyuban, A.N. 1975. The number and generation time of bacteria and production of bacterial biomass in water of the Saratov Reservoir, *Gidrobiologicheskii Zhurnal* 11:14–19.

Evison, L.M. and James, A. 1973. A comparison of the distribution of intestinal bacteria in British and East African water sources. *Journal Applied Bacteriology* 36:109–118.

Fennel, H., James, D.B., and Morris, J. 1974. Pollution of a storage reservoir by roosting gulls. *Water Treatment and Examination* 23:5–24.

Francis, J.D., Brower, B.L., Graham, W.F., Larson, O.W. III, McCaull, J.L., and Vigarita, H.M. 1982. National Statistical Assessment of rural water conditions. Executive Summary 21 pp., EPA, Office of Drinking Water, Washington, D.C.

Geldreich, E.E., Best, L.C., Kenner, B.A., and Van Donsel, D.J. 1968. The bacteriological aspects of stormwater pollution. *Journal Water Pollution Control* 40:Part 1, 1861–1872.

Geldreich, E.E. 1970. Applying bacteriological parameters for recreational water quality. *Journal American Water Works Association* 62:113–120.

Geldreich, E.E. 1972. Buffalo Lake recreational water quality: a study of bacteriological data interpretation. *Water Research* 6:913–924.

Geldreich, E.E. 1978. Bacterial populations and indicator concepts in feces, sewage, stormwater and solid wastes. pp. 51–97 in *Indicators of Viruses in Water and Food*, G. Berg (ed.). Ann Arbor Science, Ann Arbor, MI.

Geldreich, E.E., Nash, H.D., Spino, D.F., and Reasoner, D.J. 1980. Bacterial dynamics in a water supply reservoir: a case study. *Journal American Water Works Association* 72:31–40.

Geldreich, E.E. 1986. Control of microorganisms of public health concern in water. *Journal Environmental Science* 29:34–37.

Geldreich, E.E., Goodrich, J.A., Clark, R.M. 1988. Characterizing raw surface water amenable to minimal water supply treatment. *Annual Conference Proceedings, American Water Works Association*, Part 1, 545–570.

Grabow, W.O.K. and Nupen, E.M. 1972. The load of infectious microorganisms in the waste water of two South African hospitals. *Water Research* 6:1557–1563.

Greenberg, A.E. and Jongerth, J. 1966. Salmonellosis in Riverside, California. *Journal American Water Works Association* 58:1145–1150.

Grimes, D.J. 1980. Bacteriological water quality effects of hydraulically dredging contaminated Upper Mississippi River bottom sediment. *Applied Environmental Microbiology* 39:782–789.

Grimes, D.J. 1982. Bacteriological water quality effects of clamshell dredging. *Journal Freshwater Ecology* 1:407–419.

Harvey, R.W.S., Price, T.H., Foster, D.W., and Griffith, W.C. 1969. Salmonellas in sewage. A study in latent human infections. *Journal of Hygiene* (Camb.) 67:517–523.

Health and Welfare, Canada. 1981. Possible waterborne *Campylobacter* outbreak—British Columbia. *Canada Diseases Weekly Report* 7:223,226–227.

Hendricks, C.W. 1971. Increased recovery rate of salmonellae from stream bottom sediments versus surface waters. *Applied Microbiology* 21:379–380.

Herman, D. 1972. Experiences with coliform and enteric organisms isolated from industrial wastes. pp. 26–40 in *U.S. Environmental Protection Agency Seminar: The Significance of Fecal Coliforms in Industrial Wastes*, Denver Field Investigations Center, U.S. Environmental Protection Agency, Denver, CO.

Hibler, C.P., MacLeod, K., and Lyman, D.O. 1975. Giardiasis—in residents of Rome, N.Y. and in U.S. travelers to the Soviet Union. *Morbidity and Mortality Weekly Report* 24:366 and 371.

Highsmith, A.K., Feeley, J.D., Shaliy, P., Wells, J.G., and Wood, B.T. 1977. Isolation of *Yersinia enterocolitica* from well water and growth in distilled water. *Applied Environmental Microbiology* 34:745–750.

Huff, L.L. 1981. The economic analysis of health risk and the environmental assessment of revised fecal coliform effluent and water quality standards. Document No. 81/15, *Illinois Institute of Natural Resources*, Chicago, Illinois.

Kampelmacher, E.H. and Van Noorle Jansen, L.M. 1976. *Salmonella* effluent from sewage treatment plants, wastepipes of butcher's shops and surface water in Walcheren. *Zentralblatt für Bakteriologie und Hygiene, I. Abteilung* Original B 159:307–319.

Kirner, J.C., Littler, J.D., and Angelo, L. 1978. A waterborne outbreak of giardiasis in Camas, Washington. *Journal American Water Works Association* 70:35–40.

Kittrell, F.W. and Furfari, S.A. 1963. Observations of coliform bacteria in streams. *Journal Water Pollution Control Federation* 35:1361–1385.

Kowal, N.E. 1982. Health effects of land treatment: microbiological. *U.S. Environmental Protection Agency*, EPA–600/1–82–007, Health Effects Research Laboratory, Cincinnati, Ohio.

Lassen, J. 1972. *Yersinia enterocolitica* in drinking water. *Scandinavian Journal of Infectious Diseases* 4:125–127.

Lennette, E.H., Albert Balows, William J. Hausler Jr., and H. Jean Shadomy (eds.). 1985. *Manual of Clinical Microbiology*. 4th ed. American Society for Microbiology, Washington, D.C. 1149 pp.

Lindel, S.S. and Quinn, P. 1973. *Shigella sonnei* isolated from well water. *Journal of Bacteriology* 26:424–524.

Mackenthun, K.M. and Ingram, W.M. 1967. *Biological Associated Problems in Freshwater Environments*. Federal Water Pollution Control Administration, U.S. Department of the Interior, Washington, D.C. 287 pp.

McCabe, L.J., Symons, J.M., Lee, R.D., and Robeck, G.G. 1970. Survey of community water supply systems. *Journal American Water Works Association* 62:670–687.

McCall, R.G., Cotruvo, J.A., Hendricks, D.W., Jakubowski, W., Karlin, R., Logsdon, G., and Neuman, W. 1985. Waterborne *giardia*: it's enough to make you sick. *Journal American Water Works Association* 77:14, 19, 22, 26 and 84.

Mentzing, L-O. 1981. Waterborne outbreaks of *Campylobacter enteritis* in central Sweden. *Lancet* 2:(8242)352–354.

National Academy of Sciences—National Academy of Engineering. 1973. *Water Quality Criteria, 1972*, EPA–R3–73–033–3/73, Washington, D.C.

Niewolak, S. 1974. Distribution of microorganisms in the waters of the Kortowskie Lake. *Polskie Archiwum Hydrobiologii*, 21:315.

Niewolak S. 1987a. Bacteriological water quality of an artificially destratified lake. *Roczniki Nauk Rolniczyck* 101:115–154.

Niewolak, S. 1987b. Microbiological study of an artificially destratified lake. *Roczniki Nauk Rolniczyck* 101:155–172.

O'Connor, J.T., Hemphill, L., and Reach, C.D., Jr. 1982. Removal of virus from public water supplies. *U.S. Environmental Protection Agency*, EPA–600/52–82–024, Cincinnati, Ohio.

Office of Drinking Water. 1978.*Guidance for Planning the Location of Water Supply Intakes Downstream from Municipal Wastewater Treatment Facilities*. Contract No. 68–01–4473. Office of Water Supply, U.S. Environmental Protection Agency, Washington, D.C.

Olivieri, V.P., Kruse, C.W., and Kawata, K. 1977. Microorganisms in urban stormwater. *Environmental Protection Technology Series*, EPA–600/2–77–087, MERL, U.S. Environmental Protection Agency, Cincinnati, Ohio.

Peterson, M.L. 1971. Pathogens associated with solid waste processing: a progress report. *U.S. Environmental Protection Agency*, SW–49r, Cincinnati, Ohio.

Pluntze, J.C. 1984. The need for filtration of surface water supplies: viewpoint. *Journal American Water Works Association* 76:11 and 84.

Popp, L. 1974. *Salmonella* and natural purification of polluted waters. *Zentralblatt bakteriologie hygiene*, Abteilung I. 158:432–445.

Romaninko, V.I. 1971. Total bacterial number in Rybinsk Reservoir. *Mikrobiologiya* (USSR) 40:707–713.

Schiemann, D.A. 1978. Isolation of *Yersinia enterocolitica* from surface and well waters in Ontario. *Canadian Journal of Microbiology* 24:1048–1052.

Schroeder, S.A., Caldwell, J.R., Vernon, T.M., White, P.C., Granger, S.I., and Bennett, J.V. 1968. A waterborne outbreak of gastroenteritis in adults associated with *Escherichia coli. Lancet* 6:737–740.

Smith, R.J. and Twedt, R.M. 1971. Natural relationships of indicator and pathogenic bacteria in stream waters. *Journal Water Pollution Control Federation* 43:2200–2209.

Streeter, H.W. 1929. Sewage-polluted surface waters as a source of water supply. *Public Health Reports* 43:1498–1522.

Swayne, M.D., Boone, G.H., Bauer, D., Lee, J.S. 1980. Wastewater in receiving waters at water supply abstraction points. *U.S. Environmental Protection Agency*, EPA–600/2–80–044, Cincinnati, Ohio.

Symons, J.M., Carswell, J.K., and Robeck, G.G. 1970. Mixing water supply reservoirs for quality control. *Journal American Water Works Association* 62:322–334.

Taylor, D.N., McDermott, K.T., and Little, J.R. 1983. *Campylobacter enteritis* associated with drinking water in back country areas of the Rocky Mountains. *Annals of Internal Medicine* 99:38–40.

U.S. Department of Health, Education, and Welfare. 1965. *Report on Pollution of Interstate Waters of the Red River of the North (Minnesota, North Dakota)*. Public Health Service, Field Investigation Branch, Cincinnati, Ohio.

U.S. Environmental Protection Agency, Office of Water Quality. 1971. *Report on Missouri River Water Quality Studies*. Regional Office, Kansas City, Missouri.

U.S. Environmental Protection Agency, 1976. *National Interim Primary Drinking Water Regulations*. Office of Water Supply EPA–570/9–76–003. Washington, D.C.

U.S. Environmental Protection Agency. 1979. Waterborne transmission of giardiasis. Jakubowski, W. and Hoff, J.C., (eds.), EPA–600/9–79/001, Office of Research and Development, Cincinnati, Ohio, 306 pp.

Vander Velde, T.L. and Mack, W.M. 1973. Polio-virus in water supply. *Journal American Water Works Association* 65:345–348.

Van Donsel, D.J. and Geldreich, E.E. 1971. Relationships of salmonellae to fecal coliforms in bottom sediments. *Water Research* 5:1079–1087.

Vogt, R.L., Sours, H.E., Barrett, T., Feldman, R.A., Dickinson, R.J., and Witherall, L. 1982. *Campylobacter enteritis* associated with contaminated water. *Annals of Internal Medicine* 96:292–296.

Waser, E., Husmann, W., and Blockliger, S. 1934. Die Glatt Eine Systematische, Praktischen 2 weeken Dienende Flussuntersuchung in Chemischer, Bakteriologischer und Biologischer Richtung. *Berichte der Schweizerischen Botanischen Gesellschaft* 43:253–388.

Water Supply Programs Division. 1971. *Manual for Evaluating Public Drinking Water Supplies*. PB 245–006, 72 pp. U.S. Environmental Protection Agency, Washington, D.C.

Weibel, S.R., Anderson, R.J., and Woodward, R.L. 1964. Urban land runoff as a factor in stream pollution. *Journal of Water Pollution Control* 36:914–924.

Weibel, S.R., Weidner, R.B., Cohen, J.M., and Christianson, A.G. 1965. Pesticides and other contaminants in rainfall and runoff. *Journal of Water Pollution Control* 37:1075–1084.

Weiss, C.M. and Oglesby, R.T. 1960. Limnology and water quality of raw water in impoundments. *Public Works* 91:97–101.

Woodward, W.E., Hirschhorn, N., Sack, R.B., Cash, R.A., Brownlee, I., Chickadonz, G.H., Evans, L.K., Shephard, R.N., and Woodward, R.C. 1974. Acute Diarrhea on an Apache Indian reservation. *American Journal of Epidemiology* 99:281–290.

Wuhrmann, K. 1972. Stream purification pp. 119–151 in *Water Pollution Microbiology*. Vol. 1, Chap. 6, R. Mitchell (ed.), Wiley-Interscience, New York, NY.

2

Tropical Source Water

Terry C. Hazen[1,2] and Gary A. Toranzos[2]

[1]*Environmental Sciences Division, Savannah River Laboratory, Westinghouse Savannah River Company, Aiken, South Carolina and* [2]*Microbial Ecology Laboratory, Department of Biology, University of Puerto Rico, Río Piedras, Puerto Rico*

Over 2 billion people, or half of the world's population, have suffered from diseases due to drinking polluted waters (Barabas, 1986). More than 250 million new cases of waterborne disease are reported each year, resulting in more than 10 million deaths and nearly 75% of these waterborne disease cases occur in tropical areas. Indeed, nearly 50% of diarrheal disease deaths (4.6 million) occur in children under 5 years of age, living in the tropics (Snyder and Merson, 1982; Bockemühl, 1985). Many investigators and government administrators assume that the high morbidity and mortality rates simply indicate the level of contamination of water in tropical areas. However, determining biological contamination in tropical source water is much more difficult than most regulatory agencies perceive. Yet, the need to accurately determine the level of biological contamination is much greater in tropical areas than it is in temperate areas since these regions have a much greater number of waterborne diseases (Table 2.1 and 2.2). In turn, these diseases are exacerbated by the lack of adequate sewage treatment and a greater reliance on untreated (and possibly contaminated) waters as drinking water sources (Feachem, 1977; Esrey et al., 1985).

Source water quality in most tropical areas differs from that of temperate areas in three major ways: 1) physical and chemical; 2) biological; and 3) social and economic. We will examine each one of these categories, compare the standard microbial indicators in tropical source waters, and then look at techniques which may be more applicable to the tropics.

2.1 Physical and Chemical Characteristics of Tropical Waters

The physical and chemical differences between temperate and tropical source water are perhaps the most obvious, yet are quite often the most overlooked. Water temperatures in tropical areas are higher than those in temperate areas.

Table 2.1 Waterborne infectious bacteria

Organism	Disease	Infectious dose
Acinetobacter calcoaceticus	Nosocomial infections	?
Aeromonas hydrophila[a]	Enteritis, wounds	?
A. sobria[a]	Enteritis, wounds	?
A. caviae[a]	Enteritis, wounds	?
Campylobacter jejuni[a]	Enteritis	?
C. coli[a]	Enteritis	?
Chromobacterium violaceum[a]	Enteritis	?
Citrobacter spp.[a]	Nosocomial infections	?
Clostridum perfringens type C[a]	Enteritis	?
Enterobacter spp.[a]	Nosocomial	?
Escherichia coli serotypes[a]	Enteritis	$>10^6$ CFU[b]
Flavobacterium meningosepticum	Nosocomial, meningitis	?
Francisella tularensis	Tularemia	10 CFU
Fusobacterium necrophorum	Liver abscesses	10^6 CFU
Klebsiella pneumoniae[a]	Nocosomial, pneumonia	?
Leptospira icterohaemorrhagia[a]	Leptospirosis	?
Legionella pneumophila[a]	Legionellosis	>10 CFU
Morganella morganii[a]	Urethritis, nosocomial	?
Mycobacterium tuberculosis[a]	Tuberculosis	?
M. marinum	Granuloma	?
Plesiomonas shigelloides[a]	Enteritis	?
Pseudomonas pseudomallei	meliodosis	?
Salmonella enteritidis[a]	Enteritis	$>10^6$ CFU
S. montevideo B[a]	Salmonellosis	?
S. paratyphi A&B[a]	Paratyphoid fever	?
S. typhi[a]	Typhoid fever	10^5 CFU
S. typhimurium[a]	Salmonellosis	$>10^5$ CFU
Serratia marcescens[a]	Nosocomial	?
Shigella dysenteriae[a]	Dysentery	?
Staphylococcus aureus[a]	Wounds, food poisoning	?
Vibrio alginolyticus[a]	Wounds	?
V. cholerae[a]	Cholera dysentery	10^3 CFU
V. fluvialis[a]	Enteritis	?
V. mimicus[a]	Enteritis	?
V. parahaemolyticus[a]	Enteritis	10^5 CFU
V. vulnificus[a]	Wound infections	?
Yersinia enterocolitica	Enteritis	10 CFU

[a] Found in the tropics.
[b] CFU, colony forming units (see Hazen et al., 1987; Dufour, 1986; Hutchinson and Ridgway, 1977; Hawkins et al., 1985).

Table 2.2 Other waterborne pathogens

Organism	Disease	Infectious dose
Viruses		
Adenovirus[a]	Enteritis, pharyngitis	1–10 PFU[c]
Calicivirus[a]	Enteritis	1–10 PFU
Norwalk virus[a]	Enteritis	1–10 PFU
Coronavirus	Enteritis	1–10 PFU
Coxsackievirus A & B	Meningitis	1–10 PFU
Echo virus	Enteritis, Meningitis	1–10 PFU
Hepatitis A virus[a]	Hepatitis	1–10 PFU
Poliovirus[a]	Poliomyelitis	1–10 PFU
Rotavirus[a]	Enteritis	1–10 PFU
Astrovirus	Enteritis	1–10 PFU
Cyanobacteria		
Cylindrospermopsis spp.[b]	Hepatoenteritis	Bloom
Fungi		
Candida spp.[a]	Candidiasis	?
Rhinosporidium seeberi[b]	Rhinosporidiosis	?
Protozoa		
Balantidium coli[a]	Balantidiasis	?
Cryptosporidium spp.[a]	Cryptosporidiosis	?
Giardia lamblia[a]	Giardiasis	1 cyst
Entamoeba histolytica[a]	Dysentery	1 cyst
Naegleria fowleri[a]	Meningoencephalitis	?
Acanthamoeba spp.[a]	Meningoencephalitis	?
Helminths		
Schistosoma mansoni[b]	Schistosomiasis	1 cercariae
S. haemotobium[b]	Schistosomiasis	1 cercariae
S. japonicum[b]	Schistosomiasis	1 cercariae
S. intercalatum[b]	Schistosomiasis	1 cercariae
S. mekongi[b]	Schistosomiasis	1 cercariae
Fasciola hepatica[a]	Fascioliasis	1 metacercariae
Fasciolopsis buski[b]	Fasciolopsiasis	1 metacercariae
Paragonimus westermani[a]	Paragonimiasis	1 metacercaria
Clonorchis sinensis[a]	Chinese liver fluke	1 metacercaria
Diphyllobothrium latum	Pernicious anaemia	1 pleurocercoid
Dracunculus mediensis[a]	Guinea worm	1 larvae
Ascaris lumbricoides[a]	Ascariasis	1 larvae

[a] Found in the tropics.
[b] Found exclusively in the tropics.
[c] PFU, plaque forming units (see Hazen et al., 1987; Dufour, 1986; Hutchinson and Ridgway, 1977; Hawkins et al., 1985).

Water temperature rarely gets below 15°C in the tropics and may be as high as 45°C in some areas, while temperate waters range from freezing to 20°C, and under rare instances may reach 30°C (Hill and Rai, 1982). Temperature differences over the course of the year are also much greater in temperate areas than in the tropics. In Puerto Rico, water temperatures in the rain forest are always between 18° to 24°C (Carrillo et al., 1985; López et al., 1987). Inland lakes and coastal waters in South America range from 23° to 33°C (Hill and Rai, 1982; Hagler and Mendonça-Hagler, 1981). In West Africa, the annual range in water temperature is 25° to 29°C (Wright, 1986), and in East Africa the annual range in river water temperatures is 23° to 34°C (Oluwande et al., 1983). Singapore water temperatures only vary from 28° to 30°C (Jen and Bell, 1982). Even in subtropical areas like Florida, water temperatures range from 8°C during the winter to more than 30°C during the summer (Tamplin et al., 1982). The effects that these differences in temperature regimes may have upon bacterial communities may be profound. Jen and Bell (1982) found that in tropical waters, significantly greater counts of bacteria were obtained when media were incubated at 30°C for 72 h. In contrast, Holden (1970) found that 22°C for 72 h gave the highest counts in temperate waters. Hill and Rai (1982) observed that bacterial activity in central Amazonian lakes were almost entirely due to mesophiles, whereas temperate lakes had a mixture of both psychrotrophs and mesotrophs.

Thermal stratification of reservoirs and lakes is also affected by seasonal temperature changes. Nearly all lentic habitats in the tropics are oligomictic, i.e. permanently stratified (Hutchinson, 1975). Because of the small variations in surface temperatures, the relative thermal distance to mixing is nearly 3 times greater in tropical lakes as compared to temperate lakes (Hill and Rai, 1982). Since the solubility of oxygen is inversely related to temperature, tropical waters also have less dissolved oxygen at saturation and become anoxic faster than temperate areas. Tropical lakes and rivers typically have large diurnal variations in dissolved oxygen concentrations. Rivers in Nigeria were observed to change from 0.5 to 7.5 mg l^{-1} dissolved oxygen in less than 12 hours (Oluwande et al., 1983). In Puerto Rico, dissolved oxygen concentrations also varied from 1 to 8 mg l^{-1} on a diurnal basis (López et al., 1987; Carrillo et al., 1985; Pérez-Rosas and Hazen, 1988). These large changes in dissolved oxygen mean that few organisms other than facultatively anaerobic bacteria can thrive. Indeed, Carrillo et al. (1985) found higher densities of anaerobic bacteria in a pristine tropical stream than did Daily et al. (1981) on the Anacostia River, Washington, D.C., a grossly polluted, temperate river. Unfortunately, few other studies have looked at anaerobes in tropical waters. The higher average temperature of the tropics and the more constant temperature during the year are going to have profound effects on the flora and fauna found in tropical waters, including the microbial community and allochthonous species, for example, pathogens and their indicators.

Annual variations in light intensity are much smaller in the tropics (Hill and Rai, 1982). It has been well documented that bacteria vary greatly in their sensitivity to solar radiation, and some bacteria can be killed at high natural light intensities (McCambridge and McMeekin, 1981; Evison, 1988). Fujioka and Navikawa (1982) showed that the natural solar radiation in Hawaii could significantly reduce the density of coliforms and other indicators of biological pollution as well as cause injury in tropical source waters. Thus, the higher solar radiation of tropical areas could cause a significant underestimation of indicator bacteria in tropical source waters, as well as in areas receiving large amounts of solar radiation.

The constant high temperature and light intensity of tropical surface waters cause them to be in a state of hypereutrophy all year long. In temperate areas, light is quite often limiting to photosynthetic organisms during the winter, thus lowering productivity of natural communities and uptake of nutrients. In the tropics, light intensity changes vary little, with many areas having 12 h of darkness and daylight all year long. Light is usually only limiting in the tropics under dense rain forest canopies (Carrillo et al., 1985). In tropical source waters, concentrations of organics are high, while free forms of nutrients like phosphorus and nitrogen are low (Hill and Rai, 1982). More than 90% of nutrients like phosphorus and nitrogen may be tied up in the standing biomass at any given moment in tropical waters, leading to deficiencies if there are no significant allochthonous inputs (Hill and Rai, 1982). In contrast, temperate lakes vary widely in productivity during the year, owing to large changes in light intensity and temperature. In terms of nutrients and other physical chemical parameters, tropical waters are most similar to temperate waters during the late summer in eutrophic areas of the lower temperate latitudes. At this time, hypolimnetic waters are anoxic, temperatures are high, and free nutrient concentrations are low, even though standing biomass is high. Thus tropical source waters have higher densities of autochthonous bacteria than temperate waters. These higher densities of naturally occurring bacteria can cause severe limitations to assays which rely upon viable counts. This can be seen in the much greater tendency of allochthonous bacteria to overgrow standard media in tropical source waters (López et al., 1987; Santiago-Mercado and Hazen, 1987). In addition, several studies have shown that nonindicator background bacteria can produce inhibitory substances (bacteriocins) that affect the growth of indicator bacteria like *E. coli* (Means and Olson, 1981). High densities of other bacteria may also affect the differentiation of colonies on some media, for example, sheen of coliform colonies has been shown to be inhibited by high densities of nontarget bacteria (Burlingame et al., 1984). Thus, factors that influence community composition and trophic state can have a major effect on the survival and growth of indicator bacteria in tropical source water.

Water fluxes in wet tropical areas are much greater than in temperate areas. Rainfall in many tropical areas is catastrophic, with some watersheds receiving more than 61 cm in 24 h (Carrillo et al., 1985). This intensity of rainfall flushes microbes and nutrients from the soil and vegetation into the watershed, causing extreme changes in allochthonous nutrient and microbe input (Carrillo et al., 1985; Hill and Rai, 1982; Oluwande et al., 1983). In addition, many rain forest streams are literally scoured clean during such rain-falls, due to the tremendous changes in volume and velocity of the water. Subsequently, the benthos is poorly developed in these streams. Turbidity is also quite high during these catastrophic rainfalls due to washing of soil into the stream. In tropical areas where rainfall is seasonal, some streams may dry up during the dry season and flood during the rainy season; however, these seasons usually are unrelated to temperate seasons and may vary greatly in length from year to year (Wright, 1986). Hill and Rai (1982) found that pro-ductivity and total bacterial counts were inversely related to water levels in Central Amazon lakes, yet studies in Africa (Barrell and Roland, 1979; Olu-wande et al., 1983), Puerto Rico (Carrillo et al., 1985), and Hawaii (Fujioka and Shizumura, 1985) found that densities of total bacteria and coliforms in-creased with increasing rainfall. Rainfall also exerts a much greater influence on source water quality in tropical areas than it does in temperate regions.

2.2 Biological Characteristics of Tropical Waters

Wet tropical and subtropical forests alone account for more than 50.5% of the total gross production in terrestrial biospheres (Odum, 1971). As discussed above the higher light intensities and temperature increase the productivity of tropical source waters. Thus, most tropical source waters are hypereutrophic (Hill and Rai, 1982). The dominant photosynthetic species in tropical fresh-waters are nearly always cyanobacteria or other photosynthetic bacteria. Since limiting nutrients are usually phosphorus or nitrogen, the faster adsorption rate of these smaller organisms provides a favorable characteristic. Temperate source water, on the other hand, is usually dominated by diatoms or green algae in the spring, green algae in the summer, cyanobacteria in the late summer and early fall, and diatoms in the winter (Hutchinson, 1975). These differences in dominant phytoplankton will also affect the type and quantity of grazing or-ganisms and detritivores in the water. Thus, not only the quantity, but the quality of bacterial resources, that is, substrates and nutrients, are different in tropical waters. Indeed, the entire food chain is usually quite different.

The higher productivity and temperature of the tropics create an environ-ment that is high in organics and dominated by mesophilic and thermotolerant microorganisms (Hill and Rai, 1982; Santiago-Mercado and Hazen, 1987). The psychrophilic microflora of temperate waters is virtually nonexistent in tropical waters. The microflora of tropical waters is thus much more similar to the microflora of animals. In fact, it has been well demonstrated that anthropogenic protozoa, fungi, and bacteria not only survive but may grow in tropical waters. Our own studies have indicated that *Escherichia coli*, including fecal biotypes, may be a naturally occurring bacterium in tropical rain forest watersheds (Ha-zen et al., 1988; Rivera et al., 1988; Bermúdez and Hazen, 1988). The finding of *E. coli* in pristine environments is extremely unusual because this bacterium inhabits the intestine of warm-blooded animals and its presence is only expected in environments that have been exposed to recent fecal contamination. Furthermore, it seems that this bacterium is capable of surviving indefinitely in tropical environments (Carrillo et al., 1985; Valdés-Collazo et al., 1987; López-Torres et al., 1987). This suggests that *E. coli* could be a natural inhabitant in these environments and that it may be part of a previously established community.

The lack of seasonality in many tropical areas means that reproduction by plants and animals occurs all year round (Whitaker, 1975). Thus input of degradative substances (e.g. pollen, seeds, embryos, etc.) is not seasonal as it is in temperate areas. Recruitment into tropical populations and senescence are also more constant than in temperate areas where the feast or famine environment causes dramatic changes in the degradative community structure of the water, and therefore the background flora that fecal bacteria must be contrasted to and/or compete with.

In general the microbial diversity of tropical source water is greater than temperate source water, as are the floral and faunal communities within the watershed. Thus, not only is there a greater array of autochthonous microorganisms, but there is also a larger variety of allochthonous microbes and substances entering tropical waters from the surrounding milieu (Hill and Rai, 1982). Tropical microbial flora, environmental characteristics, and survival and activity characteristics of allochthonous microbes are quite dissimilar from temperate waters.

Tropical environments, as would be expected, harbor a much greater array of water borne pathogens, including many of which are totally unknown in temperate areas (Tables 2.1 and 2.2). Yet, there are few temperate waterborne pathogens that are not also found in the tropics. Thus, indicator organisms must cover a wider range and diversity of pathogens in tropical waters than in temperate waters.

2.3 Social-Economic Characteristics of Tropical Waters

More than (65%) of the world population lives in tropical areas, yet these people have less than (10%) of the world's wealth (Odum, 1971). Research and development by these nations is unaffordable and virtually nonexistent. Tropical nations have accepted regulatory recommendations of the industrialized temperate nations without any program of verification, so that temperate recommendations for fecal coliform indices in potable and recreational water are used universally in the tropics. Seldom are the underlying assumptions of temperate water-quality assays tested or even considered before being put into legislation by nations in the tropics. Yet economic hardships may be exacerbated by attempts to meet unrealistic standards. Indeed, national pride and the opinion of other peoples and nations is being severely damaged by inability to meet legislated water quality standards. Thus as stated by Odum (1971), underdeveloped nations may become "never-to-be-developed nations."

The poorness of most tropical nations has also meant that large portions of their populations do not have access to sewage disposal and potable water. There is no alternative to proper water treatment, especially in areas where quantities of source water are limited. But, treatment is a costly enterprise. Many water-treatment plants have an inconsistent supply of chlorine (Toranzos et al., 1986; Carrillo et al., 1985). Since complete treatment is only achieved when there is continuous disinfection, it would be unrealistic to expect such water treatment plants to meet the current standards for potable water. Monitoring for the presence of bacteria in raw source water and finished water is prohibitive in most cases. In fact, water distribution systems are seldom found in many tropical areas. This creates a much greater reliance on point of use surface waters (Esrey et al., 1985; Blum et al., 1987). The lack of public education as well as the persistence of primitive cultures and taboos have also perpetuated the poor health of the people, by continued use of unprotected traditional water sources that are frequently heavily contaminated with human pathogens (Feachem, 1977; Esrey et al., 1985; Blum et al., 1987; Jiwa et al., 1981). Many preventable diseases that are virtually unknown in industrialized temperate nations (Table 2.1 and 2.2) exact a horrible toll on the peoples of the tropics (Feachem, 1977; Snyder and Merson, 1982).

2.4 Indicators of Biological Pollution in Tropical Source Waters

The literature is sparse with specific references to fecal contamination in the tropics, yet there are some notable exceptions. Evison and James (1973) reviewed literature from Ceylon, Egypt, India, and Singapore and reported that densities of *Escherichia coli* in water did not seem to coincide with known sources of fecal contamination. Feachem (1974), working in rural areas of New Guinea, found fecal coliform densities from 0 to 10,000 colony-forming units (CFU) per 100 ml and fecal streptococcus levels from 0 to 6,000 CFU per 100 ml, although the mean densities for both indicators were greater than 100 CFU per 100 ml. All sites sampled were totally unacceptable for drinking water, as far as levels of indicator bacteria were concerned, and Feachem concluded that all sites were grossly contaminated with fecal material. Yet he states that fecal coliform (FC) and fecal streptococci (FS) densities were more closely correlated with the number of domestic animals than the number of people in the watershed. In fact, FC and FS densities were lowest at sites where human population densities were high. In addition, as in most studies from tropical countries, Feachem did not confirm presumptive fecal coliform isolates. In Sierra Leone, Wright (1982) reported densities of fecal coliforms from 40 to 240,000 per 100 ml using a most probable number method (MPN) and fecal streptococci densities from 7 to 64,000 CFU per 100 ml for water sources used by 29 settlements. However, no correlation was found between any of the fecal indicators that Wright measured and the presence of *Salmonella* spp., which would indicate a real health risk. Fujioka and Shizumura (1985) reported densities of fecal coliforms and fecal streptococci in tropical streams ranging from 100 to 10,000 CFU per 100 ml, including streams not known to be contaminated with sewage. Oluwande et al. (1983) found that rivers in Nigeria had total coliform counts ranging from 8 to 100,000 CFU per 100 ml. In this study it was also assumed that the high densities of fecal indicators meant that these waters were heavily contaminated with human feces, even though data showed that densities of total coliforms were quite often higher upstream from known contamination sources.

Thomson (1981) isolated *Salmonella* spp. from wells used for drinking water in Botswana and found no correlation between densities of either total coliforms, fecal coliforms, or *E. coli* and the presence of *Salmonella* spp. Lavoie and Viens (1983) reported that 95% of the traditional water sources in the Ivory Coast, West Africa had unacceptably high densities of fecal coliforms (11,421 CFU per 100 ml), yet fewer than 55% of the positive total coliform isolates were actually *E. coli* and less than 66% of the fecal coliform isolates were actually *E. coli*. However, when Lavoie (1983) examined the feces of local inhabitants he found that 92% of total coliform isolates were *E. coli* and 89% of the fecal coliform isolates were *E. coli*. Few, if any, of these studies have carefully examined and tested the underlying assumptions of the fecal-pathogen indicator being used. Bonde (1977) stated these criteria as follows:

1. The indicator must be present whenever pathogens are present.
2. It must be present only when the presence of pathogenic organisms is an imminent danger.
3. It must occur in much greater number than the pathogens.
4. It must be more resistant to disinfectants and to aqueous environments than the pathogens.
5. It must grow readily on relatively simple media.
6. It must yield characteristics and simple reactions enabling as far as possible an unambiguous identification of the group or species.
7. It should preferably be randomly distributed in the sample to be tested.
8. Its growth on artificial media must be largely independent of any other organism present.

In the late 1800's Houston proposed the idea of using three groups of bacteria (coliforms, fecal streptococci, and the gas-producing clostridia which are commonly found in the feces of warm blooded animals) as indicators of fecal pollution of waters (Hutchinson and Ridgway, 1977). He argued that since these groups could only come from fecal sources, their presence would indicate recent fecal pollution (Hutchinson and Ridgway, 1977). For nearly eighty years the coliform group of bacteria has been used as such indicators. The indicator used universally to assess biological contamination of water is *E. coli*, in both tropical and temperate countries (Barbaras, 1986).

The first drinking water regulation for microbial contamination in the U.S. was published in 1914. This was the first Public Health Service Drinking Water Standard. Subsequently, this regulation was replaced by U.S. Public Service Acts of 1915 and 1962. The current U.S. regulation comes from the Safe Drinking Water Act (Public Law 93–523, 1974). The U.S. Environmental Protection Agency proposed changes that are now being implemented (Federal Register 48:45502–45521, October, 1983). The revised law was approved in July 1986 and is currently in its first phase of implementation. The new regulation requires that there should be 0 coliforms/100 ml by any method, for any sampling frequency for drinking waters. World Health Organization (WHO) (Barbaras, 1986) allows 10 coliforms/100 ml for small community water sources. For tropical nations and tropical parts of the United States, even the old regulations may be unrealistic. As observed in Botswana (Thomson, 1981) and Sierra Leone (Wright, 1982), no correlation could be found between the presence of *Salmonella* spp. and the presence of *E. coli* in tropical source water. In Puerto Rico, densities of pathogenic yeast (Valdés-Collazo et al., 1987), *Klebsiella pneumoniae* (López et al., 1987), *Legionella pneumophila* (Ortiz-Roque and Hazen, 1987), *Vibrio cholerae* (Pérez-Rosas and Hazen, 1988), *Yersinia enterocolitica* (Elías-Montalvo et al., 1988), and *Aeromonas hydrophila* (Hazen et al., 1981) were also found to be unrelated to densities of *E. coli* in source water. Thus pathogens could be present in the absence of *E. coli* in tropical source water. It has also been demonstrated that high densities of enteric viruses, a dominant cause of waterborne disease in both temperate and tropical areas, may be found in the complete absence of fecal coliforms of *E. coli* (Berg and Metcalf, 1978; Toranzos et al., 1986a; Toranzos et al., 1986b; Keswick et al., 1985; Rose et al., 1985; Cabelli, 1983; Hejkal et al., 1982; Herrero and Fuentes, 1977). However, a large number of diverse studies has also shown that fecal coliforms and *E. coli* may become injured, such that they do not grow on standard media and/or under standard incubation conditions (Bissonnette et al., 1975; Bissonnette et al., 1977; Stuart et al., 1977). The high light intensities found in the tropics could certainly contribute to this injury phenomenon (Fujioka and Navikawa, 1982).

In Nigeria (Oluwande et al., 1983), Hawaii (Fujioka and Shizumura, 1985), New Guinea (Feachem, 1974), Puerto Rico (Carrillo et al., 1985), Sierra Leone (Wright, 1982), and the Ivory Coast (Lavoie, 1983), high densities of *E. coli* were found in the complete absence of any known fecal source. The U.S. Geological Survey reported that 54 out of 67 water sampling stations on rivers in Puerto Rico exceeded the recommended MCL/s for recreational water (< 1,000 fecal coliforms per 100 ml) during 1984 (Curtis et al., 1984). Thus only 19% of all sites sampled met the recommended MCL for recreational waters, and none of these waters could meet raw source water standards (<2 fecal coliforms per 100 ml). These findings have resulted in condemnation of sewage treatment in Puerto Rico as a source of fecal pollution of natural waters by regulatory agencies (Hazen et al., 1987). Yet, samples taken upstream from sewage-treatment plant outfalls had fecal coliform densities that were just as high as those of most downstream sites.

Recent studies have shown that *E. coli* can be isolated from pristine areas of tropical rain forests in Puerto Rico (Rivera et al., 1988; Bermúdez and Hazen, 1988). Plasmid profiles, antibiotic sensitivities, coliphage susceptibility, and physiological and biochemical characteristics confirm that even *E. coli* isolated from epiphytes in trees 15 m above the ground are very similar to clinical isolates of *E. coli* (Rivera et al., 1988). The DNA of these environmental isolates have identical mol% $G+C$ and more than 85% DNA homology with *E. coli* B (Hazen et al., 1987; Bermúdez and Hazen, 1988). Thus, tropical source waters may not only have high densities of *E. coli* in the absence of pathogens or fecal sources, but *E. coli* may be naturally occurring in some tropical areas. This observation is further corroborated by the high densities of *E. coli* in diverse tropical countries in the absence of identifiable fecal sources. *Escherichia coli* is usually found in higher densities than other enteric pathogens. However, as noted in Botswana (Thomson, 1981) and Sierra Leone (Wright, 1982), *Salmonella* spp. may be found in the complete absence of *E. coli*. In addition, *Legionella* spp. could also be found in Puerto Rico in the complete absence of *E. coli* (Ortiz-Roque and Hazen, 1987).

Gordon and Fliermans (1978) demonstrated that *E. coli* could survive for much longer periods of time in a temperate lake receiving thermal effluent. Thus, environmental factors in the tropics might also significantly affect the survival of fecal coliforms in the environment. Bigger (1937) in India first reported the growth of coliforms in tropical waters. Ragavachari and Iyer (1939) showed that coliforms can survive for several months in natural tropical river waters. In Puerto Rico, several studies (Carrillo et al., 1985; López et al., 1987; Valdés-Collazo et al., 1987; Hazen et al., 1987) have shown that *E. coli* not only survives in rain forest streams but also proliferates. Thus, once introduced into the environment, *E. coli* could remain and/or become part of the normal flora. Recent studies by Xu et al. (1982), Colwell et al. (1985), and Roszak et al. (1984) have shown that enumeration on complex media like those used to culture *E. coli* may not be reliable. These studies have demonstrated that *E. coli* and pathogens like *Salmonella enteridis*, and *Vibrio cholerae* may be able to survive and remain pathogenic and yet be unculturable on standard media. Under in situ conditions in a tropical rain forest, *Yersinia enterocolitica* (Elías-Montalvo et al., 1988), *Klebsiella pneumoniae* (López et al., 1987), *Candida albicans* (Valdés-Collazo et al., 1987), *Aeromonas hydrophila* (Hazen et al., 1982), *Salmonella typhimurium* (Jiménez et al., 1989), and *Vibrio cholerae* (Pérez-Rosas, 1984), were all shown to have shorter survival rates than *E. coli*, their presumed indicator. The inability of *E. coli* to survive in a manner similar to pathogens would further impede its ability to indicate the presence of those pathogens in waters. The time of survival of *E. coli* in source waters of temperate climates has been shown under numerous circumstances to be days, with densities decreasing by more than 90% every 60 min (Bonde, 1977). Regrowth of *E. coli* outside the intestinal tract in temperate areas has only been rarely observed (Bonde, 1977). As seen in Table 2.3, *E. coli* always survives significantly longer in situ in tropical waters as compared to temperate waters. The increased survival of *E. coli* in tropical waters suggests that temperate maximum contaminant levels based on fecal coliforms can not be justified for tropical waters.

Pagel et al. (1981) compared four fecal coliform assays in various types of freshwaters in Southern Canada. They observed that while these assays were somewhat variable in their abilities to detect fecal coliforms from environmental samples, they were all acceptable in terms of their specificity and selectivity. In similar studies, Santiago-Mercado and Hazen (1987) and Hazen et al. (1987) used the same methodology to detect fecal coliforms from freshwaters in Puerto Rico and found that the specificity of the media (determined by the ability of the medium to restrict growth of organisms other than the target bacterium) was at least 20% less than the specificity reported by the Canadian investigators (Table 2.4). Thus, all the methods gave significantly higher false-positive and false-negative errors with tropical water samples.

Table 2.3 Comparison of indicator and pathogen survival in *in situ* tropical and temperate source water

		Density	Survival (h)*	Reference
Indicators				
E. coli	Temperate	10^9	50	Gordon and Fliermans (1978)
		10^5	30.6	McFeters et al. (1974)
	Tropical	10^7	**	Carrillo et al. (1985)
		10^7	294	López et al. (1987)
		10^6	206	Valdez-Collazo et al. (1987)
S. faecalis	Tropical	10^7	226	Muñiz et al. (1988)
B. adolescentis	Tropical	10^7	92.6	Carrillo et al. (1985)
Pathogens				
V. cholerae	Temperate	10^5	13	McFeters et al. (1974)
	Tropical	10^7	198	Pérez-Rosas (1983)
Y. enterocolitica	Tropical	10^8	124	Elías et al. (1988)
C. albicans	Tropical	10^6	**	Valdés-Collazo et al. (1987)
A. hydrophila	Temperate	10^5	**	McFeters et al. (1974)
		10^7	75	Hazen and Esch (1983)
		10^8	100	Fliermans et al. (1977)
	Tropical	10^7	**	Hazen et al. (1982
S. typhimurium	Temperate	10^5	28.8	McFeters et al. (1974)
	Tropical	10^8	131	Jiménez et al. (1987)
K. pneumoniae	Tropical	10^6	125	López et al. (1987)
P. aeruginosa	Tropical	10^5	**	Cruz (1987)

* Survival time is the T_{90} time to reach 90% reduction of initial cell density.
** Survival time indefinite.

Table 2.4 Comparison of fecal coliform assay methods in tropical and temperate source waters

Performance parameter	MMB		mFC1		mFC2		mTEC	
	Temp*	Trop*	Temp	Trop	Temp	Trop	Temp	Trop
Accuracy[a]	59	67	89	84	100	105	94	93
Specificity[b]								
False positive	11	39	16	30	18	19	13	36
Undetected	4	25	1	20	2	11	2	21
Selectivity[c]	88	66	85	72	90	82	86	66
Comparability[d]								
FC recovery	26	41	41	75	48	94	45	73
Non-FC recovery	4	30	11	29	15	19	7	18
Overall rank[e]	2	4	3	2	1.5	1	1.5	3

* *Temp*, temperate (data of Pagel et al. (1981). *Trop*, tropical (data of Santiago-Mercado and Hazen, 1987). MMB = MacConkey membrane broth, mFC1 = Fecal coliform media with normal incubation, MFC2 = fecal coliform media with resuscitation at 35 °C for 2h, mTEC = mTEC media.

[a] Mean percentage number of colonies on test medium/mean number of colonies on a non-selective medium $\times$ 100, using *E. coli* (ATCC 10798 and ATCC 23848).

[b] Percentage false positive error = [false positive error + number of presumptive target colonies $-$ number of verified target colonies]/[total presumptive target colonies] $\times$ 100. Percentage false negative error = false negative counts/[verified + undetected target counts] $\times$ 100.

[c] Selectivity Index = presumptive typical colonies/[presumptive typical or target colonies + presumptive nontypical or nontarget colonies] $\times$ 100.

[d] Percentage fecal coliform (FC) recovery and percentage nonfecal coliform (NFC) recovery.

[e] Best overall efficiency of the method is given by the lowest overall rank.

Controls using known strains of *E. coli* indicated the accuracy of the methods to be the same in both studies (Table 2.4). Identification of more than 300 fecal coliform isolates from various freshwater sites around Puerto Rico showed that less than 40% of these isolates were actually *E. coli* (Santiago-Mercado and Hazen, 1987; Hazen et al., 1987). Similar studies using the same methods in the continental U.S.A. (Dufour et al., 1981), South Africa (Grabow et al., 1981), Canada (Pagel et al., 1982), and England (Evison and James, 1973) have demonstrated that more than 90% of fecal-coliform-positive isolates are identified as *E. coli*. Wright (1982) found that in Sierra Leone waters, fewer than 10% of the positive isolates from fecal coliform assays were *E. coli*. Lavoie (1983) found that less than 66% of the fecal coliform isolates from Ivory Coast well waters were actually *E. coli*. It is not surprising that the thermotolerant *E. coli* encounter more mesophilic and thermotolerant background flora in the tropics, given the higher and more constant temperatures and productivities. Santiago-Mercado and Hazen (1987) showed that high densities of mesophilic and thermophilic background flora in tropical source waters can significantly reduce the numbers of *E. coli* detected on standard fecal coliform media.

Thus *E. coli*, the indicator used by most developed nations for determining biological contamination of source waters, does not fit most of the underlying assumptions of a good indicator in tropical source waters. It is not surprising, therefore, that developing nations in tropical areas are finding it difficult (if not impossible) to meet their own legislated standards for biological contamination.

2.5   Possibilities for Tropical Standards

Fecal coliforms, as discussed above, seem to be unacceptable as indicators of recent fecal contamination in tropical source water. Since the problems with the fecal coliform assay are due in part to *E. coli* being the target organism (because of its increased survival and/or natural occurrence), changes in media and incubation would not correct these problems for tropical analyses. Thus, alternative indicators or techniques will be required to correctly access the degree of fecal contamination and/or public health importance of tropical source waters.

Several investigations of tropical source water have looked at fecal streptococcus densities. Feachem (1974) found that streams in the New Guinea highlands had densities or fecal streptococci from 0 to 6,000 per 100 ml, and the FC/FS ratio ranged from 0.20 to 1.31. Feachem interpreted these results to mean that these waters were 25 times more likely to be contaminated by pigs than humans, based upon the studies of Geldreich and Kenner (1969). However, if *E. coli* is able to survive in these waters, the ratio could be inflated. Because of their high densities, the fecal streptococci would also seem to indicate fecal contamination when none may be present. Wright (1982) found that source water in Sierra Leone had densities of fecal streptococci (7 to 64,000 per 100 ml) that were an order of magnitude lower than densities of either fecal coliforms (40 to 240,000 per 100 ml) or presumptive *E. coli* (30 to 120,000 per 100 ml). When isolates of fecal streptococci were identified, Wright (1982) found that 14 to 100% were confirmed as *Streptococcus faecalis*, the target organism for the assay. In addition, Wright (1982) could find no correlation between the incidence of *Salmonella* spp. in these source waters and the densities of fecal streptococci. Fujioka and Shizumura (1985) also found 100 to 10,000 CFU per 100 ml of fecal streptococci in streams in Hawaii, even though many of these streams were not known to have any source of fecal contamination. Densities of fecal streptococci in Hawaiian streams were also quite similar to densities of fecal coliforms. Studies in Puerto Rico have demonstrated that *S. faecalis* can survive and remain active for long periods of time in diffusion chambers in a tropical rain forest stream (Muñiz et al., 1989). Even tropical marine waters with petroleum contamination could support high densities of *S. faecalis* (Santo Domingo et al., 1989). Monitoring of Puerto Rican waters by the U.S. Geological Survey has shown that like fecal coliforms, fecal streptococci exceeded recommended MCL for recreational waters more than 80% of the time (Curtis et al., 1984). Also, as with fecal coliforms, densities of fecal streptococci were only slightly lower and occasionally higher at sampling stations upstream from sewage-treatment-plant outfalls. Evison and James (1973) suggested that *E. coli* is in much lower proportions, relative to *S. faecalis*, in human feces from tropical countries. McFeters et al. (1974) demonstrated that even in temperate waters fecal coliforms and fecal streptococci have different rates of survival. This would also invalidate the use of the FC/FS ratio for determining the origin of the fecal contamination, as would the higher survival rates of both *E. coli* and *S. faecalis* in the tropics. Though fewer studies have been done in tropical water with the fecal streptococci than with fecal coliforms, the fecal streptococci seem to satisfy only one of Bonde's eight criteria for an ideal indicator.

Evison and James (1975) suggested that *Bifidobacterium* spp. might be a more appropriate indicator for the tropics. Bifidobacteria are obligate anaerobes and thus are not likely to grow outside the intestinal tract. In addition, this genus is always found in the human gut, quite often in higher densities than *E. coli*, and is seldom observed in other animals (Resnick and Levin, 1981). With the advent of the YN-6 medium (Resnick and Levin, 1981), bifidobacteria would seem to be an excellent candidate for a fecal indicator organism in tropical source waters. Carrillo et al. (1985) showed that *Bifidobacterium adolescentis* decreased one order of magnitude each day during in situ exposure in a tropical rain forest stream in Puerto Rico. Gyllenberg et al. (1960) and Evison and James (1975) showed that densities of bifidobacteria were always greater than densities of *E. coli* in contaminated waters in temperate areas. In the Puerto Rico rain forest, however, densities of bifidobacteria-like organisms were greater than *E. coli* at all sites except those at a sewage outfall (Carrillo et al., 1985). In addition identification of isolates from the rain forest stream using the YN-6 medium showed that less than 80% were actually bifidobacteria. Bifidobacteria show promise as an indicator of recent fecal contamination in terms of lack of survival in situ and specificity as a human fecal indicator. Unfortunately, the currently available medium for enumeration (YN-6) is hampered by a lack of specificity and insufficient resolution when background anaerobic bacterial densities are high (Carrillo et al., 1985).

Clostridium perfringens, an anaerobe, has also been suggested as a suitable alternative indicator to fecal coliforms and fecal streptococci in tropical source waters. Fujioka and Shizumura (1985) found that in uncontaminated streams densities of *C. perfringens* were from 0 to 46 CFU per 100 ml, while discharge sites had from 56 to 2,100 CFU per 100 ml. The same study showed that 91% of Hawaiian isolates from membrane *C. perfringens* (mCP) medium were confirmed as *C. perfringens*. This compares favorably with the results of a study by Bisson and Cabelli (1980) in temperate waters. However, densities of *C. perfringens* in human fecal material from the tropics are much lower than any of the previously mentioned indicators (Wright, 1982). This raises serious doubts as to the ability to easily detect *C. perfringens* when pathogens are an immanent danger. Carrillo et al. (1985) found that sites which received heavy rainfall had high densities of total anaerobes (>500,000 per 100 ml). Fujioka and Shizumura (1985) made the observation that densities of *C. perfringens* increased significantly in uncontaminated sites after rainfall. This suggests that *C. perfringens* might live outside the intestinal tract under some conditions. Wright (1982) found that source waters in Sierra Leone had densities of *C. perfringens* from 40 to 1,500 per 100 ml and that they compared favorably with densities of fecal streptococci. He also found that like the fecal streptococci they were unrelated to the isolation of *Salmonella* spp. Like the bifidobacteria, *C. perfringens* seems to satisfy more, but not all, of Bonde's criteria for an ideal indicator than the fecal coliforms or the fecal streptococci. The inability of enteric nonspore-forming anaerobes to proliferate in most source waters makes them more suitable indicators of recent fecal contamination. Unfortunately, techniques involving anaerobes also require more sophisticated and expensive equipment and confirmation methods.

Coliphages have been suggested as indicators that would be more specific and not subject to the cultivation problems of *E. coli* (Berg and Metcalf, 1978; Funderburg and Sorber, 1985; Wentsel et al., 1982). Coliphages can be concentrated from large volumes of water with relative ease and high efficiency. Studies in our laboratory (Toranzos et al., 1988) have shown that when *E. coli* C3000 (ATCC 15597) is used, coliphages are detected only in water receiving sewage effluent. In the latter experiments, one liter volumes of the water samples were concentrated by the viradel technique (Goyal and Gerba, 1983) and the eluate assayed using the above strain of *E. coli*. Pristine as well as sewage-contaminated waters were sampled where indigenous *E. coli* had been isolated previously. Coliphages were consistently isolated from sites receiving sewage effluents whereas none of the other sites contained coliphages. In addition, coliphages (such as MS2, F2, T4, and T7) would not grow on more than 20 indigenous *E. coli* strains. Evison (1988), however, showed that coliphage survival in microcosms simulating temperate environmental conditions were quite variable, depending on the environmental conditions and the coliphage being used. Since the coliphage assay is specific for a particular strain of *E. coli*, it could eliminate the tropical background microflora problem seen when using standard media assays for *E. coli*. However, since the host organism for these viruses is a specific strain of *E. coli*, the viruses might reproduce in situ since this bacterium appears to survive and grow in tropical source waters. Thus, coliphage assays would be appropriate for tropical source waters only if a specific host strain of *E. coli* were demonstrated not to survive outside the intestinal tract. This seems unlikely, since all strains of *E. coli* tested to date survive in tropical freshwaters (Elías-Montalvo et al., 1988; López et al., 1987; Valdés-Collazo et al., 1987; Hazen et al., 1981; Jiménez et al., 1988; Pérez-Rosas, 1984; Carrillo et al., 1985). Coliphages that infect *E. coli* would probably be no better than *E. coli* as an indicator of fecal contamination in the tropics.

As discussed above, nonsporeforming obligate anaerobes that are only found in the intestinal tract of man or similar warm-blooded animals appear not to survive for long periods of time or proliferate in most tropical source waters. The inherent difficulty with assaying for these organisms is their requirement for complex media, anaerobic culture conditions, and differentiation from indigenous anaerobes (Carrillo et al., 1985). Assays might be developed for enumeration of bacteriophages of these bacteria that would have fewer problems than any of the assays discussed so far. Recent studies with bacteriophages of *Bacteroides fragilis* have shown that this approach is plausible (Tartera and Jofre, 1987). There is no reason to believe this approach would not also work for *Bifidobacterium* spp., a group that has already been shown not to survive for extended periods in tropical water and have better human fecal specificity than coliforms or fecal streptococci. Since these methods are hypothetical, or only at an early experimental stage, a major research and testing effort would be required before any of these techniques could be implemented in tropical source waters.

2.6   Perspective and Recommendations

The problems associated with determining fecal contamination of tropical source waters are many and varied (Table 2.5). Certainly the standard fecal indicator, *E. coli*, is unacceptable. Because few studies have reported the use of fecal indicators (other than *E. coli*) in tropical source water, objective evaluations of the efficacy of these alternate indicators is difficult. At present, ob-

Table 2.5 Problems assessing tropical microbiological source water quality

1. Regrowth of indicators in freshwater.
2. Thermotolerant and mesophilic background flora overgrow assays which rely upon the thermotolerant nature of indicators and which assume a psychrophilic background flora.
3. High false-positive natural flora using viable culture techniques.
4. Pathogens found in the absence of indicators, eg, *Salmonella* spp.
5. Higher natural productivity of environments allows grater survival and regrowth of indicators and pathogens.
6. Different flora and fauna of tropics creates different interactions and environments for indicators and pathogens.
7. Torrential rainfall in some areas causes extreme changes in allochthonous input of nutrients and microbes from surrounding environments.
8. Greater number of waterborne diseases.
9. Greater reliance on local surface waters as traditional drinking water sources.
10. Few sewage treatment and water treatment facilities and inadequate monitoring and enforcement.
11. Possible autochthonous origin of indicators.
12. Because of higher annual temperatures people spend more time in direct contact with water.

ligate anaerobes or their phages seem the best candidates for a better indicator for tropical source water, primarily due to their inability to survive outside the intestinal tract. However, all of these indicators have the inherent difficulty that they or their host may survive under some conditions and that the media used for bacterial indicator enumeration may allow the growth of false-positive background flora (Table 2.5). The viable but nonculturable phenomenon reported for many pathogens in both temperate and tropical waters suggests that indicators may only rarely be correlated with disease risk in source waters (Colwell et al., 1985; Hazen et al., 1987; Baker et al., 1983). Thus, the best indicator may be no indicator, that is direct enumeration of selected resistant pathogens. This would allow a more realistic estimation of health risk.

Immunofluorescent staining can detect densities of pathogenic bacteria as low as 10 cells per ml, a density which may give no culturable counts (Fliermans et al., 1981; Colwell et al., 1985). The use of monoclonal antibodies makes this technique specific at even the strain level for some organisms. However, as a result of cross-reactivity when using immunofluorescence (even with monoclonal antibodies), the most specific and sensitive method for detecting pathogens may be nucleic acid probes (DNA or RNA). DNA probes have already been developed and tested for enterotoxigenic *E. coli* (Bialkowska-Habrzanska, 1987; Hill et al., 1983; Moseley et al., 1982) and *Salmonella* spp. (Fitts et al., 1983). Thus, direct detection of pathogens is currently possible. Common enteric pathogens which could be enumerated are poliovirus and *Salmonella typhimurium*. Detection of either one of these in tropical source water would indicate risk of human disease. Instead of enumeration, maximum contaminant levels could be based on detection only. One potential problem with this approach is that the presence or absence of one pathogen may have little bearing on other pathogens. A multi-species test for two or more of the more resistant and common pathogens found in tropical source waters may be necessary. The public health of people living in the tropics and their economic development is dependent upon a more suitable microbiological standard being developed for tropical source waters. The currently used fecal coliform assays are unacceptable for indicating fecal contamination of tropical source waters.

References

American Public Health Association. 1985. *Standard Methods for the Examination of Water and Wastewater* 16th ed. Washington, D.C.

Baker, R.M., Singleton, F.L., and Hood, M.A. 1983. Effects of nutrient deprivation on *Vibrio cholerae Appl. Environ. Microbiol.* 46:930–940.

Barbaras, S. 1986. Monitoring natural waters for drinking-water quality. *WHO Stat Q* 39:32–45.

Barrell, R.A., and Roland, M.G. 1979. The relationship between rainfall and well water pollution in a west African (Gambian) village. *J. Hyg.* 83:143–150.

Berg, G. and Metcalf, T.G. 1978. Indicators of viruses in waters, p. 1–30. *In* Berg, G. (ed.), *Indicators of Viruses in Water and Food.* Ann Arbor Science Publications, Ann Arbor, Mich.

Bermúdez, M. and Hazen, T.C. 1988. Phenotypic and genotypic comparison of *Escherichia coli* from pristine tropical waters. *Appl. Environ. Microbiol.* 54:979–983.

Bialkowska-Hobrzanska, H. 1987. Detection of enterotoxigenic *Escherichia coli* by dot blot hybridization with biotinylated DNA probes. *J. Clin. Microbiol.* 25:338–343.

Bigger, J.W. 1937. The growth of coliform bacilli in water. *J. Pathol. Bacteriol.* 44:167–211.

Bisson, J.W. and Cabelli, V.J. 1980. *Clostridium perfringens* as a water pollution indicator. *J. Water Pollut. Control Fed.* 52:241–248.

Bissonnette, G.K., Jezeski, J.J., McFeters, G.A., and Stuart, D.S. 1975. Influence of environmental stress on enumeration of indicator bacteria from natural waters. *Appl. Microbiol.* 29:186.

Bissonnette, G.K., Jezeski, J.J., McFeters, G.A., and Stuart, D.S. 1977. Evaluation of recovery methods to detect coliforms in water. *Appl. Environ. Microbiol.* 33:590.

Blum, D., Feachem, R.G., Huttly, S.R., Kirkwood, B.R., and Emeh, R.N. 1987. The effects of distance and season on the use of boreholes in northeastern Imo State, Nigeria. *J. Trop. Med Hyg.* 90:45–50.

Bockemühl, J. 1985. Epidemiology, etiology and laboratory diagnosis of infectious diarrhea diseases in the tropics. *Immun. Infekt.* 13:269–275.

Bonde, G.J. 1977. Bacterial indication of water pollution. *Adv. Aquatic Microbiol.* 1:273–364.

Burlingame, G.A., McElhaney, J., and Pipes, W.O. 1984. Bacterial interference with coliform colony sheen production on membrane filters. *Appl. Environ. Microbiol.* 47:56–60.

Cabelli, V.J. 1983. Public health and water quality significance of viral diseases transmitted by drinking water and recreational water. *Water Sci. Technol.* 15:1–15.

Carrillo, M., Estrada, E., and Hazen, T.C. 1985. Survival and enumeration of fecal indicators *Bifidobacterium adolescentis* and *Escherichia coli* in a tropical rain forest watershed. *Appl. Environ. Microbiol.* 50:468–476.

Colwell, R.R., Brayton, P.R., Grimes, D.J., Roszak, D.B., Huq, S.A., and Palmer, L.M. 1985. Viable but nonculturable *Vibrio cholerae* and related pathogens in the environment: implications for release of genetically engineered microorganisms. *Biotechnology* 3:817–820.

Curtis, R.E., Guzman-Ríos, S., and Díaz, P.L. 1984. Water Resources Data. Puerto Rico and the U.S. Virgin Islands. *USGS Survey Water—Data Report* PR–84–1.

Daily, O.P., Joseph, S.W., Gillmore, J.D., Colwell, R.R., and Seidler, R.J. 1981. Identification, distribution, and toxigenicity of obligate anaerobes in polluted waters. *Appl. Environ. Microbiol.* 41:1074–1077.

Dufour, A.P. 1986. Diseases Caused by Water Contact, p. 23–41. In *Waterborne Diseases*. GF Craun (ed.). CRC Press Inc., Boca Raton.

Dufour, A.P., Strickland, E.R., and Cabelli, V.J. 1981. Membrane filter method for enumerating *Escherichia coli Appl. Environ. Microbiol.* 41:1152–1158.

Elías,-Montalvo, E.E., Calvo, A., and Hazen, T.C. 1989. Survival and distribution of *Yersinia enterocolitica* in tropical freshwater and food. *Curr. Microbiol.* 18:119–126.

Esrey, S.A., Feachem, R.G., and Hughes, J.M. 1985. Interventions for the control of diarrhoeal diseases among young children: improving water supplies and excreta disposal facilities. *Bull. WHO* 63:757–772.

Evison, L.M. and James, A. 1973. A comparison of the distribution of intestinal bacteria in British and East African water sources. *J. Appl. Bacteriol.* 36:109–118.

Evison, L.M. 1988. Comparative studies on the survival of indicator organisms and pathogens in fresh and sea water. *Proceedings International Conference on Water and Wastewater Microbiology* Vol. 2:50 (1–7). Newport Beach, California.

Evison, L.M. and James, A. 1975. *Bifidobacterium* as an indicator of human fecal pollution in water. *Prog. Water Technol.* 7:57–66.

Feachem, R.G. 1974. Fecal coliforms and fecal streptococci in streams in the New Guinea highlands. *Water Res.* 8:367–374.

Feachem, R.G. 1977. Infectious disease related to water supply and excreta disposal facilities. *Ambio* 6:55–58.

Fitts, R., Diamond, M., Hamilton, C., and Neri, M. 1983. DNA-DNA hybridization assay for detection of *Salmonella* spp. in foods. *Appl. Environ. Microbiol.* 46:1146–1151.

Fliermans, C.B., Cherry, W.B., Orrison, L.H., Smith, S.J., Tison, D.L., and Pope, D.H. 1981. Ecological Distribution of *Legionella pneumophila. Appl. Environ. Microbiol.* 41:9–16.

Fujioka, R.S. and Navikawa, O.T. 1982. Effect of sunlight on the enumeration of indicator bacteria under field conditions. *Appl. Environ. Microbiol.* 44:395–401.

Fujioka, R.S. and Shizumura, L.K. 1985. *Clostridium perfringens*, a reliable indicator of stream water quality. *J. Water Pollut. Control Fed.* 57:986–992.

Funderburg, S.W. and Sorber, C.A. 1985. Coliphages as indicators of enteric viruses in activated sludge. *Water Res.* 19:547–555.

Keswick, B.H., Gerba, C.P., Rose, J.B., and Toranzos, G.A. 1985. Detection of rotaviruses in treated drinking water. *Water Sci. Technol.* 17:1–6.

Geldreich, E.E. and Kenner, B.A. 1969. Concepts of faecal streptococci in stream pollution. *J. Water Pollut. Control Fed.* 41:R336–R352.

Gordon, R.W. and Fliermans, C.B. 1978. Survival and viability of *Escherichia coli* in a thermally altered reservoir. *Water Res.* 12:343–352.

Goyal, S.M. and Gerba, C.P. 1983. Viradel method for detection of rotaviruses from seawater. *J. Virol. Meth.* 7:729–285.

Grabow, W.O.K., Hilner, C.A., and Coubrogh, P. 1981. Evaluation of standard and modified mFC, MacConkey, and Teepol Media for membrane filtration counting of fecal coliforms in water. *Appl. Environ. Microbiol.* 44:192–199.

Gyllenberg, H., Niemelä, S., and Sormunen, T. 1960. Survival of bifid bacteria in water as compared with that of coliform bacteria and enterococci. *Appl. Environ. Microbiol.* 8:20–22.

Hagler, A.N. and Mendonça-Hagler, L.C. 1981. Yeasts from marine and estuarine waters with different levels of pollution in the state of Río de Janeiro, Brazil. *Appl. Environ. Microbiol.* 41:173–178.

Hazen, T.C. and Aranda, C.F. 1981. Bacteria and water quality in the Río Mameyes watershed, p. 87–111. In *Septimo Simposio de Recursos Naturales.* Commonwealth of Puerto Rico, Department of Natural Resources, San Juan.

Hazen, T.C., Fuentes, F.A., and Santo Domingo, J.W. 1988. In situ survival and activity of pathogens and their indicators. *Proceedings ISME4.* 406–411. Yugoslavia.

Hazen, T.C., Santiago-Mercado, J., Toranzos, G.A., and Bermúdez, M. 1987. What does the presence of fecal coliforms indicate in the waters of Puerto Rico? A Review. *Bol. Puerto Rico Med. Assoc.* 79:189–193.

Hejkal, T.W., LaBelle, R.L., Keswick, B., Gerba, C.P., Hofkin, B., and Sanchez, Y. 1982. Viruses in a small community water supply associated with a gastroenteritis epidemic. *J Am Water Works Assoc.* 74:318–321.

Herrero, L. and Fuentes, L.G. 1977. Estudio virologico en aguas del area metropolitana de San Jose, Costa Rica. *Rev. Lat–Amer. Microbiol.* 20:35–39.

Hill, G. and Rai, H. 1982. A preliminary characterization of the tropical lakes of the Central Amazon by comparison with polar and temperate systems. *Arch. Hydrobiol.* 96:97–111.

Hill, W.E., Madden, J.M., McCardell, B.A., Shah, D.B., Jagow, J.A., Payne, W.L., and Boutin, B.K. 1983. Foodborne enterotoxigenic *Escherichia coli*: Detection and enumeration by DNA colony hybridization. *Appl. Environ. Microbiol.* 45:1324–1330.

Holden, W.S. 1970. *Water Treatment and Examination.* London, Churchill.

Hutchinson, G.E. 1975. *A Treatise on Limnology.* Vol. 1, Part 2. John Wiley & Sons. New York.

Hutchinson, M. and Ridgway, J.W. 1977. Microbiological Aspects of Drinking Water Supplies, p. 179–218. In *Aquatic Microbiology*. FA Skinner and JM Shewan (eds). Academic Press, London.

Jen, W.-C., and Bell, R.G. 1982. Influence of temperature and time of incubation of the estimation of bacterial numbers in tropical surface waters. *Water Res.* 16:601–604.

Jiménez, L., Muñiz, I., Toranzos, G.A., and Hazen, T.C. 1989. The survival and activity of *Salmonella typhimurium* and *Escherichia coli* in tropical freshwater. *J. Appl. Bacteriol.* 66:101–109.

Jiwa, S.F., Kikrovacek, H., and Wadstrom, T. 1981. Enterotoxigenic bacteria in food and water from an Ethiopian community. *Appl. Environ. Microbiol.* 41:1010–1019.

Keswick, B.H., Gerba, C.P., Rose, J.B., and Toranzos, G.A. 1985. Detection of rotaviruses in treated drinking water. *Water Sci. Technol.* 17:1–6.

Lavoie, M.C. 1983. Identification of strain isolates as total and fecal coliforms and comparison of both groups as indicators of fecal pollution in tropical climates. *Can. J. Microbiol.* 29:689–693.

Lavoie, M.C., and Viens, P. 1983. Water quality control in rural Ivory Coast. *Trans. Royal Soc. Trop. Med. Hyg.* 77:119–120.

López-Torres, A.J., Hazen, T.C., and Toranzos, G.A. 1987. Distribution and in situ survival and activity of *Klebsiella pneumoniae* and *Escherichia coli* in a tropical rain forest watershed. *Curr. Microbiol.* 15:213–218.

McCambridge, J. and McMeekin, T.A. 1981. Effect of solar radiation and predacious microorganisms on survival of fecal coliforms and other bacteria. *Appl. Environ. Microbiol.* 41:1083–1087.

McFeters, G.A., Bissonnette, G.K., Jeseski, J.J., Thomson, C.A., and Stuart, D.G. 1974. Comparative survival of indicator bacteria and enteric pathogens in well water. *Appl. Environ. Microbiol.* 27:823–829.

Means, E.G. and Olson, B.H. 1981. Coliforms inhibition by bacteriocin-like substances in drinking water distribution systems. *Appl. Environ. Microbiol.* 42:506–512.

Moseley, S.L., Echeverria, P., Seriwatana, J., Tirapat, C., Chaicumpa, W., Sakuldaipeara, T., and Falkow, S. 1982. Identification of enterotoxigenic *Escherichia coli* by colony hybridization using three enterotoxin gene probes. *J. Infect. Dis.* 145:863–869.

Muñiz, I., Jimenéz, L., Toranzos, G.A., and Hazen, T.C. 1989. Survival and activity of *Streptococcus faecalis* in a tropical rain forest stream. *Microbe Ecol.* 16:

Odum, E.P. 1971. *Fundamentals of Ecology*. W.B. Saunders Co., Philadelphia.

Oluwande, P.A., Sridhar, K.C., Bammeke, A.O., and Okubadejo, A.O. 1983. Pollution level in some Nigerian rivers. *Water Res.* 17:957–963.

Ortiz-Roque, C. and Hazen, T.C. 1987. Abundance and Distribution of Legionellaceae in Puerto Rican Waters. *Appl. Environ. Microbiol.* 53:2231–2236.

Pagel, J.E., Qureshi, A.A., Young, D.M., and Vlassoff, L.T. 1982. Comparison of four membrane filter methods for fecal coliform enumeration. *Appl. Environ. Microbiol.* 43:787–793.

Pérez Rosas, N., and Hazen, T.C. 1988. In situ survival of *Vibrio cholerae* and *Escherichia coli* in tropical coral reefs. *Appl. Environ. Microbiol.* 54:1–9.

Pérez-Rosas, N. 1983. Survival of *Vibrio cholerae* in tropical marine environments. M.S. Thesis. Department of Biology, University of Puerto Rico, Río Piedras.

Ragavachari, T.N.S. and Iver, P.V.W. 1939. Longevity of coliform organisms in water stored under natural conditions. *Indian J. Med. Res.* 26:877–883.

Resnick, I.G. and Levin, M.A. 1981. Assessment of bifidobacteria as indicator of human fecal pollution. *Appl. Environ. Microbiol.* 42:433–438.

Rivera, S., Hazen, T.C., and Toranzos, G.A. 1988. Isolation of fecal coliforms from pristine sites in a tropical rain forest. *Appl. Environ. Microbiol.* 54:513–517.

Rose, J.B., Gerba, C.P., Singh, S.N., Toranzos, G.A., and Keswick, B.H. 1986. Isolating enteric viruses from finished waters. *J. Amer. Wat. Works Assoc.* 78:51–61.

Roszak, D.B., Grimes, D.J., and Colwell, R.R. 1984. Viable but nonrecoverable stage of *Salmonella enteritidis* in aquatic systems. *Can. J. Microbiol.* 30:334–338.

Santiago-Mercado, J. and Hazen, T.C. 1987. Comparison of four membrane filtration and MPN methods for fecal coliform enumeration in tropical waters. *Appl. Environ. Microbiol.* 53:2922–2928.

Santo Domingo, J.W., Fuentes, F.A., and Hazen, T.C. 1989. Survival and activity of *Streptococcus faecalis* and *Escherichia coli* in petroleum-contaminated tropical marine waters. *Environ. Pollution* 56:263–281.

Snyder, J.D. and Merson, M.H. 1982. The magnitude of the global problem of acute diarrhoeal disease: a review of active surveillance data. *Bull. WHO* 60:605–613.

Stuart, D.S., McFeters, G.A., and Schillinger, J.E. 1977. Membrane filter technique for quantification of stressed fecal coliforms in the aquatic environment. *Appl. Environ. Microbiol.* 34:42.

Tamplin, M., Rodrick, G.E., Blake, N.J., and Cuba, T. 1982. Isolation and characterization of *Vibrio vulnificus* in two Florida estuaries. *Appl. Environ. Microbiol.* 44:1466–1470.

Tartera, C. and Jofre, J. 1987. Bacteriophages active against *Bacteroides fragilis* in sewage eluted waters. *Appl. Environ. Microbiol.* 53:1632–1637.

Thomson, J.A. 1981. Inadequacy of *Escherichia coli* as an indicator of water pollution in a tropical climate: a preliminary study in Botswana. *S. Afr. J. Sci.* 77:44–45.

Toranzos, G.A., Gerba, C.P., Zapata, M., and Cardona, F. 1986a. Presence de virus enteriques dans des eaux de consommation a Cochabamba (Bolivie). *Revue International des Sciences de l'Eau* 2:91–93.

Toranzos, G.A., Gerba, C.P., and Hanssen, H. 1986b. Occurrence of enteroviruses and rotaviruses in drinking water in Colombia. *Water Sci. Technol.* 18:109–114.

Toranzos, G.A., Gerba, C.P., and Hanssen, H. 1988. Enteric viruses and coliphages in Latin America. *Int. J. Tox. Asses.* 3:491–510.

Valdés-Collazo, L., Schultz, A.J., and Hazen, T.C. 1987. Survival of *Candida albicans* in tropical marine and fresh waters. *Appl. Environ. Microbiol.* 53:1762–1767.

Wentsel, R.S., O'Neill, P.E., and Kitchens, J.F. 1982. Evaluation of coliphage detection as a rapid indicator of water quality. *Appl. Environ. Microbiol.* 43:430–434.

Whittaker, R.H. 1975. *Communities and Ecosystems* MacMillian Publishing Co., Inc. New York.

Wright, R.C. 1982. A comparison of the levels of faecal indicator bacteria in water and human faeces in a rural area of a tropical developing country (Sierra Leone). *J. Hyg.* 89:69–78.

Wright, R.C. 1986. The seasonality of bacterial quality of water in a tropical developing country (Sierra Leone). *J. Hyg.* 96:75–82.

Xu, H.-S., Roberts, N., Singleton, F.L., Attwell, R.W., Grimes, D.J., and Colwell, R.R. 1982. Survival and viability of nonculturable *Escherichia coli* and *Vibrio cholerae* in the estuarine and marine environment. *Microb. Ecol.* 8:313–323.

Part 2

Microbiology of Drinking Water Treatment

3

Assimilable Organic Carbon (AOC) in Drinking Water

D. van der Kooij

The Netherlands Waterworks' Testing and Research Institute, KIWA Ltd., Nieuwegein, The Netherlands

3.1 Introduction

Developments in water treatment   The removal in water treatment of microorganisms causing the so-called "water-borne" diseases and the prevention of contamination of drinking water with these organisms during storage and distribution, are major water quality objectives in water supply. These objectives affect both design criteria and operational procedures for water supply systems. The efficiency of the procedures and systems is monitored with bacteriological methods for determining the presence/absence of bacteria of fecal origin. In the past few decades, a number of developments have affected the "classical" concepts concerning water treatment and the bacteriological aspects of water quality. In the Netherlands, as in other countries, water treatment systems were adapted and new ones were constructed to cope with the increased demands for drinking water and the deteriorating quality of the raw water sources. Water treatment systems were extended with new processes, such as granular activated carbon (GAC) filtration and ozonation, particularly for the removal of persistent organic compounds affecting taste and odor or with otherwise undesirable properties.

The application of additional processes in water treatment focussed attention not only on chemical water quality parameters, but also on bacteriological water quality parameters. Increased numbers of bacteria were found in GAC filtrates (Love et al., 1974; Fiore and Babineau, 1977). Furthermore, bacterial multiplication was observed during distribution of drinking water prepared with the application of ozone as a final treatment step (Dietlicher, 1970; Snoek, 1970; Stalder and Klosterkötter, 1977). As a result, the hygienic, technical and aesthetic significance of high colony counts in drinking water and the problem of bacterial aftergrowth gained renewed attention (Geldreich et al., 1972; Müller, 1972; Victoreen, 1974; O'Connor et al., 1975; Hutchinson and Ridgway, 1977). The detection of trihalomethanes as undesirable side products of chlorination used for reducing and controlling numbers of bacteria in water during treatment and distribution (Rook, 1974), further promoted research in bacteriological processes in water treatment and distribution in the Netherlands. Drinking water prepared from surface water was studied in particular because of the presence of easily biodegradable organic compounds in surface water, the great variety of processes used for water treatment, and the need to improve water treatment by reducing the negative side effects of disinfection processes.

Water treatment in the Netherlands   The total annual production of drinking water in the Netherlands for a population of approximately 15 million, amounts to about 1000 million m^3. About two thirds of this drinking water is prepared from ground water, and one third from surface water. Table 3.1 shows the raw water types and gives examples of principal treatment systems. This table indicates that biological processes, occurring under a great variety of conditions, play a central role in water supply in the Netherlands. Treatment processes which are particularly important in this respect are: 1) storage in open surface reservoirs; 2) underground passage such as dune infiltration and river bank filtration; and 3) various types of filtration processes such as rapid sand filtration, dual-media filtration, nonsubmerged filtration, slow sand filtration and GAC filtration.

Due to the sandy character of the underground, nearly all types of ground water are exposed to aerobic and anaerobic biological processes. Underground passage in most cases results in water free from fecal pollution. Chemical disinfectants therefore generally are not used for the production and distribution of drinking water prepared from ground water, or that prepared from river water after infiltration. In some cases, for example when dune-infiltrated river water is collected in open basins, post treatment is necessary to obtain the desired bacteriological water quality. In a number of such water works, slow sand filtration is used for post treatment. Bacteriologically safe drinking water can also be produced from surface water after storage in open reservoirs without postchlorination by applying a series of physico-chemical processes in combination with slow sand filtration (Schellart, 1986).

An important effect of biological processes is the removal of biodegradable inorganic (ammonia, nitrate) and organic compounds. When applied properly, drinking water with a low concentration of compounds serving as energy sources for bacteria can be obtained. To control bacteriological processes in water treatment and distribution, The Netherlands Water Works Association (VEWIN) financed the development of a method for determining the concentration of organic compounds serving as nutrient for bacteria in (drinking) water, called the *aftergrowth potential*. The research on the aftergrowth potential method was conducted by The Netherlands Waterworks' Testing and Research Institute, KIWA Ltd. This chapter gives information about the method and about its application in water treatment and distribution.

Aftergrowth phenomena   Increasing concern about microbiological pro-

Table 3.1 Water types and water treatment in The Netherlands

Source	% of total annual production	Water type or water works	Examples of principal treatment systems — Treatment processes
Ground water	59.0	Aerobic	No treatment or removal of CO_2
	3.6	Anaerobic	Aeration followed by sand filtration for the removal of methane, ammonia, iron, and manganese.
Dune water			
Surface water direct from source	4.1	Provincial Waterworks Noord-Holland; Andijk	River Rhine water from Lake Ijssel; microstraining; break-point chlorination; coagulation/sedimentation; rapid filtration; GAC filtration; postdisinfection (Cl_2/ClO_2); microstraining; pH correction with NaOH.
after dune infiltration	14.4	Dune Water Works of The Hague	Aeration, dosage of powdered activated carbon/sedimentation, aeration, rapid sand filtration, slow sand filtration.
after bank filtration	7.0		Aeration, followed by sand filtration for the removal of methane, ammonia, iron and manganese.*
after storage in open reservoirs	10.4	Water Works of Rotterdam (Kralingen)	Coagulation/sedimentation, ozonation, dual-media filtration, GAC filtration, postchlorination.
		Water Works of Amsterdam (Weesperkarspel)	Coagulation/sedimentation in open storage reservoir, rapid filtration, ozonation, dosage of PAC, ferric chloride, coagulation/sedimentation, rapid filtration, slow sand filtration.
Additional sources	1.5		

Total annual production in 1984:1037000 m³ (VEWIN, 1987).
* In a few cases, GAC filtration is used for the removal of taste and odor compounds.

cesses in drinking water distribution systems is evident from a large number of publications in the past decade. These publications show that drinking water may harbor a large variety of bacteria, including species of the genera *Acinetobacter, Aeromonas, Arthrobacter, Bacillus, Caulobacter, Cythophaga, Flavobacterium, Pseudomonas* and *Spirillum* (Bonde, 1983; Geldreich et al., 1972; Herman, 1978; LeChevallier et al., 1980; McMillan and Stout, 1977; Reasoner and Geldreich, 1979; Veresikova and Synek, 1974, van der Kooij, 1981). The isolation of representatives of at least 31 biotypes of *P. fluorescens* and 14 biotypes of *P. putida* from various types of drinking water in the Netherlands further demonstrates the complexity and diversity of the heterotrophic bacterial flora of drinking water. These fluorescent pseudomonads constituted 1 to 10% of the heterotrophic bacterial population as determined with a standard colony count technique (van der Kooij, 1978; 1979). The presence of coliforms in drinking water during distribution recently has attracted much attention in the United States (Talbot et al., 1979; Martin et al., 1982; Wierenga, 1985; McFeters et al., 1986; Geldreich and Rice, 1987). Actinomycetes, yeasts and molds have also been isolated from water supplies (Burman, 1973; Nagy and Olson, 1985).

With electron microscopy, bacteria with unusual morphological characteristics (e.g., *Caulobacter, Hyphomicrobium* and *Gallionella*) have been observed (Allen et al., 1980; Ridgway et al., 1981; Ridgway and Olson, 1981; Brazos and O'Connor, 1985). Many of these bacteria do not contribute to colony counts on nonselective media. These microscopic observations and also cultivation methods have revealed that the bacteria may be present attached to the surfaces of the pipes and on particles. In these "biofilms," colony counts ranging from a few hundreds up to more than one million per cm^2 have been reported (Nagy and Olson, 1985; LeChevallier et al., 1987). Most likely, bacteria multiply in biofilms and are released into the passing drinking water. Disinfectants are less effective against attached bacteria than for bacteria dispersed in water (Ridgway and Olson, 1982; Herson et al., 1987; LeChevallier et al., 1988).

The negative effects of aftergrowth on drinking water quality depend on numbers and identity of the bacteria present. The following negative effects have been reported:

- Bacteriological water quality monitoring is hampered when coliforms multiply in drinking water during distribution, because these organisms should indicate undesirable external pollution. Furthermore, coliform monitoring can be hampered by the presence of large numbers of heterotrophic bacteria (Geldreich et al., 1978).

- Under certain conditions, opportunistic pathogenic bacteria can multiply in drinking water distribution systems, including plumbing installations. Examples of such bacteria are: *Legionella* spp. (Dennis et al., 1982; Tobin et al., 1981; Wadowsky, 1982), *Mycobacterium* spp. (Engel et al., 1980; Kaustova et al., 1981), *P. aeruginosa, Flavobacterium* spp., and *Aeromonas* spp. In the past few years, interest in the presence of *Aeromonas* spp. in drinking water has increased because of their recognition as potential enteric pathogens (Gracey et al., 1982; Burke et al., 1984).

- Iron bacteria can accumulate iron on their cells or sheaths, followed by formation of flocs. These flocs are visible and can give rise to consumer complaints about dirty water (de Vries, 1980; Victoreen, 1974).

- Multiplication of actinomycetes or microfungi may cause taste and odor complaints (Burman, 1965). Specific actinomycetes can attack rubber rings in pipe joints (Leeflang, 1968).

- Bacterial biomass in distribution systems can serve as a food source for animals. The presence of animals visible with the naked eye can lead to consumer complaints (Louwe Kooijmans, 1966; Smalls and Greaves, 1968; Levy et al., 1986).

- Multiplication of bacteria on the surface of construction materials may enhance corrosion (Lee, O'Connor, and Banerji, 1980).

Curative and preventive approaches for controlling aftergrowth include:

- Reduction of the concentration of organic and inorganic compounds serving as energy sources for bacteria in drinking water. Biological treatment procedures are effective for this purpose.

- Prevention of the introduction of energy sources into drinking water during distribution. Possible contamination sources include: the application of materials releasing biodegradable compounds into the water (Colbourne, 1985; Schoenen and Schöler, 1983), the access of light (promoting primary production), and pollution with contaminated soil or water during renovation or repairs of pipelines.

- The maintenance of a disinfectant residual in drinking water during distribution. However, as has been mentioned above, chlorine is relatively ineffective against bacteria attached to surfaces.

- Flushing and swabbing techniques may be used to remove detritus, bacteria, and animals from the distribution system. These methods are labor intensive, difficult to apply in a branched network, and must be repeated periodically.

For controlling and reducing aftergrowth in a practical situation, the methods mentioned above can be used, either individually or in combination with each other. Decisions about the choice of the method(s) must be based on an analysis of the situation. Such an analysis should include information about the concentration of energy sources for bacteria in water leaving the treatment facility.

3.2   Aftergrowth Potential

Measuring the aftergrowth potential   Biologists studying the presence and behavior of bacteria and animals in drinking water distribution systems concluded more than sixty years ago that the extent of this aftergrowth, as in other environments, is determined by the concentration of organic and inorganic compounds serving as nutrients for bacteria and animals (Heymann, 1928; Baylis, 1930). De Vries (1890) and Heymann (bacteriologist of the Amsterdam Waterworks) observed that growth of bacteria in distribution systems resulted in the development of animals such as *Asellus* and *Gammarus* feeding on bacteria and detritus. Heymann concluded that aftergrowth should be limited by reducing the concentration of assimilable organic compounds to low levels by the application of sand filtration. He and his colleague Von Wolzogen Kühr developed a method for determining the concentration of assimilable organic compounds in drinking water. For this purpose the water was filtered repeatedly through a filter bed (a glass cylinder containing approximately 10 liters of sand from a well-performing slow sand filter) and the decrease of the permanganate value of the water was measured (Figure 3.1). From these measurements, it was concluded that maximum reduction of the initial $KMnO_4$ value usually was obtained after two filter-bed passages and ranged from 20 to 60 percent for various water types (Heymann, 1928). However, this method has not obtained general application in the Netherlands.

In the past few years, a variety of techniques for determining the aftergrowth potential of drinking water has been described. The following principle choices have been made in these approaches:

1.   Measurement of the amount of bacterial biomass produced after incubation of the water sample. In this approach, indigenous bacteria or pure cultures of bacteria can be used. Biomass measurements include direct counts, determination of biomass constituents (e.g., DNA or ATP), or application of colony count procedures.

2.   Measurement of the growth rate of bacteria. Also here biomass must be determined.

3.   Measurement of the decrease of the concentration of organic compounds. Applied analytical procedures include $KMnO_4$ and dissolved organic carbon (DOC) determinations.

An overview of the applied techniques and the references is given in Table 3.2. Certain authors state that their method determines the total amount of biodegradable organic compounds, whereas others emphasize measuring the regrowth potential. As a general comment, it can be concluded that the sensitivity and the reproducibility of the various approaches are determined by the sensitivity and the reproducibility of the techniques for determining the concentration of biomass and organic carbon in the water. Generally, with the

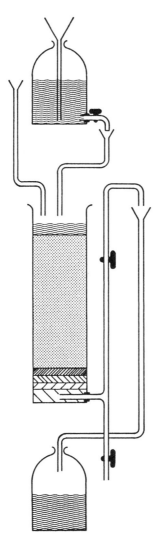

Figure 3.1 Installation as constructed by Von Wolzogen Kühr and used by Heymann (1928) for determining the concentration of assimilable organic compounds in drinking water. The water is allowed to pass through a column of sand and the difference in permanganate reactivity between the influent and effluent measured.

use of bacteriological techniques, very low densities of bacterial biomass can be determined, enabling the determination of aftergrowth potential at a level of a few micrograms of biomass per liter. In the Netherlands, a method that determines the concentration of easily assimilable organic carbon, the so-called *AOC* determination, has been developed and applied in the past decade (van der Kooij et al., 1982). The principle of this technique and results obtained by its application in water treatment and distribution are presented in this chapter.

Table 3.2 Methods for determining the aftergrowth potential of water and/or the concentration of organic compounds available for bacterial multiplication

Designation	Sample		Biomass	Process	Conditions		Measurement	Reference
	Volume	Treatment			Temp (°C)	Period (days)		
Assimilable organic compounds	A few liters	None	Adapted bacteria on sand	Sand filtration	15–20	1–2	$KMnO_4$ (mg/l)	Heymann (1928)
Assimilable organic carbon	600 ml	30 min at 60°C	Pure culture(s)	Batch culture	15	5–25	colony counts (CFU/ml)	Van der Kooij et al. (1982)
Regrowth potential	300 ml	Membrane filtration	Indigenous bacteria[a]	Batch culture	20	1–2	turbidity	Werner (1985)
dDOC[b]		Membrane filtration	Bacteria on sand	Batch culture	20	5–20	DOC (mg/l)	Joret and Levy (1986)
BDOC[c]	300 ml	Membrane filtration	Indigenous bacteria[a]	Batch culture	20		DOC (mg/l)	Servais et al. (1987)
Regrowth potential	100 ml	Membrane filtration	Surface water bacteria	Batch culture	22	2–5	ATP (ng/l)	Stanfield and Jago (1988)

[a] Reinoculation.
[b] dDOC, decrease of DOC concentration.
[c] BDOC, biodegradable DOC.

Principle and technical aspects of the AOC determination   The AOC determination is based on measuring the maximum level of growth (maximum colony count, N_{max}, in colony-forming units per ml) of a selected pure bacterial culture in a representative water sample in which the indigenous bacteria have been killed or inactivated by heat treatment. Subsequently, the AOC concentration can be calculated using the N_{max} value and the yield coefficient (Y) of the organism for a selected growth substrate which is completely utilized by the organism, even at very low concentrations (e.g., acetate or glucose). Hence,

$$\frac{N_{max}\ (CFU/ml) \cdot 1000}{Y\ (CFU/\mu g\ of\ carbon)} = AOC\ concentration\ (\mu g\ of\ carbon/1) \quad (1)$$

Essential considerations in the procedure are: 1) cleaning of materials coming in contact with the water sample; 2), sample treatment, and 3) type of organism used. The Erlenmeyer flasks and the pipettes coming in contact with the water sample must be thoroughly cleaned by an overnight contact with a 10 percent solution of potassiumbichromate in concentrated sulfuric acid. Thereafter, these materials should be rinsed with hot tap water, then with a 10 percent nitric acid solution and finally with hot tap water. Graduated pipettes should be rinsed in streaming drinking water for about four hours. Cleaned flasks and pipettes should be heated overnight at 250–300°C before use. Pipettes coming in contact with the sample must be used without cotton plugs (to avoid organic carbon contributed by the cotton).

In a typical experiment, a representative sample of the water under investigation is collected in a thoroughly cleaned glass-stoppered 1 l Erlenmeyer flask. This sample is heated at 60°C for 30 minutes to kill or inactivate the bacteria originally present in the water. After cooling, the sample is inoculated with a selected pure culture. The selected culture is adapted to growth at low-substrate concentrations by precultivation in autoclaved tap water supplied with a low concentration of a suitable substrate, e.g., 1 mg of acetate or glucose per liter. Flasks are incubated at 15°C in the dark. This temperature was selected, because it is close to the temperature observed under practical conditions in drinking water during distribution. Moreover, incubation at higher temperatures can result in a rapid decline of the colony count after having reached N_{max}. Finally, an accurate determination of the growth rate is possible at 15°C. This growth rate can give additional information about the aftergrowth potential of the water or about the growth kinetics of the applied bacteria.

The growth of the bacteria is determined by periodic colony counts. For this purpose, the spread plate technique is used. A volume of 0.05 ml of the water sample, is spread over the surface of predried plates (9 cm), which are incubated at 25°C until the colonies are large enough to be counted. The incubation period varies from 1 to three days, depending on the organism used. The cultivation medium (Lab-Lemco agar, LLA 8.0; Oxoid Ltd) contains 3 g of meat extract, 5 g of peptone and 12 g of agar per liter of demineralized water.

Properties of organisms for AOC determinations   The bacteria used must have the following properties: 1) production of cells of uniform size, preferable cocci or rods; 2) rapid growth at low concentrations of substrate; 3) rapid growth on nonselective media, producing clearly visible colonies; 4) ability to use a simple nitrogen source, preferable nitrate; 5), no requirement for vitamins or other growth factors; and 6) no changes in properties affecting growth at low solute concentrations.

The following types of organisms can be used in AOC determinations:

- Bacteria with a great nutritional versatility for the assessment of the "total" AOC concentration. For this purpose, a *Pseudomonas fluorescens* strain is used.

- Bacteria capable of using (groups of) specific compounds. A *Flavobacterium* species, which grows on carbohydrates, and a *Spirillum* species, which grows on carboxylic acids, are used for this purpose.

- Bacteria able to directly affect the bacteriological water quality, for determining the concentration of compounds present for their growth in a water type. Examples of such bacteria are: coliforms, *Aeromonas* spp., *P. aeruginosa*, *Mycobacterium* spp. etc.

Both the identity and the concentration of compounds which are potential food sources for bacteria determine the growth rate and maximum growth level. The response of a specific organism in a specific water type depends on its ability to utilize the available food sources in competition with other bacteria. To explain the observed growth rate and the maximum level of growth of a pure culture in a (drinking) water sample in terms of concentration and character of available substrate, the nutritional properties, the growth kinetics, and the yield values of the organism used must be known. With the organism used in growth measurements for AOC determinations, the following approach for determining these properties has been followed:

1. Determination of the nutritional versatility in a screening procedure, e.g., the replica plating technique as described by Stanier et al. (1966).
2. Growth measurements in drinking water supplemented with mixtures of compounds at low concentrations (1 to 10 μg of C of each compound per liter). Figure 3.2 shows the nutritional versatility of four different bacterial strains as determined using mixtures of substrates.
3. Growth measurements with individual compounds, usually at a concentration of 10 or 25 μg of C/1.
4. Growth measurements with a few selected compounds at a range of concentrations for the determination of growth kinetics and yield factors.

Growth kinetics are determined from the growth rate as observed at dif-

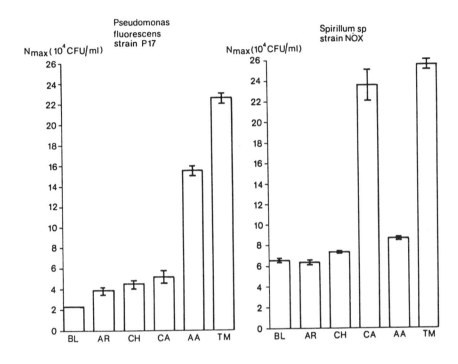

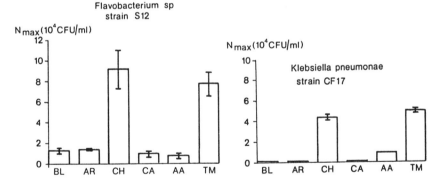

Figure 3.2   Maximum colony counts (N_{max}, CFU/ml) of four bacterial strains in drinking water (15°C) supplemented with mixtures of 18 amino acids (AA), 18 carbohydrates and polyalcohols (CH), 18 carboxylic acids (CA), 11 aromatic acids (AR), or the total mixture (TM). Individual compound concentration: 1 μg of C/l. Drinking water as used in the measurements was prepared by slow sand filtration as the final treatment step. The DOC concentration of this water varied from 3.4 to 3.6 mg per liter. Data from van der Kooij et al. (1982) and van der Kooij and Hijnen (1983, 1985, 1988).

ferent substrate concentrations. For this purpose the Monod equation is applied:

$$V = V_{max} \cdot S/(S + K_s) \qquad (2)$$

where V and V_{max} represent observed and maximum growth rates, respectively, in doublings/hour. S is the concentration of added substrate, and K_s is the substrate saturation constant, the value of S for which $V = 1/2 \, V_{max}$.

Figure 3.3 shows growth curves of a *Flavobacterium* sp. strain S12 at various concentrations of starch. The growth rates were calculated by linear-regression analysis of the slope of the curves in the exponential growth phase. V_{max} and K_s were calculated with equation 2, using nonlinear-regression analysis. The principle of using these batch experiments for the determination of the growth kinetics is based on the assumption that the value of S remains essentially unchanged as long as the observed colony count is far below the maximum colony count. The growth rate can be assessed accurately with this technique, even at concentrations of a few micrograms per liter (Figure 3.3). A further example of the growth yield determination is given in Figure 3.4, which shows that oxalate-C gives a much lower yield than acetate-C. Relevant data about the growth kinetics and yield factors for bacterial strains which have been studied in detail are given in Table 3.3.

Pseudomonas fluorescens strain P17 cannot grow on oxalate, a compound that is produced during ozonation of water (Kuo et al., 1978). To assess the concentration of this and other low-molecular-weight carboxylic acids that must be present in water after ozonation, an organism was isolated with the ability to utilize these compounds at low concentrations. This organism, *Spirillum* species strain NOX, can utilize a wide range of carboxylic acids but no carbohydrates, alcohols, or aromatic acids, and only one or two amino acids. Furthermore, this organism does not assimilate amino acids when growing on mixtures of compounds (Figure 3.2) (van der Kooij and Hijnen, 1984). For these reasons, strain NOX is used for the determination of carboxylic acids in water. Based on a similar approach, strain S12 was obtained for determining the concentration of maltose and starch-like compounds in water (van der Kooij and Hijnen, 1985).

Recent investigations showed that *Aeromonas hydrophila* strains can utilize amino acids when present as a mixture, even at concentrations as low as 0.05 μg of C per liter. Moreover, these organisms can grow at oleate concentrations below 1 μg of C/1. (van der Kooij and Hijnen, 1988). Data about the nutritional versatility and growth kinetics at these low concentrations of substrates have also been determined for other bacteria including *P. aeruginosa* (van der Kooij, Oranje, and Hijnen, 1982), another *Flavobacterium* strain (van der Kooij and Hijnen, 1981), *Klebsiella pneumoniae* (van der Kooij and Hijnen, 1988) and other coliforms (van der Kooij and Hijnen, 1985). All these bacteria can be used for determining the concentrations of available substrates in (drinking) water.

Coliforms seem less adapted to growth at low concentrations, although a *Klebsiella* strain multiplied in drinking water supplemented with mixtures of

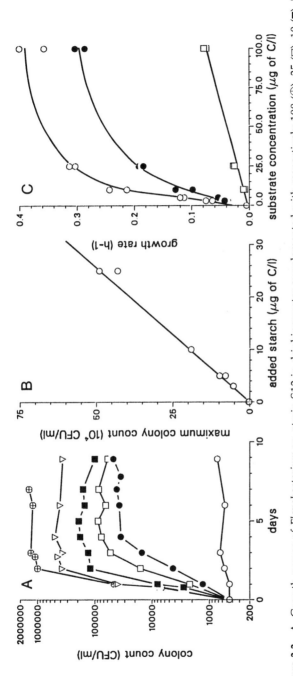

Figure 3.3   A. Growth curves of *Flavobacterium* sp. strain S12 in drinking water supplemented with respectively 100 (⊕), 25 (▽), 10 (■), 5 (□) and 3 (●) μg of starch-C per liter; blank (○). B. Growth yield for starch (○). C. Growth kinetics for starch; C. Growth kinetics for starch (○), maltose (●) and glucose (□). Data from van der Kooij and Hijnen (1985).

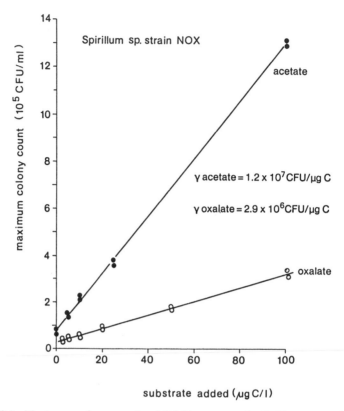

Figure 3.4 Maximum colony counts of *Spirillum* sp. strain NOX grown at different concentrations of oxalate and acetate added to drinking water. From van der Kooij and Hijnen (1984).

amino acids and carbohydrates at individual substrate concentration of 1 μg of C per liter (Figure 3.3). K_s values of coliforms for glucose and maltose are clearly above K_s values of the flavobacteria for these substrates. Growth measurements with strains of *Klebsiella pneumoniae*, *Citrobacter freundii*, and *Enterbacter cloacae* revealed that these bacteria were unable to multiply in slow sand filtrate without an added substrate. The relationship between glucose-C concentration and growth rate of two coliforms revealed that growth would not be possible at glucose-C concentrations below 3 μg of C per liter (Figure 3.5). *Aeromonas hydrophila* and other bacteria adapted to growth in drinking water have threshold values for growth below 1 μg of C per liter (van der Kooij et al., 1980; van der Kooij and Hijnen, 1988).

Combinations of strains for AOC determination   Strains P17 and NOX are used in AOC measurements either individually or in combination with each other, because strain P17 is unable to grow on a number of carboxylic acids. The following combinations can be made:

Table 3.3 Bacterial strains used in growth measurements for AOC determinations

Organism	Origin[a]	Acetate		Glucose		Oxalate		Starch		References
		K_s	Y	K_s	Y	K_s	Y	K_s	Y	
Pseudomonas fluorescens strain P17	DWS	4.0	4.1	57	ND	NG	NG	NG	NG	Van der Kooij et al. (1982a,b)
Spirillum sp. strain NOX	SSF1	ND	12	NG	NG	15.2	2.9	NG	NG	Van der Kooij et al. (1984)
Flavobacterium sp. strain S12	SSF2	NG	NG	109	18	NG	NG	8.4	20	Van der Kooij and Hijnen (1983); Van der Kooij and Hijnen (1985)
Aeromonas hydrophila M800	DWG	11.1	6.8	15.9	8.0	NG	NG	ND	ND	Van der Kooij and Hijnen (1988)

[a] DWS, drinking water prepared from surface water; SSF1, slow sand filtrate enriched with starch (100 μg C/l) and incubated at 15°C. SSF2, slow sand filtrate enriched with oxalate (25 μg C/l) and incubated at 15°C. DWG, drinking water prepared from ground water.
[b] Growth constants K_s, in μg of C per liter; Y (yield) in 10^6 CFU per μg of C. NG = no growth; ND = not determined.

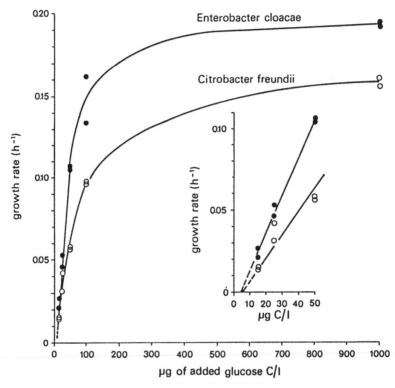

Figure 3.5   Growth rates of two coliforms in drinking water as a function of the glucose-C concentration. From van der Kooij and Hijnen (1985).

- AOC(P17) (as µg of acetate-C equivalents); only strain P17 is grown in the samples.

- AOC(NOX) (as µg of acetate-C equivalents); only strain NOX is grown in the samples.

- AOC(T) = AOC(P17) + dAOC(NOX) (as µg of acetate-C equivalent per liter). Strain NOX is grown in samples in which strain P17 had already reached a maximum value (N_{max}, CFU/ml), and dAOC (NOX) represents the AOC concentration that remained for strain NOX. In this situation, strain NOX utilizes compounds that had not been utilized by strain P17, e.g., oxalate, formate, and glyoxylate. When this combination of growth measurements is performed in water after ozonation, N_{max}(NOX) indicates the maximum oxalate concentration. This concentration can be calculated from:

$$N_{max}(NOX) \cdot 1000/2.9 \cdot 10^6$$
$$= \text{maximum oxalate-C concentration (µg of C/1)} \qquad (3)$$

- AOC(P17/NOX), (as μg of acetate-C equivalents per liter). Here, the samples are inoculated with strain P17 and strain NOX, resulting in simultaneous growth of these organisms. In this situation N_{max}(P17) and N_{max}(NOX) can still be determined by colony count determinations on the same agar plates because of differences in shape and color of the colonies. The AOC concentration is calculated from the observed N_{max} values as acetate-C equivalents by using the respective yield factors for acetate. Usually, strain NOX grows more rapidly and attains higher colony counts than strain P17. This would indicate that in these water types the majority of the AOC concentration consists of carboxylic acids. Examples of such growth curves are shown in Figure 3.6.

- AOC(NOX) + dAOC(P17) (as μg of acetate-C equivalents per liter). Here strain P17 is inoculated into the samples when strain NOX has reached its N_{max} value. In this situation, the N_{max}(P17) is an indication for the concentration of substrates other than the carboxylic acids. Such compounds may include amino acids and related compounds, carbohydrates and related compounds, aromatic acids, etc. However, in water types in which strain NOX can grow rapidly, simultaneous inoculation of strain NOX and strain P17 gives the same result as inoculation of strain P17 after strain NOX has reached its N_{max} value. An example of the effects of using either strain P17, strain NOX or combinations of these organisms in ozonated water are shown in Table 3.4. The data presented in this table demonstrate that in the water type investigated, the majority of compounds contributing to the AOC concentration were carboxylic acids. Only $6/75 \cdot 100$ (expressed as percent, 8%) of the AOC concentration observed in water after ozonation were other compounds, e.g. amino acids, carbohydrates, aromatic acids, etc.

The concentration of AOC is a fraction of the total concentration of organic carbon in water. The AOC concentration may be regarded as a measure for the biological stability of the water for heterotrophic growth and the quotient of AOC/DOC may be regarded as an indication of the biological stability of the organic compounds present in drinking water. To study the effects of water treatment on the aftergrowth potential of water, DOC measurements are needed in addition to AOC measurements.

3.3   AOC Measurements in Water Treatment and Distribution

Water treatment   As is shown in Table 3.1, drinking water in the Netherlands is prepared from river water either after infiltration in the dunes (14.4%), after bankfiltration (7.0%), after storage in open surface reservoirs

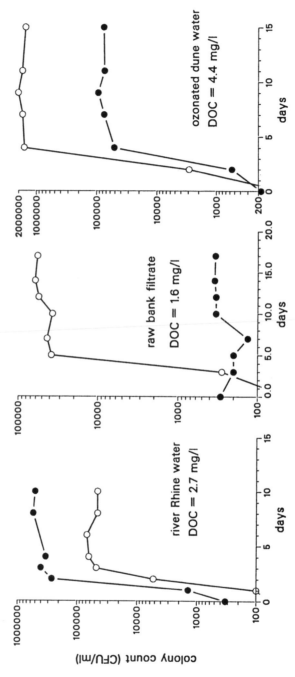

Figure 3.6 Growth curves of *Pseudomonas fluorescens* strain P17 (●) and *Spirillum* species strain NOX (○) in different water types at 15°C.

Table 3.4 AOC concentrations in ozonated water as determined with growth measurements using *Pseudomonas fluorescens* strain P17 and *Spirillum* sp. strain NOX

Organism	Maximum colony count (CFU/ml)	AOC (strain)	AOC (total)
		μg of acetate-C eq/l	
Strain P17	1.8×10^5	42	} 75
Strain NOX grown after strain P17	4.0×10^5	33	
Strain NOX	8.2×10^5	69	} 75
Strain P17 grown after strain NOX	2.5×10^4	6	

(10.4%), or directly (4.1%). The rivers Rhine and Meuse are heavily polluted with agricultural, domestic, and industrial waste water, both treated and un-treated. In the river, self-purification processes cause a permanent removal of biodegradable compounds. On the other hand, growth of algae produces organic material. Table 3.5 gives typical values of the concentrations of dissolved organic compounds, the concentrations of inorganic energy sources, the concentration of chlorophyll (as a measure of algal content), and typical values for AOC. This table indicates that DOC values in river water are relatively low, with AOC values of about one hundred μg of acetate-C equivalents per liter. Growth measurements with strains P17 and NOX in river water show that strain P17 is growing much more rapidly than strain NOX (Figure 3.6). This indicates that carboxylic acids constitute only a minor fraction of the AOC. The major part most likely is composed of compounds of biomass origin, e.g.,

Table 3.5 Organic compounds in Rhine river water

Parameter	Year	
	1983	1984
Dissolved organic carbon (DOC), mg/l	3.3	3.2
Chemical oxygen demand (COD), mg/l	15	16
Nitrate, mg/l	17	19
Ammonia, mg/l	0.84	0.67
Chlorophyll a, μg/l	23	25
AOC(P17), μg ac-C/l	126	103

Values for all parameters except AOC are averages of weekly measurements. Data from Meijers (1984) and RIWA (1986). AOC measurements are the result of duplicate samples.

amino acids, carbohydrates, and related compounds. Storage of river water in open reservoirs is attended with further biodegradation of organic compounds, but also leads to the growth of algae. Growth of algae is also observed in dune-infiltered water collected in open buffer reservoirs. Waters sampled from open reservoirs generally have AOC values of 30 to 100 μg of acetate-C per liter (Table 3.6). Incidentally, AOC values of 0.5 to 1.0 mg of C/1 have been found in situations with heavy growth of algae.

Examples of the effect of water treatment on the AOC concentration of water after storage in open reservoirs, or after dune infiltration, are given in Figure 3.7. Coagulation and sedimentation can cause a large reduction of the AOC concentration, most likely as a result of the removal of biomass components. To improve the effect of these processes at low water temperatures, starch-based coagulant aids are being used in the Netherlands at concentrations of a few ppm. These compounds are easily biodegradable and their application therefore can cause problems during storage of concentrated solutions of these products at the treatment plant (Schellart, 1986). Growth measurements with *Flavobacterium* strain S12, an organism that is specialized in the utilization of maltose and starch-like compounds, have revealed that residues of these coagulant or filtration aids and/or their degradation products can be present in drinking water (Figure 3.8). Filtration processes with biological activity can reduce these compounds to levels where growth of the *Flavobacterium* strain is no longer enhanced (van der Kooij and Hijnen, 1985).

Disinfection processes play an important role in preparing drinking water from surface water, particularly when this water is treated directly or when this water has been stored in open reservoirs. Major disinfectants applied in the Netherlands are chlorine and ozone. Ozone is not only used for disinfection, but it is also used for color removal. Chlorination has been found to cause slight increases of the AOC concentration (Figure 3.7). Also, Heymann (1928) found an increase of the concentration of assimilable organic compounds in water after chlorination. These increases are particularly important when chlorine is applied as a post-treatment step. Ozonation gives a strong increase of the AOC concentration. This increase is related to the ratio between ozone

Table 3.6 AOC concentrations in fresh water after storage in open reservoirs

Water type	AOC(P17) μg of acetate-C eq/l	DOC mg/l	AOC/DOC $\times$ 100 (%)
Dune-infiltrated river water in open collection basin	61	3.2	1.9
Lake water	47	7.4	0.6
River water after open storage	31	3.8	0.8
River Meuse water after open storage	97	4.0	2.4

Data from van der Kooij (1984).

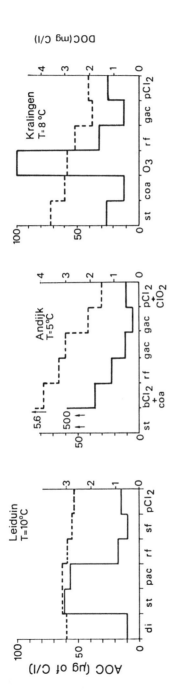

Figure 3.7   Effects of water treatment processes on the AOC(P17) concentration (solid line) and the DOC concentration (dashed line) in water from various reservoirs. Symbols: di, river water after dune infiltration; st, storage in open reservoir; coa, coagulation and sedimentation; rf, rapid filtration; pac, dosage of powdered activated carbon; gac, filtration through a bed of granular activated carbon; sf, slow sand filtration; bCl₂, breakpoint chlorination; O₃, ozonation; pCl₂, postchlorination; ClO₂, dosage of chlorine dioxide. From van der Kooij (1984).

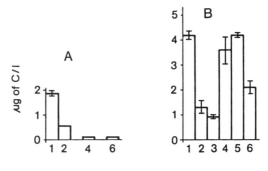

Treatment stage

Figure 3.8 Effect of water treatment on the concentration of starchlike compounds as determined by growth measurements with *Flavobacterium* species strain S12. A. Summer period; no addition of starch-based coagulant aid. B. Winter period. 1, river Meuse water after storage in open reservoirs; 2, coagulation and sedimentation; 3, ozonation (2 mg/ l); 4, addition of starch-based filter aid in winter (0.2 mg/l); 5, dual media filtration; 6, GAC filtration. From van der Kooij and Hijnen (1985).

dosage and DOC concentration. Furthermore, the nature of the organic materials present may influence the amount of AOC produced during ozonation. The production of AOC is directly related to the decrease of UV-light absorbing substances (Figure 3.9). As has been shown in Table 3.4, most compounds contributing to the AOC concentration of water after ozonation are carboxylic acids. The AOC increase caused by ozonation must be considered as an important negative side effect of this treatment process. AOC measurements can help in defining the operational conditions for ozonation in water treatment.

As has been explained above, filtration processes play a central role in water treatment in the Netherlands. Filtration through sand, granular-activated carbon, or dual media filters generally results in significant decreases of the AOC concentration (Figure 3.10). Bacteria, present in large numbers on the surface of the filtering material, utilize the compounds contributing to the AOC concentration of the water. Figure 3.10 shows that the AOC reduction clearly depends on the AOC concentration of the water and is limited at values approaching 10 μg of acetate-C equivalents per liter. Based on these observations and the AOC values as observed in ground water, water after river-bank filtration, or water after dune infiltration, it has been concluded that AOC values below 10 μg of acetate-C equivalents per liter are indicative for a biologically stable drinking water (van der Kooij, 1984).

In the past few years, increasing nitrate concentrations in ground water have become an additional problem for the water industry. The regulations of the European Community demand NO_3^- concentrations below 50 ppm. Nitrate removal using denitrifying bacteria is now being studied in a number of pilot plants in Europe, including the Netherlands. Recent observations on AOC values in filtrates from filter beds containing denitrifying bacteria have shown

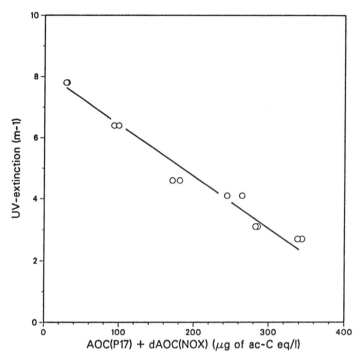

Figure 3.9   The AOC concentration in pretreated river water as a function of ultraviolet-light absorption after exposure to different concentrations of ozone. Data from van der Kooij et al. (1988).

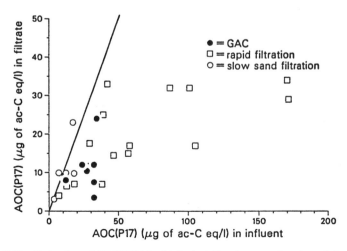

Figure 3.10   Reduction of AOC(P17) by biological filtration as a function of the influent concentration. The straight line indicates AOC influent = AOC filtrate.

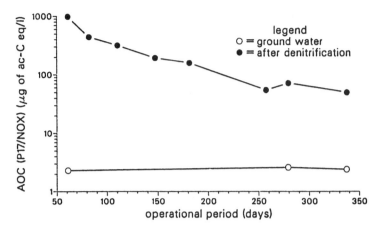

Figure 3.11   Effect of biological nitrate removal in a sulfur/limestone filter bed on the AOC(P17/NOX) concentration of ground water.

that biological nitrate removal is attended by strong increases of the AOC concentration (Figure 3.11). Hence, adequate post treatment is needed to reduce these high AOC values and the high level of bacterial biomass in water after denitrification (Hijnen et al., 1988).

Distribution systems   Prevention of aftergrowth in the distribution system is the ultimate purpose of the application of AOC measurements in water treatment. To define criteria for AOC concentrations of drinking water in which aftergrowth is strictly limited, data on aftergrowth processes in distribution systems are necessary. In a number of situations where ozonation was applied in water treatment, AOC values have been determined in water samples taken at increasing distance from the production plants. Aftergrowth was assessed with colony count determinations of heterotrophic bacteria using plates of diluted nutrient agar (LLA1.0). After spreading small volumes over the agar surface, these plates were incubated at $25°C$ during 10 to 14 days. Figure 3.12 shows AOC values and colony counts in three different water types during distribution. These data clearly indicate that AOC concentrations decrease in drinking water during distribution, but not at AOC values below 10 μg of C. Furthermore, the extent of aftergrowth as measured with heterotrophic colony counts depends on the AOC concentration of the water entering the distribution system. Further investigations in distribution systems are being conducted to support these preliminary conclusions.

3.4   Discussion and Conclusions

Controlling numbers of bacteria in drinking water during distribution by strictly limiting aftergrowth, as needed for reasons mentioned earlier in this chapter, is a challenge for both microbiologists and engineers. Practical experience and

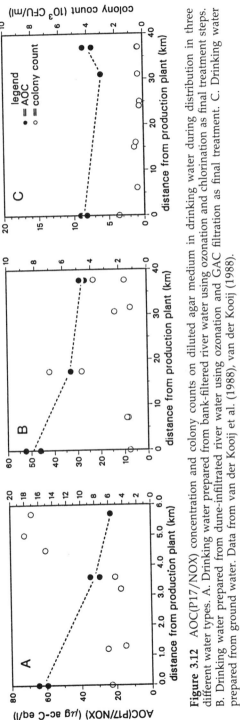

Figure 3.12 AOC(P17/NOX) concentration and colony counts on diluted agar medium in drinking water during distribution in three different water types. A. Drinking water prepared from bank-filtered river water using ozonation and chlorination as final treatment steps. B. Drinking water prepared from dune-infiltrated river water using ozonation and GAC filtration as final treatment. C. Drinking water prepared from ground water. Data from van der Kooij et al. (1988), van der Kooij (1988).

research have revealed that curative methods such as chlorination and/or pigging and swabbing are insufficiently effective in limiting aftergrowth. Bacteria survive at high concentrations of chlorine and a free chlorine residual is difficult to maintain throughout the entire distribution system, in particular when water quality promotes the formation of sediments and biofilms. Therefore, limitation of the concentration of energy sources in drinking water entering the distribution system is the most effective approach for limiting bacterial aftergrowth. Hence, microbiologists should define quality criteria for drinking water, and also for materials and chemicals in contact with drinking water, in terms of the amount of energy sources potentially available for bacteria. For the determination of an optimal sequence of water treatment processes, and for designing individual processes, AOC measurements can play a role in addition to the steadily growing number of water quality parameters. Also, ammonia and methane can serve as important energy sources, particularly in drinking water prepared from ground water sources. The concentration of these compounds can be determined with chemical methods.

The AOC concentration is defined as the concentration of easily assimilable organic compounds available for the organism under the applied test conditions. Of course, when only one or two organisms are used, this method does not include all the organic compounds that may be biodegradable for other bacteria and/or after longer test periods. However, it determines classes of compounds which are ubiquitously present because of their role in biological processes. These compounds are favorable sources of energy and carbon for a wide variety of bacteria. The data as presented in this paper indicate that a great variety of bacteria can be present in drinking water. Furthermore, many of these bacteria can grow at substrate concentrations of a few micrograms per liter. The growth rate of bacteria approaches the maximum growth rate when the substrate concentration is ten times higher than the substrate saturation constant of the involved compound. This would indicate that maximum growth rates can be attained at levels of easily assimilable organic compounds of a few tens of micrograms per liter. This has repeatedly been demonstrated with growth measurements applying the AOC technique as described in this chapter (van der Kooij and Hijnen, 1984; 1988). Therefore, the concentration of these compounds in drinking water must be very low if aftergrowth is to be limited.

Based on data obtained in the studies described above it is concluded that for limiting aftergrowth of heterotrophic bacteria, AOC concentrations in drinking water entering the distribution system should be less than 10 μg of acetate-C equivalents per liter. However, in situations with relatively high concentrations of energy sources for autotrophic bacteria, e.g. ammonia, a population of heterotrophic bacteria can also develop (Rittmann and Snoeyink, 1984), probably by using biomass constituents released by the autotrophic bacteria. Consequently, the concentrations of compounds such as ammonia and methane must also be kept to very low levels.

LeChevallier et al. (1987) have reported aftergrowth in a distribution system supplied with water with an AOC concentration above 100 μg of C per liter. In this situation, aftergrowth of coliforms was found. These bacteria multiplied in the biofilm on the walls of the pipes in the distribution system. The AOC level found in this system was similar to the AOC level observed in river Rhine water (Table 3.5). The development of a biofilm in which chlorine cannot penetrate is not surprising under these conditions. As has been shown above (Figure 3.7 and 3.9), rapid filtration of water without free chlorine strongly reduces the AOC concentration, particularly at relatively high AOC concentrations. One may expect that application of chlorine is more effective in situations where biofilm formation is limited. However, chlorination causes an increase of the AOC concentration. Thus, the question of what combination of AOC concentration and chlorine residual enables an effective limitation of aftergrowth, remains to be answered.

The investigations on AOC determinations and aftergrowth in drinking water during distribution, as conducted at KIWA Ltd, were part of the research program assigned and financed by The Netherlands Waterworks Association.

References

Allen, M.J. and Geldreich, E.E. 1978. Distribution line sediments and bacterial regrowth. Proceedings of the 5th American Water Works Association Water Quality Technology Conference, Kansas City, Mo.

Allen, M.J., Taylor, R.H., and Geldreich, E.E. 1980. The occurrence of micro-organisms in water main encrustations. *Journal American Water Works Association* 72:614–625.

Baylis, J.R., Chase, E.S., Cox, C.R., Ellms, J.W., Emerson, C.A., Knouse, H.V., and Streeter, H.W. 1930. Bacterial aftergrowth in water distribution systems. *American Journal of Public Health* 20:485–489.

Bonde, G.R. 1983. Bacteria in works and mains from ground water. *Aqua* 5:237–239.

Brazos, B.J., and O'Connor, J.T. 1985. A transmission electron micrograph survey of the planktonic bacteria population in chlorinated and non-chlorinated drinking water. Proceedings of the 13th American Water Works Water Quality and Technology Conference, pp. 275–305.

Burke, V., Robinson, J., Gracey, M., Petersen, D., and Partridge, K. 1984. Isolation of *Aeromonas hydrophila* from a metropolitan water supply: seasonal correlation with clinical isolates. *Applied and Environmental Microbiology* 48:361–366.

Burman, N.P. 1973. The occurrence and significance of actinomycetes in water supply. pp. 219–230 in Sykes, G., and Skinner, F.A. (editors), *Actinomycetales; characteristics and practical importance.* Academic Press, London and New York.

Burman, N.P., and Colbourne, J.S. 1976. The effect of plumbing materials on water quality. *Journal of Plumbing* 3:12–13.

Burman, N.P. 1965. Taste and odour due to stagnation and local warming in long lengths of piping. *Proceedings of the Society for Water Treatment and Examination* 14:125–131.

Colbourne, J.S. 1985. Materials usage and their effects on the microbiological quality of water supplies. *Journal of Applied Bacteriology Symposium Series*, 47S–59S.

Dennis, J.P., Taylor, J.A., Fitzgeorge, R.B., and Bartlett, C.L.R. 1982. *Legionella pneumophila* in water plumbing systems. *Lancet* 24:949–951.

Dietlicher, K. 1970. Wiederverkeimung ozonisierter Schnellfiltrate im Rohrnetz. *Schriftenreihe Verein fur Wasser- Boden- und Lufthygiene* 31:1–16.

Engel, H.W.B., Berwald, L.G., and Havelaar, A.H. 1980. The occurrence of *Mycobacterium kansasii* in tap water. *Tubercle* 61:21–26.

Fiore, J.V., and Babineau, R.A. 1977. Effect of an activated carbon filter on the microbiological quality of water. *Applied and Environmental Microbiology* 34:541–546.

Geldreich, E.E., Nash, H.D., Reasoner, D.J., and Taylor, R.H. 1972. The necessity of controlling bacterial populations in potable waters: community water supply. *Journal American Water Works Association* 64:596–602.

Geldreich, E.E., Allen, M.J., and Taylor, R.H. 1978. Interferences to coliform detection in potable water supplies. pp. 13–20 in Hendricks, C.W. (editor), *Evaluation of the microbiology standards for drinking water*. U.S. Environments Protection Agency, Washington.

Geldreich, E.E., and Rice, E.W. 1987. Occurrence, significance and detection of *Klebsiella* in water systems. *Journal American Water Works Association* 79:74–80.

Gracey, M., Burke, V., and Robinson, J. 1982. *Aeromonas*-associated gastroenteritis. *Lancet* ii:1304–1306.

Herman, L.G. 1978. The slow-growing pigmented water bacteria: problems and sources *Advances in Applied Microbiology* 23:155–171.

Herson, D.S., McGonigle, B., Payer, M.A., and Baker, K.H. 1987. Attachment as a factor in the protection of *Enterobacter cloacae* from chlorination. *Applied and Environmental Microbiology* 53:1178–1180.

Heymann, J.A. 1928. De organische stof in het waterleidingbedrijf. *Water en Gas* 12:61–72.

Hutchinson, M., and Ridgway, J.W. 1977. Microbiological aspects of water supplies. pp. 179–218 in F.A. Skinner and Shewan, J.M. (editors), *Aquatic microbiology*, Academic Press, London/New York.

Hijnen, W.A.M., Koning, D., Kruithof, J.C., and van der Kooij, D. 1987. The effect of bacteriological nitrate removal on the concentration of bacterial biomass and easily assimilable organic carbon compounds in ground water. *Water Supply* 6:265–273.

Joret, J.C. et Lévy, Y. 1986. Methode rapide d'évaluation du carbone éliminable des eaux par voie biologique. *Tribune de Cebedeau* 39:3–9.

Kaustova, J., Olsovsky, Z., Kubin, M., Zatloukal, O., Pelikan, M., and Hradil, V. 1981. Endemic occurrence of *Mycobacterium kansasii* in water supply systems. *Journal of Hygiene, Microbiology and Immunology* 25:24–30.

Kuo, P.P., Chian, E.S.K., and Chang, B.J. 1978. Identification of end products resulting from ozonation and chlorination of organic compounds commonly found in water. *Environmental Science and Technology* 11:1177–1181.

LeChevallier, M.W., Seidler, R.J., and Evans, T.M. 1980. Enumeration and characterization of standard plate count bacteria in chlorinated and raw water supplies. *Applied and Environmental Microbiology* 40:922–930.

LeChevallier, M.W., Babcock, T.M., and Lee, R.G. 1987. Examination and characterization of distribution system biofilms. *Applied and Environmental Microbiology* 53:2714–2724.

LeChevallier, M.W., Cawthon, C.D., and Lee, R.G. 1988. Inactivation of biofilm bacteria. *Applied and Environmental Microbiology* 54:2492–2499.

Lee, S.H., O'Connor, and Banerji, S.K. 1980. Biologically mediated corrosion and its effects on water quality in distribution systems. *Journal American Water Works Association* 72:636–645.

Leeflang, K.W.H. 1968. Biologic degradation of rubber gaskets used for sealing pipe joints. *Journal American Water Works Association* 60:1070–1076.

Levy, R.V., Hart, F.L., and Cheetham, R.D. 1986. Occurrence and public health significance of invertebrates in drinking water systems. *Journal American Water Works Association* 78:105–110.

Louwe Kooijmans, L.H. 1966. Occurrence, significance and control of organisms in distribution systems. General Report No. 3, pp. C5–C33, International Water Supply Association Congress, Barcelona.

Love, O.T., Robeck, G.G., Symons, J.M., and Buelow, R.W. 1974. Experience with activated carbon in the USA. pp. 279–295 in *Activated carbon in water treatment*, Papers and Proceedings of a Water Research Association Conference at Reading, U.K. 1973.

Martin, R.S., Gates, W.H., Tobin, R.S., Grantham, D., Sumarah, R., Wolfe, P., and Forestall., P. 1982. Factors affecting coliform bacteria growth in distribution systems. *Journal American Water Works Association* 74:34–37.

McFeters, G.A., Kippin, J.S., and LeChevallier, M.W. 1986. Injured coliforms in drinking water. *Applied and Environmental Microbiology* 51:1–5.

McMillan, L., and Stout, R. 1977. Occurrence of *Sphaerotilus, Caulobacter,* and *Gallionella* in raw and treated water. *Journal American Water Works Association* 69:171–173.

Meijers, A.P. 1984. De samenstelling van het Rijnwater in 1982 en 1983. RIWA, Amsterdam.

Muller, G. 1972. Koloniezahlbestimmungen im Trinkwasser. *Gas und Wasserfach, Wasser Abwasser* 113:53–57.

Nagy, L.A., and Olson, B.H. 1986. Occurrence and significance of bacteria, fungi and yeasts associated with distribution pipe surfaces. pp. 213–238, in Technology Conference Proceedings, American Water Works Association, Houston, Texas.

O'Conner, J.T., Hash, L., and Edwards, A.B. 1975. Deterioration of water quality in distribution systems. *Journal American Water Works Association* 67:113–116.

Reasoner, D.J., and Geldreich, E.E. 1979. Significance of pigmented bacteria in water supplies. pp. 187–196 in Proceedings Water Quality Technology Conference, American Water Works Association.

Ridgway, H.F., and Olson, B.H. 1981. Scanning electron microscope evidence for bacterial colonization of a drinking-water distribution system. *Applied and Environmental Microbiology* 41:274–287.

Ridgway, H.F., Means, E.G., and Olson, B.H. 1981. Iron bacteria in drinking-water distribution systems: elemental analysis of *Gallionella* stalks, using X-ray energy-dispersive microanalysis. *Applied and Environmental Microbiology* 41:288–297.

Ridgway, H.F., and Olson, B.H. 1982. Chlorine resistance patterns of bacteria from two drinking-water distribution systems. *Applied and Environmental Microbiology* 44:972–987.

Rittmann, B.E., and Snoeyink, V.L. 1984. Achieving biologically stable drinking water. *Journal American Water Works Association* 76:106–114.

RIWA. 1986. *De samenstelling van het Rijnwater in 1984 en 1985.* Amsterdam.

Rook, J. 1974. Formation of haloforms during chlorination of natural waters. *Journal for the Society for Water Treatment and Examination* 23:234.

Schellart, J.A. 1986. Disinfection and bacterial regrowth; some experiences before and after stopping safety chlorination by the Amsterdam Water Works. *Water Supply* 4:217–225.

Schoenen, D., and Schöler, H.F. 1983. Trinkwasser und Werkstoffe; Praxisbeobachtungen und Untersuchungsverfahren. Wasser Nr.37, DVGW-Schriftenreihe, Frankfurt.

Seidler, R.J., Morrow, J.E., and Bagley, S.T. 1977. *Klebsiella* in drinking water emanating from redwood tanks. *Applied and Environmental Microbiology* 33:893–900.

Servais, P., Billen, G., and Hascoet, M.C. 1987. Determination of the biodegradable fraction of dissolved organic matter in waters. *Water Research* 21:445–450.

Smalls, L.C., and Greaves, G.F. 1968. A survey of animals in distribution systems. *Journal Society Water Treatment Examination* 17:150–182.

Snoek, 1970. Enige chemische en bacteriologische aspecten van de toepassing van ozon bij de waterzuivering. H_2O 3:108–110.

Stalder, K., and Klosterkötter, W. 1976. Untersuchungen zur Wiederverkeimung von Trinkwasser nach Ozonbehandlung. *Zentralblatt für Bakteriologie, Mikrobiologie und Hygiene, Series B* 161:474–481.

Stanier, R.Y., Palleroni, N.J., and Doudoroff, M. 1966. The aerobic pseudomonads: a taxonomic study. *Journal General Microbiology* 43:159–271.

Stanfield, G., and Jago, P.H. 1987. The development and use of a method for measuring the concentration of assimilable organic carbon in water. Report PRU 1628-M, Water Research Centre, U.K.

Talbot, H.W., Morrow, J.E., and Seidler, R.J. 1979. Control of coliform bacteria in finished drinking water stored in redwood tanks. *Journal American Water Works Association* 71:349–353.

Tobin, J.O.H., Swann, R.A., Bartlett, C.L.R. 1981. Isolation of *Legionella pneumophila* from water systems: methods and preliminary results. *British Medical Journal* 282:515–517.

van der Kooij, D. 1978. The occurrence of *Pseudomonas* spp. in surface water and in tap water as determined on citrate media. *Antonie van Leeuwenhoek, Journal of Microbiology* 43:187–197.

van der Kooij, D. 1979. Characterization and classification of fluorescent pseudomonads from tap water and surface water. *Antonie van Leeuwenhoek, Journal of Microbiology* 45:225–240.

van der Kooij, D. 1980. Growth of *Aeromonas hydrophila* at low concentrations of substrates added to tap water. *Applied and Environmental Microbiology* 39:1198–1204.

van der Kooij, D., and Hijnen, W.A.M. 1981. Utilization of low concentrations of starch by a *Flavobacterium* species isolated from tap water. *Applied and Environmental Microbiology* 41:216–221.

van der Kooij, D. 1981. Multiplication of bacteria in drinking water. *Antonie van Leeuwenhoek, Journal of Microbiology* 47:281–283.

van der Kooij, D., Oranje, J.P., and Hijnen, W.A.M 1982. Growth of *Pseudomonas aeruginosa* in tap water in relation to utilization of substrates at concentrations of a few micrograms per liter. *Applied and Environmental Microbiology* 44:1086–1095.

van der Kooij, D., Visser, A., and Oranje, J.P. 1982. Multiplication of fluorescent pseudomonads at low substrate concentrations in tap water. *Antonie van Leeuwenhoek, Journal of Microbiology* 48:229–243.

van der Kooij, D., Visser A., and Hijnen, W.A.M. 1982. Determining the concentration of easily assimilable organic carbon in drinking water. *Journal of the American Water Works Association* 74:540–545.

van der Kooij, D., and Hijnen, W.A.M. 1983. Nutritional versatility of a starch-utilizing *Flavobacterium* at low substrate concentrations. *Applied and Environmental Microbiology* 45:804–810.

van der Kooij, D., and Hijnen, W.A.M. 1984. Substrate utilization by an oxalate-consuming *Spirillum* species in relation to its growth in ozonated water. *Applied and Environmental Microbiology* 47:551–559.

van der Kooij, D. 1984. The growth of bacteria on organic compounds in drinking water. Dissertation, KIWA, Rijswijk, The Netherlands.

van der Kooij, D., and Hijnen, W.A.M. 1985. Determination of the concentration of maltose- and starch-like compounds in drinking water by growth measurements with a well-defined strain of a *Flavobacterium* species. *Applied and Environmental Microbiology* 49:765–771.

van der Kooij, D., and Hijnen, W.A.M. 1985. Measuring the concentration of easily assimilable organic carbon in water treatment as a tool for limiting regrowth of bacteria in distribution systems. pp. 729–744 in Proceedings of the 13th American Water Works Association Water Quality Technology Conference, Houston, Texas.

van der Kooij, D. 1986. The effect of treatment on assimilable organic carbon in drinking water. pp. 317–328 in Huck, P.M. and Toft, P. (editors) *Treatment of drinking water for organic contaminants*, Pergamon Press.

van der Kooij, D., and Hijnen, W.A.M. 1985. Regrowth of bacteria on assimilable organic carbon in drinking water. *Journal Francais d'Hydrologie* 16:201–218.

van der Kooij, D., Hijnen, W.A.M., and Kruithof, J.C. 1987. The effects of ozonation, biological filtration and distribution on the concentration of easily assimilable organic carbon (AOC) in drinking water, pp D96–D113 in Proceedings 8th Ozone World Congress, International Ozone Association, Zurich, Switzerland.

van der Kooij, D. 1988. Prevention of bacterial aftergrowth in drinking water distribution systems. *Water Supply* 6:13–18.

van der Kooij, D., and Hijnen, W.A.M. 1988. Multiplication of a *Klebsiella pneumoniae* strain in water at low concentrations of substrates, Vol. 1, p. 20 (1–7) in Proceedings International Conference on Water and Waste Water Microbiology, Newport Beach, California, USA.

van der Kooij, D., and Hijnen, W.A.M. 1988. Nutritional versatility and growth kinetics of an *Aeromonas hydrophila* strain isolated from drinking water. *Applied and Environmental Microbiology* 54:2842–2851.

Veresikova, M. und Synek, M. 1974. Das Auftreten von *Sphaerotilus* und anderen Chlamydobakterien im Trinkwasser der Stadt Bratislava. *Acta Hydrochimica Hydrobiologica*. 2:261–270.

VEWIN. 1987. Waterleidingstatistiek 1984. Rijswijk, The Netherlands.

Victoreen, H.T. 1974. Control of water quality in transmission and distribution mains. *Journal American Water Works Association* 66:369–370.

de Vries, H. 1890. *Die Pflanzen und Tiere in den dunklen Raume der Rotterdamer Wasserleitung.* Gustav Fischer Verlag, Jena.

Wadowsky, R.M., Yee, R.B., Mezmar, L., Wing, E.J., and Dowling, J.N. 1982. Hot water systems as sources of *Legionella pneumophila* in hospital and non-hospital plumbing fixtures. *Applied and Environmental Microbiology* 43:1104–1110.

Werner, P. 1985. Eine Methode zur Bestimmung der Verkeimungsneigung von Trinkwasser. *Vom Wasser* 65:257–270.

Wierenga, J.T. 1985. Recovery of coliforms in the presence of a free chlorine residual. *Journal American Water Works Association* 77:83–88.

Yee, R.B., and Wadowsky, R.M. 1982. Multiplication of *Legionella pneumophila* in unsterilized tap water. *Applied and Environmental Microbiology* 43:1330–1334.

4

Effect of Starvation on Bacterial Resistance to Disinfectants

A. Matin and S. Harakeh

Department of Microbiology and Immunology, Stanford University, Stanford, California and Linus Pauling Institute of Science and Medicine, Palo Alto, California

4.1 Introduction

Public health measures aim at supplying potable water free from pathogenic agents. Such water supplies are treated with various disinfectants, most commonly chlorine, and now increasingly chlorine dioxide. For reasons of safety and economy, the minimal possible dose that is effective in inactivating certain indicator bacteria and pathogens is utilized. Bacteria are grown under laboratory conditions and are used to evaluate the performance of a disinfecting agent in the environment. The indicator organism most commonly employed in such testing is *Escherichia coli*.

Several assumptions are implicit in this practice. 1) It is assumed that the susceptibility of *E. coli* to inactivation is indicative of the susceptibility of pathogenic contaminants in the water supply. 2) The inability of *E. coli* to form colonies in laboratory media is assumed to reflect cell death. 3) The sensitivity of laboratory-grown bacteria is assumed to be a reliable measure of the sensitivity of natural populations. In this chapter, we concern ourselves entirely with critically examining the validity of the last-mentioned assumption. We first discuss the difference between growth conditions in the laboratory and in nature, and then review the evidence that indicates that bacteria grown under laboratory conditions are not necessarily a reliable index of the sensitivity of natural populations. Finally, we review studies that could have a bearing on the physiological basis of these differences.

4.2 Growth Conditions in Nature

An important difference between growth in the laboratory and in nature is the concentration of nutrients. In the laboratory, microorganisms are usually grown under rich conditions that are conducive to rapid and lavish growth. In contrast, most natural environments possess low concentration of nutrients, particularly that of organic carbon. For instance, the dissolved organic carbon in the oceans ranges between 0.3 to 1.2 mg per liter, or even lower (Morita, 1982, 1984; Van der Kooij and Hijnen, 1988). Hobbie (1966) found 6 μg glucose and 10 μg acetate per liter in Lake Erken in Sweden, and similar low estimates of available carbon have been made in soil (Gray, 1976). Consistent with these measurements are the estimated low mean generation times of microbial flora in natural environments: up to several days in aquatic environments (Jannasch, 1984; Morita, 1982; Brock, 1971), twenty days in deciduous woodland soil (Gray, 1976), and up to 48 hours for the intestinal microflora (Brock, 1971). Such findings have led microbial ecologists to agree that bacteria in nature often experience severe nutrient limitations, and that growth at submaximal rates as well as frequent starvation must be the norm for natural populations. Another marked difference between the conditions experienced by organisms in the laboratory and in nature probably concerns the extent of aggregation and surface attachment of microbes under the natural conditions. To what extent microbial populations in nature are found in aggregates and are attached to surfaces have been matters of some debate (Azam and Fuhrman, 1984; Fletcher and Marshall, 1982; Hermansson and Marshall, 1985). However, it is probable that organisms that need to be inactivated in potable water supplies do experience conditions of aggregation which probably influence their sensitivity to disinfectant agents. This has been discussed elsewhere (Hoff, 1978) and is not considered here.

4.3   The Effect of Antecedent Growth Environment

Batch Culture-grown Bacteria Versus the Natural Counterparts

Many groups of workers have found that individual bacterial populations isolated from natural environments are markedly more resistant to disinfection than the same strains after growth in laboratory batch cultures. For example, Berg et al. (1979) showed that native populations of *E. coli*, when treated with chlorine or chlorine dioxide (ClO_2) directly in the secondary sewage effluent, exhibited around 2.5 log inactivation after 10 minutes exposure to 5 mg/l of ClO_2 or chlorine. However, when cultures from this effluent were grown in nutrient broth and then subjected to the same treatment, there was an inactivation of about 7.4 log. In both situations, the total residuals of the disinfectant agents were very similar. The studies of Grieble et al. (1970) and Rhodes and Short (1970) point to the same conclusion for inactivation of *Pseudomonas aeruginosa* by acetic acid. The former group of workers found that 2% acetic acid did not eliminate *P. aeruginosa* from humidified units in a pulmonary disease ward, but the latter workers showed that broth-grown *P. aeruginosa* was inactivated with 1% acetic acid. Berg et al. (1984) similarly showed an inverse correlation between growth rate and sensitivity to ClO_2 for *Legionella pneumophila*. Their findings are consistent with the later report of Kutchla et al. that tap water-grown *Legionella* were five times more resistant to chlorine than their batch culture-grown counterparts. Similarly, *Yersinia enterocolitica* was shown to survive in chlorinated waters in spite of the fact that when grown in the laboratory it shows high sensitivity to chlorination (Wetzler, 1978).

In the above studies, the experimental protocol employed made it highly unlikely that the observed differences in sensitivity between natural populations and their batch culture-grown counterparts was the result of selection of mutants. In some of the above studies, control experiments demonstrated that passage of natural isolates through laboratory batch cultures immediately increased sensitivity. Continuous-culture studies, discussed below, point to the same conclusion, in that the sensitivity pattern of populations grown at different limiting nutrient concentrations can be manipulated in a time frame that would not permit selection of mutants.

In the studies discussed above concerning *E. coli* and *L. pneumophila*, it was also specifically excluded that increased resistance of natural populations was the result of aggregation of cells. Thus, in the case of sewage effluent, populations of *E. coli* in the native samples were blended before being exposed to the disinfectant agents.

Continuous Culture Studies   The above studies indicate that the natural populations tend to be more resistant to disinfection agents than their batch culture-grown counterparts. One explanation for this phenomenon is the difference in growth conditions in nature and the laboratory, i.e., nutrient limitation vs. nutrient excess, as discussed above. This possibility can be directly tested by making use of a chemostat in which microbial populations are exposed during growth to various degrees of nutrient limitation; the chemostat also permits establishment of steady state conditions so that the cause and effect relationships can be much more clearly determined.

The Theory and Practice of Continuous Culture   The chemostat (Figure 4.1) consists of a well-stirred culture vessel with growing microorganisms which is continuously supplied with fresh medium and in which the culture volume is kept constant by an overflow device. The inflow medium contains a known limiting nutrient, i.e., a nutrient that, during growth, is exhausted before any of the others.

A chemostat culture starts out as a batch culture, since all nutrients are present in excess, and growth occurs at μ_{max} (maximal growth rate). If no fresh medium is supplied, the depletion of the growth-limiting nutrient would eventually slow down the growth rate in accordance with the Monod equation:

$$\mu = \mu_{max} \left(\frac{s}{K_s + s} \right) \qquad (1)$$

where μ is the specific growth rate, s is the concentration of limiting nutrient, and K_s is the concentration of the limiting nutrient at which $\mu = 1/2\, \mu_{max}$.

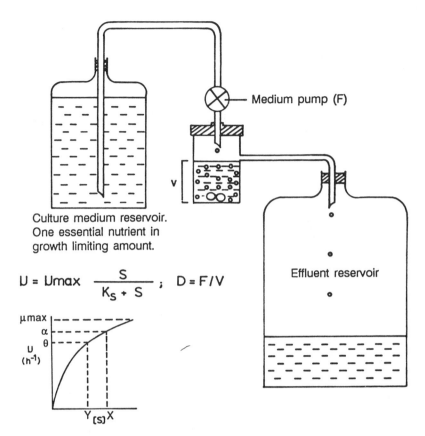

Culture medium reservoir.
One essential nutrient in
growth limiting amount.

$$U = Umax \dfrac{S}{K_S + S} \; ; \quad D = F/V$$

Figure 4.1 Schematic diagram of a chemostat vessel and (lower left) the Monod curve and the relationship between dilution rate and the concentration of a limiting nutrient. See text for explanation of symbols.

However, input of fresh medium permits growth to continue. In fact, the growth rate in the chemostat is governed by the dilution rate (D) which is given by the following equation:

$$D = F/V \tag{2}$$

where F is the flow rate of fresh medium in ml/h, and V is the culture volume in ml. The dimension of D is h^{-1}.

The change of biomass in the culture is given by:

$$\text{change} = \text{growth} - \text{output} \tag{3}$$

and since growth is μx and output is Dx:

$$\frac{dx}{dt} = \mu x - Dx = x(\mu - D) \tag{4}$$

So, if $\mu > D$, the concentration of organisms will increase, whereas the converse is true for $\mu < D$. Only when $\mu = D$ would the level of organisms remain constant with time; that is, the culture would be in a "steady state." Clearly, if $\mu > D$, consumption of substrate will be larger than the input and the substrate concentration in the growth vessel will gradually decrease, and with it μ will decrease in accordance with the Monod equation (1) until $\mu = D$; and if $\mu < D$, the converse will occur. Thus, a chemostat is a self-adjusting system which will reach a steady state at which $\mu = D$, as long as D does not exceed the critical dilution rate (D_c):

$$D_c = \mu_{max} \frac{S_R}{K_s + S_R} \tag{5}$$

where S_R is the concentration of the limiting nutrient in the inflow medium.

An easy way to visualize the working of a chemostat is in terms of the Monod curve describing the relationship between the concentration of the limiting nutrient and the growth rate (μ) (Figure 4.1). By controlling D $(=\mu)$, one automatically establishes the corresponding concentration of the limiting nutrient in the growth vessel in accordance with the Monod relationship. Thus, if the D is fixed at some value equal to α, the concentration of the limiting nutrient in the culture vessel would be X (Figure 4.1); if it is fixed at θ, Y will be the steady-state nutrient concentration in the vessel. The values of X and Y at D $= \alpha$ and θ will depend on the shape of the Monod curve for a given organism (which may change at different dilution rates). However, since most bacteria possess K_s values for individual nutrients in the μM range or less, growth at nearly all dilution rates in a chemostat involves substrate concentrations at the μM level or lower. Hence, the chemostat is an effective device for mimicking the natural environment in this respect.

Sensitivity of Nutrient-limited Bacteria in Response to Culture Dilution Rate
Matin and co-workers have conducted an examination of the effect of growth at different dilution rates in the chemostat on the sensitivity of bacteria to various disinfectant agents (Berg et al., 1982; Abou-Schleib, 1983; Harakeh et al., 1985.) This experimental approach was used to determine how the degree of nutrient limitation during growth influenced bacterial sensitivity. Results for *E. coli* and two disinfectant agents are reproduced in Figure 4.2. Similar results were obtained with *Klebsiella pneumoniae* and *Yersinia enterocolitica* (Harakeh et al., 1985). These studies were carried out at a number of growth temperatures that appear to be important in nature. At a given temperature, the culture resistant to a disinfectant agent increased with decreasing D value. At 37°C, *E. coli* grown under the limitation of an unknown component of nutrient broth exhibited three log greater resistance to ClO_2 after growth at D $= 0.06$ h^{-1} than when grown at maximal growth rate in nutrient broth batch culture. A similar phenomenon is evident at a growth temperature of 25°C (Figure 4.2A). When grown under glucose limitation, *E. coli* exhibited a similar pattern of

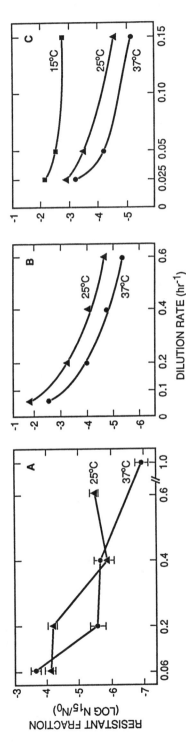

Figure 4.2 Effect of dilution rate on the susceptibility of selected organisms to different agents. The units on the ordinate express the rate of inactivation by the designated agents.

A. *E. coli* was grown in nutrient broth at 25°C and 37°C. S_R for nutrient broth was 0.16%. The highest dilution rates used were obtained using batch cultures grown in 0.16% nutrient broth at the two temperatures tested. Symbols: 25°C (▲) growth temperature; 37°C (●) growth temperature. Error bars represent one standard deviation (Berg et al., 1982).

B. *E. coli* was grown under glucose limitation and then treated with phenylphenol at 37°C (●) and 25°C (▲) (Abou-Schleib et al., 1983).

C. *Klebsiella pneumoniae* was grown under glucose limitation and then exposed to chlorine dioxide. $S_R = 0.4\%$, at three temperatures, 15°C (■), 25°C (▲), and 37°C (●) (Harakeh et al., 1985).

sensitivity to ClO_2 in relation to D, except that the highest sensitivity was noted at an intermediate D value of 0.2 h^{-1}.

In the case of phenylphenol, glucose-limited cells exhibited two orders of magnitude greater resistance as a result of 10-fold reduction in culture D value (Figure 4.2B). A similar result is evident in the case of *K. pneumoniae* (Figure 4.2C). At 37°C growth temperature, cells grown at D = 0.025 h^{-1} possessed nearly two orders of magnitude greater resistance to ClO_2 than those grown at D = 0.15 h^{-1}. *Y. enterocolitica* also exhibited the same order of magnitude difference in susceptibility in response to growth at near-μmax compared to when grown at a low D value (0.15 h^{-1} vs. 0.05 h^{-1}). A difference of the same order of magnitude was observed between batch culture grown *L. pneumophila* and the same organism grown at D = 0.06 h^{-1} (Berg et al., 1984). *Y. enterocolitica* was only marginally affected in its sensitivity by the growth temperature. In contrast, in the case of all the other bacteria tested, temperature determined the extent of susceptibility to disinfection at a given D value, with the sensitivity increasing with intensifying growth temperature. For example, *E. coli* grown in nutrient broth at 15°C at D = 0.06 h^{-1} exhibited a three-log reduction in number after 15 min treatment with phenylphenol, whereas when grown at 37°C under otherwise identical conditions, it exhibited nearly a five-log inactivation after the same treatment. (Abou-Schleib et al., 1983).

In general, *E. coli* grown under glucose limitation at a given temperature and D value tended to be more resistant than when grown under the limitation of the unknown constituent of nutrient broth (Berg et al., 1982; Abou-Schleib et al., 1983).

Sensitivity of Nutrient-limited Bacteria in Response to Culture Density   It is possible to arrange conditions in the chemostat in such a way that the growth conditions between two experiments differ only with respect to culture density; i.e., other aspects such as nutrient concentration, growth rate, temperature, aeration, etc. remain the same. This can be achieved by simply changing the concentration of the limiting nutrient in the inflow medium and then allowing the culture to attain a steady state at a given D value. This made it possible to examine the effect of culture density during growth on bacterial sensitivity to disinfection. This parameter was of interest because bacteria are typically grown to high cell densities in the laboratory, whereas in most natural environments, they attain much lower cell densities.

Three organisms have been tested in this respect, and the representative data are summarized in Table 4.1. Both *E. coli* and *Y. enterocolitica* exhibited greater sensitivity to ClO_2 after having been grown at a higher cell density, but cell density did not influence susceptibility of *K. pneumoniae*. It must be emphasized that the cell density during growth is the determining factor here rather than the density of populations subjected to disinfection. The latter was kept constant in all experiments.

Table 4.1 Effect of cell density on susceptibility of selected bacteria to chlorine dioxide

Organism	% Glucose in inflow medium	Temp (°C)	D (h^{-1})	Cell density[a]	Log N_{15}/N_0	Reference
Escherichia	0.15	25	0.40	1.10	-3.09 ± 0.13[b]	Berg et al.
coli	0.02	25	0.40	0.20	-5.06 ± 0.13	(1982)
Klebsiella						
pneumoniae	0.2	37	0.05	0.8	-4.22 ± 0.061[c]	Harakeh et al.
	0.4	37	0.05	1.8	-4.29 ± 0.045	(1985)
Yersinia						
enterocolitica	0.2	29	0.05	1.0	-4.98 ± 0.001[c]	Harakeh et al.
	0.4	29	0.05	1.6	-5.79 ± 0.017	(1985)

[a] Average optical density at 660 nm.
[b] Survival ratio after 15 min of contact with a dose of 0.75 mg of ClO_2 per liter at 23°C.
[c] Survival ratio after 15 min contact with a dose of 0.25 mg of ClO_2 per liter at 23°C.

4.4   Potential Role of Stress Proteins in Enhanced Resistance

It is not known what accounts for the increased resistance of nutritionally-deprived bacteria to disinfectant agents. There is increasing evidence, however, that such bacteria acquire increased resistance in general, and that this acquisition is dependent on the synthesis of a unique set of proteins. These aspects will be considered in this section.

Enhanced General Resistance of Nutrient-limited or Deprived Bacteria
Evidence has been accumulating over the years that bacteria grown under nutrient limitation, or those subjected to nutrient starvation, are more resistant to various deleterious agents. This was noted many years ago by Sherman and Albus (1923) and later by Elliker and Frazier (1938). Similarly, Gilbert and Brown (1978) found that Mg^{+2}-limited *P. aeruginosa* was more resistant to ethylenediaminetetracetic acid (EDTA) and polymycin B, and Green (1978) showed that *Vibrio chlorae* was more resistant to cold shock when grown under ammonium limitation.

This aspect has been recently studied rigorously by Matin and co-workers, with special emphasis on the role of starvation/stress proteins in this resistance. It was found that upon exposure to nutrient deprivation, bacteria synthesized new proteins that increased their resistance to a number of stresses. This resistance failed to develop if synthesis of starvation proteins was prevented. Peptidase-deficient mutants of *E. coli* which experienced amino acids scarcity during starvation were unable to synthesize starvation proteins. They lost viability during starvation much more rapidly than the wild type (Figure 4.3; Reeve et al., 1984a). Similarly, if protein synthesis at the onset of starvation

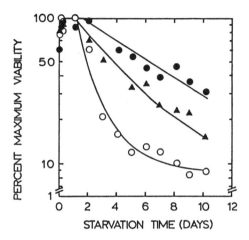

Figure 4.3 Viability of carbon-starved *E. coli*. *E. coli* K12 (●), CM17 (▲), and CM89 (○) were grown for two generations at 0.4% glucose mineral salts medium, harvested, washed, and starved for carbon. Viability was determined for each culture by serial dilution spreading on agar plus 0.02% glucose. From Reeve et al. (1984a). CM17 and CM89 are deficient in two and five peptides, respectively, and have a very low rate of protein synthesis during starvation.

was prevented by incubation at a low temperature (Schultz and Matin, 1988) or in the presence of an inhibitor of protein synthesis (Reeve et al., 1984b), the survival during starvation was greatly compromised (Figures 4.4 and 4.5). In both cases resistance to starvation increased the longer the culture was allowed to synthesize the starvation proteins. This is shown in Figure 4.5 for the chloramphenicol effect. It is clear that up to the first few hours of starvation, the sooner the protein synthesis was inhibited, the greater was the effect on the starvation survival capacity of the bacteria (Reeve et al., 1984b).

Starved bacteria were also more resistant to other stresses, and again the resistance to these stresses increased the longer the cells were allowed to synthesize starvation proteins for the first five hours after the onset of starvation. This is shown for the development of resistance to heat in Figure 4.6. This resistance developed in parallel with resistance to starvation with the maximum being reached in about five hours of protein synthesis after the onset of starvation. It is also evident that inhibition of protein synthesis, as by the addition of chloramphenicol, prevents development also of heat resistance. Precisely the same results were observed as regards the development of resistance to peroxide (Jenkins et al., 1988).

The development of enhanced resistance to starvation by nutrient deprivation-induced protein synthesis is evidently not confined to situations of complete nutrient exhaustion. Thus, Poolman et al. (1987) showed that the resistance to starvation of *Streptococcus cremoris* increased with decreasing D of the culture (Figure 4.7).

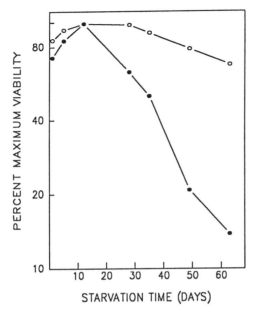

Figure 4.4   Viability of an *E. coli* K12 cell suspension after starvation at 4°C at 37°C for 10 min (■) or 5 h (O). Starvation conditions were established by allowing the culture to grow in M9 medium containing 0.025% glucose until the carbon source was exhausted. Viability was determined by plating serial dilutions on LB agar (reproduced from Schultz and Matin, 1988).

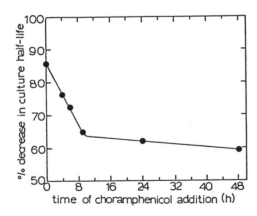

Figure 4.5   Differential effect on culture half-life of chloramphenicol added at different times during carbon starvation of *E. coli* K12. The cultures were grown to a density of 2.5×10^8 cells/ml, harvested, and then resuspended in a glucose-free medium. At the start of the experiment, the culture was divided in 50 ml aliquots; 100 μg of solid chloramphenicol was added at the times shown. The survival of the culture was monitored for 7 days. Results are expressed as the percent decrease in culture half-life compared with that of control culture (half-life, 142 h) to which no chloramphenicol was added (from Reeve et al., 1984b).

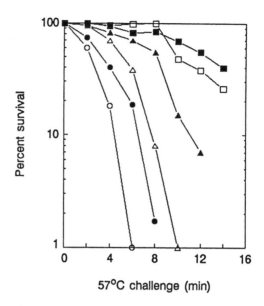

Figure 4.6   Thermal resistance induction in glucose-starved *E. coli*. Wild type *E. coli* K12 was challenged at 57°C during exponential growth (μ), or after 1 h (A), 2 h (θ), 4 h (□), or 24 h (■) following glucose depletion from the medium. The symbol (●) indicates results of a control experiment in which chloramphenicol (100 μg/ml) was added 20 min after the onset of starvation, and the culture was allowed to starve for 4 h before heat challenge. 100% bacterial counts equals ca. 3.0×10^3 cells/ml (reproduced from Jenkins et al., 1988).

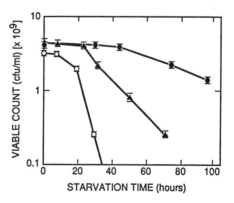

Figure 4.7   The effect of dilution rate on the viability of *S. cremoris* Wg2 cells grown during lactose limitation. The different rates tested are (●) 0.07h^{-1}, 0.11h^{-1} (▲), and 0.38h^{-1} (O) prior to starvation in the culture fluid. The reservoir medium contained 2.5% (wt/vol) of lactose. The CFUs were calculated by using six to eight agar plates (from Poolman et al., 1987).

The Nutrient-Deprivation Proteins   That bacteria exposed to nutrient deprivation synthesize unique proteins was shown directly by Groat and Matin (1986) and Groat et al. (1986). The protein synthesis pattern of growing cells and cells subjected to different nutrient deprivations was analyzed using the two-dimensional gel electrophoresis technique of O'Farrell (1975). This technique permits resolution of proteins both on the basis of their molecular weight as well as electrical charge, and thus can resolve up to a thousand different proteins. Synthesis of several polypeptides was either newly initiated or increased relative to total protein synthesis during starvation. Some 26 such polypeptides were synthesized under carbon deprivation conditions, 32 under phosphorus, and 26 under nitrogen deprivation. Although some of the starvation-induced polypeptides were unique to individual starvation conditions, several others were common to one or more of these conditions, and some 13 appeared under all three deprivation conditions.

Further experiments indicated that the different starvation proteins were synthesized at different times in the first 3 to 5 hours after the exposure of the cells to nutrient deprivation. Some were synthesized very early and their synthesis lasted for a short time, while others attained a peak level of synthesis after about one to four hours of starvation (Groat et al., 1986). This temporal distribution of synthesis of individual proteins probably accounts for the increasing degree of resistance to deleterious agents exhibited by *E. coli* with increasing period of starvation, as discussed above.

Nutrient Deprivation-Gene Expression   Synthesis of unique proteins suggests a global response of gene regulation at the onset of starvation. This possibility was directly investigated by generating a random library of *lac* Z fusions in *E. coli* using Mu dx and λ p *lac* phages (Groat et al., 1986; Schultz and Matin, 1987; Schultz et al., 1988).

These phages are believed to insert randomly in the *E. coli* genome. In proper orientation, the *lac* Z gene is controlled by the promoter of the gene in which the phage inserts, making it possible to monitor the expression of a gene by measuring β-galactosidase synthesis (which can be easily done by a colorimetric reaction). Screening of this library on low and high carbon-source media revealed the presence of several strains that synthesized β-galactosidase only during starvation. β-galactosidase synthesis by three of such fusion strains is shown in Figure 4.8 (Schultz and Matin, 1988). These strains were grown in a glucose–mineral salts medium until glucose was exhausted, and β-galactosidase synthesis was followed during growth and starvation. It is clear that the three strains induced β-galactosidase after the onset of starvation, and that they differ in the temporal pattern of this synthesis. These results are consistent with the findings discussed above that starvation leads to the induction of new protein synthesis and that these proteins fall into different temporal categories.

Although the induction of these genes has been documented primarily in the context of starvation, it appears that they are likely also expressed during

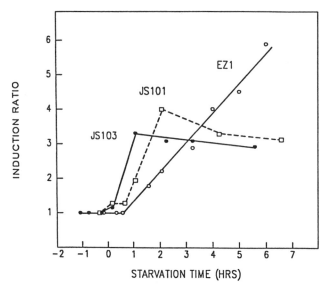

Figure 4.8 The effect of glucose depletion from M9 medium on the induction of β-galactosidase activity of *lac* fusion strains of *E. coli*. EZ1 (O) was obtained using Mu dX phage. JS101 (□) and JS103 (●) were obtained using λ p *lac*Mu9 phage as described by Bremer et al. From Schultz and Matin (1988).

nutrient limitation. (By nutrient limitation, we refer to conditions that permit growth at submaximal rates, such as those found in a chemostat, and "starvation" refers to complete absence of a nutrient.) This has been documented for one starvation gene, contained in the *E. coli* fusion strain referred to as AMS 3 (Groat et al., 1986; Schultz and Matin, 1988). This strain was cultivated in the chemostat under glucose limitation at a number of different D values and, after the establishment of steady states, cells were analyzed for their β-galactosidase content. The enzyme activity increased with decreasing D. Thus, this starvation gene became more and more activated as the concentration of glucose in the medium dwindled.

That starvation leads to the synthesis of new proteins and induction of unique genes has also been shown in *Salmonella typhimurium* (Spector et al., 1986); these proteins may also confer resistance to starvation survival in this bacterium. A number of marine bacteria also synthesize unique proteins at the onset of starvation (Amy and Morita, 1983; Jouper-Jaan et al., 1986); many of these proteins are surface proteins (Nyström et al., 1988).

Functions of Nutrient-Deprivation Proteins   It is not known what the functions of the nutrient-deprivation proteins are, and how they protect the cell against deleterious influences. It is most likely that some of them are concerned with the cells' catabolic activity/energy metabolism. This idea is supported by a variety of indirect evidence. Firstly, it is known that the synthesis of many of the starvation proteins is dependent on the cyclic adenosine monophosphate receptor protein complex (Schultz and Matin, 1988; Schultz et al., 1988), which is what would be expected if they were catabolic enzymes, or proteins concerned with cellular uptake mechanisms. Secondly, it has been shown that bacteria exposed to nutrient deprivation express new transport systems that have higher affinity for their substrates. These transport systems frequently also have a high degree of nonspecificity (Kjelleberg et al., 1987). Thirdly, it has been shown by many groups of workers that as bacteria grow under progressive nutrient scarcity, they derepress a number of their catabolic enzymes (Matin, 1979; Matin, 1981). The function of these types of starvation proteins is most likely to increase the capacity of bacteria to make better use of scarce nutrients. As such, it is unlikely that this category of proteins plays a role in the increased resistance of the cell to stresses like heat, oxidation, treatment with disinfectant agents, etc. as discussed above. This conclusion is confirmed by the finding that *cya*⁻ or *crp*⁻ mutants of *E. coli* (which cannot generate the cAMP/CRP complex) are as good as the wild type in developing resistance to various stresses when starved (Jenkins et al., 1988).

It appears that the starvation proteins whose synthesis is independent of cAMP/CRP complex confer enhanced stress resistance on *E. coli* (Schultz and Matin, 1988; Jenkins et al., 1988). Perhaps these proteins are involved in preventing degradation of essential proteins during stresses, in stabilizing DNA, or in making the cell envelope more resistant. Future research into the role of these proteins should be fruitful in providing fundamental insights and can also conceivably enhance our capacity to devise better disinfection procedures for natural bacterial populations.

Conclusions

The studies reviewed above strongly indicate that non- or slowly-growing bacteria, such as those found in most natural environments, are more resistant to deleterious agents, including disinfectants, than their lavishly-grown counterparts. It is likely that the synthesis by the former of certain stress proteins holds the key to this difference.

References

Abou-Schleib, H., Berg, J.D., and Matin, A. 1983. Effect of antecedent growth conditions on sensitivity of *E. coli* to phenylphenol. *FEMS Lett.* 19:183–186.

Amy, P.S., and Morita, R.Y. 1983. Protein patterns of growing and starved cells of a marine *Vibrio* sp. *App. Environ. Microbiol.* 45:1109–1115.

Azam, F., and Fuhrman, J.A. 1984. Measurement of bacterioplankton growth in the sea and its regulation by environmental conditions. pp. 179–196 in *Heterotrophic Activity in the Sea*, J.E. Hobbie, P.J. le B. Williams (eds.). New York: Plenum.

Berg, J., Aieta, E.M., and Roberts, P.V. 1979. Effectiveness of chlorine dioxide as a wastewater disinfectant. pp. 61–71 in *Progress in Wastewater Technology, Proceedings of the National Symposium, Cincinnati, Ohio, September 1978*, A.D. Venosa (ed.). EPA-600/9–018, U.S. Environmental Protection Agency, Cincinnati.

Berg, J.D., Matin, A., and Roberts, P.V. 1982. Effect of antecedent growth conditions on sensitivity of *E. coli* to ClO_2. *Appl. Environ. Microbiol.* 44:814–819.

Berg, J.D., Hoff, J.C., Roberts, P.V., and Matin, A. 1984. Growth of *Legionella pneumophila* in continuous culture and its sensitivity to inactivation by chloride dioxide. *Appl. Environ. Microbiol.* 49:2465–2467.

Brock, T.D. 1971. Microbial growth rates in nature. *Bacteriol. Rev.* 35:39–58.

Elliker, P.R., and Frazier, W.C. 1938. Influence of time and temperature of incubation on heat resistance of *E. coli*. *J. Bacteriol.* 36:83–98.

Fletcher, M., and Marshall, K.C. 1982. Are solid surfaces of ecological significance to aquatic bacteria? *Adv. Microb. Ecol.* 6:199–230.

Gilbert, P., and Brown, M. 1978. Influence of growth rate and nutrient limitation on the gross cellular composition of *Pseudomonas aeruginosa* and its resistance to 3- and 4-chloramphenicol. *J. Bacteriol.* 133:1062–1065.

Gray, T.R.G. 1976. Survival of vegetative microbes in soil. *Symp. Soc. Gen. Microbiol.* 26:327–364.

Green, J.A. 1978. The effects of nutrient limitation and growth rate in the chemostat on the sensitivity of *Vibrio cholerae* to cold shock. *FEMS Micro. Lett.* 4:217–219.

Grieble, H.C., Colton, F.R., and Bird, T.J. 1970. Fine-particle humidifiers: source of *Pseudomonas aeruginosa* infections in a respiratory disease unit. *New Engl. J. Med.* 282:531–535.

Groat, R.G., and Matin, A. 1986. Synthesis of unique proteins at the onset of carbon starvation in *Escherichia coli*. *J. Ind. Microbiol.* 1:69–73.

Groat, R.G., Schultz, J.E., Zychlinsky, E., Bockman, A., and Matin, A. 1986. Starvation proteins in *Escherichia coli*: kinetics of synthesis and role in starvation survival. *J. Bacteriol.* 168:486–493.

Harakeh, M.S., Berg, J.D., Hoff, J., and Matin, A. 1985. Sensitivity of chemostat-grown *Yersinia enterocolitica* and *Klebsiella pneumoniae* to chlorine dioxide. *Appl. Environ. Microbiol.* 49:69–72.

Hermansson, M., and Marshall, K.C. 1985. Utilization of surface localized substrate by nonadhesive marine bacteria. *Microb. Ecol.* 11:91–105.

Hobbie, J. 1966. *Proc. I.B.P. Symp. Amsterdam and Nieuwersluis, Koninklijke Nederlandse Akademie van Wetenschappen*, p. 245.

Hoff, J.C. 1979. The relationship of turbidity to disinfection of potable water. pp. 103–117 in *Evaluation of the Microbiology Standards for Drinking Water*, C.W. Hendricks (ed.). USEPA.

Jannasch, H.W. 1969. Estimation of bacterial growth rates in natural waters. *J. Bacteriol.* 99:156–160.

Jenkins, D.E., Schultz, J.E., and Matin, A. 1988. Starvation-induced cross protection against heat or H_2O_2 challenge in Escherichia coli. *J. Bacteriol.* 170:3910–3914.

Jouper-Jaan, Å, Dahlöf, B., and Kjelleberg, S. 1986. Changes in the protein composition of three bacterial isolates from marine waters during short-term energy and nutrient deprivation. *Appl. Environ. Microbiol.* 52:1419–1421.

Kjelleberg, S., Humphrey, B.A., and Marshall, K.C. 1983. Initial phases of starvation and activity of bacteria at surfaces. *Appl. Environ. Microbiol.* 46:978–984.

Kjelleberg, S., Hermansson, M., Mårdén, P., and Jones, G.W. 1987. The transient phase between growth and nongrowth of heterotrophic bacteria, with emphasis on the marine environment. *Ann. Rev. Microbiol.* 41:25–49.

Matin, A. 1979. Microbial regulatory mechanism at low nutrient concentrations as studied in a chemostat. pp. 323–339. In *Strategies of Microbial Life*, M. Shilo (ed.). Berlin: Dahlem Konferenzen.

Matin, A. 1981. Regulation of enzyme synthesis studied in continuous culture. pp. 69–97 in *Continuous Culture of Cells*, P.H. Calcott (ed.). CRC Press.

Matin, A., and Veldkamp, H. 1978. Physiological basis of the selective advantage of a *Spirillum* sp. in a carbon-limited environment. *J. Gen. Microbiol* 105:187–197.

Morita, R.Y. 1982. Starvation-survival of heterotrophs in the marine environment. in *Advances in Microbial Ecology*. 6:171–197.

Morita, R.Y., 1984. pp. 83–100 in *Heterotrophic Activity in the Sea*, J. Hobbie and P.J.L. Williams (eds). New York: Plenum.

Nyström, T., Mårdén, P., and Kjelleberg, S. 1986. Relative changes in incorporation rates of leucine and methionine during starvation survival of two bacteria isolated from marine waters. *FEMS Microbiol. Ecol.* 38:285–292.

O'Farrell, P.H. 1975. High resolution two-dimensional electrophoresis of proteins. *J. Biol. Chem.* 250:4007–4021.

Poolman, B., Smid, E., Veldkamp, H., and Konings, W. 1987. Bioenergetic consequences of lactose starvation for continuously culture *Streptococcus cremoris*. *J. Bacteriol.* 169:1460–1468.

Reeve, C.A., Bockman, A.T., and Matin, A. 1984a. Role of protein degradation in the survival of carbon-starved *Escherichia coli* and *Salmonella typhimurium*. *J. Bacteriol.* 157:758–763.

Reeve, C.A., Amy, P.S., and Matin, A. 1984b. Role of protein synthesis in the survival of carbon-starved *Escherichia coli* K–12. *J. Bacteriol.* 160:1041–1046.

Rhoades, E.R., and Short, S.G. 1970. Susceptibility of *Serratia, Pseudomonas,* and *Enterobacter* to acetic acid. *Antimicrobial Agents & Chemotherapy*, 498–502.

Schultz, J.E., and Matin, A. 1988. Regulation of carbon starvation genes in *Escherichia coli*. pp. 50–60 In *Homeostatic Mechanisms in Micro-organisms*, R. Whittenbury, G.W. Gould, J.G. Banks and R.G. Board (eds.). *FEMS Symposium No. 44*. Claverton Down, Bath: Bath University Press.

Schultz, J.E., Latter, G.I., and Matin, A. 1988. Differential regulation by cyclic AMP of starvation protein synthesis in Escherichia coli. *J. Bacteriol.* 170:3903–3909.

Sherman, J.M., and Albus, W.R. 1923. Physiological youth in bacteria. *J. Bacteriol.* 8:127–139.

Spector, M.S., Aliabadi, Z., Gonzales, T., and Foster, J.W. 1986. Global control in *Salmonella typhimurium*: Two-dimensional electrophoretic analysis of starvation-, anaerobiosis-, and heat shock-inducible proteins. *J. Bacteriol.* 168:420–424.

Van der Kooij, D., and Hijnen, W.A. 1989. Multiplication of a *Klebsiella pneumoniae* strain in water at low concentration substrates. In *International Conference on Water and Wastewater Microbiology*, in press.

Wetzler, T.T. 1978. *Yersinia enterocolitica* in water supplies. pp. 103–117. In *Proceedings, AWWA Water Quality Technology Conference*. Louisville, KY.

5

Microbiology of Activated Carbon

Mark W. LeChevallier and Gordon A. McFeters

American Water Works Service Company, Inc. Belleville, Illinois and Department of Microbiology Montana State University Bozeman, Montana

5.1 Introduction

The use of carbon dates back to at least 1550 B.C. where carbon is mentioned in Egyptian papyruses for medicinal uses (Hassler, 1963). Carbon has been used continuously as an absorbent since biblical times (Old Testament, Num. 19:9). Although activated carbon may be made from a number of raw materials (Capelle and de Vooys, 1983), in the 1750's toasted biscuits were put into the water of the St. Lawrence River to prevent fluxes of the Royal Navy under the command of Sir Charles Saunders (Baker, 1981). About 4 lb. of burnt biscuits were used to a hogshead of water.

In 1930, meat packing industries in Chicago first used activated carbon to correct disagreeable odors in wastewater effluents (Hassler, 1963). By 1938, over a thousand potable water plants in the U.S. were using carbon as part of the treatment process (Hassler, 1963, Baker, 1981). In 1970, the Community Water Supply Survey (McCabe et al., 1970) revealed that unpleasant tastes and odors in drinking water were the most common customer complaints. To combat taste, odor, and color problems in surface waters caused by wastes from industrial, municipal and agricultural sources, decaying vegetation, and algal blooms (Hassler, 1963; National Research Council, 1980; Symons et al., 1981), powdered activated carbon (PAC) and/or granular activated carbon (GAC) have been widely used. As a result of amendments to the National Interim Primary Drinking Water Regulations (USEPA, 1975, 1979), many water treatment systems use PAC or GAC to control organic residues such as oil, gasoline, phenols, or trihalomethanes (THM's) and their precursors (National Research Council, 1977, 1980; Symons et al., 1981, Graese et al., 1987).

Despite its long history of use, the microbiology of activated carbon and its impact on drinking water supplies has not been well characterized. A number of studies have shown that carbon particles coated with bacteria frequently penetrate treatment barriers and enter potable water supplies. Although significant advances have been reported in the literature recently, additional studies are needed to resolve the many questions regarding the microbiology of activated carbon.

Penetration of Carbon Fines Through Treatment Barriers   Concern about penetration of carbon fines through treatment barriers began in 1962 when treatment plant operators of Mt. Clemens, Michigan received customer complaints of black particles in the drinking water supply (Hansen,1970). Powdered activated carbon particles could be easily seen in toilets and left greasy, black rings in bathtubs. Robeck et al (1964) demonstrated that PAC frequently penetrated water plant filters and caused problems in distribution water, particularly when flocculation was weak. Because carbon particles do not reflect light, turbidimetric analyses are not reliable to detect carbon fine breakthrough (Hansen, 1970).

Camper et al. (1986) used a gauze filter/Swinnex procedure to collect carbon fines from 201 granular activated carbon-treated drinking water samples over a 12 month period. Examination of the filters by image analysis showed that the average sample (18,600 l of treated drinking water) contained 2,333 carbon particles ranging in size from 1.0 to $3.5 \times 10^3 \mu m$ in diameter. Stewart et al. (1988), using a modification of the Swinnex procedure (using a 10 μm Nuclepore polycarbonate filter), detected an average of 36 (range 10 to 62) carbon particles per liter in the effluent of a pilot GAC filter. The particles ranged in size from 2 to greater than 40 μm in diameter (mean diameter, 6 μm).

Evaluation of operational parameters influencing release of carbon fines showed that a deeper GAC filter bed, increased applied water turbidity, and elevated filtration rate contributed to the presence of colonized carbon fines in drinking water (Camper et al., 1987). The age of the GAC (zero to three years) material was not observed to affect the occurrence of fines in filtered water.

Studies have shown that at various times during a filter run cycle, both before and immediately after backwashing, particles enter distribution water (Amirtharajah and Wetstein, 1980). Robeck et al. (1962) demonstrated that bacterial and viral penetration through a granular filter accompanied floc breakthrough. Syrotynski (1971) observed that microorganisms, turbidity, and aluminum ions could be found in filtered drinking water at all ranges of flows, but especially at high filtration rates (1.44 to 2.52 liters/min per m^2). Preliminary studies by the American Water Works Service Co, Inc. have shown that the highest levels of carbon fines were released after filter backwash (Bill Becker, personal communication). However, penetration of carbon fines was not solely limited to the time before or after filter backwash. Camper et al. (1986) found that release of carbon fines occurred throughout the filter run.

GAC and PAC fines have not been the only filter materials released from treatment plants. Camper et al. (1987) found sand and anthracite carbon particles in 43 filter effluent samples from treatment plants not using GAC. Sand and anthracite particle counts were the same or higher than particle levels from

GAC filters. The authors noted, however, that GAC particles were significantly more populated with chlorine-protected heterotrophic plate count (HPC) bacteria than either sand or anthracite.

5.2   Enumeration of Bacteria on Carbon Surfaces

Documentation of bacterial numbers in carbon absorbers and in treated effluents is difficult to compare because of the variety of microbiological methods that have been used. Widely different microbiological media and incubation conditions (time, temperature, etc.) are known to produce varying results. In general, low bacterial counts have been reported when low temperatures and short incubation times were employed (2 to 3 days at 20 to 22°C), while higher bacterial densities have been reported on carbon particles and in carbon treated effluents when longer incubation times were used for growth of heterotrophic bacteria (5 days to 2 weeks) (Cairo et al., 1979; Schalekamp, 1979; The National Research Council, 1980; Parsons et al., 1980; AWWA Research and Technical Practice Committee on Organic Contaminants, 1981; Camper et al., 1985a). Use of a low-nutrient medium (R2A agar) has produced optimum recovery of HPC bacteria (Reasoner and Geldreich, 1985). Camper et al. (1985a) found that R2A agar incubated at 28°C for 7 days enumerated 184-times more bacteria from GAC particles than the standard spread plate procedure. Parsons (1980) incubated R2A agar for 6 days at 35°C and found that bacteria associated with GAC material ranged from a low of 1.5 times 10^4 to 1.8 times 10^8 cfu/g (wet weight). R2A agar can be particularly effective in recovering pigmented organisms. A study at Evansville, Indiana, found both virgin GAC and reactivated carbon adsorber effluents contained pigmented bacteria, even though the influent to the filters showed no detectable pigmented organisms (D. Reasoner, unpublished data, cited by Symons et al., 1981).

Methods used to recover coliform bacteria in water from carbon filters may impact recovery efficiencies. Camper et al. (1986) reported that coliform enumerations by the membrane filter (MF) technique showed only 7.0% of the GAC filter effluent samples contained attached bacteria, while coliforms were recovered from over 17% of the samples analyzed by a modified most-probable-number (MPN) procedure. It was thought that the carbon particles could alter the surface pore morphology so as to inhibit coliform growth. Other studies have demonstrated the deleterious effect of turbidity on the detection of coliforms by the MF technique (LeChevallier et al., 1981).

The basic premise of viable plate count procedures is that one colony arises from a single monodispersed microorganism (American Public Health Association, 1985). This assumption may be especially untrue when water contains suspended particulate material. Ridgway and Olson (1981) reported that 17% of turbidity particles in drinking water samples contained attached microbes, averaging between 10 and 100 bacteria per particle. The authors concluded that conventional standard plate count procedures underestimated the actual number of microorganisms by factors of 1,500 to 15,000 fold.

Much of the problem related to the enumeration of particle-associated bacteria arises from a lack of adequate techniques to desorb viable organisms so standard procedures can be used. Accurate bacterial enumeration of biofilms on GAC particle surfaces may account for conflicting conclusions derived from the research literature. Parsons (1980) recovered highest bacterial counts from GAC surfaces by sonication (20-kilohertz, 180-watt output for 4 minutes). Grinding of GAC particles in a Waring blender or tissue grinder was found to be inadequate because of cell disruption, reattachment of organisms to newly created surfaces, and simultaneous settling of bacterial cells with carbon particles. Ashidate and Geldreich (personal communication of unpublished results), however, found that sonication caused an abrupt decrease in bacterial counts after three minutes treatment. They recommended blending carbon fines in a Waring Blender for three minutes at maximum speed. This procedure was reported to be twice as effective as sonication and more than six times as effective as hand shaking for release of attached organisms.

Using controls of unattached bacteria, Camper et al. (1985a) found that sonication killed *Escherichia coli* cells at the minimum as well as 30, 50, and 75 percent of maximum power settings; even when the sample was iced (74% decrease in viable counts within nine minutes). These researchers also found that a Waring blender produce rapid temperature changes of cell suspensions. Temperatures were found to increase from 4°C to 30°C following three minutes of blending and 50°C was attained after ten minutes of treatment. Decreased bacterial viability corresponded with increased temperatures. Over 90% of the HPC bacteria were killed by temperatures produced by seven minutes blending in a Waring blender.

Camper et al., (1985a) found homogenization of carbon particles at 16,000 rpm in a chilled (4°C) buffer of Zwittergent 3-12, EGTA, Tris buffer, and peptone (Table 5.1) gave the highest recovery of plate count bacteria. The construction of the homogenizer allowed the blending vessel to be immersed in an ice bath during treatment and maintained the temperature of the cell suspensions below 10°C. Addition of peptone was necessary to prevent readsorption of the bacteria to the carbon surface. Without the peptone, more bacteria were found attached to the carbon particles than before treatment! Control experiments showed that the homogenization procedure was approximately 80 to 90 percent effective in recovering attached bacteria from activated carbon surfaces. Davies and McFeters (1988), using the homogenization procedures of Camper et al., reported recovery efficiencies of attached bacteria from carbon particles ranging from 70.5 to 106.3% (mean 81.5%, =8). The ability of the researchers to sort individual bacterial cells by differential filtration (pore sizes ranged between 0.6 and 1.0 μm) indicated that the organisms were desorbed from the particulate material.

Table 5.1 Homogenization solution for desorption of bacteria from GAC particles

Compound	Concentration
Zwittergent 3–12	10^{-6}M
Ethyleneglycol-bis-	
(β amino-ethyl ether)-	10^{-3}M
N-N-tetra acetic acid	
Peptone	0.1%
Tris buffer	10^{-2}M

Mixture was homogenized (16,000 rpm) for 3 min at 4°C. Adapted from Camper et al., 1985a.

5.3   Impact of Carbon Treatment on Water Quality

The high degree of transition porosity in activated carbon allows for an increase in the diffusion-limited rate of adsorption of materials from solutions (Mattson and Mark, 1971; Capelle and de Vooys, 1983). During the activation process, the regular array of carbon bonds in the surface of the crystallites are disrupted, yielding free valences which are highly reactive (Hassler, 1963). As a result, activated carbon is very effective for adsorbing a wide range of organic compounds (Symons et al., 1981; Capelle and de Vooys, 1983). Application of PAC or GAC has been recommended for control of certain organic compounds in potable water supplies (Suffet, 1980; Symons et al., 1981). In the United States, GAC is commonly used in place of granular media in conventional filters (GAC filter-adsorbers) primarily for removal of turbidity, organic substances, tastes, and odors. This practice is in contrast to European countries where GAC adsorbers are often used after granular media filters (post-filter adsorbers) for removal of specific trace organic substances in addition to use for taste and odor control (Graese et al., 1987).

Encouragement of bacterial growth on GAC media (sometimes termed biologically activated carbon; BAC) has been reported to increase the capacity of GAC filters for organic carbon removal, prolong bed-life of filters, remove nutrients for bacterial regrowth, and lower the chlorine demand of distribution water supplies (AWWA Committee Report, 1981; Bourbigot et al., 1982; Rittmann and Snoeyink, 1984; Pascal et al., 1986). This process has been effectively used in Europe for control of bacterial growth in distribution system waters. Van der Kooij (1987) reported that filtering water through GAC filters in which bacteria were encouraged to proliferate, could reduce assimilable organic carbon (AOC) levels by 90 percent. Because of the reduced nutrient levels, no disinfection was necessary to control bacterial regrowth in the distribution system. In the Me′ry-sur-Oise distribution system (Paris, France), treatment of the water to remove biodegradable organic substances allowed plant operators to eliminate prechlorination, reduce the ozone dosage, reduce chlorine levels in distribution water to half the previous level, and still meet European water quality standards of 100 bacteria per ml (Bourbigot et al., 1982). Considering the disinfection byproduct standards in the United States, the European practice of biological stabilization of potable water supplies appears very attractive.

Some concern, however, has been expressed about the impact of bacterial growth in GAC filters on the microbial quality of finished water supplies. If bacterial-laden particles or microbial aggregates should be sheared and released from the filter beds, these organisms might survive disinfection and enter the distribution system. Camper et al. (1986), using a gauze filter/Swinnex procedure, detected carbon fines in 201 GAC-treated drinking water samples (Table 5.2). Application of the homogenization procedure indicated that 41.4% of the samples had HPC bacteria attached to carbon particles. HPC bacteria were recovered at an average of 8.6 times higher than conventional analyses. Over 17% of the samples (MPN analyses) contained carbon particles colonized with coliform bacteria. In some instances, high levels of coliform bacteria (1,194 coliforms per sample) were recovered from the carbon fines. Significantly, 28% of the coliform organisms exhibited the fecal biotype.

Similar results have been reported by Stewart et al. (1988). Using scanning electron microscopic analysis, these investigators found that greater than 85% of the particles released from GAC filters were colonized with bacteria. Seventy-two percent of the particles were colonized with between 5 and 50 bacterial cells. Treatment of the particles with a homogenization procedure revealed that approximately 200 to 7,000 viable cells could be recovered from 1000 particles. Although few coliform bacteria were detected in their system, on one occasion a fecal coliform (*Klebsiella pneumoniae*) was found attached to carbon particles released from the GAC filter.

The significance of bacteria released from GAC filters, however, is unclear. Some of the research results indicate that release of material from water treatment filters can be a mechanism by which bacteria penetrate barriers and enter

Table 5.2 Results of analyses of particles from GAC-treated effluents

Type of analysis	Number of samples	Number showing two-fold increase*	Degree of increase	
			Mean	Maximum
Heterotrophic plate count (HPC)	198	82 (41.4)	8.6**	50.0
Coliform				
MF	201	14 (7.0)	124.3	1,194.0
MPN	191	33 (17.2)	24.5	122.2

Data adapted from Camper et al., 1986.

* Homogenized versus hand-shaken analyses

** The values give the number of times higher the viable counts are with homogenized than with hand-shaken samples.

finished water supplies. The isolation of fecal coliforms attached to carbon particles in filter effluents indicates the potential for public health problems. The survival of enteropathogenic bacteria in virgin GAC filters (see below) underscores this concern. In the early 1980's, several U.S. water utilities removed GAC from their treatment process because of concern about the impact of carbon fines on distribution system water quality. However, Camper et al. (1986) report that none of the systems they studied which released colonized carbon particles had a history of coliform problems. Clearly, additional research is necessary to clarify this point before widespread biological treatment is adopted.

5.4   Microbial Colonization of GAC Particles

Accumulation of organic compounds by GAC filters provides nutrients for bacterial attachment and growth on the carbon particles (Weber et al., 1978; Brewer and Carmichael, 1979; Cairo et al., 1979; Schalekamp, 1979; AWWA Committee on Organic Contaminants, 1981). Scanning electron microscopy of GAC fines from drinking water treatment filters have shown that particles were densely colonized by rod, coccal, and filamentous bacteria, fungi, and bacteriovorous protozoans (Weber et al., 1978; Cairo et al., 1979; Shimp and Pfaender, 1982; LeChevallier et al., 1984). Filamentous bacterial cells were observed to intertwine into the pores of the carbon particles (Figure 5.1). Acridine orange direct microscopy (AODC) has yielded similar results, showing that bacteria were predominantly found in cracks and crevices of the carbon surfaces and coated by an extracellular slime layer (LeChevallier et al., 1984).

For carbon particles released into the distribution system, laboratory studies have shown that bacterial survival may be enhanced by attachment to the activated surface (Davies and McFeters, 1988). Besides the protection from disinfection (see below), experiments have shown that the growth rate of attached *Klebsiella oxytoca* was more than 10 times faster than unattached cells in a low-nutrient medium containing glutamate, a substrate which was readily adsorbed to the carbon surface. Cellular uptake of [^{3}H]thymidine and [^{3}H]uridine (measures of DNA and RNA activity) were 5 and 11 times higher, respectively, for attached than unattached bacteria. Attachment of microorganisms to carbon particles in a nutrient-limited environment was shown to prevent a reduction in cell size. This reduction in cell size (an increase in the bacterial surface to volume ratio) is believed to allow cells to scavenge nutrients from the environment more efficiently. Davies and McFeters (1988) stated that this phenomenon indicated a higher nutrient availability at the solid/liquid interface than in the bulk water. The authors concluded that bacteria attached to carbon particles in low-nutrient drinking water have an enhanced survival strategy.

HPC organisms recovered from particles in GAC filters include: *Arthrobacter, Alcaligenes, Acinetobacter, Bacillus, Micrococcus, Corynebacterium, Pseu-*

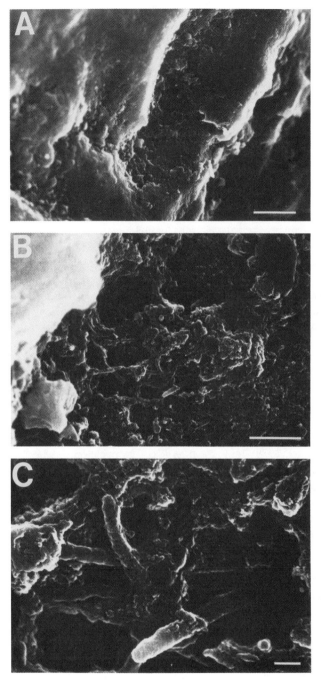

Figure 5.1   Scanning electron micrographs showing indigenous bacteria on a carbon particle. Bars: 20μm (A), 5 μm (B), and 0.5 μm (C). Reprinted with permission from LeChevallier et al., 1984.

domonas, Moraxella, Flavobacterium, and *Achromobacter* (Cairo et al., 1979; Parsons, 1980; Camper et al. 1986; Rollinger and Dott, 1987). Some of these organisms may act as secondary or opportunistic pathogens (Culp, 1969; Prier and Friedman, 1974; Lennette et al., 1985). Fungi and yeasts have been found in GAC absorbers, but usually at low levels (Parsons, 1980, Symons et al., 1981).

The numbers and types of bacteria can be influenced by the operational parameters of the GAC filters. Several researchers (Haas et al., 1983; Graese et al., 1987) have reported higher bacterial levels in GAC filters and in filter effluents when operating at high flow rates (empty bed contact time). Increased nutrient loading was thought to be responsible for stimulating bacterial growth in the filters. However, some investigators (Shimp and Pfaender, 1982) have reported lower bacterial levels in GAC filters at high flow rates. Increased hydraulic shear forces (Trulear and Characklis, 1979) and better penetration of a disinfectant residual into the filter bed were cited as possible reasons for these results. Application of ozone as a pre-disinfectant has been reported to stimulate bacterial growth on GAC by rendering organic compounds more easily biodegradable.

High levels of coliform bacteria have been found in some GAC filters. Cummins and Nash (1978) reported bacterial levels as high as 730 coliforms/ 100 ml in effluents of GAC filters at Beaver Falls, Pennsylvania even though influent waters were consistently coliform free. Coliform problems were most pronounced when water temperatures were greater than 10°C. Similar results were observed in a study at Huntington, West Virginia, where coliform bacteria were detected during a nine-week period following installation of virgin GAC. Genera including *Escherichia, Enterobacter, Klebsiella, Citrobacter, Serratia, Hafnia, Proteus,* and *Aeromonas* have been detected in GAC filter effluents (Brewer and Carmichael, 1979; Tobin et al., 1981; Camper et al., 1986). Significantly, 28% of the coliforms released from GAC filters in one study (Camper et al., 1986) exhibited the fecal biotype. Although no studies have done an exhaustive examination for pathogens in GAC filters, low levels of *Salmonella* spp. and *Escherichia coli* producing a borderline enterotoxigenic reaction in suckling mice were isolated from GAC filter beds (LeChevallier et al., 1984; Camper, et al., 1985c).

Coliform bacteria, however, have not been isolated from GAC filters in all circumstances (Fiore and Babineau 1977). Rollinger and Dott (1987) were unable to detect indicator bacteria (*Escherichia coli*, total coliforms, or fecal streptococci) after extensive examination of biologically activated carbon (BAC) filters. Cairo et al. (1979) did not detect coliform bacteria in GAC filters at Philadelphia, PA, during a six month study. Explanations for these results may be related to the coliform method employed or inhibition of indicator bacteria by coliform antagonists.

5.5   Microbial Interactions on Carbon Surfaces

The role of bacterial antagonism on GAC has indicated that this phenomenon may be an important barrier to transport of pathogens into drinking water supplies. Camper et al. (1985b) found that various waterborne pathogens (*Yersinia enterocolitica, Salmonella typhimurium*, and enterotoxigenic *E. coli*) could survive for extended periods of time provided they colonized virgin GAC (Figure 5.2). The extent of colonization by pathogens was limited by the autochthonous bacterial community. When native aquatic bacteria were added to pathogens growing on sterile GAC, the pathogens declined at rates varying from 0.08 to 0.14 log day^{-1}. When indigenous aquatic bacteria and pathogens were added simultaneously to fresh GAC, the number of pathogens declined slightly faster (0.1 to 0.22 log day^{-1}). Finally, when pathogens were added to precolonized GAC, they attached at lower levels and decreased at the fastest rates observed (0.11 to 0.77 log day^{-1}). Mechanisms of pathogen inhibition, reviewed by Savage (1972), showed that rivalry for nutrients and space, as well as production of inhibitory secondary metabolites, were some of the ways by which bacteria compete in natural environments. Means and Olson (1981) showed that 22% of bacterial isolates from chlorinated drinking water produced bac-

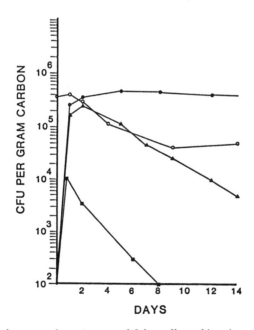

Figure 5.2   Attachment and persistence of *Salmonella typhimurium* on GAC columns. Symbols: ●, sterile river water, sterile carbon; O, continuation of above, nonsterile river water added; ▲, nonsterile river water, sterile carbon; ■, nonsterile river water, precolonized carbon. CFU, colony-forming units. Reprinted with permission from Camper et al., 1985b.

teriocin-like substances inhibitory to the growth of at least one of three selected coliforms. Other investigators (LeChevallier and McFeters 1985; Rollinger and Dott, 1987) have not detected bacteriocin-producing bacteria, but have suggested that competition for nutrients was responsible for elimination of pathogens and coliform bacteria from biofilm communities. It is likely, however, that a combination of factors exist to limit survival of allochthonous organisms in aquatic environments.

Recent experience by the American Water Works Service Company has shown that coliform bacteria can be detected in virgin GAC filters. Taking the filters out of service for 10 to 12 days to allow development of an indigenous biofilm and daily backwashing of the filters has apparently reduced the problem. Camper et al. (1985b) collected samples one week after a treatment plant had changed the carbon in GAC filters. They found that the number of coliform bacteria attached to particles in the effluent jumped from an average of 3 cells/100 ml before replacement to 1,300 cells/100 ml after the virgin carbon was added. These episodes underscore the importance of raw water pretreatment until the establishment of autochthonous microbial biofilm communities on virgin GAC (Camper et al., 1985b).

5.6   Disinfection of Bacteria on Activated Carbon

The increased coliform and HPC bacterial densities occurring during GAC treatment places critical importance on maintaining an effective disinfectant barrier. Several studies using standard bacteriological methods have determined that postfiltration disinfection adequately controlled bacterial levels (Cummins and Nash, 1978; Suffet, 1980; Symons et al., 1981; AWWA Committee Report, 1981; Haas et al. 1983). However, other studies have shown that particle-associated bacteria released from GAC filters were not significantly affected by conventional disinfection practices (Parsons et al., 1980; LeChevallier et al., 1984; Camper et al., 1986; LeChevallier et al., 1987; Stewart et al., 1988). Because particle-associated bacteria are not enumerated without application of a desorption technique, the different conclusions drawn by the various studies are largely a result of the sampling methods employed.

Disinfection experiments performed by LeChevallier et al. (1984) showed that bacteria attached to GAC were highly resistant to chlorine disinfection (Figure 5.3). Bacteria were unaffected by exposure to 2 mg/l free chlorine for 60 min. Attached cells (*E. coli*) whose protective capsule layer had been removed by centrifugation, showed no significant decrease in viability, although some cell injury was detected. Stewart et al. (1988) supported these findings by showing that bacteria attached to carbon particles were protected against disinfectant agents such as free chlorine or chloramines. They found that exposure of organisms attached to carbon particles to a 1.5 mg/l residual of either free chlorine or chloramines for 40 min did not significantly affect viable counts. Parsons et al. (1980) found that bacteria from GAC effluents were more resistant

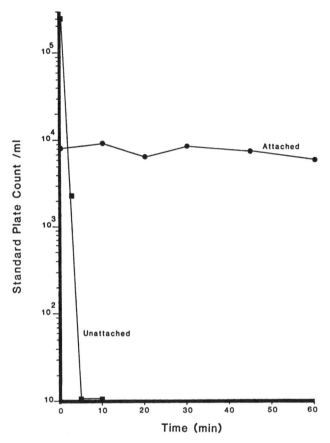

Figure 5.3   Survival of naturally occurring heterotrophic plate count bacteria exposed to 2.0 mg of free chlorine per liter for 1 h (free chlorine residual after 1 h was 1.7 mg/liter). Reprinted with permission from LeChevallier et al., 1984.

to disinfection than similar organisms from sand filters. In this study the authors did not determine the reason for increased resistance, but it is likely that resistance was due to attached organisms.

Camper et al. (1986) capitalized on the increased disinfection resistance of particle-associated bacteria to eliminate unattached organisms from their assays. They treated GAC filter concentrates with 2.0 mg of sodium hypochlorite per liter for 30 min at 4°C (pH 6.5 to 7.0) to selectively isolate attached organisms. Even with this disinfection regime, the researchers detected increased levels of coliform and HPC bacteria in 17.2 and 41.4% (respectively) of the effluent samples examined. The results underscore the importance of maintaining a chlorine residual in the distribution system, establishment of a flushing program, and regular cleaning of clearwells and storage tanks to minimize the impact of carbon fines in the distribution system (LeChevallier et al., 1984).

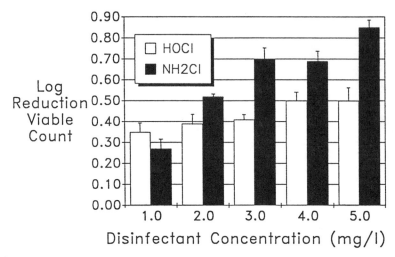

Figure 5.4 Comparison of equal weights of chlorine (HOCL) and chloramine (NH₂Cl) for disinfection of biofilms grown on GAC particles. The suspension was stirred using a magnetic stir bar for 1 h, pH 7.0, 4°C. Reprinted with permission from LeChevallier et al., 1988.

Monochloramine has been shown to be effective for control of bacterial growths in GAC filters (LeChevallier et al., 1987). Monochloramine was shown to be better able to penetrate bacterial biofilms and disinfect attached organisms than free chlorine or chlorine dioxide. Because monochloramine interacted to a lesser extent with activated carbon than free chlorine, on a weight per volume basis, monochloramine was more effective than free chlorine (mg/l as chlorine) for disinfection of GAC-associated bacteria, particularly at high (5 mg/l) disinfectant levels (Figure 5.4).

Snoeyink et al., (1981) and Voudrais (1985) have shown that the interaction of free chlorine with compounds adsorbed to the GAC surface can form a variety of products not generated in the absence of carbon. For example, adsorbed chlorophenols will react with chlorine (free or combined) to produce hydroxylated, chlorinated biphenyls. Laboratory experiments have shown that chlorine will oxidize carbon and eventually degrade particle size. Therefore, the advisability and even the practicability of applying disinfectant residuals to GAC filters requires additional investigation.

5.7   Conclusion

Operators of potable water systems are faced with a dilemma. The establishment of microbial communities on GAC particles has its benefits in the reduction of microbial nutrients and disinfectant demand, and in the antagonism of pathogen and coliform bacteria; but, also it has its hazards from the breakthrough of particle-associated opportunistic and overt pathogens. Of importance is the apparent inability of a free chlorine residual to disinfect bacterial on GAC particles. Monochloramine may be valuable for control of bacterial problems in GAC-treated water.

Activated carbon has become an important tool in the water industry for controlling a variety of water quality problems. Research in this next decade needs to focus on maximizing the benefits of activated carbon, while minimizing its hazards. Investigators need to appreciate both sides of the story to maintain a balanced perspective.

References

American Public Health Association. 1980. *Standard Methods for the Examination of Water and Wastewater*, 15th edition. American Public Health Association, Washington, D.C.

Amirtharajah, A. and Westein, D.P. 1980. Initial degradation of effluent quality during filtration. *Journal of the American Water Works Association*. 72:518–524.

AWWA Research and Technical Practice Committee on Organic Contaminants. 1981. An assessment of microbial activity on GAC. *Journal of the American Water Works Association*. 73:447–454.

Baker, M.N. 1948. The quest for pure water. *American Water Works Association*, New York, N.Y.

Bourbigot, M.M., Dodin, A., and Lheritier, R. 1982. Limiting bacterial aftergrowth in distribution systems by removing biodegradable organics. *Proceedings of the American Water Works Association Annual Conference*. Miami Beach, Florida.

Brewer, W.S. and Carmichael, W.W. 1979. Microbial characterization of granular activated carbon filter systems. *Journal of the American Water Works Association*. 71:738–740.

Cairo, P.R., McElhaney, J., Suffet, I.H. 1979. Pilot plant testing of activated carbon adsorption systems. *Journal of the American Water Works Association*. 71:660–673.

Camper, A.K., LeChevallier, M.W., Broadaway, S.C., and McFeters, G.A. 1985a. Evaluation of procedures to desorb bacteria from granular activated carbon. *Journal of Microbiological Methods*.3:187–198.

Camper, A.K., LeChevallier, M.W., Broadaway, S.C., and McFeters, G.A. 1985b. Growth and persistence of pathogens on granular activated carbon filters. *Applied and Environmental Microbiology*. 50:1378–1382.

Camper, A.K., Davies, D.G., Broadaway, S.C., LeChevallier, M.W., and McFeters, G.A. 1985c. Association of coliform bacteria and enteric pathogens with granular activated carbon. *Abstracts of the Annual Meeting of the American Society for Microbiology*. N38, p.223.

Camper, A.K., LeChevallier, M.W., Broadaway, S.C., and McFeters, G.A. 1986. Bacteria associated with granular activated carbon particles in drinking water. *Applied and Environmental Microbiology*. 52:434–438.

Camper, A.K., LeChevallier, M.W., Broadaway, S.C., and McFeters, G.A. 1987. Operational variables and the release of colonized granular activated carbon particles in drinking water. *Journal of the American Water Works Association*. 79:(5)70–74.

Capelle, A. and de Vooys, F. 1983. *Activated Carbon—A Fascinating Material*. Norit N.V., Amersfoort, The Netherlands.

Culp, R.L. 1969. Disease due to "non-pathogenic" bacteria. *Journal of the American Water Works Association*. 61:157.

Cummins, B.B. and Nash, H.D. 1978. Microbiological implications of alternative treatment. *Proceedings of the AWWA Water Quality Technology Conference.* paper 2B-1, Louisville, KY.

Davies, D.G. and McFeters, G.A. 1988. Growth and comparative physiology of *Klebsiella oxytoca* attached to GAC particles and in liquid media. Microbial Ecology. 15:165–175.

Fiore, J.V. and Babineau, R.A. 1977. Effect of an activated carbon filter on the microbial quality of water. *Applied and Environmental Microbiology.* 34:541–546.

Graese, S.L., Snoeyink, V.L., and Lee, R.G. 1987. *GAC Filter-Adsorbers.* American Water Works Association, Denver, CO.

Haas, C.N., Meyer, M.A., Paller, M.S., Zapkin, M.A., and Aulenbach, D.B. 1983. *Microbiological Alterations in Distributed Water Treated with Granular Activated Carbon.* EPA-600/S2-83-062. United States Environmental Protection Agency, Washington, D.C.

Hansen, R.E. 1970. Filter control by suspended solids determination. *Public Works.* 20:33–35.

Hassler, J.W. 1963. *Activated carbon.* Chemical Publishing Company, Inc. New York, N.Y.

LeChevallier, M.W., Evans, T.M., and Seidler, R.J. 1981. Effect of turbidity on chlorination efficiency and bacterial persistence in drinking water. *Applied and Environmental Microbiology.* 42:159–167.

LeChevallier, M.W., Hassenauer, T.S., Camper, A.K., and McFeters, G.A. 1984. Disinfection of bacteria attached to granular activated carbon. *Applied and Environmental Microbiology.* 48:918–928.

LeChevallier, M.W. and McFeters, G.A. 1985. Interactions between heterotrophic plate count bacteria and coliform organisms. *Applied and Environmental Microbiology.* 49:1338–1341.

LeChevallier, M.W., Cawthon, C.D., and Lee R.G. 1988. Inactivation of biofilm bacteria. *Applied and Environmental Microbiology.* 54:2492–2499.

Lennette, E.H., Balows, A., Hausler, W.J. Jr., and Shadomy, H.J. 1985. *Manual of Clinical Microbiology.* American Society for Microbiology. Washington, D.C.

Mattson, J.S. and Mark, H.B. Jr. 1971. *Activated Carbon.* Marcel Dekker, Inc. New York, N.Y.

McCabe, L.J., Symons, J.M., Lee, R.D., and Robeck G.G. 1970. Survey of community water supply systems. *Journal of the American Water Works Association.* 62:670–687.

The National Research Council. 1977. *Drinking Water and Health,* Vol. 1. National Academy of Sciences, Washington, D.C.

The National Research Council. 1980. *Drinking Water and Health,* Vol. 2. National Academy of Sciences, Washington, D.C.

Parsons, F. 1980. Bacterial populations in granular activated carbon beds and their effluents. Unpublished report. United States Environmental Protection Agency, Cincinnati, OH.

Parsons, F., Wood, P.R., and DeMarco, J. 1980. Bacteria associated with granular activated carbon columns. *Proceedings of the AWWA Water Quality Technology Conference.* Miami Beach, Florida.

Pascal, O., Joret, J.C., Levi, Y., and Dupin, T. 1986. Bacterial aftergrowth in drinking water networks measuring biodegradable organic carbon. Presented at the Ministere de l'Environment/US Environmental Protection Agency Franco-American Seminar, October 13–17, Cincinnati, OH.

Prier, J.E. and Friedamn, H. 1974. *Opportunistic Pathogens.* University Park Press, Baltimore, M.D.

Reasoner, D.J. and Geldreich,E.E. 1985. A new medium for enumeration and subculture of bacteria from potable water. *Applied and Environmental Microbiology.* 49:1–7.

Ridgway, H.F. and Olson, B.H. 1981. Scanning electron microscope evidence for bacterial colonization of a drinking water distribution system. *Applied and Environmental Microbiology.* 41:274–287.

Rittmann, B.E. and Snoeyink, V.L. 1984. Achieving biologically stable drinking water. *Journal of the American Water Works Association.* 76(10):106–114.

Robeck, G.G., Clarke, N.A., and Dostal, K.A. 1962. Effectiveness of water treatment processes in virus removal. *Journal of the American Water Works Association.* 54:1275–1290.

Robeck, G.G., Dostal, K.A., and Woodward, R.L. 1964. Studies of modifications in water filtration. *Journal of the American Water Works Association.* 56:198–213.

Rollinger, Y. and Dott, W. 1987. Survival of selected bacterial species in sterilized activated carbon filters and biological activated carbon filters. *Applied and Environmental Microbiology.* 53:777–781.

Schalekamp, M. 1979. The use of GAC filtration to ensure quality in drinking water from surface sources. *Journal of the American Water Works Association.* 71:638–647.

Shimp, R.J., and Pfaender, F.K. 1982. Effects of surface area and flow rate on marine bacterial growth in activated carbon columns. *Applied and Environmental Microbiology.* 44:471–477.

Stewart, M.H., Wolfe, R.L., and Means, E.G. 1988. An Assessment of the bacteriological activity on granular activated carbon particles. *Proceedings of the American Society for Microbiology Annual Meeting*, Miami Beach, Florida.

Suffet, I.H. 1980. An evaluation of activated carbon for drinking water treatment: a National Academy of Science report. *Journal of the American Water Works Association.* 72:1:41–50.

Symons, J.M., Stevens, A.A., Clarke, R.M., Gelreich, E.E., Love, O.T., and DeMarco, J. 1981. *Treatment Techniques for Controlling Trihalomethanes in Drinking Water.* EPA-600/2-81-156, United States Environmental Protection Agency, Cincinnati, OH.

Syrotynski, M. 1971. Microscopic water quality and filtration efficiency. *Journal of the American Water Works Association.* 63:237–245.

Tobin, R.S., Smith, D.K., and Lindsay, J.A. 1981. Effects of activated carbon and bacteriostatic filters on microbiological quality of drinking water. *Applied and Environmental Microbiology.* 41:646–651.

Trulear, M.G. and Characklis, W.G. 1979. Dynamics of biofilm processes. *Proceedings of the 34th Industrial Waste Conference.* Purdue University, IN.

United States Environmental Protection Agency. 1975. National interim primary drinking water regulations. *Federal Register.* 40:59568–59588.

United States Environmental Protection Agency. 1979. National interim primary drinking water regulations, control of trihalomethanes in drinking water, final rule. *Federal Register.* 44: No. 231.

Van der Kooij, D. 1987. The effect of treatment on assimilable organic carbon in drinking water. pp. 317–328 *in* Huck, P.M. and Toft, P. (editors), *Treatment of Drinking Water for Organic Compounds*, Pergamon Press, London.

Voudrais, E.A. 1985. Effects of activated carbon on the reactions of free chlorine with phenols. *Environmental Science and Technology.* 19:441.

Weber, W.J. Jr., Pirbazari, M., and Melson, G.L. 1978. Biological growth on activated carbon: an investigation by scanning electron microscopy. *Environmental Science and Technology.* 12:817–819.

6

Microbiology and Drinking Water Filtration

Gary S. Logsdon

Environmental Protection Agency Cincinnati, Ohio

Water filtration research has been undertaken for a variety of reasons. Studies have been performed to develop information for filtration theories and for design of filtration plants to remove suspended matter such as clays, algae, suspended matter in general, and asbestos fibers from water. Filtration studies related to removal of microorganisms have generally been motivated by the need to learn about the removal of pathogens or indicator organisms, or both. Reducing the risk of waterborne disease has been a goal of microbiologically related filtration research for nearly 100 years. This chapter briefly reviews that research and then discusses the results of recent investigations.

6.1 Filtration Work in the Past

In the time period when filtration research was getting under way, an incident in Europe provided dramatic evidence of the ability of slow sand filtration to remove pathogenic bacteria. As the situation was described by Hazen (1913), in 1892 the neighboring German cities of Hamburg and Altona both drew their water from the River Elbe. Hamburg employed sedimentation, whereas Altona used slow sand filtration. Altona's water supply intake was seven miles below the sewage outfalls of the cities, whereas Hamburg's was seven miles upstream but the Elbe was subjected to tidal influences, so Hamburg's raw water was sometimes contaminated by sewage. When a cholera epidemic occurred in August and September of 1892, Hamburg had a case rate of 26.31 per 1000 and a death rate of 13.39 per 1000. Altona had only 3.81 cases per 1000 and 2.13 deaths per 1000. Very few children in Altona had cholera, as compared to an earlier epidemic in 1871, suggesting that cholera was not introduced into Altona in 1892 "... by water, but by other means of infection ..." (Hazen, 1913). The Hamburg-Altona episode was probably one factor influencing Hazen's strong advocacy of slow sand filtration.

In the late 1800's and early 1900's, typhoid fever death rates in American cities were typically in the range of 30 to 100 per 100,000 (Johnson, 1916) and a considerable amount of attention was focused on improving water treatment to attempt to control typhoid fever. Hazen conducted research on removal of bacteria by slow sand filtration at Lawrence, Massachusetts in the early 1890's (Hazen, 1913). He reported that bacterial removal improved as sand size decreased, as filter bed depth increased, and as the filter run progressed. Hazen was a participant in pilot plant studies conducted for Pittsburgh in the late 1890's in which slow sand filtration and rapid sand filtration were evaluated. Removal of bacteria was studied again as a means of evaluating filter efficacy. Effective coagulation was found to be necessary for successful bacterial removal by rapid sand filtration. Later Hazen wrote, "The conditions necessary for the removal of bacteria and turbidity are very well understood, and it can be stated with the utmost confidence that no system of filtration through sand at rates many times as high as are used in ordinary (i.e. slow) sand filtration, and without the use of coagulants, will be satisfactory where either bacterial efficiency or clarification is required," (Hazen, 1913). It is remarkable that the necessity of using coagulants to attain satisfactory microorganism removal was clearly stated over 70 years ago, but in recent decades, some water utilities treating very clean surface waters have attempted to conduct rapid sand filtration without adding coagulant chemical.

The construction of water filtration plants followed the early research that showed that slow sand filtration and well-designed and well-operated rapid sand filtration plants could reduce the concentration of bacteria by 99 to 99.9 percent. Then, in the time period from World War I to the late 1920's, the bacterial removal capability of operating water treatment plants was evaluated (Streeter, 1927; Streeter, 1929). The U.S. Public Health Service undertook studies to ascertain this because municipal sewage treatment was generally inadequate, and the principal barriers to waterborne disease were natural die-off of pathogens in surface waters and water treatment processes, including disinfection and clarification. Streeter's data, developed at facilities using coagulation, sedimentation, and filtration prior to chlorination, indicated that removal of *Escherichia coli* at conventional treatment plants improved as the raw water *E. coli* concentration increased (Figure 6.1). An interesting aspect of Figure 6.1 is the better performance of plants treating muddy water (Ohio River) as compared to those treating the less turbid Great Lakes water. Streeter (1929) reported that many water purification specialists considered Great Lakes water more difficult to treat than turbid river waters. Streeter also presented data that displayed seasonal effects on bacterial removal efficiency, with greater passage of bacteria through sedimentation and filtration processes in cold weather. The percentage of plate count bacteria remaining in filter effluent increased from less than 0.1 percent to 1.0 percent under conditions of high (40,000 to 80,000 per ml) influent bacterial loadings and from 0.75 percent to 2.75 percent when raw water bacteria ranged from 1000 to 10,000 per ml (Streeter, 1927).

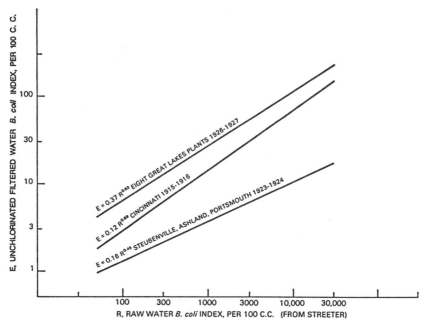

Figure 6.1   Relationship of raw and filtered coliform counts, Ohio River and Great Lakes. From Streeter (1929).

In 1933, an outbreak of amoebic dysentery took place in two hotels in Chicago during the Century of Progress Exposition. An investigation by public health authorities and others concluded that water was the mode of transmission and that cross-connections between water and sewage were involved (Jordan, 1937). After the Chicago outbreak, research on removal of amoebic cysts by rapid sand filtration was conducted at the Chicago Experimental Filtration Plant (Spector et al., 1934; Baylis et al., 1936). In several experiments involving coagulation, sedimentation, and filtration, they reported dosing from 189,000 to 870,000 cysts per gallon in the raw water, but they detected no cysts in the filtered water.

During World War II, the U.S. Army evaluated coagulation and filtration with the Portable Water Purification Unit, Model 1940, for removal of amoebic cysts. This unit consisted of a pump, chemical feeders, a pressure filter, hypochlorinator, and accessories. Two experiments were conducted with no coagulant. The report (U.S. Army and U.S. Public Health Service, 1944) stated that the experiments convincingly demonstrated that sand without coagulant did not remove the cysts of *Endamoeba histolytica* even at the reduced rate of 6.35 gallons per minute (gpm) per square foot or 15.5 m/hr (1 gpm/sf = 2.44 m³/m²/hr, or 2.44 m/hr). Thus if cysts were to be removed from water with the portable sand filter, the removal would have to be accomplished by the coagulants alone or in combination with the sand. The report also included a comment on coagulation: "It will be noted that the efficiency of the unit varies somewhat proportionally with the coagulation achieved, the efficiency of the unit being highest when the control of the coagulation is the best."

In the same study, diatomaceous earth (DE) filtration was also evaluated for water treatment for military forces. Concerning this process, the report stated: "The test results indicate that diatomaceous silica filtration is much more effective than sand filtration for the removal of amoebic cysts. A filtration unit properly designated and operated would be expected to remove completely the cysts of *Endamoeba histolytica*." Thus DE filtration was found to be very effective for removal of pathogenic cysts. In the Army-PHS work cited above, a limited amount of analysis for removal of bacteria by DE filtration was undertaken. The results showed that the highest concentrations of *E. coli* passed through the DE filter when the coarsest filter aid was used. This would be the expected result. In the usual operating procedure for DE filtration, no coagulation is practiced, so microorganism removal would be accomplished by straining, and the filter aid with the smallest particle size would remove the most bacteria.

During the 1950's concerns developed over waterborne disease caused by viruses. Water filtration research was conducted with poliovirus and reported by Robeck et al. (1962). One important finding was that at filter rates of 9.8 to 24 m/hr, when no coagulant was used, virus removal was ineffective (only 1 to 50 percent of the influent virus was removed). This is similar to the result obtained for rapid filtration of bacteria with no coagulant (Hazen, 1913) and of *E. histolytica* cysts with no coagulant (U.S. Army and U.S. Public Health Service, 1944). Robeck et al. (1962) also reported that coagulation and filtration could remove 90 to 99 percent of the influent poliovirus, and coagulation, settling, and filtration could remove more than 99.7 percent of influent viruses. These results were obtained at filtration rates of 9.8 to 15 m/hr.

Early research efforts with bacteria, cysts, and viruses all showed that effective coagulation was an essential factor in granular media filtration at rates of 2.4 m/hr or greater, when removal of a very high percentage of microorganisms was a goal of treatment. The early filtration work laid a good foundation for recent studies, many of which focused attention on DE filtration, slow sand filtration, and coagulation and filtration for removal of microbiological contaminants, particularly *Giardia* cysts.

6.2   Recent Developments in Diatomaceous Earth Filtration

During World War II, an intense effort was made to develop a filtration process that would give practically complete removal of *E. histolytica* cysts. The result was the application of the DE filtration process to potable water treatment. In DE filtration, a thin coating (3 to 5 mm) of diatomaceous earth is placed on a filter septum by recirculating a slurry of DE through the filter until essentially all DE is on the filter septum. After the filter cake is established in the pre-coating process, raw water that has been dosed with a small amount of DE is passed through the filter. Particles are removed, and they accumulate on the filter cake. Because DE (the body feed) is added to the raw water, the good hydraulic characteristics of the filter cake are maintained, and long filter runs can be attained. When a run is terminated, the filter cake is removed, disposed of, and fresh DE is used to recoat the clean septum.

Because DE filtration had proven very effective for *E. hystolytica* cyst removal, it was evaluated for *Giardia* cyst removal. Logsdon et al. (1981) used a 0.1 m² DE pilot filter and 9 μm radioactive beads as cyst models. Their results showed that removal of the cyst-sized particles could range from 90 to 99.999 percent when turbidity removal ranged from 40 to 80 percent. Turbidity removal was not a good surrogate indicator for removal of 9 μm particles. They reported that operating and maintenance procedures were important for successful DE filter operation, recommending use of 1.0 kg of precoat per m² of filter septum area, and the use of body feed at all times. In a subsequent series of tests with *G. muris* cysts, they observed removals ranging from 99.3 to 99.99 percent. Other research also proved that DE filter can be an effective barrier to passage of *Giardia* cysts. De Walle et al. (1984) observed cyst removals of 99.0% or greater in four test runs they conducted with a 0.1 m² test filter. Pyper (1985) performed one run with *Giardia* cysts using a 1.0 m² DE filter and reported 99.97 percent cyst removal. Work by Lange et al. (1986) was the most thorough evaluation of *Giardia* cyst removal by DE filtration to date. They performed 31 runs in which *Giardia* cysts were used at Colorado State University and two water utility sites. They operated the 0.1 m² DE test filter at rates of 2.4 to 9.8 m/hr, over a temperature range of 3.5°C to 15°C, and with four different grades (particle size distributions) of diatomaceous earth. Information on properties of one brand of DE can be found in Table 6.1. *Giardia* cyst removals always exceeded 99 percent, and exceeded 99.9 percent in 15 of 31 runs. *Giardia* cysts passed through the DE filter in only one of the 31 runs, during which the influent cyst concentration was 33,600 cysts/l.

Under the sponsorship of the American Water Works Association Research Foundation, Ongerth (1988) evaluated the performance of a 60 gpm DE filtration plant that was used to treat a clear creek water originating in a steep, partially wooded alpine terrain. Raw water turbidity ranged from 0.3 to 0.7 nephelometric turbidity units (NTU), while filtered turbidity varied from 0.05 to 0.5 NTU. Average turbidity removal was about 60 percent. *Giardia* cysts were detected in one of two raw water samples at a concentration of 0.13 cyst/gallon. In filtered water, cysts were found in one of three samples (about 0.01 cyst/gallon). Operation of a small DE pilot filter at the same site resulted in approximately 99 percent cyst reduction.

Lange et al. (1986) also evaluated removal of bacteria by DE filtration. Removal varied with the grade of diatomaceous earth used, ranging from 100 percent for Filter Cel®* to as low as 30 percent for Celite 545®. They could

---

*Mention of trade names or commercial products does not constitute endorsement or recommendation for use.

Table 6.1 Pore size and DE filter aid performance

DE filter aid	Median pore size μm	Median particle size μm	Flow rate	Clarity
Filter Cel	1.5	7.5	Lowest	Highest
Standard Super Cel	3.5	14		
Celite 512	5	15		
Hyflo Super Cel	7	18		
Celite 503	10	23		
Celite 535	13	25		
Celite 545	17	26		
Celite 560	22	106	Highest	Lowest

Data from Johns-Manville Filtration & Minerals Division Product Bulletin, 1981 and AWWA Manual M30, Precoat Filtration, 1988.

improve filter performance by coating the coarser DE grade with an alum precipitate, however, and reported that when this was done, total coliform removals increased to the 96 to 99.9 percent range as compared to 30 to 70 percent removal for untreated diatomaceous earth. For the coarser grades of DE, bacterial removal can be accomplished by surface attachment mechanisms when chemical conditioning is applied to the DE.

Brown et al. (1974a, 1974b) operated a DE filter and spiked the influent water with bacteriophage T2 and with poliovirus to evaluate the virus removal capability of DE filtration. In the first phase of their work, they observed average T2 bacteriophage removals ranging from 73 to 100 percent, depending on the grade of DE used. Finer filter grades were more effective. Pretreating the raw water with polymer improved bacteriophage removal from 90 percent to 98 to 100 percent for Hyflo® DE filter aid. When poliovirus was the challenge organism, removals after two hours ranged from 62 to 100 percent. Two runs of 12 hours duration were conducted. In the test with polymer-coated Hyflo®, influent poliovirus concentration was 2350 pfu/l (plaque-forming units per liter). Virus was detected in only one of 12 effluent samples (15 pfu/l) during this filter run. When polymer was added to the raw water, during a 12 hour run with 2950 pfu/l in the raw water, no poliovirus was detected in any of the 12 filtered water samples. DE filtration can reduce the concentration of virus in water, but the most effective technique requires either conditioning of the raw water or conditioning of the DE to enhance virus attachment to the DE during the filtration process.

Walton (1988) presented information, including photomicrographs, that showed that DE filters remove *Giardia* cysts by a straining or screening action. Because of the size of these cysts, their passage through the DE filter cake is physically blocked.

Because DE filtration has repeatedly proven effective for *Giardia* cyst removal, this process might be considered if filtration for removal of other protozoal pathogens such as *Cryptosporidium* oocysts was being contemplated. Information presented by Fayer and Ungar (1986) indicates that *C. parvum* oocysts are spherical to ovoid, and fully sporulated oocysts averaged 5.0 μm $\times$ 4.5μm. These microorganisms are larger than bacteria but somewhat smaller than *Giardia* cysts. Grades of diatomaceous earth that are finer in size than the coarse water treatment grades (e.g. Celite 545®) should be effective for removal of *Cryptosporidium* oocysts. The grade of diatomite that would be effective would depend upon the particular brand being used. Advice on that matter should be obtained from a technical representative of the manufacturer of the product in question. DE filtration has the capability of removing a variety of microorganisms, but it is most effective when the microorganisms are in the size range of *Giardia* or *E. histolytica* cysts. Smaller sizes of microorganisms can also be removed, however. If the removal mechanism is straining, the DE grade (particle size distribution) must be finer as the size of the target organism decreases. If surface attachment is the removal mechanism, coarse DE grades can be coated with aluminum hydroxide precipitate. The resulting positively charged diatomite can remove bacteria by surface attachment.

6.3   Recent Developments in Slow Sand Filtration

Slow sand filtration originated in Great Britain in the 1800's. In this process, uncoagulated water is applied to a sand bed about 1 meter deep at a filtration rate (approach velocity) of about 0.1 meter/hour (m/hr.) Removal of microorganisms is aided by biological processes that occur on and in the filter bed, where a microbial ecosystem becomes established with extended use. On the top of the media, a biologically active scum layer (*Schmutzdecke*) builds up and assists in filtration. As the water enters the *Schmutzdecke*, biological action breaks down some organic matter, and inert suspended particles may be physically strained out of the water. The water then enters the top layer of sand where more physical straining and biological action occur and attachment of particles onto the sand grain surfaces takes place. Also some sedimentation may occur in the pores between the media grains where velocity is sufficiently slow.

The filter must be drained and the top layer of sand removed, when the depth of the *Schmutzdecke* layer increases, clogging the top sand layer and causing the headloss through the filter to reach a predetermined level. This period of scraping is the only time during normal operations in which additional labor may be required to operate the filter. The amount of time between scrapings will depend on how much material is filtered out of the water. Thus, high turbidity levels would shorten the cycle. A run for a slow sand filter can be as short as one or two weeks when muddy or algae laden waters are treated. The runs can last for several months when very clear waters are filtered.

Slow sand filters are uncomplicated and easy to operate, but they can be used successfully only with a good quality of raw water (turbidity usually less than 10 NTU and no undesirable inorganic or organic chemicals present). In comparison to other filtration processes, slow sand filters also require large land areas per volume of water treated, and thus would probably not be used by large water utilities in the United States. Slow sand filtration is a very effective treatment process for control of microorganisms when the raw water quality is appropriate. Interest in slow sand filtration began to decline in the United States from about the time of World War I, but recently this situation has changed. Sayre (1984) wrote in the Theme Introduction to the December, 1984 issue of *Journal American Water Works Association*: "With the prospect of an imminent mandate to filter most public surface water supplies, slow-rate sand systems are likely to make a spectacular comeback during the next decade. They are simple and economical to construct, are easy to maintain with un-skilled workers, and are capable of producing safe water." This editorial comment was accurate in its description of the attributes of slow sand filtration, and it appears to be an accurate forecast with respect to use of slow sand filters, especially for small systems.

Since 1980, a number of slow sand filter investigations conducted or sponsored by the U.S. Environmental Protection Agency have been reported (Fox et al., 1984; Cleasby et al., 1984; Bellamy et al, 1985a; Bellamy et al., 1985b; Cullen and Letterman, 1985; and Pyper, 1985). Slow sand filtration studies were also sponsored by the Utah Water Research Laboratory (Slezak and Sims, 1984; McConnell, 1984), by the American Water Works Association Research Foundation (Seelaus et al., 1986), and by the town of Hundred Mile House, British Columbia (Bryck et al., 1987).

Recent slow sand filtration results for bacterial removal generally have tended to verify the results obtained by researchers in the 1890's and early 1900's. Bellamy et al. (1985a, 1985b) found that bacterial removal decreased as filtration rates increased, as water temperature decreased, as bed depth decreased, and as sand size increased. All of those findings were consistent with the older work.

One aspect of bacteria removal that was not clearly resolved by the recent research was the role of the *Schmutzdecke*, the gelatinous layer of biological matter that accumulates at the top of the sand bed. Huisman and Wood (1974) wrote that at the time of the initial startup of a slow sand filter, the living organisms necessary for effective treatment are not yet present in the sand bed. A period of ripening is necessary before the biological content of the new filter is sufficiently established. This effect was observed by Fox et al. (1984). The first stage of their initial run of a new filter lasted for 40 days, during which time 12 or 12 filtered water samples had more than 1 coliform per 100ml. The coliform counts in filtered water had been declining though, and during the following 85 days (the second stage of run number 1) none of the 24 samples tested for coliform bacteria exceeded 1 per 100 ml.

Huisman and Wood (1974) also suggested that wasting filtered water may be necessary after a filter has been cleaned, until the bacterial action of the bed is reestablished and effluent quality is satisfactory. Deterioration of water quality after scraping was noted in the pilot filters operated by Bellamy et al. (1985a). Filters operating at rates of 0.04 m/hr, 0.12 m/hr, and 0.40 m/hr all demonstrated a reduced efficiency for coliform removal after scraping. In each case, coliform removal was about one $\log_{10}$ less after *Schmutzedecke* removal. This result was not consistently observed in full-scale plants, however. Cullen and Letterman (1985) conducted evaluations of seven municipal slow sand filters located in the State of New York. On the basis of turbidity and particle counting data, they saw some evidence of ripening in four of ten filter scraping events they studied. They reported: "In most cases, results for total plate count and total coliform bacteria showed no ripening period." Even though a ripening period was not seen in all cases, conservative design of slow sand filter plants would make provision for discarding filtered water for some time period after filter scraping. In addition, conservative operating practice would include monitoring the bacteriological quality of filter effluent after scraping, with the understanding that filtering to waste should cease only after acceptable bacteriological quality was attained, or filtering to waste for a time period previously shown to be adequate.

Bellamy et al. (1985b) investigated the relationship of the biological population of the sand filter to treatment efficacy by promoting or inhibiting the growth of organisms in the filter and observing treated water quality. Three pilot-scale slow sand filters, including a control filter, were used to treat water from Horsetooth Reservoir. One filter treated raw water which had been enriched with sterile synthetic sewage in sufficient quantity to cause a dissolved oxygen drop across the filter of 3 to 4 mg/l. This nutrient addition prompted biological growth. Biological growth in another filter was inhibited by chlorination to a 5 mg/l residual between test runs. The chlorinated filter was dechlorinated with sodium thiosulfate before each test run. Reduction in total coliform counts in the filter subjected to chlorine treatment was only 60 percent. The control filter attained 97.5 percent total coliform removal, and the filter receiving nutrient supplement removed 99.9 percent of the total coliforms. Thus, the presence of a biological community within the sand bed is necessary for effective slow sand filter performance, although the necessity for a well developed *Schmutzdecke* is less certain.

Virus removal by slow sand filters was evaluated in London (Poynter and Slade, 1977; Slade, 1978) because of that city's reliance on slow sand filtration to treat water from the River Thames and River Lee. Using pilot filters, Poynter and Slade (1977) concluded that *Escherichia coli* removals were similar to or slightly less than poliovirus removals. The similarity of the results obtained for bacteria and poliovirus was consistent over a wide range of temperatures and flow rates. They wrote: "If this relationship can be confirmed on a full scale working filter it will be very useful because it implies that the normal bacteriological control of slow sand filtration can be assumed to apply to enteroviruses, thus obviating any need for further virological controls." Slade (1978) followed up on the first study with a project at the Thames Water

Authority's Coppermills Works. Two 0.337 hectare filter beds were studied, and Slade concluded: "The removal of bacteria was generally similar to that of the viruses, *E. coli* reduction being parallel to but slightly less than that of the viruses. This confirms earlier reports that *E. coli* removal can be used as a guide to virus reduction by slow sand filters." The English work from the late 1970's could lead to the conclusion that a recently developed test for total coliform and *E. coli* (Edberg, Allen, and Smith; 1988) might be very useful for assessing both the bacteria removal efficacy and the virus removal efficacy of slow sand filters. For this to be done, water would have to be sampled before and after filtration, with both of those samples prior to disinfection, so that microbiological quality changes would be the result of filtration only. Slade (1978) also noted, however, that the absence of *E. coli* could occur in water samples that did contain viruses and explained that this could be related to the small volume (100 ml) of water tested for *E. coli* vs. the much larger volume of water (15 to 200 l) tested for viruses. Thus, an apparent 100 percent removal of *E. coli* might be an artifact caused by the detection limit of 1 per 100 ml.

Uncertainty exists about the mechanism of virus removal by slow sand filtration. Robeck et al. (1962) reported that removal of poliovirus by clean sand beds ranged from 22 to 96 percent. Poynter and Slade (1977) reported quarterly averages for poliovirus removal ranging from 98.25 to 99.997 percent when pilot scale slow sand filters were used. Poynter and Slade attributed virus removal to biological activity. They wrote: "The reduction in efficiency at low temperatures, the adverse effects of drainage, and the phenomenon of maturation are all consistent with the view that the removal of bacteria and viruses by slow sand filtration is essentially a biological process. "Concerning physical processes, such as adsorption, they explained that such processes were important but added that experiments with small sand filters had demonstrated that viruses were not removed by clean, sterile sand when operated at normal flow rates (0.2 m/hr).

Taylor (1970) reported that it was considered possible that the large surface area presented by a 0.6 m depth of sand might be capable of adsorbing large numbers of virus particles. Adsorption would be expected to have its greatest effect under conditions of low flow rates and low temperatures when microbial efficacy was minimized, and in the absence of dissolved salts and organic substances. Taylor reported on experiments conducted with a 100 mm diameter filter containing clean sand sterilized with chlorine to a depth of 600 mm. Distilled water containing 100 PFU poliovirus/ml was filtered at a rate of 0.1 and 0.2 m/hr at temperatures of 20°C and 5°C. Samples taken before and after treatment showed no significant reduction in the virus concentration. Taylor concluded that adsorption did not account for most of the virus removal accomplished in slow sand filters. In contrast with results reported by Robeck et al. (1962) and Taylor (1970), McConnell (1984) used reovirus and did not show any effects related to the maturing of the sand bed. She detected no reoviruses in slow sand filter effluent in tests conducted at 1, 13, and 21 weeks of filter operation. Using [125]Iodine-labelled reovirus, McConnell reported that influent concentrations were reduced by greater than 5 $\log_{10}$ concentration units in slow sand filters having either clean sands or aged sand.

Lance et al. (1976) presented results that they said could be explained by adsorption of virus onto a sandy soil at depths below 5 cm in the soil column, but they interpreted a two $\log_{10}$ reduction of virus in the top 5 cm to factors other than adsorption. The results of Poynter and Slade (1977) and Slade (1978) both showed that slow sand filters were less efficient for removing viruses when water temperature was low. This behavior would be expected for biological processes but not for adsorption. McConnell's work suggests that reovirus may adsorb onto sand very readily, but the other slow sand filter research reviewed in this chapter indicates that biological action is also involved. Additional research on virus removal mechanisms may be warranted, especially if slow sand filters are used to treat waters that receive sewage effluent from municipal treatment plants or septic tanks, and are thus possibly subject to contamination by human viruses.

Extensive research on removal of *Giardia* cysts by slow sand filtration has been conducted. Research conducted at Colorado State University was reported by Bellamy et al. (1985a, 1985b). Results from the first phase of the work (Bellamy et al., 1985a) with three 0.30 m diameter pilot filters operated at 0.04, 0.12, and 0.40 m/hr indicated that slow sand filtration was very effective for *Giardia* cyst removal. Removals exceeded 98 percent for all operating conditions tested, including new filter sand. After a *Schmutzdecke* was established, cyst removal approached 100 percent. Filtration rate did influence cyst removal slightly with some decline observed at the 0.40 m/hr rate. Average cyst removals for all filter bed conditions exceeded 99.99 percent for the 0.04 and 0.12 m/hr rates, but was slightly over 99.98 percent for the 0.40 m/hr filter rate.

The second phase of research (Bellamy et al., 1985b) showed that sand sizes from 0.13 mm to 0.62 mm were effective for *Giardia* cyst removal even though coliform removal declined as sand size increased. Cyst removal approached 100 percent for filters operated continuously at 17°C and at 5°C. The authors concluded that slow sand filtration was an effective technology for *Giardia* cyst removal.

Pyper (1985) evaluated *Giardia* cyst removal at a small municipal slow sand filter plant. This facility had two filter beds, each with an area of 37 m². The plant operated at a rate of 0.08 m/hr. Because *Giardia* cysts and sewage were spiked into the raw water in order to define the limits of treatability, all of the water not needed for sampling purposes was wasted. At this plant, cyst removal efficacy was influenced by water temperature. When the raw water temperature ranged from 7.5°C to 21°C, cyst removal was 99.98 to 99.99 percent. When temperature dropped to below 1.0°C (0.5° to 0.75°), cyst removal was somewhat less effective—from 99.3 to 99.91 percent. During one experiment at 0.5°C, both *Giardia* cysts and sewage were spiked in the raw water at the same time. The biological capacity of the filter seemed to have been stretched to the limit, as cyst removal dropped to 93.7 percent. Treatment of both sewage and *Giardia* cysts would not be expected to be a problem for plants treating pristine waters, but siting slow sand filters downstream from sewage treatment plants, sewer mains, or septic tanks could lead to drinking water quality problems if a treatment failure or main break took place. This supports the concept that a sanitary survey ought to be undertaken in conjunction with plant design studies.

Seelaus et al. (1986) reported that a slow sand filter was built and operated at Empire, Colorado in the mid-1980's, following a 1981 outbreak (110 cases) of waterborne giardiasis in this small community caused by unfiltered water. Their paper presented engineering information on the design of the plant, which began operation in January, 1985. *Giardia* cysts were detected in five raw water samples collected between March 16 and April 27, 1985. The cyst numbers ranged from 14 to 48 cysts, in sample sizes ranging from 1892 to 3860 liters. Filtered water samples collected on the same dates contained no *Giardia* cysts. In four of five cases, filtered water sample volume was similar to or greater than raw water sample volume. These winter results showed that even when the water was so cold that a thin layer of ice formed on the water surface within the filter plant, cysts were effectively removed from this clear mountain water.

Bryck et al. (1987) reported on pilot testing, design, construction, and evaluation of a slow sand filter at the community of 100 Mile House, British Columbia. The facility had three filters, each 43 m long and 6 m wide. This plant operated at an average rate of 0.08 m/hr and a maximum of 0.19 m/hr during its first year of operation (November 1985 to November 1986). In a water quality evaluation program, *Giardia* cysts were detected in 11 of 21 raw water samples. No *Giardia* cysts were detected in any of the 22 filtered water samples. This confirmed the earlier work indicating slow sand filtration was very effective for *Giardia* cyst removal.

Tanner (1987) conducted sanitary surveys at thirteen slow sand filters in Idaho. Four filters exceeded the 0.4 m/hr filtration rate that was suggested by Huisman and Wood (1974) as the normal upper limit. Nine did not conform to recommended (Huisman and Wood, 1974) sand specifications of an effective size in the range of 0.15 to 0.35 mm, a uniformity coefficient of less than 3, and a filter bed depth of at least 0.7 meter. Four of the systems studied had experienced coliform maximum contaminant level (MCL) violations during 1986. Those four also had filter beds that were not deep enough, or had excessive filtration rates, or both. The initial filter bed depth is determined by the design engineer, but if the sand in the filter bed is not augmented on a periodic basis, several years of scraping could result in removal of so much sand that the depth would become inadequate. In the case of filtration rate, problems could occur if the original design called for excessive flow rates. Tanner's study showed the importance of following proper design standards and proper operating procedures.

Slow sand filtration has proven to be a very effective process for microorganism removal during one and one half centuries of use and is being recommended for use in communities in developing nations by the World Health Organization. It has been shown in pilot plant studies and in full scale plant evaluations to be capable of attaining very high (three to four $\log_{10}$) removals of bacteria, virus, and protozoan cysts.

6.4   Recent Developments in Coagulation and Filtration

About 1900, engineers in the United States began to recommend the use of rapid sand filters for treatment of muddy surface waters like those found in the Ohio River and Mississippi River valleys. These filters treat water that has been conditioned by coagulation, flocculation, and sedimentation to remove most of the particulate matter, and this results in longer filter runs. Rapid sand filters typically were operated at rates of 5 m/hr in the first half of this century. Since about 1960, use of dual media (coal and sand) or mixed media (coal, sand, and garnet or ilmenite) has permitted operation of rapid rate filters at rates as high as 10 to 24 m/hr. For clear waters (10 NTU or lower) the sedimentation and sometimes flocculation processes may be omitted, and the process is known as direct filtration. In direct filtration, all removal of particulate matter occurs within the filter. Effective coagulation is essential for successful operation of rapid rate filters, with or without sedimentation, because surface attachment is the mechanism by which particle removal occurs.

In the opinion of many engineers, the capability of coagulation and filtration to remove bacteria from raw water had been established in the first third of the twentieth century, so attention was focused on other microbial contaminants. Pilot plant studies for removal of poliovirus were conducted by Robeck et al. (1962), who operated the process train in such a way as to produce very low filtered water turbidity.* They used a microphotometer that could measure turbidity as low as 0.01 Jackson unit, and reported that such values "... actually occurred during parts of many runs." Poliovirus removals of 90 to 99 percent for direct filtration, and greater than 99.7 percent for coagulation, flocculation, settling, and filtration were reported. Their paper has figures that show several runs in which filtered water turbidity was less than 0.1 TU and virus passage was less than 5% (removal of 95% or higher). Also shown were several episodes of rising filtered water turbidity from about 0.1 TU to turbidities in the 0.6 to 2.8 TU range. These increases were accompanied by increases in virus passage through the filters. One phase of this work differed from the rest. Generally raw water turbidity ranged from 10 to 50 TU, but one set of tests was conducted with low turbidity water. The authors wrote, "...

---

*In a review of turbidity standards and calibration, Heer (1987) explained that turbidity measurement originally was based on a suspension of diatomaceous earth in distilled water and was sometimes reported as "parts per million, silica turbidity." Later the unit of turbidity was called the Turbidity Unit (TU). Turbidity measurements made by a nephelometer are expressed in terms of Nephelometric turbidity Units (NTU's), whereas the term Jackson Turbidity Unit (JTU) is used when measurement is made with a Jackson Candle Turbimeter.

to see whether turbidity was really necessary to help form floc, a run was made at 2 gpm/sq ft without adding any local clay." The raw water turbidity was 0.2 TU, and filtered water turbidity was below 0.01 TU. In spite of the low turbidity, virus penetration occurred after 8 or more hours of filter operation, when head loss exceeded 6 feet. The low turbidity condition (a blend of hard ground water and demineralized water) was the exception to the successful removal of viruses when filtered water turbidity was below 0.1. Robeck et al. (1962) concluded that a floc breakthrough was usually accompanied by an increase in virus penetration when turbid water was treated, even though the finished water turbidity remained below 0.5 TU. However, when clear water was treated, the virus penetration could increase without a corresponding increase in filtered water turbidity. Robeck et al. (1962) stated that a series of conventional water treatment processes, well operated, should collectively prevent passage of poliovirus in drinking water.

Guy et al. (1979) conducted pilot plant studies for removal of bacteriophages and enteroviruses. Their flocculation and sedimentation unit removed 99.9 percent of poliovirus and 99.7 percent of bacteriophage T4 that had been added to the raw water. The incremental virus removal by filtration was not consistent, ranging from 18.8 to 37.5 percent for poliovirus and from 0 to 87 percent for bacteriophage T4. The very high removals by sedimentation exceeded those reported earlier by Chang et al. (1958) whereas the incremental removal by filtration was less than Robeck et al. (1962) observed when water was coagulated and flocculated but not settled before filtration.

Stetler et al. (1984) evaluated an operating water filtration plant treating a turbid river water in Michigan by alum coagulation, rapid mixing, flocculation, sedimentation, and filtration. During the study, the point of chlorination was changed from before the filters to after the filters for a period of 8 hours of operation preceding any sampling for microorganisms. This was done once per month for 13 months, and both raw and filtered water samples were analyzed for turbidity and viruses (see Table 6.2). In 7 of 13 sampling periods (every month from September through March) some viruses were recovered from the filtered water, so actual removal percentages could be calculated and they ranged from 72 to 94 percent. In 6 of 7 instances, the percentage reduction of viruses exceeded the percentage reduction of turbidity. Virus reduction was also similar to or greater than turbidity reduction in 3 of 5 cases in which no viruses were recovered, if a zero value in finished water is not called 100 percent removal, but rather only a percentage that exceeds the percent removal that would be observed if a single plaque-forming unit had been detected. The filtered water turbidities greatly exceed the values associated with good virus removal in the research of Robeck et al. (1962). Cold water temperature might have had some influence on the ability of the filtration plant to remove viruses during the November to March time period, but surface waters are not especially cold in southern Michigan in September and October, so the temperature effect would not completely explain the passage of viruses by this plant.

Table 6.2 Virus and turbidity removal at a Michigan water treatment plant

	Turbidity			Viruses, PFU/380 l			
Date	Raw NTU	Filtered NTU	% Reduction	Finished NTU	Raw	Filtered	% Reduction
5/81	38	2.6	93	1.1	3	0	>67
6/81	42	1.8	96	0.6	14	0	>92
7/81	69	1.2	98.3	0.5	23	0	>95
8/81	18	1.3	93	0.4	3	0	>67
9/81	28	3.0	89	0.4	90	9	90
10/81	53	3.8	93	1.1	54	15	72
11/81	11	2.5	77	1.0	13	2	85
12/81	8.3	2.0	76	0.9	64	6	91
1/82	15.0	1.8	88	2.0	48	3	94
2/82	6.0	1.0	83	1.0	293	22	92
3/82	7.0	3.0	57	1.0	71	4	94
4/82	20	7.3	63	0.6	5	0	>80
5/82	14.0	1.0	93	0.8	0	0	

Stetler et al. (1984).
NTU, nephelometric turbidity units. PFU, plaque-forming units.

In another study of water treatment for virus removal, Rao et al. (1988) conditioned the raw water in pretreatment so that filtration resulted in water having low turbidity. This research involved testing with poliovirus type 1, SA 11 rotavirus, and hepatitis A virus. In initial jar test work, when the reduction of turbidity by sedimentation was 90 percent or greater, poliovirus reduction ranged from 87 to over 99 percent, rotavirus reduction was 99 percent or greater, and hepatitis A virus reduction ranged from 93 to 99 percent. Three test runs with a continuous flow pilot plant (one run for each virus) demonstrated excellent removal of both turbidity and viruses. Results are given in Table 6.3 and are consistent with the findings reported by Robeck et al. (1962) for virus removal in a pilot plant.

As the extent of the waterborne giardiasis problem became apparent in the latter 1970's, a considerable amount of research was undertaken on cyst removal, beginning in the late 1970's and continuing into the late 1980's. Cyst removal by coagulation, flocculation, and sedimentation was evaluated by Arozarena (1979), whose jar test results showed that at pH 6.5, at an alum dose of 5 mg/l, cyst removal was about 96 percent, whereas at doses of 10 mg/l or greater cyst removal exceeded 99 percent. This established the importance of using an adequate coagulant dose to attain effective coagulation.

Direct filtration pilot plant research for *Giardia* cyst removal (Logsdon et al., 1981) suggested that obtaining low turbidity filtered water would aid in *Giardia* removal. Useful information about unusual filter conditions was obtained. In three of six filter runs with above-normal turbidity at the beginning

Table 6.3 Virus and water turbidity removal in pilot plant studies

Virus	Virus assays[a]					Water turbidities[b]				
	Virus input	Virus in settled water	% Removal	Virus in filtered water	% Removal	Source water	Settled water	% Removal	Filtered water	% Removal
Poliovirus	5.2×10^7	1.0×10^6	98	8.7×10^4	99.84	12	2.0	83.3	0.08	99.4
Rotavirus	9.3×10^7	4.6×10^6	95	1.3×10^4	99.987	24	2.8	88.3	0.07	99.71
Hepatitis A[c]	4.9×10^{10}	1.6×10^9	97	7.0×10^8	98.6	26	3.9	85.0	0.07	99.73

Rao et al. (1988)
[a] Plaque-forming units per 200 l sample.
[b] Nephelometric turbidity units.
[c] Physical particles per 250 l as determined by electron microscope analysis of virus inocula prior to addition to pilot plant water.

of the run, cyst concentrations during the first half-hour or so were higher than those measured after the turbidity had declined. Also, in one run with weak floc, a filtration rate increase resulted in a turbidity increase by a factor of 4 (from about 0.25 NTU to 1.0 NTU), while the cyst concentration increased by a factor of 25. During another filter run, cysts were dosed into the raw water continuously. About two hours later, after cyst dosing had ceased, the filtration rate was increased, and turbidity increased from 0.3 NTU to about 1.0 NTU. The cyst concentration in filtered water after the rate increase was 2900 cysts/l, higher than the concentration had been in the raw water when cysts were being added to the raw water. This work showed that cysts could be accumulated in filters and later discharged. Filter disturbances that caused turbidity increases were likely to cause the passage of cysts into filtered water.

Some indication of the importance of attaining low turbidity for cyst removal was obtained in research performed at the University of Washington (De Walle et al., 1984). Direct filtration runs were performed with unfiltered surface water (Seattle tap water) typically having a turbidity of 1.0 NTU or lower. When filtered water turbidity was in the 0.02 to 0.04 NTU range, cyst reduction was on the order of 2 to 3 $\log_{10}$ (99% to 99.9%). When turbidity was in the 0.19 to 0.52 NTU range, cyst reduction ranged from less than 1 $\log_{10}$ to 3 $\log_{10}$ (less than 90% to 99.9%), a much more variable performance. The low-turbidity filtered water appeared to be associated with both higher cyst removal percentages and more consistent performance.

Research on *Giardia* cyst removal by direct in-line filtration (coagulation and filtration) was conducted at Colorado State University. Most of the research was done with low turbidity water (about 5 NTU or lower). Because treating cold water is an important consideration for several months each year in parts of Colorado, some work was done at low temperatures. Al-Ani et al. (1986) presented data for treatment of 3° to 4°C water with optimum doses of alum and a cationic polymer. With alum and the polymer, cyst removals ranged from 97.6 percent to greater than 99.9 percent, while turbidity removals ranged from 82.4 to 92.7 percent. The authors concluded: "With proper chemical pretreatment, removals can be expected to exceed 70 percent for turbidity, 99 percent for bacteria, and 95 percent for *Giardia* cysts." Attaining a substantial turbidity reduction was important, even for low turbidity water. They wrote: "If for example, raw water turbidity is 0.5 NTU and if removal of turbidity is 70 percent or greater, the probability is 0.85 that the removal of *Giardia* cysts will exceed 99 percent."

In further work, the Colorado State University researchers used a 0.61 m × 0.61 m dual media pilot filter (Mosher and Hendricks, 1986; Hendricks et al., 1988) and confirmed the earlier findings of Al-Ani et al. (1986). Concerning the need to attain very low filtered water turbidity, they concluded: "*Surrogate Indicators*. Percent reduction of turbidity is an effective and useful indicator of efficient filtration of low turbidity waters. If turbidity is reduced at least 70 percent, as reported by Al-Ani, et al. (1986), the percent reduction in *Giardia* cysts should be greater than 99 percent." As an example, they suggested that turbidity levels in finished water should be 0.1 NTU or less if raw water turbidity was 0.4 NTU. Microscopic analysis of cartridge filter samples, comparing raw water and finished water, also was recommended as an effective indicator of efficient filtration of low turbidity waters.

During this study, the researchers conducted 12 in-line filtration runs when water temperature ranged from 0.2° to 0.4°C. They reported that when proper chemical pretreatment was practiced with cold, low-turbidity raw water, turbidity removals ranged from 84 to 96 percent, *Giardia* cyst removal was virtually 100 percent, and total coliform bacteria removal was 97 percent or greater.

In another research study, Horn and Hendricks (1986) evaluated a 20 gpm Culligan Multi-Tech® filtration system (a two stage coagulation-filtration system operating in pressure vessels). They reported that the filtration system was effective in removing *Giardia* cysts and total coliform bacterial (E. *coli*) at the 99 percent level under low-turbidity raw water conditions, using proper chemical pretreatment with Nalco 8109®. Turbidity removal percentages exceeded 90 percent and were believed to serve as a good surrogate indicator of filtration efficiency.

In another study, Rivers and Hendricks (1985) evaluated a pilot scale Automatic Backwash (ABW®) Filter. Dual media (0.25 m of anthracite over 0.28 m of sand) and single media (0.28 m of sand) configurations were tested. They reported that overall removals of turbidity, total coliform bacteria and *Giardia* cysts were 70 to 99 percent, 86 to 99 percent and 99.8 to 100 percent, respectively, with proper coagulation. The lower end of each range represented results from a single media filter and the upper portion of the range was based on dual media filtration.

Pilot plant research on *Giardia* cyst removal by coagulation, flocculation, sedimentation, and filtration was conducted by Logsdon et al. (1985), using Ohio River water with a turbidity range of 8 to 32 NTU. The authors concluded that removal of *Giardia* cysts by sedimentation ranged from 65 to 93 percent, and was generally similar to percentage removals observed for turbidity. Concerning filtered water quality, they wrote: "The concentration of *Giardia* cysts was influenced by relatively small turbidity changes (0.2 to 0.3 NTU), increasing greatly as turbidity increased." The quality of filtered water was related to both coagulation practice and filter media. When alum was used as the coagulant, with no polymer added to strengthen the floc, filtered-water turbidity was higher from the 0.9 mm effective size anthracite media filter than from the sand filter and the dual media filter, both of which contained at least a portion of filtering material with an effective size in the range of 0.4 to 0.5 mm. When polymer was used to strengthen the alum floc, the effluent filtered-water turbidity and cyst removal percentage for the anthracite filter were comparable to the dual media and the sand filter results for alum alone.

After several pilot scale investigations had been conducted, an evaluation was made of eight water filtration plants in the Rocky Mountains to learn whether pilot plant results obtained by various researchers could be produced at full-scale facilities (Hendricks et al., 1988). *Giardia* cysts were found in the raw water at seven of the eight plants. *Giardia* cysts were found in finished water at two plants, and cartridge-filter analysis of treated water samples from those two plus one other plant had elevated counts of microscopic particles in the effluent, as compared to the clean samples from well-operated filter plants.

The cartridge filter analyses were done by C.P. Hibler of Colorado State University or were done under his direction. The technique has been described by a former Hibler student (Howell, 1987). The presence of cyst-sized particles in filtered water can indicate inability of a filtration plant to remove *Giardia* cysts. Cellular plant material, i.e. fragments of vegetation typical of fecal matter from beavers or muskrats, is particularly indicative of potential for presence of *Giardia* cysts.

Hendricks et al. (1988) related plant performance to coagulation practice. The three plants that passed particulate matter (including two that passed cysts) either did not add coagulant chemical or did not coagulate the water properly to achieve effective particle destablization. During the time period included in this study, two of those three plants changed their pretreatment practice and coagulated properly. Under the latter circumstances, even though *Giardia* cysts were present in the raw water, they were not detected in the filter effluent. The authors concluded:

> The data show that without proper chemical pretreatment *Giardia* cysts will pass the filtration process. Lack of chemical coagulation or improper coagulation was the single most important factor in design or operation of those rapid rate filtration plants where *Giardia* cysts were found in finished water.
> The study has underlined the importance of proper chemical pretreatment in rapid rate filtration. The data indicate that with proper chemical coagulation, the finished water should be free of *Giardia* cysts, have few microscopic particles, and have turbidity levels of less than 0.1 NTU.

Thus their field evaluation supported the cumulative results of earlier pilot plant research.

In another study, Ongerth (1988) evaluated plant performance at two facilities. Two-week field studies at each site were conducted, along with pilot plant studies, to develop information on *Giardia* cyst removal.

A 2,200 m^3/day package plant designed to provide in-line flash mixing, flocculation, sedimentation, and filtration was being used to treat water from an impoundment on a western upland watershed. Raw-water turbidity averaged 0.3 NTU, with a range of 0.25 to 0.6 NTU. Even though no coagulant chemical was used, the filtered-water turbidity averaged 0.13 NTU, but peaks as high as 0.3 to 0.5 NTU were observed upon filter startup, just after backwash. The effects of improper operation (no coagulation) were apparent in the cyst removal results; only about 40 percent reduction of *Giardia* cysts was found. Pilot filter operation at this site produced filtered water turbidity averaging 0.03 NTU with proper coagulation, but some cysts were found in filtered-water samples immediately after backwash.

Ongerth (1988) also evaluated a 15,000 m³/day in-line direct filtration plant treating water from a creek that drained a high mountain watershed that included a marshy meadow. Facilities consisted of chemical feed equipment, piping, filters, and pumps. Chemical mixing was accomplished by adding chemical to the raw water pipe ahead of the filter. This plant used 0.15 mg/l of a cationic polymer as a filter aid. Raw water averaged 0.3 NTU and filtered water turbidity averaged 0.1 NTU. Cyst removal, based on average influent and effluent concentrations, was about 85 percent.

Operation of a pilot plant, with 16 mg/l alum and 0.05 mg/l polymer used as a filter aid, produced filtered water having a turbidity of about 0.05 NTU. Two samples of filtered water were analyzed, and no cysts were found. The pilot plant results suggest that improper coagulation practice was employed at the full scale plant. Using a low dose of cationic polymer as a filter aid, instead of using an adequate dose of the polymer as the primary coagulant for purposes of destabilizing the particles in the raw water, is not likely to provide the optimum in treatability. This plant also had room for improvement in the coagulation practice.

In another field study, Howell (1987) used Hibler's microscopic particulate examination procedure to evaluate water quality and operating procedures at 27 surface water treatment plants in Montana. Howell reported that 10 of the 27 had problems removing cyst-sized particles. She pointed out the need to monitor filter performance during all phases of the filter run, not just while quality is optimum. One plant could produce 0.02 to 0.03 NTU filtered water, but the clearwell turbidity was 0.8 NTU. This plant had short runs (1 to 1.5 hours), occurring at an interval of about every two hours. In addition, several filter stops and starts might occur before a filter would be backwashed. At this plant, the turbidity setpoint for initiation of backwash was lowered from 1.0 to 0.5 NTU, and arrangements were made to lengthen the filter runs. These actions improved effluent quality.

Howell reported that filtered water turbidities above 0.7 NTU always were related to water quality that indicated a risk of *Giardia* breakthrough based on the microscopic particulate analysis. Turbidities below 0.7 NTU were given filter evaluation results ranging from poor to excellent. Howell wrote: "As the above mentioned examples illustrate, systems with less than 0.1 NTU effluent during optimal performance can be at risk for *Giardia* contamination if breakthrough due to shear and backwash are not controlled." Her recommendations for more effective plant operation included ". . . the use of optimized chemical pretreatment, reducing shock to filters by slowing the opening of valves after backwash, reducing breakthrough during filtration runs by initiating backwash at lower turbidity levels, and reducing the number of short filtration runs between backwash cycles."

The *Giardia* research conducted over the past decade has been particularly useful because it has explored conditions that favored effective cyst removal as well as those that hindered it. In addition, the field studies that were conducted to evaluate plant performance also contained the plant survey and evaluation aspect that is essential to understanding why the plants performed in the way they did, as well as to explaining how to improve plant performance.

Research on microorganism removal by coagulation and filtration has not been limited to viruses and *Giardia* cysts. McCormick and King (1982) performed direct filtration studies with a pilot plant that included coagulation, flocculation, and filtration. They used raw waters from five sources in Virginia. Total coliform counts ranged from 0 to 2300 per 100 ml. They stated: "When turbidities in the filtered water were reduced to 0.10 NTU or less, coliform bacteria removal was practically 100 percent." They also reported that raw water in a temperature range of 4° to 6°C could be successfully treated.

Cleasby et al. (1984) operated a direct in-line filter at a quarry site near Iowa State University, spiking the influent with a small amount of sewage to increase the coliform count. When the water in the gravel pit was covered with ice, water temperature varied from 2° to 4°C. During this situation, total coliform removal for three runs averaged 77.7 percent during the first hour (initial quality improvement occurring) and 89.7 percent during the remainder of the runs. During a period of snowmelt, water temperature was 2° to 3°C. A filter run with alum resulted in 93 percent removal of total coliforms for the first hour and 96 percent during the rest of the run. A run with cationic polymer resulted in corresponding total coliform removals of 72 and 89 percent during the initial phase and remainder of the run. The authors concluded: "The performance of direct in-line filtration was not impaired by water temperature as low as 2°C. In fact, when the good-quality raw water was treated during winter ice cover, excellent filtrate and long filter runs were obtained."

This work was continued at the same site by Hilmoe and Cleasby (1986) who performed pilot plant studies with in-line direct filters operated in a constant rate mode and a declining rate mode. Among the water quality characteristics they measured were turbidity, particle count, total coliforms, and temperature. These experiments were conducted in September and October, 1982, during which time the water temperature ranged from 17.5° to 24°C. As in the earlier work, the sewage was added to the raw water to increase the total coliform concentration. In this phase of the work, filtered water turbidity ranged from a low of 0.16 NTU to a high of 1.22 NTU (means varied from 0.20 to 0.34 NTU) for the constant rate filters. Turbidity ranged from a low of 0.17 NTU to a high of 0.86 NTU for the declining rate filters, and means varied from 0.20 to 0.35 NTU. Hilmoe and Cleasby stated that particle and coliform counts followed the turbidity trends closely. In this series of experiments, total coliform counts ranged from 165 to 625 per 100 ml and from 14 to 96 per 100 ml in filtered water after each run had been completed. During the first hour, coliform reduction ranged from 68.8 to 97.1 percent. Water quality improved after the first hour in every run.

Research with a 7 liter/minute conventional treatment train pilot plant (Logsdon and Rice, 1985) suggested that the condition of bacteria before they enter a water treatment process train can influence the results of treatment. Fresh-cultured *Klebsiella pneumoniae* was easily removed by conventional treatment, with reductions as high as 99.999 percent (five $\log_{10}$) being observed. On the other hand, *Klebsiella* naturally present in the sewage added to the raw Ohio River water used for their studies was not so readily removed, as the reductions slightly exceeded 99 percent, just two $\log_{10}$. In the latter situation, the river water with added sewage was held for three to five days to allow the bacteria to acclimate to "river" conditions. This simulated the presence of a sewage discharge about three to five days' travel time upstream. If addition of bacteria is necessary to assess the efficiency of filtration, use of organisms originating in sewage appears to give more realistic results. Logsdon and Rice also showed that higher numbers of bacteria passed through filters immediately after backwashing, during the initial improvement period. This was especially pronounced when dechlorinated tap water was used to backwash the filters.

An extensive review of filter operation as related to backwashing, initial improvement, and filter-to-waste (wasting the filtered water for a period of time immediately after backwash) was conducted by Bucklin and Amirtharajah (1988). These researchers evaluated water quality at two treatment plants in Montana. They observed turbidity peaks in filtered water after filter runs were started, and noted that higher than usual coliform counts were detected during the initial portion of filter runs. Because chlorination was practiced at the plants, and the backwash water was chlorinated, many of the bacteria were injured. Use of MT-7 agar to detect injured coliforms gave consistently higher counts than m-Endo agar. High bacterial counts and turbidity associated with filter washing and start-up could be more effectively minimized by treatment manipulation than by inclusion of a filter-to-waste period. In particular, they emphasized that proper chemical pretreatment of the raw water, maintenance of a free chlorine residual in the backwash water and slow, gradual, careful increase of the filtration rate were techniques that would yield better water quality during the initial improvement period.

A different perspective on removal of bacteria by water filtration was presented by O'Connor et al. (1985) and by Brazos et al. (1987). Both projects reported upon by these authors involved evaluation of water quality at treatment plants. These authors focused primarily on surface water treatment plants in Missouri. The first phase of the work (O'Connor et al. 1985) was conducted in both summer and winter and included analysis for turbidity, total coliforms, heterotrophic plate count, and direct cell count of total bacteria by acridine orange stain and fluorescence microscopy. The authors reported that on the average, surface water filtration plants removed less than one order of magnitude of bacterial cells in winter but two orders of magnitude in the summer. Their data showed four plants attaining total bacterial cell removals of 91.0 to 95.0 percent in the winter, with two plants removing 31.3 percent and 76.8 percent. During the summer, removals at two plants were less than two $\log_{10}$ (89.7 and 97.1 percent). Physical removals at the other four plants ranged from 99.78 to 99.98 percent—that is, close to or exceeding three $\log_{10}$. Averaging the results of six plants lowered the apparent performance by about one log. A majority (4) of the surface water treatment plants exceeded a one $\log_{10}$ total bacterial cell reduction in winter and approached or exceeded three $\log_{10}$ reduction of total bacterial cells in the summer.

Brazos et al. (1987) conducted a long term monitoring project of the Jefferson City water filtration plant. They measured about two $\log_{10}$ removal of total bacterial cells from May to October, and about one $\log_{10}$ removal from December to March. The authors discussed plant performance and wrote: "The filtration process had little ability to compensate for ineffective removal by pretreatment. In fact, only when pretreatment was working well did filtration appear to accomplish significant additional bacterial removals." The authors also noted that turbidity removals were good, and the finished water turbidity averaged about 0.3 NTU. Even so, the total number of bacterial cells per ml ranged from 10^4 to 10^6 (10^7 to 10^9 cells per liter).

Some of the observations of O'Connor et al. (1985) and particularly those of Brazos et al. (1987) seem to be very different from many other results reviewed earlier in this chapter, but appearances may be deceiving. The investigations of Logsdon et al. (1981), DeWalle et al. (1984) and Hendricks et al. (1988) presented examples of pilot plant operation under conditions of improper coagulation (insufficient coagulant dose of improper pH) or no coagulation. Logsdon et al. and DeWalle et al. observed erratic and less than optimum removal of *Giardia* cysts under such conditions. Al-Ani et al. (1986) observed erratic removal of *Giardia* cysts, total coliform bacteria, and standard-plate-count bacteria when no coagulant was added to raw water prior to rapid rate filtration. Hendricks et al. (1988) and Howell (1987) performed plant evaluations that included assessments of plant operation. The above studies have clearly established that improper or inadequate coagulation could enable *Giardia* cysts and microscopic particulate matter to pass through filtration plants. Even though the filtered water turbidity at Jefferson City averaged 0.3 NTU, well below the 1 NTU Maximum Contaminant Level, this low turbidity may have been attained because of the effect of a combination of lime softening and coagulation followed by two stages of sedimentation, with the total time for sedimentation exceeding usual times in single-stage treatment. In the absence of investigations that confirm that coagulation was optimized at Jefferson City, it should not be assumed that coagulation practice could not be improved there.

In contrast to one log removal of total bacterial cells, water filtration plants on Lake Superior have demonstrated a capability to remove large numbers of asbestos fibers—particles even smaller than bacteria (Logsdon et al., 1983). The direct filtration (flocculation included) plant at Two Harbors, Minnesota had amphibole asbestos fiber removals ranging from 89 to 99.9 percent, including a January, 1981 removal of greater than 98 percent. The Silver Bay, Minnesota direct filtration (flocculation included) plant had amphibole fiber removals ranging from greater than 78 percent to 99.3 percent. Silver Bay also removed greater than 98 percent of the influent fibers when the plant was monitored in January, 1981. The Duluth, Minnesota plant, in December through March, 1979, typically removed 3 to 4 $\log_{10}$ of amphibole fibers when operated in a conventional mode. Temperature data were not given by the authors, but in winter water temperature of Lake Superior definitely would not exceed 4°C. The results of plant operation for asbestos removal on Lake Superior show that with careful control of coagulation, direct filtration plants can remove one to two $\log_{10}$ of small particles (asbestos fibers)and conventional plants can remove three to four $\log_{10}$.

6.5 Summary

All of the treatment processes reviewed in this chapter can remove viruses, bacteria, and *Giardia* cysts from raw water with proper facility design, necessary maintenance, and appropriate operational procedures. Some of the information in this chapter indicates that full-scale treatment plants do not always perform as efficiently as pilot plants, but in other cases, pilot plant and full scale results are in agreement. Although some may view pilot plant results as representative of idealized conditions, in fact pilot plants are good predictors of the actual capability of well-designed, well-maintained, and carefully operated filtration plants.

Filtration plant performance can be evaluated by a number of methods, including turbidity, particle count by electronic instrument, total microscopic count of bacteria, microscopic analysis of particulate count from cartridge filters, and analysis for viruses, bacteria, or protozoan cysts. When culture techniques are used to evaluate microorganism removal in filtration, disinfection must not be practiced before filtered water samples are collected; otherwise the actual degree of physical removal can not be determined. Simultaneous use of more than one evaluation technique may be appropriate, not only for research but also for plant operation.

The need for additional work at full-scale plants that display lower-than-expected microorganism removals is apparent. Plant performance can be improved by conducting a thorough evaluation of facilities and operating procedures, and then implementing changes that are needed (Logsdon, 1987). Upgrading filtered water quality can at times be less expensive than anticipated, especially if operating changes rather than capital equipment and facilities changes are the recommended solution. The results of numerous studies conducted over the decades, though, indicate that water filtration plants can be effective barriers to the passage of microorganisms into finished drinking water. Achieving this goal at all plants will require the cooperation and assistance of design engineers, regulatory officials, and water utility staff members. With hard work by all, it can be done.

References

Al-Ani, M.Y., Hendricks, D.W., Logsdon, G.S., and Hibler, C.P. 1986. Removing *Giardia* cysts from low turbidity water by rapid rate filtration. *Journal American Water Works Association* 78(5):66–73.

Al-Ani, M.Y., McElroy, J.M., Hibler, C.P., and Hendricks, D.W. 1985. *Filtration of Giardia Cysts and Other Substances: Volume 3. Rapid Rate Filtration.* EPA/600/2-85/027. U.S. Environmental Protection Agency, Cincinnati, Ohio.

U.S. Army and U.S. Public Health Service. 1944. Report 834. *Efficiency of Standard Army Water Purification Equipment and of Diatomite Filters in Removing Cysts of Endamoeba histolytica from Water.*

Arozarena, M.M. 1979. *Removal of Giardia muris Cysts by Granular Media Filtration.* Unpublished Master's Thesis, University of Cincinnati, Cincinnati, Ohio.

Baylis, J.R., Gullans, O., and Spector, B.K. 1936. The efficiency of rapid sand filters in removing the cysts of the amoebic dysentery organisms from water. *Public Health Reports* 51:1567–1575.

Bellamy, W.D., Silverman, G.P., Hendricks, D.W., and Logsdon, G.S. 1985a. Removing *Giardia* cysts with slow sand filtration. *Journal American Water Works Association* 77(2):52–60.

Bellamy, W.D., Hendricks, D.W., and Logsdon, G.S. 1985b. Slow sand filtration: influences of selected process variables. *Journal American Water Works Association* 77(12):62–66.

Brazos, B.J., O'Connor, J.T., and Lenau, C.W. 1987. Seasonal effects on total bacterial removals in a rapid sand filtration plant. pp 795–833 in *Proceedings of the 1986 Water Quality Technology Conference.* American Water Works Association, Denver, Colorado.

Brown, T.S., Malina, J.F., Jr. and Moore, B.D. 1974a. Virus removal by diatomaceous earth filtration—Part 1. *Journal American Water Works Association* 66(2):98–102.

Brown, T.S., Malina, J.F., Jr., and Moore, B.D. 1974b. Virus removal by diatomaecous earth filtration—Part 2. *Journal American Water Works Association* 66(12):735–738.

Bryck, J., Walker, B., and Mills, G. 1987. *Giardia* removal by slow sand filtration—pilot to full scale. pp. 49–58 in *Proceedings AWWA Seminar on Coagulation and Filtration: Pilot to Full Scale*, American Water Works Association, Denver, Colorado.

Bucklin, K., and Amirtharajah, A. 1988. *The Characteristics of Initial Effluent Quality and Its Implications for the Filter to Waste Procedure.* American Water Works Association Research Report. American Water Works Association, Denver, Colorado.

Chang, S.L. and Kabler, P.W. 1958. Removal of coxsackie and bacterial viruses in water by flocculation, Part 1. *American Journal of Public Health* 48:51–61.

Cleasby, J.L., Hilmoe, D.J., and Dimitracopoulos, C.J. 1984. Slow sand and direct in-line filtration of a surface water. *Journal American Water Works Association* 76(12): 44–55.

Cullen, T.R. and Letterman, R.D. 1985. The effect of slow sand filter maintenance on water quality. *Journal American Water Works Association* 77(12):48–55.

DeWalle, F.B., Engeset, J., and Lawrence, W. 1984. *Removal of Giardia lamblia Cysts by Drinking Water Treatment Plants.* EPA-600/2-84-069. U.S. Environmental Protection Agency, Cincinnati, Ohio.

Edberg, S.C., Allen, M.J., and Smith, D.B. 1988. National field evaluation of a defined substrate method for simultaneous enumeration of total coliforms and *Escherichia coli* from drinking water: comparison with the standard multiple tube fermentation method. *Applied and Environmental Microbiology* 54:1595–1601.

Fayer, R. and Ungar, B.L.P. 1986. *Cryptosporidium* spp. and Cryptosporidiosis. *Microbiological Reviews* 50:458–483.

Fox, K.R., Miltner, R.J., Logsdon, G.S., Dicks, D.L., and Drolet, L.F. 1984. Pilot plant studies of slow rate filtration. *Journal American Water Works Association* 76(12):62–68.

Guy, M.D., McIver, J.D., and Lewis, M.J. 1977. The removal of virus by a pilot treatment plant. *Water Research* 11:421–428.

Hazen, A. 1913. *The Filtration of Public Water Supplies, 3rd Ed.* John Wiley & Sons, New York.

Heer, J.M. 1987. Turbidity Standards, Calibration, and Practice. *Water World News* March/April 18–22.

Hendricks, D.W., Seelaus, T., Janonis, B., Mosher, R.R., Gertig, K.R., Williamson-Jones, G.L., Jones, F.E., Alexander, B.D., Saterdal, R., and Blair, J. 1988. *Filtration of Giardia Cysts and Other Particles Under Treatment Plant Conditions*. American Water Works Association Research Foundation Research Report, American Water Works Association, Denver, Colorado.

Hilmoe, D.J. and Cleasby, J.L. 1986. Comparing constant-rate and declining-rate filtration. *Journal American Water Works Association* 78(12):26–34.

Howell, D.G. 1987. The Evaluation of Montana Surface Water Treatment Plants for Their Ability to Remove *Giardia*. pp. 437–448. In *Proceedings of the 1986 Water Quality Technology Conference*, American Water Works Association, Denver, Colorado.

Huisman, L. and Wood, W.E. 1974. *Slow Sand Filtration*. World Health Organization, Geneva, Switzerland.

Johnson, G. 1916. The typhoid toll. *Journal American Water Works Association* 3:249–326 and 791–868.

Jordan, H.E. 1937. Epidemic amebic dysentery—The Chicago outbreak of 1933. *Journal American Water Works Association* 29:133–137.

Lance, J.C., Gerba, C.P., and Melnick, J.L. 1976. Virus movement in soil columns flooded with secondary sewage effluent. *Applied and Environmental Microbiology* 32:520–526.

Lange, K.P., Bellamy, W.D., Hendricks, D.W., and Logsdon, G.S. 1986. Diatomaceous earth filtration of *Giardia* cysts and other substances. *Journal American Water Works Association* 78(1):76–84.

Logsdon, G.S. 1987. Evaluating treatment plants for particulate contaminant removal. *Journal American Water Works Association*, 79(9):82–92.

Logsdon, G.S. and Rice, E.W. 1985. Evaluation of sedimentation and filtration for microorganism removal. pp. 1177–1197 In: *Proceedings of the 1985 American Water Works Association Annual Conference*, American Water Works Association, Denver, Colorado.

Logsdon, G.S., Symons, J.M., Hoye, R.L. Jr., and Arozarena, M.M. 1981. Alternative filtration methods for removal of cysts and cyst models. *Journal American Water Works Association* 73:111–118.

McConnell, L.K. 1984. *Evaluation of the Slow Rate Sand Filtration Process for Treatment of Drinking Water Containing Viruses and Bacteria*. Unpublished Master of Science Thesis. Utah State University, Logan, Utah.

McCormick, R.F. and King, P.H. 1982. Factors that affect the use of direct filtration in treating surface waters. *Journal American Water Works Association* 74:234–242.

Mosher, R.R. and Hendricks, D.W. 1986. Rapid rate filtration of low turbidity water using field-scale pilot filters. *Journal American Water Works Association*. 78(12):42–51.

O'Connor, J.T., Brazos, B.J., Ford, W.C., Plaskett, J.L., and Dusenberg, L.L. 1985. Chemical and Microbiological Evaluations of Drinking Water Systems in Missouri: Summer Conditions. pp. 1199–1225 in: *Proceedings of the 1985 American Water Works Association Annual Conference*, American Water Works Association, Denver, Colorado.

Ongerth, J. 1988. A study of water treatment practices for the removal of *Giardia lamblia* cysts. pp. 171–175 in: Proceedings of Conference on Current Research in Drinking Water Treatment, EPA/600/9-88/004, U.S. Environmental Protection Agency, Cincinnati, Ohio.

Poynter, S.F.B. and Slade, J.S. 1977. The removal of viruses by slow sand filtration. *Progress in Water Technology* 9:75–88.

Pyper, G.R. 1985. *Slow Sand Filter and Package Treatment Plant Evaluation: Operating Costs and Removal of Bacteria, Giardia and Trihalomethanes*. EPA-600/2-85-052, U.S. Environmental Protection Agency, Cincinnati, Ohio.

Rao, V.C., Symons, J.M., Ling, A., Wang, P., Metcalf, T.G., Hoff, J.C., and Melnick, J.L. 1988. Removal of hepatitis A virus and rotavirus by drinking water treatment. *Journal American Water Works Association* 80(2):59–67.

Robeck, G.G., Clarke, N.A., and Dostal, K.A. 1962. Effectiveness of water treatment processes for virus removal. *Journal American Water Works Association* 54:1275–1290.

Sayre, I.M. 1984. Filtration. *Journal American Water Works Association* 76(12): 33.

Seelaus, T.J., Hendricks, D.W., and Janonis, B.A. 1986. Design and operation of a slow sand filter. *Journal American Water Works Association* 78(12):35–41.

Slade, J.S. 1978. Enteroviruses in slow sand filtered water. *Journal of the Institution of Water Engineers and Scientists* 32:530–536.

Slezak, L.A. and Sims, R.C. 1984. The application and effectiveness of slow sand filtration in the United States. *Journal American Water Works Association* 76(12):38–43.

Spector, B.K., Baylis, J.R., and Gullams, O. 1934. Effectiveness of filtration in removing from water, and of chlorine in killing, the causative organism of amoebic dysentery. *Public Health Reports* 49:786–800.

Stetler, R.E., Ward, R.L., and Waltrip, S.C. 1984. Enteric virus and indicator bacteria levels in a water treatment system modified to reduce trihalomethane production. *Applied and Environmental Microbiology* 47:319–324.

Streeter, H.W. 1927. Studies of the Efficiency of Water Purification Processes. *Public Health Bulletin No. 172*, U.S. Public Health Service, Government Printing Office.

Streeter, H.W. 1929. Studies of the Efficiency of Water Purification Processes. *Public Health Bulletin 193*. U.S. Public Health Service, Washington, D.C.

Tanner, S.A. 1987. *Evaluation of Sand Filters in Northern Idaho*. Unpublished Master's Thesis, University of Washington, Seattle, Washington.

Taylor, E.W. 1969–1970. *Forty-Fourth Report on the Results of the Bacteriological, Chemical and Biological Examination of the London Waters for the Years 1969–1970*. Metropolitan Water Board, London, England.

Walton, H.W. 1988. Diatomite filtration: how it removes *Giardia* from water. pp. 113–116 In: Wallis, P.M. and Hammond, B.R. (editors), *Advances in Giardia Research*, University of Calgary Press, Calgary, Alberta.

7

Home Treatment Devices and Water Quality

Edwin E. Geldreich and Donald J. Reasoner

Environmental Protection Agency, Cincinnati, Ohio

Drinking water quality is not uniform in its characteristics. While it may meet current government regulations for a safe supply, there can be pronounced differences in hardness, taste, odor, and other properties. Furthermore, there is growing evidence that groundwater quality in some small systems may have undesirable microbial, organic, inorganic, or radionuclide contaminants related to the source water characteristics of the supply. Periodic spills of a variety of pollutants into surface waters may also challenge the effectiveness of existing treatment processes in some public water supplies.

When informed of these contaminants in drinking water, public reaction is often one of loss of confidence in the municipal supply, prompting a search for an alternative drinking water source. Alternative solutions are perceived to be spring water, private wells, bottled water, and home treatment devices. Among these alternatives, home treatment devices, known as point-of-use devices, receive much interest because of advertising claims for treatment of specific problems and consumers' fears of exposure to carcinogenic agents. In the United States, the Water Quality Association (Cirola, 1985) has reported that there are at least 325 manufacturers of home water treatment devices that are marketing a wide range of these appliances through 5,000 retailers, distributors and wholesalers. Annual sales are nearly $1 billion and these products are serving more than 5 million families and more than 250,000 commercial, industrial and institutional customers. These statistics suggest that there is public acceptance of this approach to refining the quality of public water supply. The present discussion will consider whether home treatment devices are the solution to water quality problems or possibly another aspect of the problem.

7.1 Home Treatment Appliances

There are a number of devices that have been developed for home treatment of public water supplies. Such devices use a variety of basic process concepts (filtration, adsorption, ion exchange, reverse osmosis and distillation) to achieve a desired contaminant reduction (Table 7.1). In some designs, the treatment package consists of several processes in series to achieve better removal efficiencies and control microbial of quality. An overview of these processes is important for better understanding of their effectiveness and limitations in a small packaged configuration for use in the home.

Filtration devices have been designed to remove turbidity, particulate materials (such as asbestos), and certain types of colloidal color in water. These filters are often made as a spool of spirally-wound fibers, or string or acrylic filaments and can be used to remove particles and dirt from the water. Pleated filters made of paper or nonwoven fabric material will also remove much of the larger particulates but the efficiency of all of these materials is subject to wide variation in effectiveness. Use of membrane materials with controlled pore size in the pleated version will provide more uniform removal of some predetermined particle sizes. Precoating the fibrous fluted material with activated carbon can provide additional taste and odor removal while a careful application of a uniform diatomaceous earth precoating over a fibrous base (to avoid cracks in the filter barrier) may restrict passage of *Giardia* cysts.

Adsorption by granular activated carbon is used to remove tastes and odors associated with chlorine residuals, certain organic chemicals, and chemical compounds produced by various microorganisms including actinomycetes, iron and sulfur bacteria, fungi, and algae.

Carbon filter units vary in several aspects that may account for observed differences in use-life and selective absorption characteristics. Carbon material may be produced from animal bone, pecan or coconut shells, wood, or coal. To improve adsorption, the carbon product is activated by exposing the material to high temperatures and steam in the absence of oxygen. This process results

Table 7.1 Point-of-use methods for reduction or removal of contaminants in water supply

Method	General Removal Applications
Water Treatment Devices	
Filtration	Turbidity, particulates, color
Adsorption	Chlorine, organic substances, odors
Ion exchange	
Cationic	Calicum, Mercury, Magnesium, Iron
Anionic	Arsenic, Selenium
Distillation	Inorganic substances, dissolved solids
Reverse osmosis	Metals, total dissolved solids
Water softening	Barium
Lime Softening	Cadmium
Water Purifiers	
Filtration barrier	*Giardia*
Disinfection barrier	Bacteria, viruses

in the development of minuscule pores within the material, thereby increasing the surface area for absorption and particulate entrapment. Most of the material is granular in structure, and is used to pack into cartridges that may contain as little as 30 g granulated activated carbon in faucet mounted devices or over 3,000 g carbon in the larger devices installed on line under the kitchen sink. In an effort to avoid problems of channel formation in the carbon bed material that would reduce effective filtration, several manufacturers have recently introduced a pressed carbon cake that is said to be a more effective adsorption medium over an extended time.

Ion exchange resins are used to exchange their "soft" sodium ions for the calcium, magnesium, and iron ions in "hard" waters. The ion exchange treatment process requires recharging of bed material when all of the sodium ions are gone from the resin. In the reconditioning process, new sodium ions are introduced into the resins as the "hard" ions are discarded in a reverse exchange. These systems may also remove nitrates and sulfates by freeing the water of mineral cations and anions.

Distillation units heat the water to a vapor, vent volatile impurities off, and separate solids; cooling condenses the purified water vapor back to liquid. Care must be exercised not to overdrive these systems above the boiling temperature because of concern for flash-over of volatile organic substances, particulates with imbedded bacteria, and possibly chlorine into the product water. Heating the water to boiling and inclusion of an air gap between the raw supply and the product water provides an effective barrier to microbial migration from contaminated source into the processed water.

Reverse osmosis (RO) is another popular treatment principle in which the dissolved solids are separated from the water supply by applying a pressure differential across a semi-permeable membrane. The semi-permeable membrane allows the water to flow through, but prevents dissolved ions, molecules, and solids from passing through. The differential pressure forces the water through the membrane, leaving the dissolved particles behind, thus imparting a higher quality to the product water. The membrane may also provide barrier protection against many organisms. However, if water pressure or microbial degradation of the material causes a breakdown of the membrane, treatment effectiveness is lost.

Water purifiers are treatment devices that must remove all types of pathogenic organisms from the water so that the processed water is safe for drinking. To accomplish this objective, devices have been designed to include one or more of the following: membrane filters with small pore sizes that are a barrier to organism passage, silver salts impregnated in granular activated carbon (GAC) to serve as a bacterial inhibitor, halogen based disinfectants imbedded in contact resins, ultraviolet light exposure or generated ozone contact of product water processed by point-of-use treatment units.

7.2   Installation Practices

Home water treatment devices are available in several different package configurations depending on the placement in the plumbing system, the intended use, and the unit cost. Faucet add-on units are inexpensive and the easiest to install but contain so little carbon that their effectiveness is short-lived. A more popular home treatment package is the in-line device that is plumbed into the cold water line under the sink. This type may be configured to treat all water going to the mixing faucet at the kitchen sink or have a separate faucet outlet on a cold water line bypass to be used only for drinking water use. Point-of-entry treatment devices are also available to process the entire home water supply. Point-of-entry units approach the size of a home hot water tank and require more carbon or other treatment agents to process a larger volume of water. Installations of under-the-sink or point-of-entry home water treatment devices are generally done by a professional plumber as contrasted to the simple faucet add-on units.

Selection of the home treatment unit must relate to the water supply contaminant of concern. According to a Cornell University study of small water supplies (Francis, et al, 1984), the most frequent contamination concerns are bacteria, iron, manganese, selenium (in the north central and western regions of the U.S.), mercury, particulates, organic materials, and disagreeable tastes and odors. Recognizing these water quality problems, a variety of home treatment devices have been developed to remove a general class of contaminants. Water softeners to remove hardness, and particulate filters to remove turbidity have been available to the public for many years. More recently, manufacturers of such treatment devices have turned their attention to the development of processes that would be effective in reducing a variety of physical and chemical contaminants to below the critical limits specified in primary and secondary water quality standards.

Treatment devices to produce microbiologically safe drinking water have also appeared on the market to address the needs of backpackers, campers, military organizations, emergency control agencies, homes using individual wells or cisterns, water supplies at foreign embassies, and short term solutions to small water supply deficiencies. Most of the devices identified as water purifiers need to be carefully evaluated for their capability to remove bacteria, viruses, and protozoan cysts in test waters that might represent worst case situations (pH, turbidity, and microbial challenge). The Guide Standard and Protocol of the U.S. Environmental Protection Agency for testing microbiological water purifiers (Table 7.2) specifies *Klebsiella terrigena* to be representative of coliform bacteria, Poliovirus 1 and Rotavirus as typical environmental viruses, and *Giardia* cysts as the protozoan pathogen of concern.

The effective use of point-of-use devices in specific treatment applications is an area not clearly understood by the public. There is a limit to the service life of filter cartridges in these home treatment units which will vary with the characteristics of the water being processed, daily water use, and the treatment

Table 7.2 Challenge organisms and microbial reduction requirements for evaluating water purifiers[a]

Organism	Influent challenge[b]	Minimum required reduction	
		Log	%
Bacteria:			
Klebsiella terrigena (ATCC-33257)	10^7/100 ml	6	99.9999
Virus:			
Poliovirus 1 (LSc) (ATCC-VR-59) and,	1×10^7/l	4	99.99[c]
Rotavirus (Wa or SA-11) (ATCC-VR-899 or VR-2018)	1×10^7/l		
Cyst (Protozoan):[d]			
Giardia muris or *Giardia lamblia* or	10^6/l	3	99.9
As an option for units or components based on occlusion filtration: particles or spheres, 4–6 μm	10^7/l	3	99.9

[a] Guide Standard and Protocol for Testing Microbiological Water Purifiers (Office of Drinking Water Task Force Report, U.S. Environmental Protection Agency, April 1987)
[b] The influent challenges may constitute greater concentrations than would be anticipated in source waters, but these are necessary to properly test, analyze, and quantitatively determine the indicated log reductions.
[c] Virus types are to be mixed in roughly equal 1×10^7/l concentrations and a joint 4 log reduction will be acceptable.
[d] It should be noted that new data and information with respect to cysts (i.e., *Cryptosporidium* or others) may in the future necessitate a review of the organism of choice and of the challenge and reduction requirements.

capacity of the unit selected. To illustrate this issue, our initial laboratory study (Taylor et al; 1979) of four point-of-use carbon-filter devices demonstrated limited removal of chlorinated organic chemicals (predominantly chloroform). All four units contained GAC but the quantity ranged from 226 g to 593 g, and the effective size of the carbon ranged between 0.38 mm and 0.65 mm. Installed in the cold water line, these point-of-use devices initially removed 55 to 100% of the chlorinated organics (chloroform) from the influent water when new filter cartridges were first put on-line. After 6 to 8 weeks, removal was reduced to <20% as a result of decreased effectiveness of the filters. By the end of the 20 week test period, organics were passing directly through the filter or were displaced (unloaded) by other adsorbed organics. This unloading of slugs of chlorinated organics is an undesirable condition and is related directly to the amount and type of carbon in the filter device and to the volume of water treated. The removal of brominated organics, generally a much smaller portion of the total trihalomethanes, was more consistent and did not decrease nearly as rapidly during the test period. During the same study, total organic

Table 7.3 Organic chemical removal by point-of-use treatment units

Unit Type	No.	Estimated contact, sec.	GAC, grams	THM, % Reduction	NPTOC, % Reduction	Service LIFE, Gallons
Faucet, bypass	5	1.6–36.9	27–895	6–69	6–31	40–4000
Faucet, no bypass	1	0.9	16	6	2	200
Line bypass	14	5.0–185.0	146–3402	23–99	6–87	720–4000
Stationary	5	1.2–4.5	208–398	15–46	7–12	3000–3400
Portable pour-through	3	14.1–43.8	30–97	19–74	6–30	300–2000
Portable, other	1	0.9	n.a.*	4	0	200

Data from Bell et al. (1984)
* n.a. not applicable.
GAC = granular activated carbon
THM = trihalomethane
NPTOC = non purgeable total organic carbon.

carbon (TOC) removal appeared to be minimal and short-lived, probably because influent TOC levels were always less than 2 mg/l and effluent concentrations were within 0.1 to 0.3 mg/l of the influent levels one day after filter installation. No TOC reduction was detected after 2 to 4 weeks. These observations were also generally supported by a study on point-of-use devices done by the Gulf South Research Institute (Bell et al., 1984). In that study, performed on 29 different units, removal of selected specific halogenated organic compounds ranged from a low of 2% to a high of 99% (Table 7.3).

Removal of chlorine was generally excellent, ranging from about 53% to 65% in one run and 77% to 97% in a second run. These removal rates accounted for the perceived improvement in taste and odor of the processed water. Chlorine removals >95% for a period of several months were not unusual, even when the free chlorine levels of the influent were >1.0 mg/l. The efficiency of chlorine or organic removal and the length of time this efficiency could be maintained under normal use depended on the volume of water passed through the filter, the residual chlorine or organic concentration, the amount of GAC in the filter cartridge, and the environmental (air, water) temperatures.

7.3   Bacterial Quality From Full-service Units Used In The Kitchen

The bacterial quality of water produced by home treatment devices will differ from that delivered to the faucet by the public utility (Wallis et al., 1974; Taylor et al., 1979, Geldreich et al., 1985). Bacterial densities in the product water from point-of-use carbon filters can be expected to increase by 1 to 2 log over the numbers detected in public water sampled at the tap (Taylor et al, 1979).

Microbial proliferation relates to the species present in the tap water influent to the filter device, the presence or absence of a free-chlorine residual, seasonal changes in water temperatures, ambient air temperatures around the device (generally located under the kitchen sink), and the service duration for a given carbon cartridge.

Adventitious organisms passing through the distribution system may become transient colonizers in a carbon filter for varying periods of time (Geldreich et al., 1985). Monitoring the dechlorinated tap water source revealed that coliform bacteria were released from one or more carbon filter devices during periods ranging from one to 10 months, even though the filters were not challenged with any coliforms during a twelve month test run (Figure 7.1). Three coliform species were detected in single samples at various times rather than for prolonged periods: *Enterobacter aerogenes*—four times, *E. cloacae*—four times, and *Klebsiella pneumoniae*—one time. These observations suggest that occasional coliform occurrences in a public water supply may be more readily detected by monitoring the passage of tap water over time through a carbon filter device than from scheduled grab samples taken periodically from distribution sampling sites. Other organisms detected in the dechlorinated tap water and the number of times isolated during the 12-month study were: *Alcaligenes* spp.—two times, *Serratia marcescens*—four times, *S. rubidaea*—two times, and on different one-day occasions *Pseudomonas cepacia*, *P. fluorescens*, and *P. maltophila*. Not all of these organisms were isolated from each of the filter units at the same sampling time and no individual unit yielded isolates of all of the organisms. Only *E. cloacae*, *Alcaligenes* spp., and *P. maltophila* were found in all the units at some time during the 12 month period, and none of the isolates

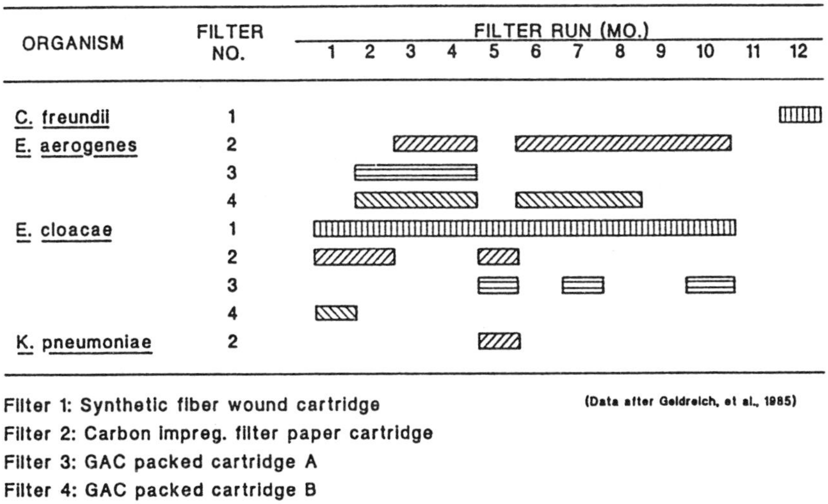

ORGANISM	FILTER NO.	1	2	3	4	5	6	7	8	9	10	11	12
						FILTER RUN (MO.)							
C. freundii	1												▥
E. aerogenes	2					▨▨		▨▨▨▨▨					
	3				═══								
	4			▨▨▨			▨▨▨						
E. cloacae	1		▥▥▥▥▥▥▥▥▥▥▥▥▥▥▥▥▥▥▥▥										
	2		▨▨			▨							
	3						═		═			═	
	4	▨											
K. pneumoniae	2					▨							

Filter 1: Synthetic fiber wound cartridge
Filter 2: Carbon impreg. filter paper cartridge
Filter 3: GAC packed cartridge A
Filter 4: GAC packed cartridge B

(Data after Geldreich, et al., 1985)

Figure 7.1   Coliforms isolated from unchallenged point-of-use unit product water over a 12 month period.

were found to be continuously present. Since the point-of-use device had been disinfected prior to the beginning of this study period, the organisms isolated must have been present in the dechlorinated distribution water influent to the test system and were able to survive and multiply to some extent in the individual point-of-use filter cartridges. The dechlorinating filter on the tap water influent to the test units served to remove free-chlorine from the tap water and introduce a significantly higher density of colonized heterotrophic bacteria than was detected in the public supply. Thus, reasonable worse case conditions were set up for the tests; i.e., no free-chlorine residual in the water introduced to the carbon filters and a heterotrophic bacterial density greater than that found in the treated municipal water supply.

Minimizing bacterial proliferation in point-of-use carbon filters should be an important consideration in the design of such devices. For example, treatment unit designs must avoid the use of carbon-impregnated paper filters, since microbial activity quickly degrades the cellulose fibers and releases large numbers of bacteria into the product water. While the densities of heterotrophic bacteria in product water from carbon-packed canisters initially increased by 10 to 100 times over those of dechlorinated influent (tap water), these effluent densities tended to stabilize after several months in contrast to the microbial releases observed with paper filters.

Static water conditions over night or for longer intervals also provide an opportunity for continued growth of organisms colonizing carbon filters. Under no-flow conditions, water temperatures can increase to room ambient temperature and no chlorine residual that might suppress bacterial growth reaches the carbon sites. Examination of the bacterial quality changes that might result from growth accumulated overnight in the device and be entrained in the product water, revealed that test units frequently had higher bacterial densities in morning samples (Table 7.4). In the winter, the carbon filter used on a chlorinated tap water and the acrylic fiber-wound carbon filter on the non-chlorinated source water produced water with lowest bacterial densities. However, warm water and ambient air temperatures were important factors that contributed to higher populations during summer months on all filters. The carbon-impregnated paper filter product water had bacterial densities that were often higher than for any other device tested and showed little improvement with cyclic flushings during the day.

The problem of water quality deterioration in the product water becomes even more severe with long-term static conditions that might be encountered during vacation periods. For an extreme example, water passage was discontinued through five point-of-use devices (including the control unit) for a six week period after these filters had been in use for six weeks. Heterotrophic plate count (HPC) values for each filter were averaged for the six week period prior to the no-flow test and then compared to HPC results obtained from the product waters released within the first few seconds of the restart of operation (Table 7.5). As anticipated, static conditions provided opportunity for bacterial growth which increased by several log over densities associated with overnight

Table 7.4 Daily fluctuations in heterotrophic bacterial densities from five point-of-use home treatment devices

| Average count per ml | Nonchlorinated source | | | | Chlorinated tap water |
	Acrylic fiber-wound carbon	Carbon-impregnated paper	Carbon #1	Carbon #2	Carbon #3 (control)
Winter*					
A.M.	280	4,700	1,300	13,000	380
P.M.	160	5,800	770	12,000	100
Summer**					
A.M.	8,200	200,000	7,300	7,500	8,500
P.M.	6,200	340,000	3,800	11,000	5,000

 * Test period Jan. 1981 to May 1981; average water temp 10.5°C; average 26 samples.
** Test period June 1981 to Sept. 1981; average water temp 23.8°C; average 25 samples.

no-flow conditions. Ambient air temperatures that translated into similar static water temperatures, coupled with available nutrients in the carbon particles and no flushing, were prime factors that proved favorable to increased microbial growth. Therefore, following a static period of several hours, it is important to flush any point-of-use device by running the water to waste for a brief interval; at least 30 seconds at full flow. After a long term (two days or more) stagnation period, flushing should be extended to full flow for two to three minutes to break up biofilm development.

Table 7.5 Microbial quality deterioration related to static periods without flushing*

| | Nonchlorinated source | | | | Chlorinated tap water |
	Acrylic fiber-wound carbon	Carbon impregnated paper	Carbon #1	Carbon #2	Carbon #3 (control)
Average count per ml**	29	25,000	260	2,600	11
6 week static test	693,000	1,100,000	103,000	30,000	29,000
Cyclic flushing resumed	130	11,000	500	5,100	190

 * Test period December, 1981 to March, 1982.
** Heterotrophic Plate Counts (48 hrs; 35°C) averaged over a 6 week period.

Bacterial amplification or colonization of point-of-use devices is another potential health related concern. To examine this aspect, several home treatment devices were subjected to challenge water that contained microorganisms that might be found in contaminated public water supplies as a result of cross-connections, line breaks, or backsiphonage. Three groups of organisms were selected: coliforms, opportunistic pathogens, and primary pathogens. The coliform group included *Klebsiella pneumoniae, Enterobacter aerogenes, Enterobacter cloacae, Citrobacter freundii,* and *Escherichia coli.* Of the opportunistic pathogens frequently found in drinking water, *Pseudomonas aeruginosa, Serratia marcescens, Aeromonas hydrophila,* and *Legionella pneumophila* were selected. Waterborne primary pathogens were represented by *Salmonella typhimurium* and *Yersinia enterocolitica.*

The challenge inoculum of a candidate organism was adjusted to a dose of about 200 to 300 cells per test unit and introduced into dechlorinated tap water supplied to each point-of-use device (Reasoner et al., 1987). The single dose challenge was then monitored for release in the product water within 15 seconds after inoculation and thereafter during the first flow cycle in the morning on a daily basis for four weeks. The purpose was to determine if the challenge organisms had colonized the filter units. Results indicated some challenge organisms passed directly through the filters, with little or no retention and no subsequent colonization, whereas other bacterial species evidently adsorbed onto and colonized the carbon. Over an extended period of time these colonizing organisms released varying numbers of viable cells into the product water (Figure 7.2). Among the coliforms, *E. coli* and *E. cloacae* quickly passed through the carbon filters within a matter of minutes of the initial challenge. Those coliform species often associated with biofilm development in some water distribution systems (Geldreich, 1988) were found to be capable of varying persistence in the test devices. The length of retention time ranged from 5 days (*C. freundii*) to several months (*Enterobacter aerogenes*), depending on the coliform species. Among the opportunistic pathogens tested, *Pseudomonas aeruginosa* and *Serratia marcescens* were the most persistent bacteria (156 days) in product waters after the carbon filters received a challenge dose.

Carbon filters alone were not a barrier to the passage of either a *Salmonella* or *Yersinia* strain, nor were the test devices installed on the kitchen cold water line colonized by these waterborne pathogens. Both pathogens flushed through the test filters within minutes of the initial challenge and could not be isolated from one liter samples taken several hours later or on succeeding days.

7.4   Bacterial Quality of Third Faucet Treatment Devices

Third-faucet units are intended to treat water only for drinking and cooking requirements rather than inclusion of large volume demands in dishwashing. Bacterial colonization could be a more serious problem with such third-faucet

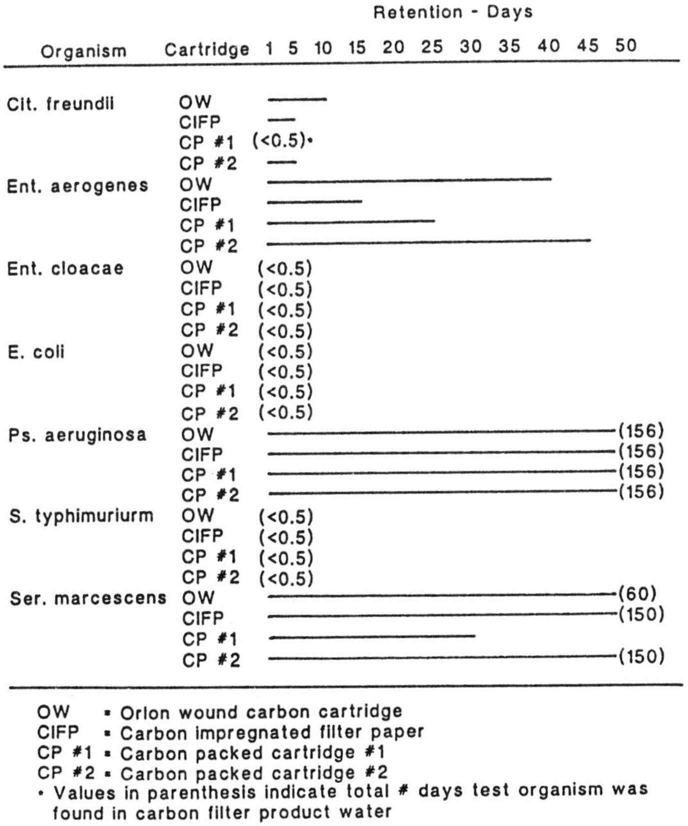

Retention - Days

Organism	Cartridge	1	5	10	15	20	25	30	35	40	45	50
Cit. freundii	OW	——										
	CIFP	—										
	CP #1	(<0.5)•										
	CP #2	—										
Ent. aerogenes	OW	————————————————										
	CIFP	———————————										
	CP #1	—————————										
	CP #2	————————————————										
Ent. cloacae	OW	(<0.5)										
	CIFP	(<0.5)										
	CP #1	(<0.5)										
	CP #2	(<0.5)										
E. coli	OW	(<0.5)										
	CIFP	(<0.5)										
	CP #1	(<0.5)										
	CP #2	(<0.5)										
Ps. aeruginosa	OW	——————————————————————(156)										
	CIFP	——————————————————————(156)										
	CP #1	——————————————————————(156)										
	CP #2	——————————————————————(156)										
S. typhimuriurm	OW	(<0.5)										
	CIFP	(<0.5)										
	CP #1	(<0.5)										
	CP #2	(<0.5)										
Ser. marcescens	OW	——————————————————(60)										
	CIFP	——————————————————————(150)										
	CP #1	————————————										
	CP #2	——————————————————————(150)										

OW   • Orlon wound carbon cartridge
CIFP   • Carbon impregnated filter paper
CP #1 • Carbon packed cartridge #1
CP #2 • Carbon packed cartridge #2
• Values in parenthesis indicate total # days test organism was found in carbon filter product water

Figure 7.2   Microbial challenge for carbon filters.

point-of-use treatment devices because of flow rate reduction and larger contact beds of carbon particles. Third faucet units are generally designed to restrict the pass-through (flow rate) of water to 1.5 gallons per minute and incorporate larger carbon beds, with multi-element components that may include prefilters for particulate removal and silver as a bacteriostatic agent (Reasoner et al., 1987). Seven test units of varying design, number of cartridge elements and carbon capacity were plumbed into the laboratory tap water supply. Table 7.6 describes some of the critical features of the third faucet point-of-use devices selected for this study. The test cycle provided a simulated daily-use cycle that consisted of seven flow periods that ranged from 28 seconds to 3.5 minutes, and approximately a 13 hour non-flow period overnight. Total flow time per test unit during each 24 hour period was about 12.9 minutes at a maximum of 1.5 gallons per minute (5.7 liters per minute), or a total of about 19.3 gal/day/unit (73.1 l/day/unit), depending on the test unit.

Variability in the bacterial quality of product water relates to when the

Table 7.6 Characteristics of some third-faucet point-of-use treatment units

Unit[a]	Number of cartridges	Rated capacity gal (l)	Carbon weight (g)	Measured flow[b] gpm (l/min.)
1. Aqualux, Model CB-2[c]	1 microporous, 2 GAC[d]	2000(7571)	1150	0.79(3.0)
2. Culligan, Model SG-1	1 GAC	4000(15,142)	854	0.84(3.20)
3. Hurley Town & Country	1 GAC	4000(15,142)	895	0.95(3.60)
4. Seagull IV	1 GAC	1000(3785)	300	0.37(1.40)
5. Everpure, QC4 and THM[e]	1 GAC 1 PAC[d]	1000(3785)	765	0.65(2.48)
6. Aquacell Bacteriostatic	2 GAC[d]	2000(7571)	417	0.33(1.24)
7. Continental, Model 350	1 GAC	720(2726)	3402	0.60(2.28)
8. Sears Water Filter Model 3464 (dechlorinating filter)	1 GAC	3420(12,946)	398	2.3(8.7)

[a] Units 1,2,4,5,6 and 7 installed as cold water by-pass to a third faucet; unit 3 installed downstream of water flow control solenoid so that plastic influent line was not subjected to water system pressure during non-flow periods; unit 8 was installed in cold water line as a stationary unit. All filters installed according to manufacturer's recommendations.
[b] Flow rate as measured by collecting water in a one-liter graduated cylinder for 15 seconds, multiplied by 4. Units are gallons per minute (gpm) or liters per minute (lpm).
[c] Manufacturers: 1-Aqualux Water Processing Co., Norwalk, CT 06850; 2-Culligan USA, Northbrook, IL 60062; 3-Hurley Chicago Co., Inc., Worth, IL 60482; 4-General Ecology, Inc., Paoli, PA 19301; 5-Everpure, Inc., Westmont, IL 60559; 6-Aquatron International, Inc., Gretna, LA 70053; 7-Continental Water Conditioning Co., New Orleans, LA; 8-Sears, Roebuck and Co., Chicago, IL.
[d] Impregnated with silver as a bacteriostatic agent.
[e] One GAC (THM-1) cartridge followed by one bacteriostatic powdered activated carbon (QC 4-DC) cartridge.

sample is taken (morning vs. afternoon), seasonally with respect to water temperature, periodically as a response to filter replacement frequency, and individually as a result of unit design. Overnight static water conditions permit bacterial regrowth in the moist environment of a packed carbon cartridge. Therefore, it was not surprising to observe higher heterotrophic bacterial counts in the first morning sample as contrasted to afternoon samples from the same test unit. Results from treatment unit number 4 (Table 7.7) typify results obtained from all of the filter units in the third-faucet point-of-use investigation with regard to the morning vs. afternoon variation in the microbial quality of the product water produced, although this unit generally yielded the highest heterotrophic bacterial densities of all the test units. All of the third-faucet treatment devices produced a morning water quality that had one log higher

Table 7.7 Heterotrophic bacterial density (monthly mean), comparisons for A.M. and P.M. samples.*

	(Organisms per 1 ml)							
	1984**		1985**		1986		1987**	
Month	A.M.	P.M.	A.M.	P.M.	A.M.	P.M.	A.M.	P.M.
Jan.	—	—	140,000	130,000	230,000	190,000	71,000	35,000
Feb.	470,000	190,000	130,000	120,000	260,000	200,000	110,000	72,000
Mar.	270,000	150,000	110,000	88,000	230,000	200,000	100,000	73,000
Apr.	220,000	130,000	190,000	110,000	270,000	180,000	110,000	120,000
May	460,000	260,000	240,000	63,000	330,000	220,000	170,000	140,000
June	410,000	210,000	120,000	47,000	230,000	160,000	120,000	81,000
July	150,000	42,000	150,000	48,000	220,000	170,000	64,000	97,000
Aug.	15,000	14,000	360,000	43,000	220,000	140,000	4,900	7,800
Sept.	46,000	41,000	5,500	6,600	150,000	100,000	27,000	34,000
Oct.	84,000	49,000	19,000	19,000	140,000	130,000	91,000	83,000
Nov.	140,000	110,000	140,000	110,000	150,000	92,000	140,000	140,000
Dec.	150,000	93,000	270,000	250,000	210,000	120,000	—	—
Overall mean	220,000	117,000	156,000	86,000	220,000	158,000	92,000	80,000
A.M. to P.M. Decrease	−47%		−45%		−28%		−13%	

* Data for third faucet "unit 4" treatment device.
** Filter replacements: July 6, 1984; Sept. 9, 1985; Aug. 8, 1987.

bacterial densities than observed with under the sink full service units. The most likely explanation of these significant increases in bacterial density for third-faucet treatment units is that water flow is more restricted, there is more carbon per cartridge, and contact time of the water with the carbon is longer. The slow water flow, and greater amount of carbon with more surface area available for nutrient adsorption and organism attachment, provide better opportunities for increased bacterial colonization and biofilm development. Following overnight stagnation, some bacterial growth would be washed out of the units during the flow periods but not to the extent observed with full-service units operating at high flow-through rates.

Some of the variation in heterotrophic bacterial densities may be attributed to the influence of seasonal changes in water temperature. This can be seen in the trend of the monthly mean bacterial density for the afternoon sample of Unit 1 (Figure 7.3). The high monthly mean for March, 1984, is attributed to the fact that two new GAC filter cartridges were installed in Unit 1 and apparently there was a surge of bacterial growth. Beginning with a relatively low heterotrophic bacterial density in April, 1984, the densities rose to a peak in September and then decreased monthly through December. The effect of

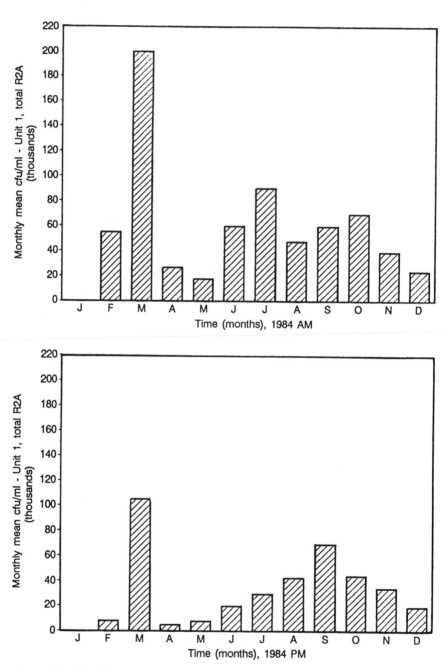

Figure 7.3 Monthly mean heterotrophic plate count (R2A medium) variation for Unit 1 (see Table 7.6). Data are for the year 1984. Upper, AM and lower, PM samples.

increasing water temperature from April to September was to accelerate the rate of bacterial growth on the nutrients adsorbed by the carbon particles.

Service life of carbon filters was also investigated as a factor that could influence the microbial quality of product water from third-faucet treatment devices. As anticipated, the heterotrophic bacterial densities in product water from new filters increased rapidly within the first week after being placed in service and then became relatively stable at densities that ranged from 10^4 to 10^5 organisms per ml. Unit 2 (see Table 7.7), for example, showed a 3-log increase in a period of one week, and the heterotrophic bacterial densities in the product water from that unit remained high for the remaining five weeks of the comparison period. Unit 5, with new filter cartridges, showed the second highest bacterial levels, whereas with the old filter cartridges, it consistently yielded heterotrophic bacterial densities as low as, or lower than, the dechlorinated tap water. These surprising results may indicate that the powdered activated carbon precoat filter in this unit required some time to stabilize. Alternatively, it may be that since the powdered carbon has more surface area than granular activated carbon, more bacteria were able to colonize and grow at this barrier. The long-term trend for the old filters in Unit 5 was toward heterotrophic bacterial densities similar to those in dechlorinated tap water. Thus, microbial colonization and growth in point-of-use carbon filter cartridges can occur quickly and to very high densities of bacteria.

No two home treatment devices showed identical trends in heterotrophic bacterial variation throughout a year or over the entire study period. It is likely that unit size, construction materials, internal design, and amount or type of carbon contributed to the heterotrophic bacteria density differences observed. Unit 1 (Table 7.6), for example, was largely constructed of plastic and had two granular activated carbon cartridges; Unit 4 was the smallest device, contained a packed powdered carbon filter and had a stainless steel filter housing; and Unit 7 was a single, large, wound fiberglass/epoxy canister filled with granular activated carbon. Metal cartridge housings such as present on Units 2, 3, 4 and 5 may contribute to more rapid equilibration of the water and GAC within the cartridges to ambient room temperatures. Any increase in water and filter cartridge temperature would result in increased growth of bacteria on the carbon, resulting in higher bacterial densities in the product water.

The occurrence of pigmented bacteria (forming yellow, orange, pink, brown, or black colonies) in the product water was of interest because significant populations of these bacteria are often found in treated distribution water (Reasoner and Geldreich, 1979). These organisms generally appear to be part of the normal bacterial flora of clean water, most often occurring in higher numbers in waters low in nutrients. In the case of point-of-use treatment devices, the pigmented bacteria were considered as potentially useful markers that might aid in interpreting the changes in the microbiological quality of the product water.

The mean heterotrophic bacterial density of the afternoon sample of dechlorinated tap water from Unit 8 usually contained the highest proportions of pigmented bacteria (Table 7.8) but not necessarily the highest densities. Seasonal trends in the proportion of pigmented bacteria did occur in the product water. All of the point-of-use devices tested modified the density of pigmented bacteria introduced in the influent water with the mean percent of change ranging from less than 4% to as high as 74%, depending on whether the samples were taken in the morning or afternoon. Unit 4, with the highest heterotrophic bacterial densities throughout the study (Table 7.7) usually had the lowest percentage of pigmented bacteria. The significance of these differences is occurrence between test units is not clear but may relate to carbon source (wood, animal, pecan shell, coconut shell, etc.) used by the manufacturer or by design configurations that crated some impact on available attachment sites.

Silver has been incorporated in some third-faucet treatment devices in an attempt to control bacterial activity (Reasoner et al., 1987; Tobin et al., 1981). In this study, treatment devices 1, 5 and 6 contained silver as a bacteriostatic agent (Table 7.6). The effectiveness of silver in controlling the bacterial population was apparently limited, however, because two of the three units containing silver (Units 1 and 6) developed bacterial densities as high as those units that did not contain silver. The bacterial flora of the treatment devices containing granular activated carbon impregnated with silver appeared to be different, both in colony appearance and types of organisms recovered compared to the bacterial flora of other carbon filters without silver. Apparently silver served as a selective agent, allowing silver tolerant bacterial strains to grow.

Bacterial challenges using input water contaminated with biofilm associated coliforms and waterborne bacterial pathogens were investigated because these organisms might colonize home treatment units if attached to some unsafe water supply. *Klebsiella pneumoniae* was selected to represent a biofilm coliform,

Table 7.8 Pigmented bacteria in afternoon samples for third-faucet treatment units

	Study	Point-of-use treatment unit*							
Year	quarter	1	2	3	4	5	6	7	8
1985	Winter	6,300	580	1,800	2,200	780	2,600	480	730
	Spring	7,800	1,200	3,900	3,000	700	2,600	1,800	1,900
	Summer	3,600	4,800	3,000	1,700	320	2,500	1,700	3,100
	Fall	3,800	1,700	7,800	1,000	2,100	3,200	2,100	900
1986	Winter	2,500	960	50,000	2,000	4,000	4,300	910	1,500
	Spring	11,000	2,900	14,000	3,600	6,900	6,700	1,800	4,800
	Summer	12,000	3,200	15,000	2,800	7,000	5,900	2,100	3,500
	Fall	11,000	900	4,900	7,700	3,700	6,300	1,300	1,500

* See Table 7.6 for characteristics of the various units.
** Mean bacteria densities per ml.

Aeromonas hydrophila and *Legionella pneumophila*, as typical opportunistic pathogens, and *Campylobacter jejuni*, as a primary pathogen. Only *K. pneumoniae* colonized the test filters for an extended period of time (Reasoner et al., 1987). *Aeromonas hydrophila* colonized the point-of-use filters during a warm water period (20°C, October, 1984), but not during a cooler water period (12°C, February, 1986). *L. pneumophila* apparently did temporarily persist in some of the point-of-use test filters, but not in a consistent fashion (Table 7.9). The recovery methodology for *L. pneumophila* is not efficient and lacks sensitivity at low cell concentrations and may account for failure to recover this organism in some experiments. Cool or cold water temperatures and the presence of any disinfectant residual in the water probably mitigate against colonization by

Table 7.9 Survival/colonization of *Legionella pneumophila* in a challenge to four point-of-use treatment units

1987 Date	Cell input dosage (CFU)	Point-of-use unit	Time period	Effluent concentration, CFU/liter			Duration (days)
				Start	Maximum	End	
April 30	14,000,000	Unit 2	A.M.	<1	<1	<1	—
			P.M.	3,200	3,200	3,200	0.5
May 5	—		A.M.	330	330	330	5.0
			P.M.	330	330	330	5.5
May 6	—		A.M.	<1	<1	<1	0
			P.M.	<1	<1	<1	0
May 13	—		A.M.	<1	<1	<1	0
			P.M.	<1	<1	<1	0
April 30	14,000,000	Unit 4	A.M.	5,200,000	5,200,000	<1	0.1
			P.M.	5,000	5,000	5,000	0.5
May 5	—		A.M.	170	170	170	5.0
			P.M.	<1	<1	<1	0
May 6	—		A.M.	<1	<1	<1	0
			P.M.	<1	<1	<1	0
April 30	14,000,000	Unit 5	A.M.	<1	<1	<1	—
			P.M.	39,000	39,000	39,000	0.5
May 5	—		A.M.	170	170	170	5.0
			P.M.	50	50	50	5.5
May 6	—		A.M.	<1	<1	<1	0
			P.M.	<1	<1	<1	0
April 30	14,000,000	Unit 8	A.M.	130,000	130,000	<1	0.1
			P.M.	1,500	1,500	1,300	0.5
May 5	—		A.M.	<1	<1	<1	0
			P.M.	<1	<1	<1	0
May 6	—		A.M.	<1	<1	<1	0
			P.M.	<1	<1	<1	0

Data are colony-forming-units (CFU) for input and effluent. For characteristics of the individual units, see Table 7.6.

this organism in carbon filter cartridges when the organism is introduced at low cell concentrations. However, warm water temperatures (>25°C), large numbers of heterotrophic bacteria, and low or no disinfectant residual for prolonged periods of time would probably encourage growth of *L. pneumophila* or other *Legionella* species. Therefore, it is possible that *L. pneumophila*, under appropriate conditions, and at high inoculum levels, could become established in point-of-use filter devices.

The waterborne pathogen *Campylobacter jejuni* (ATCC 29428) was used in a challenge to third faucet home treatment devices (Table 7.10). High levels of inocula were used in these experiments because of the difficulty of recovering *C. jejuni* at low concentrations. Also, the total chlorine residual present in the influent water to the point-of-use test units caused some reduction of viable *C. jejuni* in flask experiments that compared water neutralized with sodium thiosulfate to tap water that had passed through a carbon filter for chlorine removal. The reduction in viability was not significant in the 15 to 20 minutes of experimental exposure of the tests organisms, but prolonged exposure (several hours) would undoubtedly have resulted in considerable die-off. The overall poor recoveries of both *L. pneumophila* and *C. jejuni* from the dechlorinated tap water indicate the difficulties of recovering these organisms in the presence of large numbers of other heterotrophic bacteria, and of the sensitivity of each organism to a disinfectant residual (free or combined).

Growth and persistence of pathogens on granular activated carbon filters is possible (Camper et al., 1985). Laboratory experiments using pure cultures of *Y. enterocolitica*, *Salmonella typhimurium*, and an enterotoxigenic *Escherichia coli* in sterile river water readily demonstrated that all three organisms colonized sterile GAC and maintained high populations for at least several days. However, when GAC had an established biofilm of river water bacteria, the

Table 7.10 Survival/colonization of *Campylobacter jejuni* (ATCC 29428) used to challenge point-of-use treatment units

Date	Inoculum (CFU)	Experiment number	Unit number	Effluent concentration (CFU/liter)			Duration (days)
				Start	Maximum	End	
9/17/87	3.87 × 10⁶	1	1–8	0	0	0	0
9/23/87	3.23 × 10⁶	2	1–8	0	0	0	0
9/29/87	>3.0 × 10⁹	3	1–8	0	0	0	0
10/6/87	1.63 × 10⁶	4	1–8	0	0	0	0
10/13/87	1.17 × 10⁶	5	8	3	3	0	0.5
10/21/87	6.33 × 10⁶	6	8	0	0	0	0
10/28/87	3.5 × 10⁸	7	8	2	2	0	0.5
11/17/87	3.72 × 10⁹	8	8	17	17	0	0.5
11/19/87	3.0 × 10⁸	9	2,4	9,4	9,4	0	0.5

pathogens attached at a lower rate and decreased in concentration rapidly. It must be noted that in these studies the concentration of pathogens introduced into the GAC was higher than would be expected from a contamination event in a potable water system. It appears, on the basis of data presently available, that most of the pathogenic or opportunistic bacteria that have been tested will generally not colonize and/or persist long in the carbon filter devices that already have a biofilm population of nonpathogenic heterotrophic bacteria. However, some organisms such as *K. pneumoniae, A. hydrophila,* and *L. pneumophila* can colonize and could prove to be a health hazard for some consumers.

Summary

Various studies have shown that home treatment devices that use a carbon filter are effective for the removal of contaminants in drinking water, including chlorine disinfectant residuals, as well as organic and particulate substances, but the effectiveness may be short-lived. For this reason, filter cartridges need to be replaced at intervals determined by service life measured in volumes of water processed. The microbial quality of the product water produced by these devices is extremely variable, being a function of input water flora, service life of the filter, water temperature, and frequency of static water conditions. Whether or not individuals who consume water from a biologically active filter suffer adverse health effects is dependent upon the bacteria involved, the number of bacteria ingested, and the individual's general health and resistance to the particular organism(s). Incorporation of barrier filters or the use of bound-resin disinfectant material has been included in the design of some carbon filter units to minimize these microbial growths. However, their effectiveness as a protection from waterborne pathogens in contaminated water supply has not yet been evaluated. Inclusion of silver as a bacteriostatic agent appears to be of limited benefit in controlling the bacterial quality of the product water. Particulate accumulation around the contact agent, silver, and the development of silver resistant biofilms may be important deterrents to the effectiveness of this bacteriostatic agent in point-of-use treatment devices.

Amplification of the heterotrophic bacterial population in carbon filter treatment devices is of concern. If opportunistic pathogens are present in this microbial flora, their growth may reach an infective dose level for some individuals. Those people at greatest risk are those whose health is compromised as a result of immunosuppressive therapy, infection with AIDS, or old age. Obviously, epidemiological studies are needed on water supplies that may have a high heterotrophic bacterial population (over 1,000 organisms per ml). Potable water with this bacterial characteristic would provide the opportunity for a challenge by numerous opportunistic pathogens at levels that may not only colonize the filter bed material but possibly amplify to densities approaching an infective dose level for compromized individuals. However, the potential for adverse human health effects by the user population from ingestion of large numbers of heterotrophic bacteria in the product water appears to be low. There have been no documented and verified reports to date of waterborne illness resulting from consumption of product water from any point-of-use devices.

Along with these microbial quality changes is the problem of carbon bed material unloading adsorbed organic compounds into the product water at higher concentrations as selective adsorption shifts from one group of organic compounds to another group or as the total adsorption capacity is reached for the mass of granular activated carbon material available in the cartridge.

In an effort to minimize the deterioration of water quality by point-of-use treatment devices, the user should apply the following recommendations:

1. Use the point-of-use device only on a microbiologically safe water supply, unless specifically recommended by the manufacturer for other applications.

2. After a prolonged quiescent period (several hours or overnight), the home treatment device should be allowed to run to waste for 30 seconds or longer at full flow. Longer flushing is desirable after a prolonged non-use period such as a vacation.

3. Change the filter cartridge(s) at least as frequently as recommended by the manufacturer.

4. Adhere to the manufacturer's maintenance recommendations and specific instructions relative to changing the filter cartridge(s).

References

Bell, F.A. Jr., Perry, D.L., Smith, J.K., and Lynch, S.C. 1984. Studies on home water treatment systems. *Journal American Water Works Association*, 76:126–130.

Camper, A.K., LeChevallier, M.W., Broadaway, S.C., and McFeters, G.A. 1985. Growth and persistence of pathogens on granular activated carbon filters. *Applied Environmental Microbiology* 50:1378–1382.

Cirola, D. 1985. Background on the point-of-use water quality improvement industry. Presented at the Point-of-Use Water Quality Improvement Industry: Regional Seminar, Chicago, Ill., Nov. 7.

Francis, J.D., Brower, B.L., Graham, W.F. Larson III, O.W., and McCall, J.L. *National Statistical Assessment of Rural Water Conditions. Vol. 1*, EPA-570/9-84-004A. U.S. Environmental Protection Agency, Office of Drinking Water, Washington, D.C.

Geldreich, E.E. 1988. Coliform non-compliance nightmares in water supply distribution systems. In: *Water Quality: A Realistic Perspective*, Univ. of Mich. Press, Ann Arbor, MI.

Geldreich, E.E., Taylor, R.H., Blannon, J.C., and Reasoner, D.J. 1985. Bacterial colonization of point-of-use water treatment devices. *Journal American Water Works Association* 77:72–80.

Reasoner, D.J., Blannon, J.C., and Geldreich, E.E. 1987. Microbiological characteristics of third-faucet point-of-use devices. *Journal American Water Works Association* 79:60–66,.

Reasoner, D.J. and Geldreich, E.E. 1979. Significance of pigmented bacteria in water supplies. pp 187–196 in *Proceedings American Water Works Association, Water Quality Technology Conference,* Philadelphia, PA.

Taylor, R.H., Allen, J.J., and Geldreich, E.E. 1979. Testing of home use carbon filters. *Journal American Water Works Association* 71:577–579.

Tobin, R.S., Smith, D.K., and Lindsay, J.A. 1981. Effects of activated carbon and bacteriostatic filters on microbiological quality of drinking water. *Applied Environmental Microbiology* 41:646–651.

Wallis, C., Stagg, C.H., and Melnick, J.W. 1974. The hazards of incorporating charcoal filters into domestic water systems. *Water Research* 8:111–113.

8

Domestic Water Treatment for Developing Countries

Malay Chaudhuri and Syed A. Sattar

Department of Civil Engineering, Indian Institute of Technology, Kanpur, India and University of Ottawa, Ottawa, Ontario, Canada

Nearly 75% of the present global population lives in the developing parts of the world and since the rate of increase in the population in many such areas is also higher than that in industrialized countries, the share of the Third World population will have increased to almost 80% by the year 2000 (WHO,* 1981). Moreover, many developing regions either suffer from chronic shortages of fresh water or the readily accessible water resources available there are heavily polluted, mainly with domestic wastes. The larger proportion of the population in developing countries lives in rural and suburban areas (WHO, 1981) and conventionally treated drinking water is generally unavailable in such settings. Even though urban centers in these countries have centralized facilities for conventional treatment of drinking water, the quality of such treated water is often suspect, either because of improper treatment or as a result of its contamination during distribution or storage. This lack of sufficient quantities of fresh water, and the consumption of unsafe water, are known to be responsible for a large proportion of the disease burden in these regions (Briscoe, 1987; Esrey and Habicht, 1986; Grant, 1987). Therefore, provision of adequate quantities of safe water for the growing population of the developing world has become a challenging task. As shown in Table 8.1, the International Drinking Water Supply and Sanitation Decade (IDWSSD), which was launched in 1981, is aimed at addressing this problem (Deck, 1986; Lowes, 1987). The traditional approach prior to the launching of the IDWSSD was to build large and capital-intensive water treatment facilities in urban areas of the developing world (Bourne, 1984). Provision of safe water in the home or within 15 minutes walking distance is one of the objectives of primary health care in the Global Strategy for Health for All by the Year 2000 (WHO, 1982).

   Safe water is defined as water that does not contain harmful chemical substances or microorganisms in concentrations that could cause illness in any

---

*WHO, World Health Organization

Table 8.1 Developing world's population with water supply and the International Drinking Water Supply and Sanitation Decade's targets for 1990

Geographic regions as defined by the World Health Organization	Population $\times$ 10^3 (% served)			
	1970	1980	1983	1990
Africa				
urban	28,732	44,508	63,662	54,797
	(70)	(62)	(61)	(83)
rural	121,223	168,551	202,190	120,473
	(10)	(24)	(26)	(60)
Americas				
urban	156,298	219,757	213,172	277,513
	(76)	(77)	(85)	(88)
rural	117,939	119,510	112,026	81,678
	(24)	(41)	(40)	(54)
Eastern Mediterranian				
urban	63,786	71,661	44,234	60,368
	(80)	(83)	(86)	(99)
rural	140,546	133,941	91,800	138,740
	(22)	(30)	(26)	(72)
South-East Asia				
urban	149,418	232,601	252,999	309,156
	(51)	(64)	(66)	(89)
rural	668,684	787,360	822,889	921,659
	(9)	(31)	(43)	(90)
Western Pacific				
urban	31,839	56,403	34,250	50,117
	(80)	(81)	(70)	(87)
rural	62,354	101,788	108,713	121,030
	(27)	(40)	(45)	(91)

Adapted from Deck (1986).

form; *an adequate water supply* is one that provides safe water in quantities sufficient for drinking, and for culinary, domestic, and other household purposes so as to make possible the personal hygiene of members of the household (WHO, 1987). A sufficient quantity of safe water should be available on a reliable, year-round basis near to or within the household where the water is to be used. The daily per capita consumption of drinking water in the rural areas of the developing world ranges between 35 and 90 liters (Kalbermatten, 1983); in urban areas where house connections are available the water consumption rate may be as high as 150 liters per day (Banks, 1983).

8.1   Basic Considerations in Domestic Water Treatment

The most desirable drinking water supply is one that requires no treatment at all. Unfortunately, the wide-spread pollution of water has rendered readily accessible sources of water unsuitable for human consumption without some degree of treatment; this is also true of groundwater in many areas of the world. In fact, in certain tropical countries the combined influence of high population density, rampant pollution, and drought has so seriously affected the quality of available raw water that its treatment for drinking virtually amounts to wastewater reclamation (Evison and James, 1977).

The primary objective of any drinking water treatment method is to render the water safe for human consumption by eliminating from the raw water potentially dangerous microorganisms and chemicals. In view of the alarming number of cases of water-borne infections in developing countries, the effective removal and inactivation of harmful infectious agents during treatment assumes primary importance. However, the presence of color, odor, or turbidity could make an otherwise safe water supply aesthetically unappealing (Chambers, 1978; Huisman et al., 1982). Therefore, the general acceptability of a treatment process is based on how well it maintains or improves the aesthetic quality of the water being treated. The basic methods for the removal of color, odor, turbidity and harmful microorganisms from water have remained unchanged for the past few years and these are based primarily on the processes of sedimentation (with or without coagulation/flocculation of suspended matter), filtration, and disinfection.

The available technology for drinking water treatment that is suitable for developed countries and urban areas in the Third World is often inappropriate for villages and small communities in developing regions because of the limitations of funds and skilled manpower. Urban dwellers in developing countries are also becoming increasingly aware of the potential dangers of consuming drinking water which is either improperly treated in centralized facilities or becomes contaminated during distribution or storage. There is, therefore, renewed interest in the development and testing of those simple and inexpensive devices or systems of drinking water treatment which show promise for application in small communities or individual households in developing countries. This chapter will focus on recent developments in such systems and devices, with particular emphasis on their demonstrated or potential capability in improving the microbiological quality of potable waters. Information on the design and engineering aspects of such systems is available in a number of other publications (Dangerfield, 1983; Hofkes, 1983; IDRC,* 1981; IRCCWS,** 1982; Pescod, 1983; Pickford, 1977; Reid, 1982; Schiller and Droste, 1982; Schulz and Okun, 1984; WHO, 1987).

---

*International Development Research Center
**International Reference Center for Community Water Supply

8.2 Coagulation/Flocculation

Commercially available coagulants/flocculants and coagulation/flocculation aids are routinely used in the treatment of drinking water in urban areas in industrialized as well as developing countries. With the possible exception of alum, these aids are not readily available in many parts of the developing world and several indigenous substances may serve as low-cost alternatives (Schulz and Okun, 1984). In fact, the traditional use of certain types of plant materials and clays in the clarification of drinking water by villagers in parts of Africa, India, and South America has been in practice for several hundred years (Jahn, 1981). However, systematic studies on the comparative efficiency of these traditional substances in the removal of turbidity and microorganisms from water began only recently (Jahn, 1981, 1984; Okun and Schulz, 1983; Schulz and Okun, 1984).

Villagers in south India use seeds from the *nirmali* plant (*Strychnos potatorum*) for treating water for drinking. The seeds are rubbed on the inner surface of earthen vessels and overnight storage of raw water in such vessels clarifies it by the coagulation/flocculation and subsequent settling of suspended materials. Under experimental conditions, extracts/scrapings from these seeds could clarify tap water made turbid by the addition of kaolinite or bentonite clay (Tripathi et al., 1976; Abel et al., 1985). Tripathi et al. (1976) reported 94% and 40 to 50% removal of turbidity and plate count bacteria, respectively, from samples of canal water (turbidity 500 nephalometric turbidity units (NTU); standard plate count 1400 colony-forming units (CFU)/ml) using a *nirmali* seed dose of 2 mg/l.

The seeds from two other plants, *Moringa oleifera* and *Moringa stenopetala*, are used to clarify river water in certain parts of Sudan. Feeding experiments in rats have shown these seeds to be nontoxic (Berger et al., 1984).

Rauwaq, a form of bentonite clay, is also traditionally used by Sudanese villagers for the clarification of water for drinking. Olsen (1987) compared the efficiency of the pulp from *M. oleifera* seeds and *rauwaq* in removing cercariae of *Schistosoma mansoni* from experimentally contaminated samples of tap water and river water. Separate portions of the water were treated with the pulp and samples of the clay from two different sites (Alti and Kutranj) in a manner used by Sudanese villagers. All these flocculants proved to be effective water clarifiers, but the seed pulp and the clay from Kutranj also removed >90% of the cercariae from the water phase between one and three hours. The clay from Alti was less efficient in this regard, suggesting that among the natural flocculants, moringa seeds may be more reliable.

Flocculation of experimentally-contaminated water by *rauwaq* from the banks of River Nile was able to remove 3 to 4 $\log_{10}$ of a coxsackie-, a human adeno- and a bovine parvovirus from experimentally contaminated samples of water; infectious virus particles associated with the floc became undetectable when the floc was dried for 24 hours (Lund and Nissen, 1986).

In the dissolved air flotation (DAF) process, microscopic air bubbles (20 to 80 μm in diameter) are generated when a small quantity of water saturated with air is introduced into the water being treated; the flocs with air bubbles attached to or trapped in them float to the top. Folkard et al. (1986) described a low-cost DAF system for use in the developing world and have also demonstrated that its application together with *M. oleifera* seeds holds promise in drinking water treatment for rural communities. The removal of microorganisms during such treatment remains to be tested.

8.3 Filtration

Sand filters   Sand, which is cheap, inert, durable, and widely available, makes a highly desirable filter medium for the treatment or pretreatment of potable water (Huisman et al., 1982) and slow sand filters (SSF) are used in many rural settings because they are relatively simple to build and operate (Droste and McJunkin, 1982). Such filters consist of a 60 to 75 cm deep column of fine sand to permit a filtration rate of no more than 200 liters/m²/hour of water (IRCCWS, 1973). SSF are generally satisfactory in the retention of bacteria and viruses and they can effectively remove cysts, ova, and cercariae from the water being treated. Complete removal of *Salmonella typhi* by SSF was observed as long back as 1890 at Lawrence, Mass., by the Massachusetts State Board of Health (McCarthy, 1975) and according to a report by Hazen in 1913 (Logsdon and Lippy, 1982), SSF removed 98.31 to 99.41 percent of the applied bacteria (see also Chapter 6). Tests on working filters by Huisman and Wood (1974) showed reduction of total bacteria by a factor of 1000 to 10000 and that of *Escherichia coli* by a factor of 10 to 1000. Studies carried out on pilot SSF in India showed 85 percent of the filtered water samples to be free from *Escherichia coli* under all operating conditions (Joshi et al., 1982), whereas a pilot study by Fox et al. (1984) showed 4 to 5 log reduction of total coliforms and coliforms in the filtered water were <1/100 ml. For enteroviruses, in experimental filters, Poynter and Slade (1977) observed that poliovirus was removed with an efficiency (99.68 to 99.99 percent) similar to that of naturally occurring bacteria (coliforms and plate count bacteria), whereas McConnell et al. (1984) observed a 4-log reduction of reovirus. Pilot plant studies indicated virtually 100 percent removal of *Giardia* cysts by SSF (Bellamy et al., 1985). Sridhar et al. (1985) have also shown that a relatively simple sand filter, which does not require any specialized materials, skills or tools to construct, can be used at the village level to remove water fleas (*Cyclops*), which can carry guinea worm disease.

SSF are a very desirable means of pretreating water in rural setting in the developing world and must not be regarded as an old-fashioned method (Li, 1982; WHO, 1987). Rapid sand filters, which can have filtration rates 50 times higher than those of SSF, are more complicated to build and operate and are, therefore, unsuitable for most rural communities in the Third World (Huisman et al., 1982).

Turbidity reduction in raw water by plain sedimentation (Huisman et al., 1982; Ahmad et al., 1984) or extended storage (Myhrstad and Haldorsen, 1984) prior to slow sand filtration can substantially improve the life and operation of SSF. Such pretreatment can also improve the microbiological quality of raw water through the inactivation of microorganisms during the retention period and the settling of parasite eggs and particle-associated bacteria and viruses with other solids. Studies in the Kigoma Region of Tanzania have shown that raw waters containing >500 fecal coliforms and fecal streptococci/100 ml require a minimum storage of four days for improving the microbiological quality of water (Myhrstad and Haldorsen, 1984). However, certain climatic and geological factors as well as extended retention times required for turbidity removal limit the application of plain sedimentation in many settings. Horizontal-flow roughing filters can be used for the pretreatment of raw water to be passed through SSF and recent improvements and simplification in the design of such horizontal-flow devices make them more suitable for application in the developing world (Wegelin and Schertenlieb, 1987).

Rice hull ash filters   Rice hulls, which are readily available in many parts of the developing world, are of low nutritional and agricultural value. The original structure and porosity of the hulls is retained even after their incineration and such ash, containing almost 90% silica, has been found to be a suitable media for the filtration of drinking water (Barnes and Mampitiyarachichi, 1983). Filters made out of such ash removed turbidity better than sand filters at an optimal flow rate of about 1 $m^3/m^2/h$, which is intermediate between the flow rates of slow and rapid sand filters. When water contaminated with *Escherichia coli* was passed through the ash filter, there was 90 to 99% removal of the bacteria as compared to 60 to 96% removal by a sand filter under similar test conditions.

Coal-based filtration/adsorption media   Studies were undertaken at the Indian Institute of Technology, Kanpur, India, to evaluate the potential of coal-based filtration/adsorption media for removal of turbidity and bacteria from water (Prasad, 1986). The media included Giridih bituminous coal (GBC), both raw as well as pretreated/impregnated with either alum, ferric hydroxide, lime, or manganese dioxide. The results of downflow column studies (5 cm bed depth; 5 min bed contact time), using dechlorinated tap water spiked with *E. coli* (ca. 50 CFU/ml) and natural turbidity (50 NTU), indicated the potential of Alum-GBC as a filtration/adsorption media for domestic water filters (Table 8.2).

   Alum-GBC was subjected to laboratory tests at the University of Ottawa to study its potential in removing enteric viruses (Sabin strain of poliovirus type 1 and the Wa strain of human rotavirus) from artificially contaminated dechlorinated tap water (Chaudhuri and Sattar, 1986 and 1988). In batch sorption tests, with input poliovirus concentration of 2.34 to 2.83 $\times$ 10^6 PFU/1 and sorbent concentration of 10 g/l, Alum-GBC removed 96 percent of the

Table 8.2 *Escherichia coli* and turbidity removal by coal-based filtration/adsorption media

Media	Effluent turbidity, NTU			% *Escherichia coli* removal		
	1 hr	2 hr	5 hr	1 hr	2 hr	5 hr
GBC	11.0	12.0	12.0	30	20	20
Alum-GBC	5.0	6.0	11.0	97	100	100
Ferric hydroxide-GBC	11.0	10.0	10.0	94	100	88
Lime-GBC	7.0	7.0	7.0	59	81	64
Manganese dioxide-GBC	8.0	6.0	6.0	23	30	25

Adapted from Prasad (1986).
NTU, nephelometric turbidity units.

virus when the pH of the water was between 6.3 and 8.9. Under similar conditions, with input rotavirus concentrations of 2.01 to 2.65 $\times$ 10^6 plaque-forming units (PFU)/l, the removal was >99.9 percent; with input rotavirus concentration of 1.18 to 2.71 $\times$ 10^7 PFU/l and sorbent concentration of 4 g/l, virus removal was >95.6 percent in the pH range of 7.5 to 7.7. Virus sorption was rapid and a plateau was attained in 15 to 30 min. Compared with activated carbon (Filtrasorb 400, Calgon Corporation, Pittsburgh, PA), Alum-GBC exhibited greater sorption energy and about one log higher poliovirus sorption capacity. Downflow column tests (8 cm bed depth; 0.09 l/h flow rate), using dechlorinated tap water spiked with poliovirus (1.54 $\times$ 10^5 PFU/l), further indicated the potential of Alum-GBC as a filtration/adsorption media for domestic water filters (Table 8.3).

Results of the laboratory studies on removal of turbidity, bacteria, and enteric viruses have indicated the potential of alum-pretreated coal as a filtration/adsorption media for domestic water filters. A collaborative study between the Indian Institute of Technology, Kanpur, India, and the University of Ottawa, Canada, funded by the International Development Research Centre, Ottawa,

Table 8.3 Alum-GBC column study for poliovirus removal

Throughput volume, ml	% Virus removal
25	96.39
50	97.60
100	93.97
200	98.90
300	93.97
500	93.97

Adapted from Chaudhuri and Sattar (1986).

Canada, is presently underway to further investigate the feasibility of using coal-based media in domestic water filters for developing countries.

The UNICEF filter The upward-flow filter developed by the United Nations Children's Fund (UNICEF) is a very simple and inexpensive device for providing clean drinking water in rural settings (Childers and Claasen, 1987). It consists of two tanks placed one on top of the other. The upper tank is used for raw water storage. The lower tank, which is larger, contains a layer of crushed charcoal sandwiched between two layers of fine sand. Raw water enters the base of the lower tank, pushes upward through the filter and accumulates above the filter for collection via a delivery tube. Such a device with a capacity of 40 l/day can provide clean drinking water for 10 persons and, depending on the quality of the raw water, is expected to require cleaning only once a year. Properly designed studies are required to assess how well this filter is able to remove harmful microorganisms from the water being treated.

8.4 Disinfection

Chlorine Although chlorine solutions of various strengths are available, calcium hypochlorite (bleaching powder) containing 30% available chlorine is cheap and more appropriate for Third World situations. Many remote villages use hand-dug wells or stored rain water for their domestic water supply needs. Diffusion-type chlorinators can be used for the disinfection of such raw water. A mixture of one part bleaching powder and two parts sand is placed in a corrosion-resistant porous or perforated container and it is suspended in the water to allow the chlorine to diffuse. The diffusion rate should maintain a free chlorine residual of 0.2 to 0.5 mg/l of the water over a period of at least one week and the volume of the water and its rate of withdrawal should permit a contact time of no less than 30 minutes.

The basic design and operation of diffusion chlorinators are simple and the vessels for holding the hypochlorite mixture can be made out of indigenously available clay pots. However, their proper application still requires a careful survey of the local conditions, water-use practices and the nature and quality of source water. An understanding of these factors will determine the type and number of chlorinators to be used for a given well or water storage tank and how frequently the hypochlorite mixture will need to be changed (Mann, 1983).

Venkobacher and Jain (1983) have described a device to disinfect water in rural areas of developing countries. It is based on sorbents indigenously available in India. Activated rice husk, activated coconut shell, and a resin (De Actinine-N resin) are used as sorbents in a fixed-bed disinfector with chlorine or iodine. Such materials could act as a filter and a disinfectant. Chlorine, which is a strong oxidizing agent, reacted irreversibly with the sorbents and was not released in the water to be disinfected. Iodine (a milder oxidizing agent), on the other hand, was released from the sorbents for water disinfection. The use of iodine might also be beneficial in areas where goiter is a problem. Comparative tests showed the resin and activated rice husk to be more promising. Iodinated sorbents were able to inactivate *Escherichia coli* in experimentally contaminated samples of well water, whereas chlorine-treated sorbents failed to do so. The bactericidal activity was due to the released iodine and not the contact of bacteria with the sorbent surface. Water with turbidity levels of up to 55 NTU could be readily disinfected and the presence of a certain level of turbidity appeared to help in controlling the release of iodine and improved disinfection activity.

The bacteriostatic and bactericidal properties of silver have been used in the development of many devices for the treatment of drinking water. More recently, the Karachi laboratories of the Pakistan Council of Scientific and Industrial Research (PCSIR, 1985) have developed and tested several silver-based devices for the treatment of potable water for use in the developing parts of the world. The one most suitable for use in rural settings consists of nylon bags with silver-coated sand. The bags can be placed in earthenware pitchers used for storing water in village households. Preliminary tests with *Salmonella typhi, Escherichia coli, Shigella dysenteriae,* and *Staphylococcus spp.* have indicated that there is a 90% decrease in the number of viable bacteria when samples of experimentally-contaminated water are filtered through the bags (placed near the inside of the faucet opening) at a flow rate of 1 l/min. Further testing will, however, be necessary to (a) distinguish between the bacteriostatic and bactericidal effects of the levels of silver released from the bags, (b) test for the development of silver-resistant strains of bacteria and (c) determine the influence of water quality on the disinfectant action of silver.

Work at the Indian Institute of Technology, Kanpur, has also examined the usefulness of incorporating silver into Alum-GBC in order to improve its performance in eliminating bacteria from water (Jayadev, 1987). Column studies (5 cm bed depth; 5 min bed contact time) using water from a canal (turbidity 12–50 NTU; total coliforms 34–160 CFU/ml; heterotrophic plate count 350–730 CFU/ml) showed high bacterial removal (96% total coliforms and 82% plate count bacteria) by silver-incorporated alum-pretreated Giridih bituminous coal (Alum-GBC-Ag). Batch sorption-desorption tests, using *Escherichia coli,* showed marked reduction in the number of viable bacteria following silver incorporation. Further long duration (4–8 days) column studies (10 cm bed depth; 10 min bed contact time) using a river water (turbidity 12–23.5 NTU; fecal coliforms 14–130/100 ml), showed the capacity of Alum-GBC-Ag in producing low levels of turbidity (1.5–2.5 NTU) and fecal coliforms (0–3/100 ml) in the effluent (Osman, 1988).

Virtually nothing is known about the virucidal activity of silver. Mahnel and Schmidt (1986) examined two commercially available silver-based compounds (Certisil and Micropur) for their virucidal potential. A picornavirus of animal origin was the only enteric virus out of the five viruses tested. The lability of all the test viruses was increased by silver but it required 48 hours or longer for a 3 $\log_{10}$ reduction in their infectivity titre. Other compounds (Certisil-Combina and Sanosil) which contained a chemical disinfectant together with silver proved to be slightly better in the inactivation of the tested viruses, with a 99.9% reduction in virus infectivity titre in 24 hours. For further discussion of the use of silver in treatment of drinking water, see Chapter 7.

8.5   Commercial Devices

The popularity of point-of-use commercial devices for drinking water treatment is rapidly increasing in urban areas of many developing countries. However, no hard data are as yet available from field studies to indicate that the regular use of these devices leads to a reduction in the number of cases of water-borne infections. Concerns over the possibility of negative health and environmental impacts due to large-scale use of these devices also need to be addressed.

Two commercial domestic water treatment devices in India are tap-attachable water filters and candle filters (Kulkarni et al., 1980). In the tap-attachable filters, the water passes through either a cotton wad held on a coarse sintered silica crucible or a sponge type wad placed over a microporous bed held on a porous Buchner type funnel. According to the manufacturers, these filters are capable of filtering out fine particulates and clusters of bacteria. The candle filters are more popular and comprise a top chamber with one or more microporous ceramic filter candles and a bottom chamber with a tap. It is claimed that these filters trap microorganisms, dirt, and other suspended impurities.

A series of three Zero-B (zero bacteria) water purifiers incorporating ion exchange technology has been recently introduced. These purifiers are claimed to eliminate all water-borne bacteria and produce a water totally free of diarrhea, dysentery, gastroenteritis, cholera, typhoid, and other disease-causing bacteria. The three purifiers are:

1. *Tumbler*: Provides bacteria-free water at office or during travel. Recommended life is one year for one person or 700 liters of water filtered.

2. *Filter tap attachment*: A device with refill cartridges that can be directly attached to a candle filter tap. Recommended life of cartridge is 6 months for a family of 6 or 3,000 liters of water filtered.

3. *Basin*: A basin with refill cartridge which purifies water while it is poured. Recommended life of cartridge is 6 months for a family of 6 or 3,000 liters of water filtered.

Another device is a tap-attachable ultraviolet water purifier (Alfa Unipure) with an electronic eye to indicate any malfunction. It is claimed to produce 100% bacteria-free water @ 2 liters/min.

8.6   Laboratory Evaluation of Commercial Devices

Chapter 7 provides detailed evaluations of point-of-use devices as they are employed in the United States but there is very limited published information on the evaluation of point-of-use devices in developing countries. The National Environmental Engineering Research Institute, Nagpur, India, conducted a laboratory study to test the performance of the four most popular advertised water filters in terms of turbidity and coliform bacteria removal (Kulkarni et al., 1980). Two tap-attachable filters were tested for turbidity removal only using well water with Black Cotton clay turbidity (20–32 NTU). Both filters performed poorly. The filtrate turbidity for filter A (cotton wad held on a coarse sintered crucible) always remained above 3.5 NTU (range 3.5–20.0 NTU) while that for filter B (microporous bed held on a porous Buchner type funnel) varied in the range 6.0–20.0 NTU. Two candle filters (C and D) were tested for turbidity as well as coliform bacteria removal. For both filters, the filtrate turbidity remained above 3.0 NTU during the first five hour run when the influent turbidity varied in the range 15.0–24.0 NTU. During the second 7 hour run with influent turbidity in the rate 8.0–20.0 NTU, filtrate turbidity decreased from 0.50 NTU in the first hour to 0.15–0.20 NTU at the end. Data for removal of coliform bacteria by the candle filters indicated considerable improvement of the bacteriological quality of the artificially contaminated well water (Table 8.4).

According to the test reports supplied by the manufacturer, the tap-attachable ultraviolet purifier produced 100% bacteria-free water. The test reports are summarized below:

1. Viable count of *Escherichia coli* in influent: 7.8×10^4 colony forming units (CFU)/ml; no *E. coli* detected in the effluent.

2. Influent: 10^6 CFU/ml of *E. coli* and *Pseudomonas aeruginosa*; no *E. coli* of *P. aeruginosa* detected in effluent after 48 hours of incubation.

Table 8.4 Coliform bacteria removal by candle filters

Raw water coliform/100 ml	Filter C		Filter D	
	Run 1	Run 2	Run 1	Run 2
5.4×10^6	8	240	22	33
1.6×10^5	17	170	49	240
2.4×10^4	2	13	6	49
2.4×10^4	49	5	13	17
9.2×10^2	49	17	33	33

Adapted from Kulkarni et al. (1980).

8.7   Concluding Remarks

The general lack of sufficient quantities of safe drinking water in the developing world continues to be a serious problem; this is particularly true for rural and suburban communities. As Bourne (1984) has pointed out, 1000 or 2000 years ago people living in what are now developing regions had better systems of water supply than they do today; this is not only a reflection of the high priority ancient civilizations placed on this issue, but it is also an example of development moving backwards.

Although concerted efforts are now underway to correct the situation through the IDWSSD, the unabated increase in the population is making it difficult to meet the Decade's initial objective of providing coverage for 85% of the rural population of the developing world (WHO, 1986); this means that more than one billion persons will still be without access to safe water by 1990. At the same time, the increasing awareness of the importance of water in economic development is putting ever greater pressures on the best available water resources. Since new water resources are not available in many areas, there are fears of an impending water crisis (Biswas, 1984).

Even if resources could be made available, the introduction of conventional methods of drinking water treatment in villages and small communities in the developing world would not be a viable approach. There is, therefore, a pressing need for the development and introduction of simple and inexpensive ways of making safe potable water available to such communities (Chambers, 1978; Erbel, 1983; Henry, 1978).

Water treatment methods (e.g. slow sand filtration) already in use by rural communities are being examined more carefully and technology is also being adapted to make it suitable for rural settings which have limited resources and skills (Commonwealth Science Council, 1982; Dangerfield, 1983; Gregory et al., 1983; Hofkes, 1983; IDRC, 1981; Schiller and Droste, 1982; Schulz and Okun, 1984; WHO, 1987). But very little information is available thus far on the removal and inactivation of harmful microorganisms by processes and devices which appear to show some promise. For example, we do not yet know how well the UNICEF's upward-flow filter (Childers and Claasen, 1987) can improve the microbiological quality of water and how efficient is the DAF system (Folkard et al., 1986) in eliminating potentially dangerous microorganisms from water.

For the foreseeable future, the use of chlorine (in the form of calcium hypochlorite) is likely to continue as the most appropriate form of drinking water disinfection in rural communities of the developing world because of its relatively low cost, residual effect and broad-spectrum of activity (Mann, 1983). Sophisticated equipment and training are required for proper water disinfection by chlorine gas in imported cylinders and the technology for the on-site production of chlorine is still unsuitable for routine application in such settings (WHO, 1987).

Many communities boil their water for drinking. Whereas this is a satisfactory method of destroying a wide variety of infectious agents, it usually requires 1 kg of wood to boil 1 liter of water (IRCCWS, 1973). The increasing shortage of burning wood makes this method undesirable unless alternate sources of energy are available. The use of solar energy shows some promise in this regard (Ciochetti and Metcalf, 1984; Sundaresan et al., 1982).

In industrialized countries, considerable progress has been made in the development (Gerba et al., 1984 and 1988; Thurman and Gerba, 1988) of point-of-use and point-of-entry devices and the testing (Bell et al., 1984; Tobin and Armstrong, 1980; U.S.E.P.A., 1986) of their capacity to remove and inactivate infectious agents (see Chapter 7). The findings (Dufour, 1988) in the U.S. also indicate that the prolonged use of such devices is not harmful to human health. As mentioned above, the use of a variety of devices for the in-home filtration and/or disinfection of water is rapidly increasing in many parts of the developing world and greater efforts are required to control their quality and assess their suitability for tropical settings.

Providing safe drinking water can immediately and dramatically improve the health of many communities and can also lead to the elimination of serious diseases such as dracunculiasis (Hopkins, 1983). However, any attempts at providing safe and *adequate* quantities of water to the developing regions of the world must be properly integrated with other aspects of development such as sanitation and education (McGarry, 1977; Wood, 1983). Improvements in the quality of potable water, particularly in remote rural settings in developing countries, in themselves do not appear to make a substantial difference in the incidence of many types of gastrointestinal diseases (Esrey et al., 1985; Feachem, 1984). This was again illustrated by a relatively recent study in a Brazilian village of about 200 inhabitants (Kirchoff et al., 1985). There were no sanitation facilities in this village and water for domestic use was collected in cans from a nearby pond which was found to have a mean fecal coliform count of 9700/100 ml. Introduction of an in-home water chlorination program resulted in a notable decrease in the fecal coliform count in treated water, but there was no reduction in the incidence of diarrhea with the use of the disinfected water.

The preparation of this chapter was made possible by a grant from the International Development Research Centre, Ottawa, Canada.

References

Abel, R., Kawata, K., and El-Sebaie, O.D. 1985. Coagulant properties of an indigenous seed: home water treatment in Tamil Nadu. *Tropical Doctor* 15:45–47.
Ahmad, S., Wais, M.T., and Agha, F.Y.R. 1984. Appropriate technology to improve drinking water quality. *Aqua*, no. 1:26–31.
Banks, P.A. 1983. Rural water supply and sanitation. pp. 239–261 in Dangerfield, B.J.

(editor), *Water Supply and Sanitation in Developing Countries*, The Institute of Water Engineers and Scientists, London, England.

Barnes, D. and Mampitiyarachichi, T.R. 1983. Water filtration using rice hull ash. *Waterlines* 2:21–23.

Bell, F.A. 1984. Studies on home water treatment systems. *Journal of the American Water Works Association* 77:127–130.

Bellamy, W.D., Silverman, G.P., Hendricks, D.W., and Logsdon, G.S. 1985. Removing *Giardia* cysts with slow sand filtration. *Journal of the American Water Works Association* 77:52–60.

Berger, M.R., Habs, M., Jahn, S.A.A., and Schmahl, D. 1984. Toxicological assessment of seeds from *Moringa oleifera* and *Moringa stenopetala*, two highly efficient primary coagulants for domestic water treatment of tropical raw waters. *East African Medical Journal* 61:712–716.

Biswas, A. 1984. Role of water in economic development. pp. 115–134 in Bourne, P.G. (editor), *Water and Sanitation: Economic and Sociological Perspectives*, Academic Press, Inc., New York.

Bourne, P.G. 1984. Water and sanitation for all. pp. 1–20 in Bourne, P.G. (editor), *Water and Sanitation: Economic and Sociological Perspectives*, Academic Press, Inc., New York.

Briscoe, J. 1987. A role for water supply and sanitation in the child survival revolution. *PAHO Bulletin* 21:93–105.

Chambers, R. 1978. Identifying research priorities in water development. *Water Supply and Management* 2:389–398.

Chaudhuri, M. and Sattar, S.A. 1986. Enteric virus removal from water by coal-based sorbents: development of low-cost water filters. *Water Science and Technology* 18:77–82.

Chaudhuri, M. and Sattar, S.A. 1988. Removal of human rotavirus from water by coat-based sorbents. Presented at the *First Biennial Water Quality Symposium: Microbiological Aspects*, August, 1988, Banff, Alberta, Canada.

Childers, L. and Claasen, F. 1987. UNICEF's upward-flow water filter. *Waterlines* 5:29–31.

Ciochetti, D.A. and Metcalf, R.H. 1984. Pasteurization of naturally contaminated water with solar energy. *Applied and Environmental Microbiology* 47:223–228.

Commonwealth Science Council. 1982. *Report of the Regional Workshop on Rural Drinking Water Supply*, May, 1982, Madras, India; Commonwealth Secretariat, London, England.

Dangerfield, B.J. 1983. *Water Supply and Sanitation in Developing Countries*. The Institute of Water Engineers and Scientists, London, England.

Deck, F.L.O. 1986. Community water supply and sanitation in developing countries, 1970–1990: An evaluation of the levels and trends of services. *World Health Statistics Quarterly* 39:2–31.

Droste, R.L. and McJunkin, F.E. 1982. Simple water treatment methods. pp. 101–122 in Schiller, E.J. and Droste, R.L. (editors), *Water Supply and Sanitation in Developing Countries*, Ann Arbor Science, Ann Arbor.

Dufour, A.P. 1988. Health risks posed by heterotrophic bacteria colonizing granular activated carbon filters. p. 151 in Janauer, G. (editor), *Proceedings of the Fourth Conference on Progress in Chemical Disinfection*, April, 1988, Binghamton, New York.

Erbel, K. 1983. Low cost water supply—an appropriate technology for rural areas and urban fringes? *Water Supply* 1:1–7.

Esrey, S.A., Feachem, R.G., and Hughes, J.M. 1985. Interventions for the control of diarrhoeal diseases among young children: improving water supplies and excreta disposal facilities. *Bulletin of the World Health Organization* 63:757–772.

Esrey, S.A. and Habicht, J.-P. 1986. Epidemiologic evidence for health benefits from improved water and sanitation in developing countries. *Epidemiologic Reviews* 8:117–128.

Evison, L.M. and James, A. 1977. Microbiological criteria for tropical water quality. pp.

30–51 in Feachem, R., McGarry, M., and Mara, D. (editors), *Water, Wastes and Health in Hot Climates*, John Wiley & Sons, London.

Feachem, R.G. 1984. Interventions for the control of diarrhoeal disease among young children: promotion of personal and domestic hygiene. *Bulletin of the World Health Organization* 62:467–476.

Folkard, G.K., Sutherland, J.P., and Jahn, S.A.A. 1986. Water clarification with natural coagulants and dissolved air flotation. *Waterlines* 5:23–26.

Fox, K.R., Miltner, R.J., Logsdon, G.S., Dicks, D.L., and Drolet, L.F. 1984. Pilot plant studies of slow sand filtration. *Journal of the American Water Works Association* 76:62–68.

Gerba, C.P., Hou, K., and Sobsey, M.D. 1988. Microbial removal and inactivation from water by filters containing magnesium peroxide. *Journal of Environmental Science and Engineering* 23:41–58.

Gerba, C.P., Janauer, G.E., and Costello, M. 1984. Removal of poliovirus and rotavirus from tapwater by a quaternary ammonium resin. *Water Research* 18:17–19.

Grant, J.P. 1987. *The State of the World's Children.* UNICEF, New York, 110 pages.

Gregory, R., Mann, H.T., and Zabel, T.F. 1983. The roles of upflow filtration and hydraulic flocculation in low-cost water technology. *Water Supply* 1:97–112.

Henry, D. 1978. Designing for development: what is appropriate technology for rural water and sanitation? *Water Supply and Management* 2:365–372.

Hofkes, E.H. 1983. *Small Community Water Supplies: Technology of Small Water Supply Systems in Developing Countries*, John Wiley & Sons, New York.

Hopkins, D.R. 1983. Dracunculaisis: an eradicable scourge. *Epidemiological Reviews* 5:208–219.

Huisman, L., Van Gorkum, N., and Kempenaar, K. 1982. Water supplies in rural areas of developing countries. pp. 147–185 in Reid, G.W. (editor), *Appropriate Methods for Treating Water and Wastewater in Developing Countries*, Ann Arbor Science, Ann Arbor.

Huisman, L. and Wood, W.E. 1974. *Slow Sand Filtration*, WHO, Geneva, Switzerland.

International Development Research Centre. 1981. *Rural Water Supply in Developing Countries*, IDRC, Ottawa, Canada.

International Reference Centre for Community Water Supply. 1973. *The purification of water on a small scale.* Tech. Paper No. 3. 19 pages. IRC, The Hague, The Netherlands.

International Reference Centre for Community Water Supply and Sanitation. 1982. *Practical Solutions in Drinking Water Supply and Wastes Disposal for Developing Countries*, Technical Paper No. 20, IRC, The Hague, The Netherlands.

Jahn, S.A.A. 1981. *Traditional Water Purification in Tropical Developing Countries.* German Agency for Technical Cooperation. Eschborn, West Germany.

Jahn, S.A.A. 1984. Traditional water clarification methods using scientific observation to maximize efficiency. *Waterlines* 2:27–28.

Jayadev, S. 1987. *Silver Incorporated Filtration/Adsorption Media for Removal of Bacteria from Water.* M. Tech. dissertation. Indian Institute of Technology, Kanpur, India.

Joshi, N.S., Kelkar, P.S., Dhage, S.S., Paramasivam, R., and Gadkari, S.K. 1982. Water quality changes during slow sand filtration. *Indian Journal of Environmental Health* 24:261–276.

Kalbermatten, J.M. 1983. Water supply and sanitation in developing countries. pp. 1–5 in Dangerfield, B.J. (editor), *Water Supply and Sanitation in Developing Countries*, The Institute of Water Engineers and Scientists, London, England.

Kirchhoff, L.V., McClelland, K.E., Pinho, M.D.C., Araujo, J.G., De Sousa, M.A., and Guerrant, R.L. 1985. Feasibility and efficacy of in-home water chlorination in rural North-eastern Brazil. *Journal of Hygiene* 94:173–180.

Kulkarni, D.N., Tajne, D.S., and Parshad, N.M. 1980. Performance of domestic water filters. *Indian Journal of Environmental Health* 22:30–41.

Li, K.G. 1982. Methods of choice: small wells, slow sand filtration, and rapid sand filtration for water supplies in developing countries. pp. 186–206 in Reid, G.W.

(editor), *Appropriate Methods for Treating Water and Wastewater in Developing Countries*, Ann Arbor Science, Ann Arbor.

Logsdon, G.S. and Lippy, E.C. 1982. The role of filtration in preventing waterborne diseases. *Journal of the American Water Works Association* 74:649–655.

Lowes, P. 1987. Half time in the Decade. pp. 16–17. *Developing World Water*, Grosvenor Press International, Hong Kong.

Lund, E. and Nissen, B. 1986. Low technology water purification by bentonite clay flocculation as performed in Sudanese villages: virological examinations. *Water Research* 20:37–43.

Mahnel, H. and Schmidt, M. 1986. Über die Wirkung von Silberverbindungen auf Viren in Wasser. *Zentralblatt für Bakteriologie und Hygiene B.* 182:381–392.

Mann, H.T. 1983. Water disinfection practice and its effect on public health standards in developing countries. *Water Supply* 1:231–240.

McCarthy, W.J. 1975. Construction and development of slow sand filters at Lawrence, Mass. *Journal of the New England Water Works Association* 89:36.

McConnell, L.K., Sims, R.S., and Barnett, B.B. 1984. Reovirus removal and inactivation by slow sand filtration. *Applied and Environmental Microbiology* 48:818–825.

McGarry, M.G. 1977. Institutional development for sanitation and water supply. pp. 195–212 in Feachem, R., McGarry, M. and Mara, D. (editors), *Water, Wastes and Health in Hot Climates*, John Wiley & Sons, New York.

Myhrstad, J.A. and Haldorsen, O. 1984. Drinking water in developing countries—The minimum treatment philosophy. A case study. *Aqua*, no. 2:86–90.

Okun, D.A. and Schulz, C.R. 1983. Practical water treatment for communities in developing countries. *Aqua*, no. 1:23–26.

Olsen, A. 1987. Low technology water purification by bentonite clay and *Moringa oleifera* seed flocculation as performed in Sudanese villages: effects on *Schistosoma mansoni* cercariae. *Water Research* 21:517–522.

Osman, S. 1988. *Silver Incorporated Alum Treated Bituminous Coal Media for Removal of Bacteria and Turbidity from Water*. M. Tech. Dissertation. Indian Institute of Technology, Kanpur, India.

Pakistan Council of Scientific and Industrial Research. 1985. Annual Report 1983–1984. Scientific Information Division, PCSIR, Karachi, Pakistan.

Pescod, M.B. 1983. Low-cost technology. pp. 263–294 in *Water Supply and Sanitation in Developing Countries*, Dangerfield, B.J. (editor), The Institute of Water Engineers and Scientists, London, England.

Pickford, J. 1977. Water treatment in developing countries. pp. 162–191 in Feachem, R., McGarry, M., and Mara, D. (editors), *Water, Wastes and Health in Hot Climates*, John Wiley and Sons, New York.

Poynter, S.F.B. and Slade, J.S. 1977. The removal of viruses by slow sand filtration. *Progress in Water Technology* 9:75–88.

Prasad, V.S. 1986. *Development of Filtration/Adsorption Media for Removal of Bacteria and Turbidity from Water*. Ph.D. dissertation. Indian Institute of Technology, Kanpur, India.

Reid, G.W. 1982. The problem, interface for decision and appropriate technology-science technology, technology transfer and utilization. pp. 1–25 in Reid, G.W. (editor), *Appropriate Methods for Treating Water and Wastewater in Developing Countries*, Ann Arbor Science, Ann Arbor, Michigan.

Schiller, E.J. and Droste, R.L. 1982. *Water Supply and Sanitation in Developing Countries*, Ann Arbor Science, Ann Arbor, Michigan.

Schulz, C.R. and Okun, D.A. 1984. *Surface Water Treatment for Communities in Developing Countries*, John Wiley & Sons, New York.

Sridhar, M.K.C., Kale, O.O., and Adeniyi, J.D. 1985. A simple sand filter to reduce guinea worm disease in Nigeria. *Waterlines* 4:16–19.

Sundaresan, B.B., Paramasivam, R., and Mhaisalkar, V.A. 1982. Technology choice for rural water supply. pp. 33–42 in *Workshop on Rural Drinking Water Supply*, Commonwealth Science Council, London, England.

Thurman, R.B. and Gerba, C.P. 1988. Characterization of aluminum metal on poliovirus. *Journal of Industrial Microbiology* 3:33–38.

Tobin, R. and Armstrong, V.C. 1980. *Assessing the Effectiveness of Small Filtration Systems for Point-of-use Disinfection of Drinking Water Supplies.* Publication No. 80-EHD-54, National Health and Welfare, Ottawa, Canada.

Tripathi, P.N., Chaudhuri, M., and Bokil, S.D. 1976. *Nirmali* seed—a naturally occurring coagulant. *Indian Journal of Environmental Health* 18:272–281.

U.S. Environmental Protection Agency. 1986. *Report of the Task Force on Guide Standard and Protocol for Testing Microbiological Water Purifiers.* EPA, Washington, D.C.

Venkobacher, C. and Jain, R.K. 1983. Studies on development and performance of mixed bed disinfector. *Water Supply* 1:193–204.

Wegelin, M. and Schertenleib, R. 1987. Horizontal-flow roughing filtration for turbidity reduction. *Waterlines* 5: 24–27.

Wood, W.E. 1983. Who will look after the village water supply? *Waterlines* 2:17–20.

World Health Organization. 1981. *Global Strategy for Health for All by the Year 2000.* WHO, Geneva, Switzerland.

World Health Organization. 1982. *Seventh General Programme of Work Covering the Period 1984–1989.* WHO, Geneva, Switzerland.

World Health Organization. 1986. *International Drinking Water Supply and Sanitation Decade Directory.* WHO, Geneva, Switzerland.

World Health Organization. 1987. *Technology for Water Supply and Sanitation in Developing Countries.* World Health Organization, Geneva, Technical Report Series No. 742.

9

Microbiology of Drinking Water Treatment: Reclaimed Wastewater

W. O. K. Grabow

Department of Medical Virology, University of Pretoria, Pretoria, South Africa

The water reclamation plant in Windhoek (South West Africa/Namibia) represents a milestone in water reuse technology (Figure 9.1). Prompted by water shortage, the plant was commissioned in 1969 as the first, and to date probably still the only, plant in the world which directly reclaims drinking water from wastewater for the supplementation of a city's drinking water supply (Stander and Clayton, 1977).

The 7 megaliter/day plant functions on the principle of a multiple-barrier treatment train. The design is based on that of the Stander Wastewater Reclamation Plant in Pretoria which was used exclusively for research purposes. During the course of time various process units have been modified or replaced to facilitate operation, refine performance with regard to the physical/chemical quality of the water, and reduce the cost of water reclamation. Long-term operation of the Windhoek plant, experimental runs on the Stander plant under various operational conditions for research purposes, and the progressive sequence of modifications, offered an ideal opportunity for research on the behavior of health-related microorganisms in advanced water treatment processes.

The primary objective of the microbiological research was to establish the safety of the reclaimed water. Equally important, however, was the development of methods for reliable routine quality surveillance by means of tests which can be carried out at high frequency and relatively low cost in laboratories with limited expertise and facilities. High frequency testing with results available in the shortest possible time was of major importance because the history of disease outbreaks associated with treated water supplies shows that if contamination of the water or plant failure is known early enough, most, if not all, of the infections can be prevented (Grabow, 1986a).

The unit processes of the water reclamation system which had the most

185

Fig. 9.1.   The reclamation plant in Windhoek: milestone in the history of drinking water treatment.

important impact on the microbiological quality of the water were the following: Conventional sewage purification terminated by maturation pond (in the case of the Windhoek plant) or activated sludge (in the case of the Stander plant) treatment—primary clarification by high pH lime or ferrichloride flocculation in combination with polyelectrolytes—primary chlorination—secondary polyelectrolyte clarification—sand filtration—disinfection by intermediary breakpoint chlorination (ozonation in some experimental runs on the Stander plant)—activated carbon filtration—final chlorination. Technical details of the Windhoek and Stander reclamation plants have been described elsewhere (Grabow et al., 1980; Grabow and Isaäcson, 1978).

9.1   Indicator and Pathogenic Organisms

Indicator organisms   Since the selection of practical and reliable indicator organisms was one of the prime objectives of the study, a clear understanding of the concept is essential. The most important features generally expected of indicator organisms are that they should be: 1) Present whenever pathogens are present; 2) present in the same or higher numbers than pathogens; 3) specific for fecal or sewage pollution; 4) at least as resistant as pathogens to water treatment and disinfection processes; and 5) preferably nonpathogenic and detectable by simple, rapid and inexpensive methods (Grabow, 1986a). For the purposes of routine quality surveillance of water reclamation processes, the counts of indicator organisms are not expected to correlate with those of pathogenic organisms through the whole or part of a treatment train. Of fundamental importance, however, is that the absence of the indicator in any given volume of water should give a reliable indication of the absence of pathogens in at least the same volume of water.

Viruses   The term "human viruses" is used here to refer to those human enteric viruses which produce a cytopathogenic effect in primary vervet monkey kidney cells. This comprises polioviruses (all three types), coxsackie A viruses (types 7, 9 and 16 of the 24 types), coxsackie B viruses (all six types), echoviruses (all 34 types), and reoviruses (all six types). As in most other studies on viruses in water, the term does not include the viruses responsible for the great majority of waterborne viral infections. Among these are the hepatitis A virus (HAV), and the wide variety of gastroenteritis viruses including rota, norwalk, adeno types 40 and 41, parvo, corona, astro, and some others which have not yet been classified. Known waterborne viruses such as adeno types 3 and 4, which cause conjunctivitis, as well as the epidemic non-A, non-B hepatitis virus, are also not detectable (Grabow, 1986b).

The limited range of viruses detectable by methods based on a cytopathogenic effect in monkey kidney cells implies that the value of conventional virological testing of water is limited to that of an indicator system. In addition, virus is a poor indicator system because: 1) the tests are in the order of ten to a hundred times more expensive than coliform tests (Grabow et al., 1980); 2) the tests require sophisticated laboratory facilities and highly skilled expertise; 3) results are available only after one to four weeks (Grabow and Nupen, 1981; IAWPRC Study Group on Water Virology, 1983); 4) the need to process large volumes of water for the recovery of small numbers of viruses is inconvenient and increases the risk of sample contamination (Grabow et al., 1984b); 5) the efficiency of adsorption/elution methods generally used for the recovery of viruses from water rarely exceeds 50% for some viruses and is much lower for others (Grabow et al., 1984c); 6) the numbers, behavior, and resistance of the detectable "indicator" viruses may differ extensively from those of infectious viruses commonly transmitted by water (Grabow, 1986b; Grabow et al., 1984a; IAWPRC Study Group on Water Virology, 1983). These shortcomings imply that conventional technology for the virological analysis of water is hardly suitable for any purpose other than research.

The virological tests used in this study had the following distinct advantages over those used in most other research on viruses in water: 1) The recovery of viruses by ultrafiltration using flat membranes had an average efficiency of 94% for treated water which is much higher than that of conventionally used adsorption/elution methods (Grabow et al., 1984c; Nupen et al., 1980); 2) the primary vervet kidney cells used for the detection of viruses were more susceptible than conventional cell lines such as BGM (Grabow and Nupen, 1981); 3) the detection of viruses by the 50% tissue culture infections dose (TCID50) method using roller tubes incubated for 21 days is more sensitive than conventional plaque assays incubated for about seven days and detects slow-growing viruses such as reovirus (Grabow and Nupen, 1981).

To date a total volume of more than 13,000 liters of reclaimed water collected after intermediary breakpoint chlorination and final chlorination has been analysed for viruses. This volume consisted of 10 liter samples collected at regular intervals, and a series of 100 liter samples. Viruses were not detected in any of these samples (Grabow, 1986a; Grabow et al., 1978a, 1980). This implies that the water was well within the most stringent virological quality limits recommended in the international literature for directly reclaimed drinking water (Grabow et al., 1980; 1984c: IAWPRC Study Group on Water Virology, 1983), and that the direct reclamation of virologically safe drinking water from wastewater is feasible.

The average count of viruses in maturation pond effluent was of the order of 0.08/10 liter, and in activated sludge effluent 3,000/10 liter (Table 9.1). These numbers were rapidly reduced in early treatment stages, and viruses were rarely detected after sand filtration. An important feature of viruses in a multiple-barrier treatment system is that, contrary to many other indicators, they cannot multiply in any of the process units.

Coliphages   These bacterial viruses meet more of the requirements of indicators for viruses than any of the other organisms used for this purpose because: 1) Some of them resemble human viruses in size, morphology, struc-

Table 9.1 Average counts of microorganisms at certain treatment stages in the Windhoek and Stander water reclamation plants

Treatment stage	Plate count per ml	Total coliforms per 100 ml	Fecal coliforms per 100 ml	Streptococci per 100 ml	Acid-fast bacteria per 100 ml	Pseudomonas aeruginosa per 100 ml	Coliphages per 100 ml	Human viruses per 10 l
Maturation pond effluent[a]	30,000	5,000	1,400	80	not done	not done	320	0.08
Activated sludge effluent[b]	44,000	600,000	140,000	7,400	500,000	90,000	70,000	3,000
Lime treatment	250	50	not done	200	not done	0	1	0
Sand filtration	6,000	200	1	2	200,000	260	2	0
Breakpoint chlorination	10	0	0	0	1	0	0	0
Carbon filtration	2,750	1,000	1	0	90	490	0	0
Final chlorination	20	0	0	0	4	0	0	0

[a] Raw water intake of Windhoek reclamation plant.
[b] Raw water intake of Stander reclamation plant.
For further details on counts of microorganisms after individual treatment stages see Grabow et al., 1978a, 1978b, 1980, 1984b.

ture, and composition (particularly male-specific phages such as MS2); 2) generally they are at least as resistant to water treatment and disinfection processes; 3) they usually outnumber human viruses in water by a factor of up to 1,000 or more; and 4) they are nonpathogenic and detectable by simple and economic methods which yield results within 8 to 18 h (Grabow, 1986a; Grabow et al., 1984b; IAWPRC Study Group on Water Virology, 1983).

In research on the optimum utilization of coliphages as indicators for reclaimed drinking water, a sensitive and direct plaque assay for the detection of coliphages in 100 ml samples was developed (Grabow and Coubrough, 1986). A comparison of a variety of *Escherichia coli* host strains revealed that generally the highest counts of coliphages were obtained by means of *E. coli* strain C (ATCC 13706) (Grabow et al., 1984b). Analysis of water samples collected after individual process units of the Windhoek reclamation plant, as well as samples from other water environments, showed that somatic coliphages outnumbered male-specific coliphages and were equally resistant to treatment processes (Grabow et al., 1980). This implies that plaque assays using *E. coli* C as host proved the method of choice for routine quality surveillance of water reclamation processes.

The value of coliphages as indicators of the absence of human viruses was illustrated by an analysis of a series of samples of raw and treated drinking water derived by conventional methods or the direct reclamation of wastewater (Grabow et al., 1984b). Of the 558 samples tested, 211 were positive for coliphages and negative for human viruses, while only two were positive for human viruses and negative for coliphages, even though only 10 ml volumes were tested for coliphages and 10 liter volumes for human viruses. In addition, there was good reason to believe that the human virus isolates were laboratory contaminants.

Maturation pond effluent contained in the order of 32 coliphages per 10 ml and activated sludge effluent 7,000/10 ml (Table 9.1). This exceeded the count of human viruses by a factor of more than 1,000. The ratio of coliphages to human viruses increased through the various treatment processes of reclamation plants, which counts in favor of the value of coliphages as indicators for the absence of human viruses. The increase in the ratio is due to the release of coliphages by coliform bacteria in certain process units such as sand and activated carbon filters, and probably also to higher resistance and less efficient removal of coliphages (Grabow et al., 1978b, 1980). Coliphages were generally detectable after most treatment processes except intermediary breakpoint chlorination. Using the highly sensitive direct plaque assay on 100 ml samples, they were occasionally detected in final reclaimed water. These coliphages must have been released by host organisms growing in the activated carbon filters.

Heterotrophic plate count   Determined by conventional pour plate methods (Grabow and Isaäcson, 1978), this bacterial count proved the most practical and sensitive indicator of the removal and inactivation of microorganisms in individual process units of reclamation systems, as well as the microbiological safety of final treated water (Grabow, 1986a; Grabow and Isaäcson, 1978; IAWPRC Study Group on Water Virology, 1983). Important features which contributed to the value of this indicator include: 1) Counts of the organisms in the raw water intake were much higher than those of any pathogens; 2) counts increased in certain process units such as sand and activated carbon filtration (Grabow et al., 1980) which facilitated assessment of the efficiency of subsequent treatment processes; 3) the count included organisms, particularly bacterial spores (Grabow et al., 1969, 1978b), which are more resistant to the treatment processes concerned than most pathogens including viruses (IAWPRC Study Group on Water Virology, 1983); 4) the count can be determined by simple and inexpensive methods which yield results within two days (Grabow and Isaäcson, 1978).

Some specifications for drinking water derived by conventional methods include recommended limits for heterotrophic plate counts of 100/ml or 500/ml. These limits are often disregarded because many water suppliers find it too difficult to consistently produce and deliver water which conforms to them (Grabow, 1981; IAWPRC Study Group on Water Virology, 1983). In the case of drinking water directly reclaimed from wastewater, however, the heterotrophic plate count is considered an essential determinant in routine quality assessment (Grabow, 1986a; IAWPRC Study Group on Water Virology, 1983).

The heterotrophic plate count of maturation pond effluent was in the order of 30,000/ml and that of activated sludge effluent 44,000/ml (Table 9.1). Despite increases in certain process units, the count was well within the 100/ml limit after intermediary breakpoint chlorination and final chlorination.

Total coliform bacteria   These organisms, which are used as indicators of the general sanitary quality of treated drinking water supplies, proved most useful for reclaimed water because of the following reasons: 1) When detected in advanced treatment stages, isolates can be purified and identified, and the presence of *Escherichia coli* renders almost conclusive evidence of fecal pollution (Grabow, 1978); 2) the count includes organisms such as *Klebsiella* and *Aerobacter* species which are generally more resistant to disinfection processes than most pathogens including viruses (Grabow, 1982; Grabow et al., 1969, 1978b, 1981); 3) the organisms multiply in certain process units, which facilitates assessment of the efficiency of subsequent units (Grabow et al., 1978a); and 4) counts can be determined by simple and inexpensive methods which yield results within 24 h (Grabow and DuPreez, 1979).

The term "coliform bacteria" represents a vaguely defined group of organisms which have a long history in water quality assessment. Over the years a confusing variety of definitions and methods of detection have been used (Grabow, 1981; Grabow and DuPreez, 1979). Research on the applications of the determinant in the routine quality surveillance of directly reclaimed drinking water has revealed the following important issues: 1) Cytochrome oxidase-positive organisms (mainly *Aeromonas* and *Pseudomonas* species) should not be excluded because these organisms include potential pathogens and indicate inadequate treatment; 2) membrane filtration techniques using agar media are more practical than membrane filtration techniques using saturated pads, or most probable number procedures; 3) special procedures such as enrichment steps were not found necessary because the slightly higher counts did not justify the additional labor, cost, and technical know-how required; 4) in a comparison of a variety of growth media and detection methods, membrane filtration using m-Endo LES agar and incubation at $35 \pm 0.5°C$ for 22 to 24 h yielded the highest counts and proved the method of choice (Grabow, 1981; Grabow and DuPreez, 1979). In the latter tests, membranes were sometimes overgrown by organisms such as *Pseudomonas* species and Gram-positive bacilli. This tended to impair the counting of coliforms, but disqualified the water in any case because the presence of these organisms reflected inadequate disinfection (Grabow and DuPreez, 1979).

Originally, coliform bacteria were considered as harmless indicator organisms, many of which are normal inhabitants of the human gastrointestinal tract. Gradually, however, their status changed and today they should be regarded as potential pathogens. The reason is that the incidence of members of the group, such as *E. coli*, which carry plasmids coding for transferable enteropathogenicity, enterotoxicity, and resistance to antimicrobial chemotherapeutic agents, has been increasing rapidly (Grabow et al., 1974). This is another important reason for including total coliforms in the assessment of the quality of directly reclaimed drinking water.

The total coliform count of maturation pond effluent was in the order of 5,000/100 ml and that of activated sludge effluent 600,000/100 ml (Table 9.1). These counts were reduced by lime treatment and clarification processes, but increased in sand and activated carbon filters. Breakpoint and final chlorination reduced counts to levels well within drinking water limits.

Fecal coliform bacteria   This group of bacteria, which is more closely associated with fecal pollution than total coliforms, did not prove of particular value in the assessment of the quality of reclaimed water. Except for the activated carbon filters, which occasionally had a few fecal coliforms in their effluents, the organisms failed to multiply in any of the process units. Counts were reduced rapidly in early stages of the treatment train, after which they were no longer detectable (Grabow et al., 1980). The organisms were not particularly resistant to any of the treatment processes.

The fecal coliform test had distinct advantages for the convenient and rapid detection of fecal pollution. In a survey of 312 samples of raw and various stages of treated water, the presence of *E. coli* among 1,139 isolates varied from 51% to 93% (Grabow et al., 1981). The remainder consisted primarily of *Klebsiella pneumoniae*, which ranged from 4% to 32%. The test is relatively simple and inexpensive. Membrane filtration using conventional M-FC agar without rosolic acid, and incubation in a water bath at $44.5 \pm 0.2°C$ for 22 to 24 h, proved superior to a variety of other test methods (Grabow et al., 1981). Colonies were more distinctive in the absence of rosolic acid, and modifications such as the addition of a resuscitation top layer or pre-incubation at lower temperature did not yield higher counts.

The fecal coliform count was primarily used for the assessment of raw water quality. Maturation pond effluent had counts in the order of 1,400/100 ml and activated sludge effluent 140,000/100 ml (Table 9.1). These counts were reduced to an average of 1/100 ml after sand filtration. Fecal coliforms occasionally detected after activated carbon filtration were probably *Klebsiella* species of nonhuman origin.

Fecal streptococci   The detection of these organisms by membrane filtration using M-Enterococcus Agar was relatively simple and inexpensive, and they are known to be highly specific for fecal pollution (Grabow, 1986a; Grabow and Isaäcson, 1978). The highly selective growth medium yields distinctive colonies which are easy to count, while contaminant organisms rarely interfere. The organisms also proved more resistant than coliform bacteria, and possibly even coliphages and viruses, to various treatment and disinfection processes (Grabow, 1982, 1986a; Grabow et al., 1969, 1978a). Despite these attractive features, fecal streptococci were not considered particularly useful for testing reclaimed water because they were completely removed in early treatment stages and failed to multiply in subsequent process units, and results were available only after 48 h (Grabow et al., 1980).

The fecal streptococci count of maturation pond effluent was in the order of 80/100 ml, and that of activated sludge effluent 7,400/100 ml (Table 9.1).

Pseudomonas aeruginosa   Outstanding indicator features of this bacterium were relatively high resistance to disinfection, and ability to multiply in process units such as sand and activated carbon filters which facilitated the monitoring of subsequent treatment processes. *Pseudomonas aeruginosa* was even more resistant to ozonation than acid-fast bacteria (Grabow et al., 1980). The absence of *P. aeruginosa* is not only important from an indicator point of view, but also because this species is an opportunistic pathogen often associated with waterborne transmission.

A most probable number method yielded high counts of *P. aeruginosa*, but the procedure was relative cumbersome, time-consuming, of uncertain sensitivity, and yielded results only after a number of days (Grabow, 1978). Activated sludge effluent had a *P. aeruginosa* count in the order of 90,000/100 ml. Counts were relatively high after sand and activated carbon filtration (Table 9.1).

Acid-fast bacteria   In the tests carried out in this study, this vaguely defined group of bacteria consisted primarily of rapid-growing acid-fast organisms such as *Mycobacterium fortuitum* and *M. phlei*. Counts were done by a membrane filtration procedure which included incubation on a modified Middlebrook 7H9 medium and incubation for nine days at 37°C (Grabow et al., 1980).

Shorter incubation yielded lower counts, while longer incubation resulted in higher counts as slow-growing strains developed colonies.

The most important indicator feature of acid-fast bacteria is their exceptional resistance to chlorination (Grabow, 1982; Grabow et al., 1980, 1983, 1984a), and other disinfection processes such as high-pH lime treatment (Grabow et al., 1969). There was no indication that pathogenic microorganisms, including viruses, can survive disinfection processes which inactivate acid-fast bacteria.

Another valuable indicator feature of acid-fast bacteria is their ability to multiply in sand and activated carbon filters, which facilitated the testing of subsequent disinfection processes. An analysis of isolates from sand filter effluents revealed that they consisted mainly of rapidly growing, scotochromogenic mycobacteria of the *Mycobacterium parafortuitum* complex (Tsukamura et al., 1983). Some of these isolates had not previously been described, and one cluster was named *Mycobacterium austroafricanum*.

The count of acid-fast bacteria in activated sludge effluent was in the order of 500,000/100 ml (Table 9.1). Throughout the reclamation treatment train, the counts of acid-fast bacteria were exceeded only by those of the heterotrophic plate count, and by total coliforms or *P. aeruginosa* after certain process units.

Clostridium perfringens   Although the spores of this organism are more resistant to disinfection processes than most pathogens (Grabow, 1986a; Grabow et al., 1969), they proved of little indicator value in reclamation plants because their numbers were generally reduced to low levels in treatment processes preceding disinfection units (Grabow and Isaäcson, 1978). Indicator advantages include the high specificity of *C. perfringens* for fecal pollution, and detection by relatively simple methods which yield results within 24 h (Grabow, 1986a).

Staphylococcus aureus   These bacteria were reduced to low levels in early treatment stages and proved of little indicator value in reclamation systems (Grabow, 1978; Grabow and Isaäcson, 1978).

Legionella pneumophila   Preliminary results of a recently initiated study indicate that these organisms multiply in activated carbon filters (unpublished observations).

Miscellaneous bacterial pathogens   A series of tests for salmonellae, shigellae, and *Yersinia enterocolitica* revealed that these organisms were present in low numbers in the raw water and that they were reduced to undetectable levels in early treatment stages (Grabow et al., 1980).

Candida albicans   This organism has attractive indicator features for certain purposes, but in reclamation treatment trains it proved of little value because its numbers were reduced to undetectable levels in early stages of treatment (Grabow et al., 1980). The *C. albicans* count in activated sludge effluent was about 100/100 ml.

Parasite ova   A series of tests in which 10 liter samples of water were analysed for the ova of human intestinal parasites by means of a filtration procedure showed that they were reduced to undetectable levels by sedimentation, clarification, and filtration processes (Grabow, 1978).

Endotoxins   Using a commercial *Limulus* amoebocyte lysate assay, the average concentration of endotoxins was found to decrease during various treatment stages, but to increase during activated carbon filtration from 72 to 133 endotoxin units/ml (Grabow and Burger, 1985). Final chlorination and mixing with a conventional supply reduced this level to 84 units/ml. Although there is no reason to believe that this concentration may constitute a meaningful health risk, it was higher than that of spring water (3 units/ml), well water (10 units/ml), and a drinking water supply derived by conventional treatment of surface water (15 units/ml).

9.2   Effect of Individual Treatment Processes

Conventional sewage treatment   The microbiological quality of maturation pond effluent (raw water intake of the Windhoek reclamation plant) was superior to that of activated sludge effluent by a wide margin (Table 9.1). In fact, the quality of the raw water intake of the Windhoek reclamation plant was well within the total and fecal coliform limits of 20,000/100 ml and 2,000/100 ml, respectively, which have been considered acceptable for the treatment of drinking water by conventional methods (Committee on Water Quality Criteria, 1972).

High pH lime treatment   This process reduced counts of microorganisms by 97% or more (Grabow et al., 1969, 1978a, 1978b). Acid-fast bacteria, spore-formers, fecal streptococci, and *S. aureus* were the most resistant. Counts of viruses were generally reduced to undetectable levels.

Primary ferrichloride-polyelectrolyte clarification   The reduction in counts of microorganisms ranged from 32% to 88% (Grabow et al., 1980).

Secondary chlorination in combination with pre-chlorination
Counts were reduced by 99% or more (Grabow et al., 1980).

Sand filtration   Heterotrophic plate counts and counts of coliforms (including fecal coliforms), *P. aeruginosa* and acid-fast bacteria increased to various extents, while those of fecal streptococci, *C. albicans*, coliphages, and viruses decreased (Grabow et al., 1980). In one series of tests, acid-fast bacteria grew

the most vigorously and their counts increased from an average of 10/100 ml
to 218,000/100 ml.

Intermediary breakpoint chlorination   The operational conditions for
breakpoint chlorination were specified as: 1–2 mg/liter free chlorine residual
for 1–2 h at a pH of less than 8 and turbidity of less than 1 NTU (Grabow et
al., 1978a). This disinfection procedure is generally accepted as more than
adequate for the final polishing of drinking water supplies, and no pathogens
are known to survive this treatment (Grabow, 1982; IAWPRC Study Group on
Water Virology, 1983). The efficiency of the disinfection process is confirmed
by the absence of viruses and the consistent conformation to drinking water
quality standards (Grabow, 1986a).

In laboratory experiments, the hepatitis A virus (HAV) proved more sensitive to free chlorine residuals at pH 6 and 8 than *M. fortuitum*, coliphage
MS2, coliphage V1 (representative of coliphages common in sewage-polluted
water), reovirus, poliovirus, *S. faecalis*, and *E. coli* (Grabow et al., 1983). At pH
10, however, HAV was more resistant to free chlorine residuals than coliphage
MS2, reovirus, and *E. coli*. HAV was more resistant than the SA-11 rotavirus
to free chlorine residuals at all pH levels. In the case of combined chlorine
residuals, on the other hand, HAV was more resistant than *E. coli*, *S. faecalis*,
poliovirus, reovirus, and coxsackie B5 virus, while it was less resistant than
only SA-11 rotavirus, MS2 coliphage, and *M. fortuitum* (Grabow et al., 1984a).
These survival data indicate that viruses and coliphages tend to be relatively
more resistant to combined chlorine residuals than bacteria, while bacteria tend
to be relatively more resistant to free chlorine residuals than viruses and coliphages. This implies that bacteria such as coliforms are reliable indicators of
the inactivation of viruses by free chlorine residuals as applied in the reclamation plants, but not of the disinfection of water by combined chlorine residuals as in the case of the chlorination of wastewater or the chloramination
of drinking water supplies (Grabow et al., 1984a).

The hepatitis B virus does not constitute a risk in reclaimed water. The
hepatitis B surface antigen (HBs Ag) was not detected in the feces and urine
of patients with HBs Ag antigenemia. Laboratory experiments revealed that
this was due to disintegration of spherical and tubular HBs Ag particles, as
well as the coats and cores of Dane particles, by unidentified *P. aeruginosa*
antagonists and enzymes including carboxypeptidase A and two types of sub-
tilisin enzymes (Grabow et al., 1975). These findings explain the failure of
detecting HBs Ag in sewage by means of a sensitive and specific affinity
chromatography method (Grabow and Prozesky, 1973), and the rare association of infections with transfer other than parenteral (Grabow et al., 1975).

Activated carbon filtration   The heterotrophic plate count and counts of
total coliforms, *P. aeruginosa*, and acid-fast bacteria increased considerably
(Grabow et al., 1980). Counts of fecal coliforms and coliphages increased to
a limited extent. The behavior of other determinants could not be investigated

because they were rarely detected in either the influent or effluent of the filters. However, other research has shown that activated carbon filters reduce counts of viruses by up to 86% (Grabow et al., 1978a).

The prolific growth of certain organisms in activated carbon filters has quality implications which require careful operation and post-disinfection. Some of these organisms, such as *P. aeruginosa* and *Legionella* species, have meaningful health implications. Others may be aesthetically unacceptable, contribute to growth in distribution systems, and eventually play a role in taste and odor problems. The relatively high level of endotoxins in the filtered water is a reflection of the heavy growth of Gram-negative bacteria in the filters. Careful management of activated carbon filters is also important because breakthroughs in saturated filters may release loads of microorganisms beyond the disinfection capability of post-chlorination systems (see also Chapter 6).

Final chlorination   This final disinfection step was operated so as to consistently produce water of a quality well within generally accepted microbiological drinking water specifications (Grabow et al., 1978a, 1980; IAWPRC Study Group on Water Virology, 1983).

Overall reclamation plant performance   The combined effect of the various process units in the multiple barrier reclamation systems conformed to the most stringent requirements for the microbiological safety of directly reclaimed drinking water, including the capability of reducing viral counts by 12 log units (Grabow et al., 1980). These results show that the technology to directly reclaim microbiologically safe drinking water from wastewater is available. Similar findings have been reported for advanced water treatment systems in other parts of the world (Grabow et al., 1978a, 1980; IAWPRC Study Group on Water Virology, 1983).

9.3 Epidemiology of Infectious Diseases in Consumers of Directly Reclaimed Drinking Water

Relatively speaking, Windhoek offered an ideal opportunity for epidemiological research on the consumers of directly reclaimed drinking water. It is a fairly isolated community with limited resident fluctuation. The drinking water supply of only certain parts of the city was supplemented with reclaimed water for intermittent periods, which made it possible to eliminate mutual sources of infection other than the water supply. Results of initial surveys were reported by Grabow and Isaäcson (1978), and those of an intensive study over the period 1974 to 1983 by Isaäcson et al. (1986). During the ten years of the latter study the population of Windhoek increased from approximately 75,000 to 100,000.

The study embraced both potential short term (mainly infectious) and long term (mainly noninfectious chronic) effects. Surveillance of short term effects included potentially waterborne infectious diseases with particular reference to diarrhoea and viral hepatitis. More than 15,000 cases of diarrhoeal diseases and about 1,000 cases of viral hepatitis were investigated and statistically analyzed with regard to incidence among consumers of directly reclaimed drinking water and control groups.

The results of the study indicate that the consumption of directly reclaimed drinking water did not cause any epidemiologically detectable adverse health effects.

9.4   Quality Criteria and Indicator Systems

The history of waterborne diseases shows that infections associated with treated water supplies could never be ascribed to shortcomings in technological know-how regarding the treatment of water or the assessment of the quality of water. Waterborne infections have always been due to the absence of treatment, the choice of treatment, the operation of treatment processes, plant failure, secondary contamination of treated water supplies, or human negligence (Grabow, 1986b; IAWPRC Study Group on Water Virology, 1983). This implies that the challenge in supplying safe water resides in the selection of appropriate treatment systems, appropriate operation, and appropriate water quality surveillance. Economic considerations do, of course, play a major role in this selection.

Based on certain interpretations of data on minimal infectious doses, the efficiency of water treatment processes, the theoretical possibility of undetectable low level transmission of diseases, and epidemiological findings, the absence of viruses from 1,000 US gallons (3,785 liters) of drinking water has initially been recommended. As more accurate details on the causes and epidemiology of waterborne diseases became available, these proposals have been revised, and the latest recommendations of the World Health Organization and the European Economic Community specify the absence of viruses from 10 liter samples of drinking water (Grabow, 1986b).

The wide variety of microbiological determinants used in the study on water reclamation played a valuable and essential role in the assessment of the quality of the reclaimed water and in research on the mechanisms of action and the efficiency of individual process units. However, all these determinants are not necessary for the routine quality surveillance of an established multiple-barrier treatment train of which the efficiency and capabilities of individual process units are known. Determinants and quality specifications used for routine monitoring of the Windhoek reclamation plant have, therefore, been revised as more details became available.

At one stage the following quality criteria were recommended for directly reclaimed drinking water (Grabow et al., 1978a, 1980): Absence of viruses from 10 liters, a heterotrophic plate count of less than 100/ml, and absence of total coliforms, acid-fast bacteria, *P. aeruginosa*, and coliphages from 100 ml samples after at least two process units in the treatment train. In view of additional information and experience presently available, some of these criteria can be eliminated with confidence. The safety of the water with regard to viruses and other pathogenic microorganisms can reliably be monitored without the tests for viruses, acid-fast bacteria, and *P. aeruginosa* (Grabow, 1986a; Grabow et al., 1980; IAWPRC Study Group on Water Virology, 1983). Experience has also shown that despite satisfactory plant operation, some indicators are occasionally detected in treated water. This may be due to accidental sample contamination, experimental error, or the multiplication of organisms in certain process units (Grabow et al., 1984b).

In view of the above considerations, the following guidelines are recommended for the routine quality surveillance of an established multiple barrier treatment system of proven efficiency such as the Windhoek reclamation plant:

1. In 95% of samples collected daily after at least two treatment stages, the heterotrophic plate count should not exceed 100/ml, and total coliforms and coliphages should be absent from 100 ml samples.

2. At least one disinfection process should consistently conform to the following or equivalent specification: A free chlorine residual of 1–2 mg/liter for 1–2 h at a pH of less than 8.0 and turbidity of less than 1.0 NTU.

These guidelines are in agreement with internationally accepted recommendations (Grabow, 1986b; Grabow et al., 1984c), and require a quality of water which is considerably better than that of most conventional drinking water supplies (Grabow, 1978, 1986a; Grabow et al., 1984b). For instance, a survey carried out in the United States revealed that 36% of 2,600 water treatment plants supplied water which did not conform to conventional coliform limits, while some 56% of the plants experienced failure, and 33% bypassed water without treatment for periods of up to 300 days (Grabow and Isaäcson, 1978).

The results of a survey of typical drinking water supplies in Windhoek and Pretoria (Grabow, 1986a) illustrate that the heterotrophic plate count limit is most frequently exceeded and the virus limit least frequently (Table 9.2). The bacterial and coliphage tests have the advantage that, if for any reason necessary, their sensitivity can easily be increased by testing larger volumes of water, and they will still remain practical tests (IAWPRC Study Group on Water Virology, 1983).

9.5 Future Research Requirements

Although technology and expertise for the direct reclamation of drinking water from wastewater, and assessment of the microbiological safety of the reclaimed water, are well established, various aspects could benefit from further refine-

Table 9.2 Samples of drinking water in Windhoek and Pretoria which exceeded quality limits

		Samples which exceeded limits	
Determinant	Limit	Number/number tested	%
Heterotrophic plate count	100/ml	48/582	8.2
Total coliforms	0/100 ml	17/582	2.9
Fecal coliforms	0/100 ml	5/582	0.9
Coliphages	0/10 ml	3/188	1.6
Human viruses	0/10 liter	1/242	0.4

ment (Grabow, 1986b). These include the time required for obtaining results. Ideally, the entire process should be developed into a rapid-response automatic alarm system. Progress in research along these lines is currently being made in various laboratories. Presently developed nucleic acid hybridization techniques for the detection of virtually any viruses should be applied in research on the incidence and behavior in reclamation systems of viruses not previously studied. Indications are that the techniques could eventually be developed into simple and inexpensive test kits suitable for routine monitoring (Grabow, 1986b; see also Chapter 18). Accumulating evidence regarding the proliferation of health-related microorganisms in activated carbon filters would seem to require further attention, particularly with regard to the operation of the process and the post disinfection of the water (see also Chapter 5).

More details on the role of water in the transmission of disease are of major importance. This information is required for the better understanding of health risks, and the clear definition of acceptable health risks and water quality standards (Grabow, 1986b; IAWPRC Study Group on Water Virology, 1983). Progress along these lines would have far reaching implications on reclamation technology, the quality assessment of reclaimed water, and the cost of water reuse.

9.6 Conclusions

Comprehensive analysis of the microbiological quality of water treated by two multiple-barrier reclamation systems demonstrated that the direct reclamation of safe drinking water from wastewater is technologically feasible and economically competitive under certain circumstances. Analysis of the efficiency of individual unit processes illustrated the safety margin and precautions against breakdown or failure incorporated in the systems concerned.

The safety of the directly reclaimed drinking water was confirmed by the failure to detect adverse health effects on consumers in detailed epidemiological studies over more than ten years.

Research on the incidence and behavior of a wide variety of health-related microorganisms in the raw water and individual treatment processes revealed that routine quality surveillance of the final water as well as the efficiency of individual units is possible by means of microbiological indicator tests which are relatively simple, inexpensive, and fast. These features are important because it is considered essential to carry out routine quality surveillance of directly reclaimed drinking water at the highest possible frequency.

Evidence has been presented that in the case of established multiple-barrier treatment systems in which the efficiency of individual processes is known a combination of indicators consisting of the heterotrophic plate count, total coliforms, and coliphages is sufficient for routine quality surveillance. In a system of this kind, the final water and the water at one intermediary stage should have a heterotrophic plate count of less than 100/ml, and total coliform and coliphage counts of 0/100 ml, using specified methods of detection.

Additional microbiological safety is incorporated by the specification that the treatment train should have at least one automatically controlled disinfection stage with a free chlorine residual of 1–2 mg/liter for 1–2 h at a pH of less than 8 and turbidity of less than 1 NTU. In addition, final disinfection should yield a product which conforms to microbiological drinking water specifications. Disinfection of similar or superior efficiency and reliability would, of course, also be acceptable.

Various areas of improvement and further research have been identified. Among these are the microbiological routine quality surveillance procedures which could be even faster and more convenient, ideally in the form of continuous, automatic alarm systems. The results revealed reasons for further research on the operation and control of activated carbon filters because they tend to serve as breeding ground for a variety of organisms which may have health and other quality implications.

References

Committee on Water Quality Criteria. 1972. *Water Quality Criteria*, p. 58, US Government Printing Office, Washington, D.C. 20402.

Grabow, W.O.K. 1978. South African experience on indicator bacteria, *Pseudomonas aeruginosa*, and R+ coliforms in water quality control. pp. 168–181 in Hoadley, A.W. and Dutka, B.J. (editors), *Bacterial Indicators/Health Hazards Associated With Water*, ASTM STP 635, American Society for Testing and Materials, Philadelphia.

Grabow, W.O.K. 1981. Membrane filtration: a South African view. pp. 355–392 in Dutka, B.J. (editor), *Membrane Filtration: Applications, Techniques, and Problems*, Marcel Dekker, Inc., New York.

Grabow, W.O.K. 1982. Disinfection by halogens. pp. 216–260 in Butler, M., Medlen, A.R. and Morris, R. (editors), *Viruses and Disinfection of Water and Wastewater*, Print Unit, University of Surrey, Guildford.

Grabow, W.O.K. 1986a. Indicator systems for assessment of the virological safety of treated drinking water. *Water Science and Technology* 18:159–165.

Grabow, W.O.K. 1986b. Recent developments in water virology and future research needs. pp. 679–683 in Megusar, F. and Gantar, M. (editors), *Perspectives in Microbial Ecology*, Slovene Society for Microbiology, Ljubljana.

Grabow, W.O.K., Bateman, B.W., and Burger, J.S. 1978a. Microbiological quality indicators for routine monitoring of wastewater reclamation systems. *Progress in Water Technology* 10:317–327.

Grabow, W.O.K. and Burger, J.S. 1985. Endotoxins in drinking water supplies. *Annual Congress of the South African Societies of Pathologists and Microbiologists, Johannesburg, South Africa, 1–3 July*.

Grabow, W.O.K., Burger, J.S., and Nupen, E.M. 1980. Evaluation of acid-fast bacteria, *Candida albicans*, enteric viruses and conventional indicators for monitoring wastewater reclamation systems. *Progress in Water Technology* 12:803–817.

Grabow, W.O.K. and Coubrough, P. 1986. Practical direct plaque assay for coliphages in 100-ml samples of drinking water. *Applied and Environmental Microbiology* 52: 430–433.

Grabow, W.O.K., Coubrough, P., Hilner, C., and Bateman, B.W. 1984a. Inactivation of hepatitis A virus, other enteric viruses and indicator organisms in water by chlorination. *Water Science and Technology* 17:657–664.

Grabow, W.O.K., Coubrough, P., Nupen, E.M., and Bateman, B.W. 1984b. Evaluation of coliphages as indicators of the virological quality of sewage-polluted water. *Water S.A.* 10:7–14.

Grabow, W.O.K. and DuPreez, M. 1979. Comparison of m-Endo LES, MacConkey, and Teepol media for membrane filtration counting of total coliform bacteria in water. *Applied and Environmental Microbiology* 38:351–358.

Grabow, W.O.K., Gauss-Müller, V., Prozesky, O.W., and Deinhardt, F. 1983. Inactivation of hepatitis A virus and indicator organisms in water by free chlorine residuals. *Applied and Environmental Microbiology* 46:619–624.

Grabow, W.O.K., Grabow, N.A., and Burger, J.S. 1969. The bactericidal effect of lime flocculation/flotation as a primary unit process in a multiple system for the advanced purification of sewage works effluent. *Water Research* 3:943–953.

Grabow, W.O.K., Hilner, C.A., and Coubrough, P. 1981. Evaluation of standard and modified M-FC, MacConkey, and Teepol media for membrane filtration counting of fecal coliforms in water. *Applied and Environmental Microbiology* 42:192–199.

Grabow, W.O.K. and Isaäcson, M. 1978. Microbiological quality and epidemiological aspects of reclaimed water. *Progress in Water Technology* 10:329–335.

Grabow, W.O.K., Middendorff, I.G., and Basson, N.C. 1978b. Role of lime treatment in the removal of bacteria, enteric viruses, and coliphages in a wastewater reclamation plant. *Applied and Environmental Microbiology* 35:663–669.

Grabow, W.O.K. and Nupen, E.M. 1981. Comparison of primary kidney cells with the BGM cell line for the enumeration of enteric viruses in water by means of a tube dilution technique. pp. 253–256 in Goddard, M. and Butler, M. (editors), *Viruses and Wastewater Treatment*, Pergamon Press, Oxford.

Grabow, W.O.K., Nupen, E.M., and Bateman, B.W. 1984c. South African research on enteric viruses in drinking water. *Monographs in Virology* 15:146–155.

Grabow, W.O.K. and Prozesky, O.W. 1973. Isolation and purification of hepatitis-associated antigen by affinity chromatography with baboon antiserum. *Journal of Infectious Diseases* 127:183–186.

Grabow, W.O.K., Prozesky, O.W., Appelbaum, P.C., and Lecatsas, G. 1975. Absence of hepatitis B antigens from feces and sewage as a result of enzymatic destruction. *Journal of Infectious Diseases* 131:658–664.

Grabow, W.O.K., Prozesky, O.W., and Smith, L.S. 1974. Drug resistant coliforms call for review of water quality standards. *Water Research* 8:1–9.

IAWPRC Study Group on Water Virology, 1983. The health significance of viruses in water. *Water Research* 17:121–132.

Isaäcson, M., Sayed, A.R., and Hattingh, W.H.J. 1986. *Studies on Health Aspects of Water*

Reclamation during 1974 to 1983 in Windhoek, South West Africa/Namibia. WRC-Report No. 38/1/86. Water Research Commission, Pretoria. 82 pp.

Nupen, E.M., Basson, N.C., and Grabow, W.O.K. 1980. Efficiency of ultrafiltration for the isolation of enteric viruses and coliphages from large volumes of water in studies on wastewater reclamation. *Progress in Water Technology* 12:851–863.

Stander, G.J. and Clayton, A.J. 1977. Planning and construction of wastewater reclamation schemes as an integral part of water supply. pp. 383–391 in Feachem, R., McGarry, M., and Mara, D. (editors), *Water, Wastes and Health in Hot Climates*, Wiley, London.

Tsukamura, M., Van der Meulen, H.J., and Grabow, W.O.K. 1983. Numerical taxonomy of rapidly growing scotochromogenic mycobacteria of the *Mycobacterium parafortuitum* complex: *Mycobacterium austroafricanum* sp. nov. and *Mycobacterium diernhoferi* sp. nov., nom. rev. *International Journal of Systematic Bacteriology* 33:460–469.

Part 3

Microbiology of Drinking Water Distribution

10

Bacterial Distribution and Sampling Strategies for Drinking Water Networks

A. Maul,[1] A.H. El-Shaarawi,[2] and J.C. Block[1]

[1]Centre des Sciences de l'Environnement, Metz, France and
[2]National Water Research Centre, Burlington, Ontario, Canada

A major objective of public health microbiology is to ensure bacteriologically safe and reliable drinking water. Sampling programs designed to monitor the bacterial density in water distribution systems usually involve the collection of a number of water samples at various locations and over an extended period of time. The results of the bacterial counts provided by the sampling program are then compared with bacterial water quality guidelines in order to determine the potability of the water analyzed.

Most bacterial water quality guidelines or regulations thus specify: 1) a maximum proportion of samples exceeding a particular value of indicator bacteria (e.g., coliforms); and/or 2) a threshold value of the mean sample count averaged over a specified period (Health and Welfare Canada, 1978; U.S. environmental Protection Agency, 1976). However, the efficiency of a sampling program for monitoring the microbiological quality of the water largely depends on both the dispersion patterns of the bacterial populations in the system sampled and the characteristics of the sampling design itself.

This chapter is concerned with discussing some basic bacteriological and statistical issues involved in the design of sampling strategies for monitoring the health of a drinking water network both spatially and temporally. Specifically, the issues considered include: 1) the origin and nature of bacteria found in the network; 2) the estimation of the frequency of sampling both spatially and temporally; and 3) the sequential and nonsequential approaches to monitoring. All these issues are examined for systems of varying degrees of homogeneity, starting from the very simple and small system to the very large and complex one.

10.1   The Origin of Bacteria in the Network

It is well established that the distribution of bacteria within drinking water networks exhibit very heterogeneous and complicated patterns (Christian and Pipes, 1983; El-Shaarawi et al., 1985; and Maul et al., 1985a,b). Certain patterns can be predicted and explained in terms of the physical characteristics of the network and the chemical and biological properties of the water supply. The unexplained patterns are usually attributed to randomness and their characteristics are summarized by using probabilistic models such as the Poisson and the negative binomial distributions. It is helpful when modelling the variations of bacteria in the network to discriminate between macro and micro scales of variabilities as well as physiological heterogeneity of the bacteriological population. Figure 10.1 illustrates the macro scale of variability, with dramatic changes in the bacterial densities among the network zones (Zone 1, 10 colony-forming units (CFU)/ml; Zone 2, 100 CFU/ml; Zone 3, 1000 CFU/ml; and Zone 4, 10,000 CFU/ml). On the other hand, the microspatial heterogeneity is caused by the patchiness in the distribution of bacteria (approximately 50% of the bacteria in the water sample may be found as 5 μm clumps) in the flowing water and in the irregular adsorption/accumulation/desorption of bac-

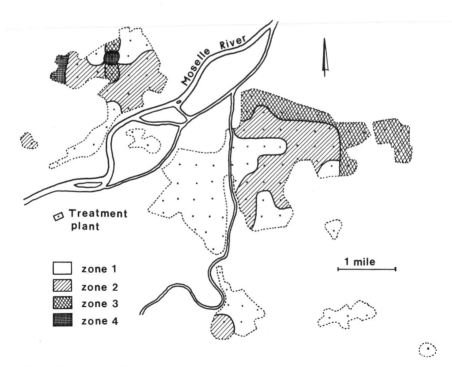

Figure 10.1   Zones of bacterial density for the Metz water distribution system. Numbers identifying the zones correspond to increasing levels of the bacterial density: (Zone 1, 10 CFU/mL; Zone 2, 100 CFU/mL; Zone 3, 1000 CFU/mL, Zone 4, 10,000 CFU/mL).

teria onto/from the pipe walls (approximately 10^7 cells/cm^2). Finally, the physiological heterogeneity of the bacterial population is caused by a complex mixture of dead, dormant, injured, and healthy bacteria. As a result, it is common to find in the finished water samples: 1) total bacterial counts (epifluorescence technique) around 10^5/ml cells; 2) actively respiring bacterial count (iodo-nitro triphenyl tetrazolium or INT technique) around 10^3 cells/ml; 3) plate count (15 days of incubation at 20 °C) around 10^2 CFU/ml; and 4) plate count (72 hours of incubation at 20 °C) around 1 CFU/ml.

Furthermore, the microorganisms of the drinking water network can be classified into indigenous (autochthonous) and exogenous (allochthonous) populations. The indigenous microorganisms are well adapted to their environment and represent a stable ecosystem. They are hard to eradicate. Examples of these microorganisms include *Pseudomonas*, *Flavobacterium*, and protozoa. On the other hand, exogenous microorganisms such as fecal indicators and algal cells are transported by the flowing of water in the network. Although the algae are of course unable to grow in the pipes, some of the fecal indicator (intestinal bacteria) can grow under certain conditions.

In addition, there are two major origins for the bacteria found in the water samples from any sampling point in the network. The first type is introduced directly into the network from the water treatment plant or from the source, since all water sources contain bacteria. The second origin is from bacteria growing on the pipe walls which are later released into the flowing water by shear loss or by erosion. The ratio between the densities of the two types (bacteria introduced by the water route/released bacteria from the walls) decreases from the water treatment plant to the most external zones of the network.

It is not surprising then to observe the pattern shown in Figure 10.1 for the distribution of bacteria in a water network. Hence, a well-designed sampling program will allow not only the optimal determination of the number and the location of samples to be taken, but it will also help to discriminate between the parameters governing bacterial proliferation in drinking water distribution systems.

10.2   Statistical Aspects of Drinking Water Quality Monitoring

Prior to the preparation of a sampling program for monitoring the quality of drinking water, answers to the following three questions are usually required: 1) how many samples should be taken? 2) where should the samples be collected? and 3) when should the samples be taken? For any sampling design of this nature, two erroneous decisions are possible. The first is declaring that the quality of the water in the system is violating the regulation when it is not (i.e., producers' risk, α) and the second is declaring compliance with the regulation when, in fact, this is not true (i.e., consumers' risk, β). Therefore, the basic objective of the microbiological monitoring of drinking water should consist of controlling the risk β at a specific level. Suppose that the regulation specifies that the quality of water in the network is acceptable when the unknown actual mean bacterial density, λ, expressed as the number of bacteria contained in one unit-volume of water in the distribution system sampled is less than μ. This can be problematic since there is always the risk of arriving at the wrong conclusion because only a small portion of the water involved is analyzed for bacteria. The situation which leads to making each of the two kinds of erroneous decisions, assuming different levels for the mean bacterial density of the sample analyzed, $\bar{r}$, is shown in Table 10.1

The probability that bacterial water quality guidelines are violated (Table 10.1, column headed by $\bar{r} \geq \mu$) during monitoring programs depends on:

1. The true mean bacterial density of the water (λ).

2. The number of samples collected.

3. The variability of the distribution of bacteria (Esterby, 1982; Pipes and Christian, 1982).

This shows that relating the number of samples to population density as in the U.S. Water Quality Regulations (U.S. Environmental Protection Agency, 1976) is not always advisable. As noted by Muenz (1978), an increase in the number of samples with the size of the population served is not desirable on statistical grounds, since the precision of estimating the condition of water quality depends very little on the volume of water in the network (El-Shaarawi et al., 1985). The most important parameter in setting a sampling program to monitor drinking water quality is the dispersion patterns of the bacteria in the water and not the size of the population. Clearly the size of the population might be correlated with the degree of bacterial heterogeneity in the water system and in this case, it represents an indirect factor in setting the sampling design.

Table 10.1 Characterization of the risk of sampling, assuming different levels for λ and $\bar{r}$ as compared with the standard μ

	Mean bacterial density of the sample analyzed ($\bar{r}$)	
Actual mean bacterial density of the network (λ)	$\bar{r} < \mu$ The regulation is met	$\bar{r} \geq \mu$ The regulation is violated
$\lambda < \mu$	Right decision	Wrong decision (Producers' risk α)
$\lambda \geq \mu$	Wrong decision (Consumers' risk β)	Right decision

10.3   Characterization of the Bacterial Dispersion Pattern

A common approach for studying the dispersion of bacteria in water is to describe the data by means of a probability distribution in order to characterize the bacterial dispersion pattern. This can be achieved, for example, by analyzing the historical information available for the system and, in particular, summarizing this information in the form of a frequency distribution which could then be fitted to an appropriate probability model.

Models for the Dispersion of Bacteria in Water   The variation of the number of bacteria in equal sample volumes from a completely homogeneous material follows the Poisson distribution and the probability of finding r organisms in a sample is given by:

$$P(r) = e^{-\lambda} \lambda^r / r! \tag{1}$$

where λ is the mean of the distribution. The value of λ completely specifies the probability of observing a particular value of r. Also, the mean and the variance of the distribution are equal to λ. The goodness of fit of the Poisson distribution to a data set can be tested using Fisher's index of dispersion (Fisher et al., 1922).

When the Poisson model is not adequate (i.e., the assumption of homogeneity is inappropriate), the natural model to consider is the negative binomial distribution (Fisher, 1941), which is given by the formula:

$$P(r) = \frac{(k + r - 1)!}{r! \, (k - 1)!} \frac{p^r}{(1 + p)^{k+r}} \tag{2}$$

This distribution is specified by the parameters p and k. The mean and variance of the distribution are $\lambda = pk$ and $p(1 + p) k = \lambda + \lambda^2/k$, respectively. The parameter k is inversely related to the variation of λ, and it can be shown that when $k \to \infty$ and $p \to 0$ such that $pk = \lambda$ is constant, the limiting form of the negative binomial is the Poisson distribution (El-Shaarawi et al., 1981). Further, if the number of bacteria in the samples is assumed to follow a negative binomial distribution, the parameter k is independent of the volume of the samples analyzed. Therefore, the value of the parameter k of the negative binomial distribution, which can range between 0 and plus infinity, can be considered a measure of bacterial heterogeneity in the water system: as k approaches infinity the negative binomial distribution simplifies to the Poisson and, inversely, the smaller the value of k the greater the bacterial heterogeneity of the system. The negative binomial distribution has been shown to give a good fit to many actual bacterial frequency distributions and is widely used to represent bacterial counts from water samples collected over time and/or space (Christian and Pipes, 1983; El-Shaarawi et al., 1981; El-Shaarawi et al., 1985; Maul et al., 1985a,b; Muenz, 1978; Pipes et al., 1977; Thomas, 1955). The parameters of the probability distribution are then used to summarize the information in the data and to assist in designing future data collections for monitoring the bacteriological water quality (El-Shaarawi et al., 1985; Esterby, 1982; Maul et al., 1985b; Pipes and Christian, 1982).

Parameter Estimation and Goodness-of-fit Test   These two topics, fitting a probability distribution and estimating parameters, are not separable but their order of application to a set of data is logically determined. The adequacy of the negative binomial distribution to fit the data can be examined by calculating the chi-square goodness-of-fit statistic (Snedecor and Cochran, 1967).

The procedure starts by estimating the parameters of the distribution. An estimate $\hat{k}$ for k can be obtained by using the method of maximum likelihood; thus $\hat{k}$ satisfies the equation:

$$n \ln \left(1 + \frac{\bar{r}}{k}\right) - \sum_{i=1}^{n} \sum_{j=1}^{r_i} \frac{1}{(k + r_i - j)} = 0 \qquad (3)$$

where $\bar{r}$ is the arithmetic mean of $r_1, \ldots, r_n$ which are the bacterial counts from n independent samples. Equation 3 can be solved numerically on a computer using the Newton-Raphson method. The maximum likelihood estimate for p is $\hat{p} = \bar{r}/\hat{k}$. It should be pointed out that, in this case, $\bar{r}$ is also the maximum likelihood estimate for the mean of the distribution $\lambda = pk$.

The next step consists of grouping the n observations into g classes, with the restriction that the classes are formed such that the expected frequency of each class is not too small. The chi-square goodness-of-fit statistic is based on the comparison of both the observed and the expected frequencies corresponding to each of the g classes. Thus, the quantity:

$$\chi^2 = \sum_{j=1}^{g} \frac{(f_j - e_j)^2}{e_j}$$

where f_j and e_j represent the observed and expected frequencies in class j respectively, has an approximate chi-square distribution with $g-3$ degrees of freedom. A large and significant value of χ^2 suggests that the negative binomial distribution is not an adequate model for representing the dispersion of bacteria in the water distribution system considered. Such an outcome suggests further investigations and computations, as outlined in Section 10.5.

Hence a water distribution system will be called a simple heterogeneous system if the negative binomial distribution provides an adequate fit to the bacterial counts from samples collected under given sampling conditions, whereas a system showing the unsuitability of the negative binomial for representing the data will be called a compound heterogeneous system.

Assessment of the Bacterial Variability in a Simple Heterogeneous System   The degree of bacterial heterogeneity, which plays an essential role in the design of the strategy for sampling, can be assessed (in terms of parameter k) by analyzing the historical bacteriological information available for any water distribution system modelled by the negative binomial distribution. When the available data consist of bacterial counts from water samples collected during *l* bacteriological surveys which were performed on the water system considered, then it is of interest to determine a common value (k_c) of the *l* sets of data. This common value may next be considered a measure of bacterial heterogeneity, assuming the constancy of the value of the parameter k in the system.

Thus, if each of these *l* sets of data is represented by a negative binomial distribution with parameters p_i and k_i ($i = 1,2, \ldots ,l$), the maximum likelihood estimate, $\hat{k}_c$, for k_c satisfies the equation (under the assumption of the equality of the k_i's)

$$\sum_{i=1}^{l} m_i \ln \left(1 + \frac{\bar{r}_i}{k_c} \right) - \sum_{i=1}^{l} \sum_{j=1}^{m_i} \sum_{t=1}^{r_{ij}} \left(\frac{1}{k_c + r_{ij} - t} \right) = 0 \qquad (4)$$

where $\bar{r}_i$ is the arithmetic mean of $r_{i1}, r_{i2}, \ldots ,r_{im_i}$; which are the bacterial counts from the m_i samples of the ith set of data. This equation, which is an extension of equation 3 to *l* sets of data, can likewise be solved numerically on a computer by the Newton-Raphson method.

Note that if the *l* values $\hat{k}_i$ ($i = 1,2, \ldots ,l$) corresponding to the *l* surveys, are somewhat discrepant from each other so that the hypothesis of the equality of the k values may not be accepted, a more realistic approach for assigning a specific value for k to characterize the bacterial variability in the system under specific sampling conditions was given by Maul et al. (1985b).

10.4   Sampling Strategy for Monitoring the Bacteriological Water Quality in a Simple Heterogeneous System

When the dispersion of bacteria in the water system can be represented by the negative binomial distribution and the government regulation specifies a limit on the mean value of bacterial density (i.e., maximum contaminant level), it is possible to develop an efficient sampling design for monitoring the bacteriological water quality in the network. It must be emphasized that the methods described hereafter apply or may be adapted to any bacteriological water quality data (i.e., any particular indicator bacterium), provided that the above mentioned conditions are fulfilled.

Choice of Sampling Design   Two different sampling designs will be considered. The first approach presented here is based on the collection of a predetermined number of samples over a period of time which is also prespecified. The intent of such a sampling design is to achieve the collection of the samples over a very short period of time. This approach, which allows an "instantaneous" assessment of the bacteriological water quality in the system, will be called the one-run sampling design.

On the other hand, if the results of a monitoring program in a simple heterogeneous system are to be reported for specific intervals of time (say, monthly), then sequential sampling offers another way of monitoring the level of bacterial density in the water system. Sequential sampling allows a decision to be made on the mean bacterial density as soon as the accumulated data provide a strong indication to determine whether or not the system is in compliance with the regulations. In some cases, sequential sampling may show increased efficiency in comparison with the one-run design.

The One-run Sampling Design   If a water distribution system is modelled by the negative binomial distribution with parameters p and k ($\lambda = pk$) corresponding to given sampling conditions, the probability (PA) of accepting that the system is in a state of compliance (i.e., $\bar{r} < \mu$, see Table 10.1) is approximated as:

$$\text{PA} = \Phi \left(\frac{\mu - \lambda}{(\lambda/n)^{1/2} (1 + \lambda/k)^{1/2}} \right) \tag{5}$$

where $\Phi(z)$ is the area under the standard normal distribution from $-\infty$ to z (El-Shaarawi et al., 1985).

Note that if the actual mean bacterial density of the network (λ) exceeds the standard (μ), the probability PA represents the consumer's risk (β). Hence, the estimate of the number of samples (n) to be collected, when PA is set at a prespecified level, β, is given as

$$n \geqslant \frac{\lambda \, z_0^2 \, (1 + \lambda/k)}{(\mu - \lambda)^2} \tag{6}$$

where z_0 is the normal variable corresponding to the probability level β; in other words, β is the area under the standard normal distribution from $-\infty$ to z_0.

From equation 6 and the specification of z_0 it appears that n is a function of λ, β, and k. Moreover, the minimum number of samples required (n) is inversely related to the values of λ, β and k. The incidence of bacterial heterogeneity (i.e., in terms of parameter k) on the sampling design is illustrated in Figure 10.2, which gives n as a function of k, assuming different levels for β and λ. (The standard has been set at $\mu = 100$ for the purpose of this example.) Reciprocally, equation 5 can be used to calculate the risk β associated with a given number of samples n. The level of the consumers' risk can also be obtained graphically. For instance, Figure 10.3 gives β as a function of n for different values of k when $\lambda = 200$ and $\mu = 100$. Curves like those presented in Figure 10.2 or 10.3 may be easily obtained from equations 5 or 6 considering their specifications of the maximum contaminant level.

Sequential Sampling   If the bacterial dispersion pattern can be fitted to

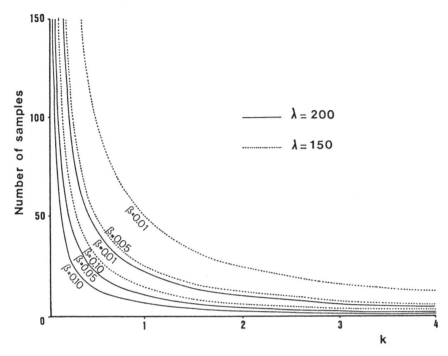

Figure 10.2   Number of samples to be collected as a function of k, assuming different levels for the bacterial density λ and the risk β.

the negative binomial distribution, and the degree of nonrandomness as measured by the parameter k can be considered constant during the sampling period, then sequential sampling can be used.

The sampling continues until a decision is made from the accumulated data so that prior estimation of adequate sample number is unnecessary. This may avoid going through such a strenuous sampling program as using a large predetermined number of samples collected in a single run.

The first task in sequential sampling is to set out two opposing hypotheses about the mean of the population level (λ), namely:

$$H_0: \lambda \leq \lambda_0$$

and:

$$H_1: \lambda \geq \lambda_1$$

Each hypothesis is subject to two types of sampling error:

1. α is the probability of accepting H_1 when H_0 is true.
2. β is the probability of accepting H_0 when H_1 is true.

For example, one can carry out a sequential sampling for deciding whether

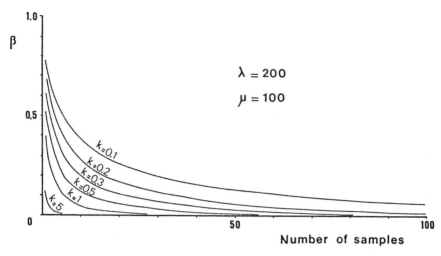

Figure 10.3   Consumers' risk β as a function of the sample size.

the mean bacterial density (λ) is less than the guideline μ (the H_0), or greater than μ by 100 percent (the H_1), with error risks $\alpha = \beta = 0.05$.

For a given step of the procedure, the cumulative number of bacteria (T_n) found in the first n samples is compared to the quantities A(n) and B(n) given as

$$A(n) = \frac{\ln (\beta/(1 - \alpha))}{\ln[(\lambda_1(k + \lambda_0))/(\lambda_0(k + \lambda_1))]} - \frac{nk \ln((k + \lambda_0)/(k + \lambda_1))}{\ln[(\lambda_1(k + \lambda_0))/(\lambda_0(k + \lambda_1))]}$$

$$B(n) = \frac{\ln ((1 - \beta)/\alpha)}{\ln[(\lambda_1(k + \lambda_0))/(\lambda_0(k + \lambda_1))]} - \frac{nk \ln((k + \lambda_0)/(k + \lambda_1))}{\ln[(\lambda_1(k + \lambda_0))/(\lambda_0(k + \lambda_1))]}$$

(7)

The sampling process has to be continued until T_n falls outside the interval determined by A(n) and B(n). If T_n is as $T_n < A(n)$, H_0 is accepted with error risk β and if T_n is as $T_n > B(n)$, H_1 is accepted with error risk α.

When an estimate for k has been calculated from the results of preliminary sampling programs, the lines defined by A(n) and B(n) can be plotted. In the actual study, the sampling is continued until either the cumulative number of bacteria has moved outside the uncertainty region between the two lines or until some predesignated sample size has been reached. Furthermore, it is even possible to combine several tests on the same graph so that different categories of water quality can be characterized (Green, 1979). Figure 10.4 shows, for example, the curves obtained from equation 7 when both α and β are set equal to 0.05. These curves are given for different levels of the two opposing hypotheses, assuming k = 0.44.

The method is illustrated in Figure 10.4, which also displays the path of the cumulative bacterial counts obtained from a sequential sampling design

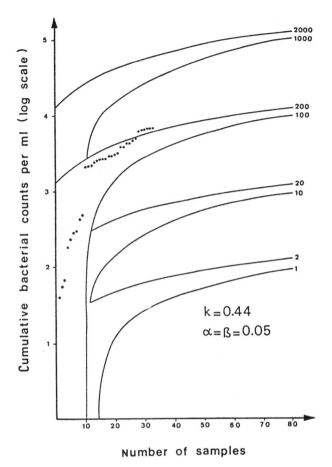

Figure 10.4. Sequential sampling illustrated for the Metz water distribution system using total counts.

conducted on the drinking water distribution system of the city of Metz (France). Since a threshold value of 100 has been adopted for μ, the process was continued until the value for ln T_n fell outside the uncertainty region delimited by the curves associated with 100 and 200. As it appears in Figure 10.4, the violation of the standard could be observed from the 28th sample onwards for the data used in this example.

Sequential sampling may show increased efficiency in comparison with the one-run design, especially when the true mean bacterial density is far enough above or below the standard. Nevertheless, if the cumulative sample T_n remains between A(n) and B(n) for some large sample size, the true mean bacterial density probably lies somewhere between λ_0 and λ_1.

In sequential sampling, both the producers' risk (α) and the consumers' risk (β) are considered and the graphical procedure illustrated in Figure 10.4

allows further classification of the water analyzed into different classes of bacterial concentration.

Moreover, a major advantage of the procedure is that sequential sampling can be performed on any water system modelled by the negative binomial distribution or its limiting form (the Poisson distribution). This can be done even when there is no reliable estimate for k available. The method allows a new estimate of k (see equation 3) to be calculated after each step of the sampling and, thus, the graph corresponding to the last estimated value of k can be visualized on a monitor screen by means of a computer. This offers an interesting way of following the process until a decision is made on the mean bacterial density of the water system sampled.

The steps of the sequential sampling method, summarized in Figure 10.5, include automatic calculation of k, criteria for identifying the model (i.e., Poisson or negative binomial), plotting of the decision lines, the cumulative bacterial counts (T_n), and printing of the outcome corresponding to the stage of the process considered.

10.5 Sampling Strategy for Monitoring the Bacteriological Water Quality in a Compound Heterogeneous System

Although the negative binomial distribution is a rather versatile model which appears to provide an adequate fit to a wide range of bacteriological data, the goodness-of-fit test presented in Section 10.3 may sometimes lead to the conclusion that the negative binomial distribution is not suitable for representing the data (Maul et al., 1985a). The cause of such an outcome may be related to the size of the network sampled, since the bacterial counts might follow a multimode distribution relevant to various zones of bacterial density, especially in large water systems. Such zones have been observed by Maul et al. (1985a), who identified a structured pattern in the spatial dispersion of heterotrophic bacteria in the network of the city of Metz. The authors observed a bacteriologically stable water system and the consistency of this structured spatial dispersion pattern of bacteria was explained by some physical and chemical characteristics of the system.

Hence, if the frequency distributions of microorganisms obtained from the water system sampled cannot be modelled by the negative binomial distribution, then the underlying assumption of a structured spatial heterogeneity should be examined. The success of such an attempt, which aims at giving a clear picture of the spatial variation of bacterial populations in the network, is useful for: 1) understanding the mechanisms of the incidence and the spreading of bacteria in the network; 2) determining problem areas; and 3) designing future monitoring sampling programs.

The procedure should start by examining the possibility of dividing the

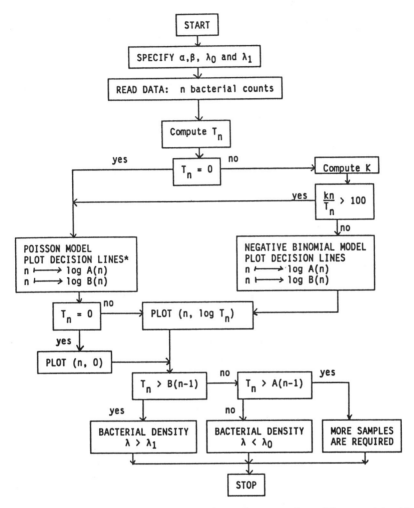

Figure 10.5 Flow-diagram of the sequential sampling procedure. *The quantities A(n) and B(n) corresponding to the Poisson model are obtained from equation 7 when $k \to \infty$. Hence they are calculated as

$$A(n) = \frac{\ln\left(\beta/(1-\alpha)\right)}{\ln \lambda_1/\lambda_0} - \frac{(\lambda_0 - \lambda_1)n}{\ln \lambda_1/\lambda_0}$$

$$B(n) = \frac{\ln((1-\beta)/\alpha)}{\ln \lambda_1/\lambda_0} - \frac{(\lambda_0 - \lambda_1)n}{\ln \lambda_1/\lambda_0}$$

entire network into several distinct zones by means of an objective classification of the water system. The next step considers the hypothesis that the bacteriological data can be represented within the zones by the negative binomial distribution so that the compound heterogeneous system may be considered as being composed of several simple heterogeneous subsystems. The successive stages of the procedure are discussed and illustrated by examples in the following sections.

Preliminary Sampling   If the historical bacteriological information available is too sparse or inadequate for determining the spatial and temporal variation of bacterial populations in the network, then there is no substitute for preliminary sampling and it is not advisable to skip it. To provide a consistent image of both the spatial and the temporal changes of bacterial density in the water system, we recommend carrying out several surveys following the same approach (i.e., collection of water samples and sample processing). In particular, water samples should be collected during a short period of time from a number of sites spread over the area of the entire network according to a systematic sampling design (Maul et al., 1985a).

Clustering Method   If the same general pattern can be seen for all the surveys, thus reflecting a bacteriologically stable water distribution system, it is appropriate then to divide the network into zones corresponding to different levels of bacterial density.

A nonhierarchical nearest-centroid clustering method can be used to divide the sampling stations in the water distribution system into sets on the bases of the data from the surveys. When these sets are given on the map of the distribution system, they are called zones. A clustering method which is well adapted for this purpose has been given in detail by Anderson et al. (1984). As an example, Figure 10.1 shows the zones of bacterial density as they were observed for the water distribution system of the city of Metz. The application of the clustering method to the heterotrophic bacterial counts from water samples collected at 100 stations during six surveys adequately divided the water system into four zones which appeared to be rather densely built-up. Further, the arrangement of the zones shows that high densities of heterotrophic bacteria were more likely to occur in peripheral locations far from the treatment plant rather than close to it.

One-run Sampling Design   If the entire water system has been divided into l zones with the negative binomial distribution representing a suitable model for the dispersion of bacteria in each zone (i.e., system composed of several simple heterogeneous subsystems) and a specific value for k has been assigned to each zone, then the patterns of bacterial heterogeneity can be used for:

1. A determination of the location and the number of stations that need to be sampled for a given level of the risk β; or

2. An optimal allocation of the stations to the different zones of bacterial density, assuming the total number of stations is prespecified.

If β is fixed for all the zones, then the number (n_i) of samples needed from the ith zone can be calculated directly from equation 6, which is applicable to any simple heterogeneous system. Seen another way, the total number of samples (N) to be collected might be fixed for administrative purposes or practical constraints and the problem consists of determining how to divide N among the zones. The optimal number of stations (n_i) that should be allocated to each of the l zones is given as:

$$n_i = N\frac{1 + \lambda/k_i}{\sum\limits_{i=1}^{l} (1 + \lambda/k_i)} \tag{8}$$

where k_i is the parameter associated with zone i.

The risk β corresponding to such a design is given as:

$$\beta = \Phi \left(\frac{\mu - \lambda}{(\lambda/N)^{1/2} \left(\sum\limits_{i=1}^{l} (1 + \lambda/k_i) \right)^{1/2}} \right) \tag{9}$$

or may also be obtained from equation 5 for any of the l zones.

The two different ways for performing a one-run sampling design on a compound heterogeneous system (i.e., if β or N is fixed) are illustrated here, using the data on heterotrophic bacterial counts from the Metz water distribution system. Table 10.2 presents the value for k that could be assigned to each zone of the network (see Figure 10.1) following the approach outlined in Section 10.3. With estimates of k in hand, it is possible to determine the number of stations that should be allocated to each zone, provided that the values for β or N are given. Thus, assuming the guideline μ and the level of λ are set at 100 and 200, respectively, the column headed by ñ in Table 10.2 gives the allocation corresponding to $\beta = 0.05$, while the column headed by n̂ presents the optimal allocation of 100 samples among the zones ($\beta \simeq 0.0235$). The results given in Table 10.2 show, for example, that at least 70 samples are needed from the Metz water distribution system in order to be able to detect

Table 10.2 Optimal allocation of sampling stations to the water distribution system of the city of Metz

Zone	k	Number of sampling stations	
		ñ	n̂
1	0.43927	24.70	35.96
2	0.50514	21.48	31.28
3	0.51945	20.89	30.42
4	7.01250	1.59	2.33

violations of the water quality with a probability above 0.95 when $\mu = 100$ and $\lambda = 200$.

As to the location of the stations allotted to the different zones of a water system, they can be randomly selected among a number of sites spread over the network area according to a spatial grid pattern.

Sequential Sampling   The procedure presented in Section 10.4, which continually takes new information into account as it becomes available, can also be performed on a compound heterogeneous system. Thus, sequential sampling is applicable to each zone taken separately, provided that the negative binomial distribution is a suitable model to represent bacterial counts within the zones. Monitoring the bacterial density within the different zones taken apart may be advisable since the temporal variation in the bacterial population does not necessarily occur in the zones in the same manner. In other words, the zones may not be affected uniformly by changes in bacterial density and may thus behave rather independently from each other (Maul et al., 1985a). Consequently, considering zones separately within the network may be useful for determining problem areas. Furthermore, examining the zones separately will facilitate taking effective remedial action.

Summary

The origin and the characteristics of the dispersion patterns of bacteria in a water distribution system are examined and discussed. Appropriate mathematical models, in terms of a probability distribution, are considered for representing the dispersion of bacteria in water. The parameters of these models and the general configuration of bacterial heterogeneity in the network are then used to assist in designing sampling programs for monitoring the microbiological water quality. Two different sampling designs are developed: sequential sampling and the one-run sampling design. The latter allows an "instantaneous" assessment of the water quality in the system. In both cases, emphasis is laid on the risk of sampling (i.e., making the wrong decision).

References

Anderson, J.E, El-Shaarawi, A.H., Esterby, S.R., and Unny, T.E. 1984. Dissolved oxygen concentrations in Lake Erie (U.S.A.-Canada). 1. Study of spatial and temporal variability using cluster and regression analysis. *Journal of Hydrology* 72:209–229.

Christian, R.R. and Pipes. W.O. 1983. Frequency distribution of coliforms in water distribution systems. *Applied and Environmental Microbiology* 45:603–609.

El-Shaarawi, A.H., Esterby, S.R., and Dutka, B.J. 1981. Bacterial density in water determined by Poisson or negative binomial distributions. *Applied and Environmental Microbiology* 41:107–116.

El-Shaarawi, A.H., Block, J.C., and Maul, A. 1985. The use of historical data for estimating the number of samples required for monitoring drinking water. *The Science of the Total Environment* 42:289–303.

Esterby, S.R. 1982. Fitting probability distributions to bacteriological data. Considerations for surveys and for water quality guidelines. *Journal Français d'Hydrologie* 13:189–203.

Fisher, R.A. 1941. The negative binomial distribution. *Annals of Eugenics* 11:182–187.

Fisher, R.A., Thornton, H.G., and MacKenzie, W.A. 1922. The accuracy of the plating method of estimating the density of bacterial populations. *Annals of Applied Biology* 9:325–359.

Green, R.H. 1979. *Sampling Design and Statistical Methods for Environmental Biologists.* Wiley, New York. 257 pp.

Health and Welfare Canada. 1978. *Guidelines for Canadian Drinking Water Quality.* Canadian Government Publishing Centre, Supply and Services Canada, Hull, Quebec.

Maul, A., El-Shaarawi, A.H., and Block, J.C. 1985a. Heterotrophic bacteria in water distribution systems. 1. Spatial and temporal variation. *The Science of the Total Environment* 44:201–214.

Maul, A., El-Shaarawi, A.H., and Block, J.C. 1985b. Heterotrophic bacteria in water distribution systems. 2. Sampling design for monitoring. *The Science of the Total Environment* 44:215–224.

Muenz, L.R. 1978. Some statistical considerations in water quality control. pp. 49–56 in Hendricks, C.W. (editor), *Evaluation of the Microbiology Standards for Drinking Water,* U.S. Environmental Protection Agency, Washington, D.C.

Pipes, W.O. and Christian, R.R. 1982. *Sampling Frequency—Microbiological Drinking Water Regulations—Final report.* U.S. Environmental Protection Agency, EPA R-805-637/9–82-001, Washington, D.C. 159 pp.

Pipes, W.O., Ward, P., and Ahn, S.H. 1977. Frequency distributions for coliform bacteria in water. *Journal of American Water Works Association* 69:644–647.

Snedecor, G.W. and Cochran, W.G. 1967. *Statistical Methods,* 6th edition. Iowa State University Press, Ames, Iowa, 649 pp.

Thomas, H.A., Jr. 1955. Statistical analysis of coliform data. *Sewage and Industrial Wastes* 27:212–228.

U.S. Environmental Protection Agency, Office of Drinking Water. 1976. *National Interim Primary Drinking Water Regulations,* EPA-570/9–76-003, Washington, D.C.

11

Invertebrates and Associated Bacteria in Drinking Water Distribution Lines

Richard V. Levy

Millipore Corporation, Bedford, Massachusetts

11.1 Introduction

In March of 1975, the United States Environmental Protection Agency (USEPA) under the provisions of the Safe Drinking Water Act (SWDA, 1974), proposed the National Interim Primary Drinking Water Regulations (NIPDWR, 1977) based upon the concept that water destined for human consumption should be free from physical, chemical, radiological and microbiological contamination in accordance with drinking water standards set by the United States Public Health Service (USPHS) in 1962. Amendments to the NIPDWR were added in 1976, 1979, and 1980 for organic and inorganic chemicals, radionuclides, and microbiological contaminants.

The NIPDWR recommended that surface waters should be low in microbiological contaminants and turbidity prior to disinfection and at the consumers' tap (see Chapter 1). Rationale for these provisions include the following:

1. The presence of particulate matter interferes with effective disinfection at the point of entry of water into the distribution system.
2. Because suspended matter may exert a chlorine demand, the maintenance of low turbidity throughout the distribution system will facilitate adequate chlorine residual.
3. Regrowth of microorganisms in a distribution system is stimulated if organic matter is present.
4. Research has shown that organisms such as nematodes could ingest enteric pathogenic bacteria as well as viruses and that nematode-borne bacteria were completely protected from inactivation with chlorine, even when 90% of the nematodes were completely immobilized.

5. Demonstration that most of the bacteria recovered from chlorinated drinking water could be associated with particles.

6. Knowledge that invertebrates contribute to the phenomenon of biological magnification, and that bottom-dwelling organisms may concentrate waterborne contaminants such as pesticides, biocides, metals, and radioactive materials in their tissues or within their bodies.

Therefore, higher organisms such as invertebrates entering a drinking water distribution system may be contaminated or may become contaminated within the system with materials or microbiological agents which could affect the public health.

The occurrence and persistence of invertebrates in public drinking water supplies is a common complaint concerning both unfiltered and filtered water supplies and distribution systems. These invertebrates include crustaceans (amphipods, copepods, isopods, ostracods), nematodes, flatworms, water mites, and insect larvae such as chironomids. Invertebrates enter water treatment plants and most are trapped and removed by various processes. Some invertebrates, however, may go on to breed in the treatment plant and release either adults or juveniles into the system. In unfiltered systems, invertebrates pass through coarse screening and penetrate the disinfection barrier to enter the distribution system. Many of these microorganisms are inactivated immediately, others die slowly, while others may actually live and reproduce in the system. These invertebrates may harbor and protect bacteria from disinfection with chlorine and monochloramines at concentrations normally present in distribution systems. Associated bacteria may survive even when the invertebrates are totally inactivated. Invertebrates may therefore:

1. Constitute a health hazard if they are carrying or become exposed to and associate with bacterial pathogens;

2. Contribute significantly to the organic load entering many distribution systems;

3. Contribute to the continued colonization of distribution systems by invertebrate associated bacteria in spite of residual chlorine;

4. Decrease the effectiveness of chlorination by increasing the chlorine demand of the water.

Open reservoirs and rivers are often used as a source of potable water yet they are often rich in flora and fauna typical of natural aquatic ecosystems. These water sources may contain opportunistic pathogens if watershed protection is marginal and/or if the water source is used for sewage effluent disposal. Organic matter (living and dead) contaminated with pathogenic organisms derived from these waters may enter drinking water supply distribution systems. In the case of invertebrate animals, adsorption and ingestion of enteric bacteria can provide considerable protection from chlorine disinfection (Levy et al., 1984a). Levy et al. (1984b) demonstrated bacterial attachment to both external and internal surfaces of amphipods. The association of bacteria with other aquatic invertebrates has been reported (Atlas et al., 1982; Boyle and Mitchell, 1981; Ducklow et al., 1979; Ducklow et al., 1981; King et al., 1988; Klug and Kotarski, 1980; Sarai, 1976; Sochard et al., 1979).

Since contamination of potable waters may occur both from within (back-siphonage, cross-connections, leaks and repairs, deficient filtration and disinfection) and from without (animal defecation and contamination, pasture run-off and improper waste disposal), invertebrates could come into contact with microbiological contaminants at multiple locations. These invertebrates may be the cause of tastes and odors, water discoloration, point-of-use filter and/or water meter clogging, high turbidity, loss of water authority credibility, and public health concerns (Levy et al., 1986).

To better understand invertebrate occurrences and infestations, it is necessary to consider the role invertebrates play in impoundment (reservoir) ecology. These principles and life styles apparently carry over into the distribution system, and explain how some invertebrates enter, continue to survive, and even reproduce within the harsh conditions found in pipe networks. Historically, of those invertebrates commonly found in surface waters, amphipods and chirononmids (fly larvae) have received the most attention from scientists and consumers of potable water. Amphipods are benthic, omnivorous crustaceans found to be common and abundant in the more permanent freshwaters of North and Central America (Pennak, 1935; Edmondson, 1973). They burrow through beds of aquatic vegetation, feeding upon the epiphytic growth that covers the rooted aquatics and are known to feed upon dead animal and plant matter, such as filamentous algae (Cooper, 1965). For example, *Hyalella* is both a grazer and deposit feeder (Strong, Jr., 1972). These data further confirm that *Hyalella* would be most likely to inhabit littoral areas of surface reservoirs. Therefore, amphipods are less likely to be transported into the distribution system as an adult, but probably enter as juveniles when they are less able to resist water currents. Amphipods have been shown to harbor bacteria (Atlas et al., 1982), and to pass viable bacteria (including *Bacillus cereus*) through the gut tract and out with the fecal materials (Willoughby et al., 1982).

Chironomid adults (midges) resemble small mosquitoes and are commonly seen flying in great numbers near water. There are over 200 species of midges in North America alone. Adult female midges lay a mass of eggs in the water. These eggs hatch into larvae that require from one to two months or more to reach pupation. During this time the fly larvae wriggle in and swim near the bottom sediment. Some attach themselves to plants or rocks. Near the end of the larval stage, they migrate to the water surface, pupate, and stay there a few days before emerging as adults. The larval stages of chironomids frequently enter municipal drinking water systems. Many communities report these larvae in their drinking water during the spring and summer months of the year. These occurrences have led to a series of published reports concerning the occurrence of chironomids and other invertebrates in water systems throughout the world.

Members of other taxonomic groups of invertebrates have been demonstrated to occur not only in surface reservoirs, but also in drinking water distribution systems. Nematodes (Nematoda) are roundworms which occur in moist soil, freshwater, estuarine, and marine habitats. They are generally aerobic and are found in benthic habitats where they may consume bacteria, algae, other nematodes, and invertebrates. Most adult forms are about 1 mm long (Nicholas, 1975; Croll, 1977). Nematodes found in untreated drinking waters originate from animals drifting into the water distribution system (Mott and Harrison, 1983). In treated drinking water, the source of nematodes is likely to be either drifting or colonization and breeding within the water treatment plant (Cobb, 1918; George, 1966; Tombes et al., 1979). Larval and adult worms have been demonstrated to survive in such diverse habitats as aerobic water and wastewater plants (Chang and Kabler, 1962). Furthermore, nematodes have been found to occur in public drinking water supplies (Chang et al., 1960a; Levy et al., 1984b). Since nematodes are known to ingest bacteria, they should be considered potential carriers of enteric bacterial pathogens.

Copepods may be found as part of the plankton, although some species may be frequent the bottom substrate. *Cyclops* is the most common American genus, living in lakes and ponds (Pennak, 1935). Their diet consists of microscopic plants and animals captured while filtering the water. Many copepods exhibit diurnal vertical migrations within the water column, and are most likely transported by water currents into drinking water distribution systems. Since marine copepods have been reported to harbor bacteria attached to their external surfaces, including egg sacs (Huq et al., 1983), they are of interest.

Cladocerans, e.g., water fleas or *Daphnia*, live year round in freshwaters throughout America and Europe where algal and bacterial foods are present. These minute crustaceans (0.2 to 3.0 mm in length) are common at the bottoms of lakes and pools where they forage among the dead plant and animal debris, or are active in the water column as members of the plankton. Their food is primarily protozoa and algae, although some species consume organic debris and bacteria; therefore they might transport other microorganisms into treated waters protected within their gut tracts. Since parthenogenic development is a common method of reproduction during most of the year, once inside a distribution system they could reproduce without mating. The young undergo continued development in the brood chamber and hatch into young which look like the adult. Most females lay eggs which become attached to debris or are carried within the shell for a brief period of time. Special winter eggs, produced by sexual reproduction, survive the more rigorous conditions, while summer eggs hatch quickly (Pennak, 1935; Edmondson, 1973; Palmer and Fowler, 1973).

Ostracods are sometimes called mussel or seed shrimp. They are small crustaceans found in marine and fresh water habitats. The ostracod has evolved a special shell not unlike that of a bivalve clam. They may be carnivorous, herbivorous, scavengers, or filter feeders. Algae, smaller crustaceans, snails, and annelid worms are the most common sources of food, although they may gain nutrition by ingesting detritus particles (Barnes, 1980). They are frequently collected in the reservoirs and from hydrant samples where they appear to be browsing in pipe debris.

Mites (Hydracarina) are the only members of the Class Arachnoidea which have become adapted to fresh waters. They are particularly active during the late spring and early autumn in and near wooded ponds and impoundments, but may be collected year round, even under the ice. The majority of mites are carnivorous or parasitic. Their food consists of small insects, worms, or dead animal tissues. They can be cannibalistic. Some species are excellent swimmers, while others are strictly creepers and crawlers on the benthic sediments. Those collected from the distribution system are of the later type. Mites remain active throughout the winter.

In conclusion, representatives of almost all taxonomic groups of invertebrates occur naturally in freshwater ecosystems (Pennak, 1935; Edmondson, 1973; Palmer and Fowler, 1973). Other invertebrate groups (not described here) which have been demonstrated to occur in water supplies drinking water distribution lines include annelids, coelenterates, other crustaceans, flatworms (platyhelminthes), insects, and bivalve mollusks.

11.2   Review of the Literature

The first recorded data concerning macroinvertebrate infestations of a water system was by Malloch (1915). He reported that chironomid larvae were found in the public water works supply of Boone, Iowa. Malloch also reported that he had obtained a specimen of the genus *Chironomus* from a reservoir supply of Champaign, Illinois. Bahlman (1932) noted that a similar problem existed in the water supply of Cincinnati, Ohio, while Brown (1933) and Arnold (1936) reported that Chironomidae infestations occurred in a number of California systems. Other cities reported similar infestations: Charlottesville, Virginia and Elizabeth City, N.C. (AWWA, 1929); Champaign, Illinois, for a second time (Amsbary, 1928); Peoria, Illinois (Flentje, 1945a), and Alexandria, Virginia (Flentje, 1945b); Washington Suburban Sanitary District (Hechmer, 1932); Washington, D.C. (Brown, 1933); Lake St. Clair, Detroit, Michigan (Hudgins, 1931); Ohio river (Chang et al., 1959); Indianapolis (Crabill, 1956); and several locations in England (Floris, 1935; Kelley, 1955) and Germany (Buchmann, 1933).

Martin E. Flentje (1945b) prefaced his paper entitled "The Control and Elimination of Pest Infestations in Public Water Supplies" with the statement that "numerous infestations of public water supplies by such pests as midge fly larvae and crustacea have been reported in the water works literature." Invertebrate colonization of potable water distribution systems has since been substantiated. Levy et al. (1984a) and others (Luczak et al., 1980; Small and

Greaves, 1968) have suggested that certain invertebrates live and reproduce within distribution systems, while others are temporarily suspended in the water column within the pipe network. Recent studies on the nature of the encrustations and tubercles within pipelines suggest that there are adequate food sources (direct or indirect) so that certain "benthic" invertebrates may actually infest sections of a distribution network that provides a favorable habitat (Luczak et al., 1980; Allen et al., 1980;, Tuovinen et al., 1980).

English (1958) summarized the biological problems of water main infestation. He defined infestation to be the existence of a colony or population of organisms which were actually growing or reproducing within the distribution system. Organisms carried along with the water flow within the system did not represent an infestation. These situations are most appropriately termed occurrences. Smalls and Greaves (1968) did the most extensive ecological survey of animals within distribution systems. They used hydrant sampling to survey a distribution system for the presence of animals. Over 100 different types of animals were found in the distribution systems they sampled. They concluded that the population density of invertebrates depends on the following:

1. Recruitment of individuals of a given species via treatment plant and service reservoirs;
2. Maintenance of sufficient food supplies;
3. Suitability or adaptability of the species to the water distribution system environment;
4. The species reproductive capacity.

Furthermore, they reported that seasonal reductions in the number of individuals observed sometimes leads to the erroneous conclusion that the infestation no longer exists. Some invertebrates can exist for longer periods of time in resting stages. None of the animals they collected were thought to have public health significance, since the chance of ingesting even a nematode containing viable pathogens was, in their view, remote.

Luczak et al. (1980) carried out hydrobiological investigations to determine qualitatively and quantitatively the aquatic invertebrates present in the network of the central waterworks in Warsaw, Poland. They concluded that the methods used by the waterworks were not sufficient to prevent the penetration of invertebrates into the distribution network. They collected live flagellated protozoa, rotifers, and nematodes. Although many invertebrates and plant species were not always found to be alive upon collection, there was no relationship between chlorine residuals and the occurrence of live invertebrates. Differences in animal distribution within the water supply network were found to be most closely correlated to seasons.

Invertebrate infestations are frequently the cause of consumer complaints of dirty and discolored water, even when the invertebrates themselves go unnoticed. Walker (1983) reported that infestations with *Asellus* were manifested by the appearance of blackened water due to the fecal debris which becomes suspended in the water column. These "frasse" (as they are known) were shown to be high in iron and manganese, indicating a diet of metal-rich water main deposits. On occasion, frasse were found to contain lead or copper, indicating that the animals were grazing within older, corroded plumbing or that the fecal deposits had been present in waters rich in those elements before appearing in the consumers water.

Levy et al. (1984a) examined organisms which were found in a distribution reservoir and the drinking water distribution system of Worcester, Massachusetts. They collected invertebrates to determine the general circumstances surrounding the occurrence of such animals. Subsequently, they cultured and isolated bacteria associated with those invertebrates. Within this drinking water distribution system, invertebrates were collected aseptically from point-of-use water filters or hydrants. During the second year of the study, hydrant samples were subdivided into initial effluent and effluent samples to determine those invertebrates trapped in the hydrant itself versus those present in the water column. This study permitted the location of the areas within the distribution system in which invertebrates were most likely to be planktonic versus those areas where invertebrates were actually colonizing the pipelines. Invertebrates were found to be asymmetrically distributed within the distribution system. During the period of May through October of 1983 and 1984, a total of 264 hydrant samples were taken. Less than 20 of the same hydrants were sampled twice during one season, those that were sampled more than once were expected to yield invertebrates. In 1983 and 1984, the predominate invertebrates present in the Worcester distribution system were copepods (48 and 55 percent, respectively). Amphipods (19 and 15 percent) and fly larvae of several species (15 and 15 percent) were present in approximately equal numbers during the study. Surprisingly, nematodes were relatively common in hydrant samples (4 and 5 percent). Cladocerans (*Daphnia*), water mites, and ostracods were less common. There was very little variation in the frequency of occurrence between the two years. The slight increase in the frequency of copepod occurrence may be attributed to extensive cleaning and relining of pipelines which facilitates flow within the system. The decrease in amphipod occurrence may be attributed to active flushing and pipe rehabilitation in areas which frequently reported occurrences.

11.3   Bacterial Association with Invertebrates

Bacteriology of Natural Invertebrate Associations: Surface Reservoirs

Chang et al. (1960b) demonstrated that nematodes were capable of protecting *Salmonella* from chlorination; however no data were available on the "natural flora" of invertebrates associated with both raw and treated drinking waters. Levy (1985) characterized the bacteria associated with amphipods collected from one of the surface water reservoirs supplying drinking water to the city of Worcester. A variety of bacteria were recovered from amphipods collected during the late spring, summer, and fall. During the winter months only a small proportion of the amphipod population survived, apparently retreating to deeper waters, and their bacteria flora was reduced. Bacteria isolated from amphipods varied seasonally; *Pseudomonads* dominated the invertebrate-associated flora during most of the year. Coliform bacteria, e.g. *Enterobacter* and *Klebsiella*, were recovered from ambient reservoir waters, but not in association with amphipods. Members of the genus *Serratia* were recovered infrequently. Bacterial species recovered from amphipods were also resident in the water column, though not always concurrently, prompting speculation that invertebrates could serve as reservoirs for bacteria unable to survive in the planktonic environment.

Bacteriology of Natural Invertebrate Associations: Distribution Systems   Levy (1985) collected and isolated bacteria from four taxonomic groups of invertebrates sampled from distribution systems (see Table 11.1). The results of the bacteriological studies indicate that all of the invertebrates collected from the distribution systems were colonized by bacteria. Of particular note was the lack of coliform bacteria associated with these animals. Similarly, Levy et al. (1986) were unable to recover coliform bacteria from distribution

Table 11.1 Bacteria associated with invertebrates collected from a drinking water distribution system, Worcester, Massachusetts, 1982–1984

Bacterium	Invertebrate Type			
	Amphipod	Copepod	Fly Larva	Nematode
Acinetobacter sp.	+			
Achromobacter xylooxidans	+			
Aeromonas hydrophila			+	
Bacillus sp.	+			
Chromobacter violaceum	+			
Flavobacterium meningosepticum		+		
Moraxella sp.	+	+		
Pasteurella sp.	+	+		
Pseudomonas diminuta		+		
Pseudomonas cepacia			+	+
Pseudomonas fluorescens	+		+	
Pseudomonas maltophilia	+			
Pseudomonas paucimobilis	+	+		
Pseudomonas vesicularis	+			
Serratia sp.	+			
Staphyloccocus sp.	+			

Levy et al. (1986).

specimens. Unfortunately, Levy did not collect data on reservoir copepods, fly larvae, or nematodes. These data suggest that invertebrate-associated coliforms are less able to colonize invertebrates, and those which succeed are more sensitive to disinfection, and do not survive as well as other, better adapted bacterial flora. Unpublished data suggested that the heterotrophic flora of distribution system amphipods qualitatively paralleled that of those found in drinking water reservoirs.

Scanning electron microscopy (SEM) was used to verify the presence of bacteria associated with all the invertebrates (see Figure 11.1A,B,C). Bacteria were detected on the surfaces of all invertebrates except copepods (Figure 11.1D). Levy (1985) speculated that bacteria associated with copepods were located internally, while bacteria flora of other invertebrates were located both internally and externally. The gut tracts of amphipods were examined and they proved to be generally devoid of flora except in one sampling when staphyloccocci were isolated and visualized by SEM. Because of the proportionally larger size and life span of amphipods, their bacterial flora tended to be greater than that of the copepods, fly larvae, and nematodes recovered from either the reservoirs or the distribution system.

From additional studies completed by Levy (1985), it can be predicted that invertebrate-associated bacteria collected from the distribution systems will be

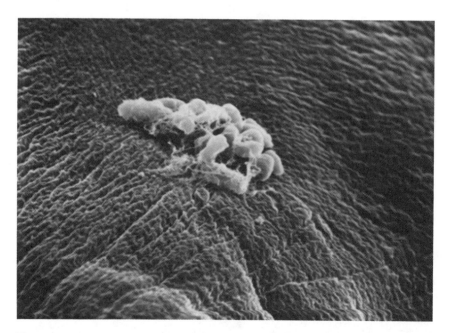

Figure 11.1   Scanning electron micrographs of invertebrates collected from the drinking water distribution system of Worcester, Massachusetts, showing the presence of bacteria on the external surfaces. A (above), and B, amphipods; C, nematodes; D, copepod.

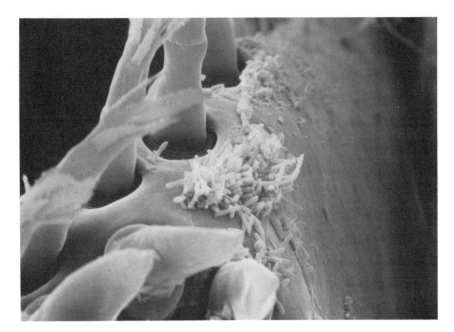

Figure 11.1B

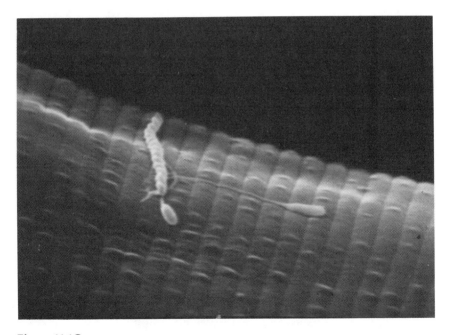

Figure 11.1C

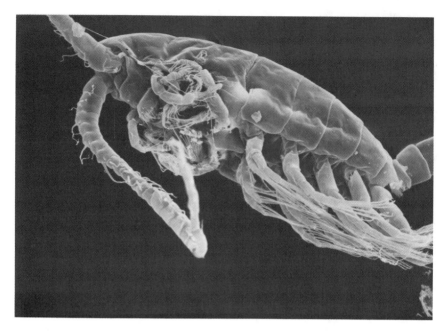

Figure 11.1D

similar to bacteria found associated with invertebrates in water supply reservoirs before chlorination. Quantification of the associated bacterial flora demonstrates that the numbers of bacteria found in association with various invertebrate taxons may be expected to vary. This variance may reflect the following considerations:

1. Size as well as physical and biochemical characteristics of the invertebrate's external surface exposed to the ambient water environment must be considered.

2. Life style of an invertebrate may predispose it to bacterial colonization. Those which actively browse in and among the bottom sediments of reservoirs and pipelines, e.g., amphipods, would be exposed to higher numbers of bacteria for longer periods of time when compared to invertebrates such as copepods that live in the water column, where bacterial cells are more diffuse.

3. Life span of an invertebrate should play a role in determining the number of permanent associations formed with ambient bacteria, since contact time appears to be factor in attachment.

11.4 Experimental Association and Persistence of Bacteria with Invertebrates

Invertebrates may come in contact with potential pathogens before or after entering the distribution system. Bacteria could associate with invertebrates under these conditions and possibly form permanent associations. To evaluate this further, Levy (1985) reported the following experiments.

Approximately 250 amphipods were transferred to a 4 l flask containing approximately 3 l of filter-sterilized (0.45μm) reservoir water which had been collected and stored at 4°C until shortly before use. At the same time, another 4 l flask was prepared with only filter-sterilized reservoir water as a control. 2.0 ml of an *Enterobacter cloacae* suspension was added to each flask, yielding a concentration of suspended cells inside the test flasks of approximately 1.0 $\times$ 10^6 cells per ml at time zero. The flasks were incubated at 20°C.

At selected time intervals (as close to 6 hr. as possible), samples were taken from the control flask containing only bacteria and from the experimental flasks containing both suspended bacteria and amphipod-associated (seeded) bacteria.

Similarly the persistence of associated bacteria under a different set of conditions was tested by seeding amphipods with *Enterobacter*. The persistence of associated bacteria under conditions where the amphipods were no longer exposed to high densities of suspended *Enterobacter* was evaluated. Experiments revealed that amphipods do adsorb *Enterobacter* under laboratory conditions. Maximum adsorption occurred at approximately 30 hours incubation time. The number of *Enterobacter* adsorbed was always less than 1.0 $\times$ 10^3 bacteria per amphipod. *Enterobacter* associated with *Hyalella* increased up to the 30 hour time (251 colony-forming-units (CFU)/amphipod) and then declined linearly. When amphipods were removed from the original flask and placed in a milieu without bacteria, the number of associated *Enterobacter* declined linearly and more quickly. Monitoring of the bacteria numbers in the newly replaced sterile medium indicated that bacteria were shed from the amphipods. Control unassociated bacteria declined linearly, at a rate similar to that of associated bacteria in the presence of high numbers of suspended bacteria.

The data cited above were evaluated by linear regression and used to project the time it would take for the number of bacteria in each situation to reach 1% survival. Estimated 1% survival of associated *Enterobacter* was found to be similar to unassociated *Enterobacter* under the same conditions: 263 hours and 268 hours respectively. However, unassociated bacteria in the control flask were found to survive better and estimated to survive longer (1% survival after 598 hours) than any other experimental situation. The apparent lower survival of associated bacteria placed in a sterile ambient medium was due, in part, to the shedding of bacteria from the amphipod into the medium.

11.5   Control of Invertebrates

Control of invertebrates within drinking water distribution systems is essential to the maintenance of potable water aesthetics. Historically, invertebrates have been controlled by chemical, physical, or biological means, with varying degrees of success. A summary of selected invertebrate occurrences and control methodologies applied to those situations may be found in Sands (1968) and in Levy (1986).

Chlorination did not prove effective in the elimination of midge fly larvae (*Chironomidae*). Flentje (1945b) reported that these larvae could withstand concentrations as high as 50 mg/l with a contact period of 24 hours. Copper sulfate was also shown to be ineffective. However, when both chemicals were introduced at the same time, killing times were shortened. Copepods genus (*Cyclops*) were controlled using cuprichloramine application over a three-week period. This type of treatment killed *Cyclops* emerging from eggs which had passed through filtration systems (Crabill, 1960). Ward and DeGraeve (1978) reported that Cladocerans (*Daphnia magna*) were very sensitive to residual chlorine. Total concentrations of 0.22 mg/l and 0.07 mg/l were lethal to 3 day old *D. magna* in 5.5 and 10.5 hours, respectively. In a 48 hour experiment run with *D. magna* specimens which were less than one day old, an LD_{50} of 0.017 mg/l total chlorine residual was noted. It was concluded that low levels of chlorinated effluents (from a water treatment facility) may adversely affect the survival of some invertebrates. Mitcham et al. (1983) reported that in treated lowland river water, chloramination was preferred to free residual chlorine for inactivating adult *Cyclops* (Copepoda), while free chlorine proved more effective in eliminating *Bosmina* (Cladocera).

Lorenz et al. (1986) demonstrated that nematodes could survive 0.5 mg/ l at 10°C for 43 hours at pH 6.9 and 72 hours at pH 7.8. Inactivation took two to three times longer at 4°C than at 24°C at 3.0, 5.0 and 10.0 mg/l free chlorine. Nematodes exhibited $C \times T$ (concentration $\times$ time) coefficients of between 450 and 2550 milligrams minutes per liter. Nematodes are 3 to 4 orders of magnitude more resistant to free chlorine than bacteria and are of equal resistance to *Giardia* cysts.

DDT treatment, when used in previous infestations, has proved effective (Flentje, 1945c). This is not surprising since the excellent performance of DDT may be attributed (at least partially) to its long half-life (15 years). Since such techniques were first used, DDT has been shown to produce high residuals in the environment, to be nonbiodegradable, and to concentrate in body tissues (fatty tissue). On January 1, 1973, its use was prohibited for most pest problems in the United States.

Pyrethrin dosing, using a 2 percent alcoholic solution, was practiced in England for the control of *Asellus aquaticus* in water distribution systems. The process involved the addition of 0.001 mg/l continuously for 77 hours (Hechmer, 1932). Mitcham and Shelley (1980) reported the use of a synthetic pyrethroid, permethrin, to control animals within water mains. Results in both studies were considered very successful. The addition of pesticides to drinking water has not found widespread acceptance.

Physical means of control have also been attempted. Weiss (1971) described the following considerations with reference to the location of water intakes:

1. "The natural movement of planktonic organisms, both vertically and spatially."
2. "Seasonal fluctuations of species and seasonal changes in population density of organisms associated with tastes and odors."
3. "Changes in water chemistry due to shifts in oxidation-reduction potential resulting from biological consumption of oxygen."
4. "Vertical motion of water masses due to oscillations resulting from seiches or from internal waves generated at depths of marked density gradient."

Weiss further suggested that the installation of multi-level or modifiable intake structures could be used effectively only if its use was mediated by intelligent monitoring of the above listed factors. This technique has what appears to be added benefits because it could be used to control the supply of nutrients available to organisms that may inhabit the pipes and dead ends, and could result in less fouling of the insides of pipes.

Macroinvertebrates can be excluded from entering a distribution system by the installation of occluding screens. The use of excelsior filters placed over intake pipes is recommended. Flentje (1945b) reported that filters 6 feet $\times$ 3 feet $\times$ 3.5 feet effectively stopped crustaceans and other organisms from entering the system. These filters had a life span of at least 4 months. Kelley (1955) stated that after examination and experimentation with control mechanisms, specifically sand filters and microstrainers, "there appeared to be no way of organizing slow sand filters to prevent the seasonal recurrence of the troubles. The solution of the problem lay either in the replacement of slow sand filters by a rapid sand plant or in the installation of microstrainers between the filters and the chlorination contact tank. The former solution was abandoned because systems already using rapid sand filtration plants showed evidence of colonization of filters and continued infestations. Kelley further proposed the use of microstrainers (10 foot diameter) with apertures of 23 μm. Thus, it would be likely that only small larval organisms (which have a higher sensitivity to chlorine) would pass through such a filter. However, insect and crustacean eggs and protozoan cysts, which are known to be resistant to chlorination, were ignored in their study and not monitored.

Microstrainers were installed as an alternative method of algae control in a Connecticut water supply (Wilbur, 1961; Mackenthun and Keup, 1970). Installation costs of $43,500/million gallons per day (mgd) and operating costs of $2/mgd were reported. When combined with proper chlorination, this system was reported to be effective. No follow-up was possible, since the water system was later converted to filtration.

Similarly, microstrainers (35 μm) were installed on the outlet of service reservoirs in Langford, England (Williams, 1974). Following a rigorous program to eradicate chironomid larvae from service reservoirs and the distribution system flushing, this installation caused some 99% removal of larvae from the final water. However, continued reports of larvae in the system occurred, leading researchers to discover that these larvae were reproducing parthenogenically (without males of the species and without passing through an adult stage). Continued infestion led to the use of synthetic pyrethrins; the results were inconclusive.

In 1969, the City of New York (1969) systematically attacked an infestation problem. They continually monitored (chemically and biologically) their reservoirs and other water sources, and reduced the number of invertebrates entering the distribution system by switching water sources on a regular basis. This practice, coupled with regular flushing of hydrants, chlorination of gate houses and stagnant areas, and the systematic relining of pipes produced good results.

Reservoir water level fluctuations may be used to control certain benthic (Chironomidae) and periphytic (Crustacea) populations by killing not only the organisms themselves (and their eggs and larvae), but also their biotic foods and substrate (aquatic plants) (MacKenthun and Keup, 1970). Winter lowering of water levels to expose littoral areas induces freezing and desiccation of rooted aquatic plants and algae species. Increasing water depths (shortly after spring growth begins to renew plant populations) may restrict the amount of light sufficiently to kill seedlings. Since many invertebrates, e.g. *Hyalella*, require such aquatic plants, this action may seriously affect crustacean populations by limiting their numbers. However, fish populations may also be affected because they feed on such invertebrates.

Biological controls have also been explored. Flentje (1945b) presented methods which may still be practiced today with varying degrees of success:

1. "Keeping down vegetation in the vicinity of the reservoir. Adult midges seem to prefer higher weeds in which to rest during the daytime. At Peoria [Illinois] such a procedure has apparently aided in discouraging midge flies from remaining in the vicinity of the reservoir."

2. "Removing one of the food supplies of the larvae by keeping down algae growths."

3. "Stocking the reservoir with fish."

Other possible methods could include the stocking of littoral areas with macroinvertebrate predators. Larvae of Odonata (damselflies and dragonflies) are known to prey upon other invertebrates, including *Hyalella* and Chironomidae.

Although the diets of such predators have been documented, the use of biological control methods has not been reported in the literature.

Survival of Invertebrates in Distribution Systems   In sampling the Worcester distribution system, it was noted that amphipods (*H. azteca*) whether collected from filters or hydrants, were usually alive (Levy, 1985). In fact, they were dramatically larger (greater than 1.0 cm) and considerably darker in color than their reservoir counterparts. The dark color suggested a diet which included the iron-rich, bacterially colonized pipe sediments. The age of the animals was estimated to be in excess of one year, (suggesting that these invertebrates were overwintering in the pipe system.

When placed in a laboratory microcosm, *H. azteca* collected from the Worcester drinking water distribution system were able to survive for extended periods of time without changing water or adding food sources. At 4°C, these amphipods exhibited a 10% survival after 8 months in captivity. Amphipods were observed to be browsing in the rust and fecal debris in the bottom of the microcosm. When an animal died, the others were observed to be cannibalistic, which may explain the large numbers of body parts and exoskeletons collected in several regions of the Worcester distribution system. These data suggest that amphipods can survive under harsh environmental conditions for extended periods of time, particularly in areas of reduced flow, high concentrations of organic debris and low chlorine residuals.

Toxicity of Free Available Chlorine to *H. azteca*   Susceptibility tests of *H. azteca* to free available chlorine (FAC) were conducted under chlorine-demand-free conditions using μ-Carrier Flasks described previously (Levy et al., 1984a). *Hyalella* was exposed to 1.0 mg/l free available chlorine for periods of 0, 60, 120, 180, 240, and 300 minutes at both 4°C and 20°C in a chlorine-demand-free system. Because the amphipods exerted a slight chlorine demand, chlorine levels were adjusted to be greater than 1.0 mg/l at the start of the experiment. Typically, chlorine levels dropped 0.2 mg/l over the longest contact time, while mean chlorine levels were always 1.0 ± 0.2 mg/l for the contact period.

Amphipods proved to be resistant to chlorine applied at 1.0 mg/l. At 4°C they exhibited a 63 percent survival after 300 minutes. There was 58 percent survival at 20°C after the same contact period. Generally, FAC was less effective at 4°C than at 20°C. These data support the viewpoint that adult amphipods can withstand chlorine concentrations commonly applied at gate houses and routinely encountered in the distribution system. More data on the effects of chloramines would be useful. Data from Levy et al. suggested that there would be limited advantage in the application of chloramines over free available chlorine. Observations that most of the copepods and cladocerans in the Worcester distribution system are dead or dying are substantiated by disinfection work done by Mitcham et al. (1983). They reported that total chlorine (54 to 89 percent free chlorine) at concentrations which ranged from 1.15 to 1.85 mg/l was effective in killing *Bosmina* (Cladocera). They also reported that chloramines produced an even greater percent kill of adult *Cyclops* (Copepoda) than free chlorine, which produced unpredictable results. Chloramines were also more effective than free chlorine in inactivating nauplii of *Cyclops*. The results of this study appear to be of less value because the exact measurements of FAC available and chloramines were not reported. In Worcester, where combined chlorine levels were frequently below 0.5 mg/l, nematodes, fly larvae, and amphipods were the only invertebrates commonly found alive within the distribution system. Copepods and cladocerans were almost always inactivated, indicating that these levels of chlorine were sufficient to inactivate those crustaceans.

Toxicity of Copper Sulfate to *H. azteca*   The addition of copper sulfate ($CuSO_4$) to municipal water supplies is a widely accepted method for the control of algae and other nuisance organisms (Muchmore, 1978). To determine the 96 hour LD_{50} of *H. azteca* (amphipods) in relation to varying concentrations of $CuSO_4$, Levy (1985) performed bioassays which were executed in the manner prescribed by *Standard Methods* (APHA, 1985). Previous toxicity experiments performed on the amphipod *Gammarus pseudolimnaeus* indicated that the LD_{50} would be approximately 0.02 mg/l (Arthur and Leonard, 1970). In these experiments, the LD_{50}, calculated by linear estimation, was 0.06 mg/l. The sensitivity of *H. azteca* to low concentrations of copper sulfate suggest that this chemical should be an effective control agent. One would expect $CuSO_4$ to be more effective under our experimental conditions since the reservoir water used was relatively soft and there was no chemical demand present in the test vessels except the amphipods. Since $CuSO_4$ is relatively inexpensive, easy to apply, and a proven algicide, this chemical appears to be a viable choice for amphipod control within the reservoir and distribution system. He speculated that other chemicals (e.g. cuprichloramines) might be more desirable for invertebrate control in distribution systems.

11.6   Control of Invertebrate-associated Bacteria

Experimental Seeding and Disinfection of Associated Bacteria: Chlorine and Chloramine Inactivation Kinetics   Chlorine and monochloramine inactivation experiments were conducted as reported by Levy et al. (1984a). Disinfection of associated *Enterobacter*, *Klebsiella*, and *Salmonella* at concentrations of 0.5, 1.0 and 2.0 mg/l free available chlorine and monochloramines at pH 6.0 and 8.0 respectively was conducted. The chlorine-to-nitrogen weight-ratio was 5:1 in all of the monochloramine experiments, since this ratio had been found to give the greatest rate of inactivation (Ward et al., 1984). Results of these inactivation experiments were summarized in Levy (1986).

Bacteria associated with *Hyalella* always survived better than similarly inactivated unassociated bacteria, as reported in other studies (Butterfield et al., 1943; Wolfe et al., 1984) This was true at both 4°C and 20°C. FAC at equivalent contact times and concentrations was always more effective in inactivating associated bacteria than monochloramine, regardless of temperature. *Enterobacter* and *Salmonella* behaved similarly in the presence of FAC at all concentrations and temperatures, while *Klebsiella* was the most susceptible to inactivation with FAC at both 4°C and 20°C. *Enterobacter* was more resistant to inactivation with monochloramines than the other bacteria at all concentrations and temperatures. *Enterobacter* exhibited interesting inactivation kinetics after disinfection with monochloramines for 2 hours at 20°C and 4°C the number of associated bacteria which survived either decreased only slightly, or increased in these experiments.

These data indicate that at least in the laboratory, recently adsorbed test bacteria are able to escape inactivation by either FAC or monochloramines, and survive better than the same bacteria unassociated in the water column. The inactivation kinetics data do not support the notion that laboratory strains are less resistant to inactivation with chlorine and chloramines than environmental strains. This work strongly supports the work of LeChevallier et al. (1985) on the inactivation of surface-associated bacteria: bacteria associated with particulate granular activated carbon were afforded enhanced protection from inactivation with FAC and monochloramines when compared to unassociated bacteria under similar conditions (see Chapter 5). Therefore, the association of bacteria with invertebrates affords protection to recently attached bacteria as well as to their natural bacterial flora.

11.7 Evaluation of the Public Health Significance of Invertebrates and Their Associated Bacteria.

The potential public health significance of invertebrates was demonstrated by Chang et al. (1960a; 1960b) who established that nematodes were present in drinking water supplies. They reported that these nematodes could ingest enteric pathogens and protect them from disinfection with free chlorine, but they were unable to verify the release of viable pathogens within the excreta of the animals.

Tracy, Camarena, and Wing (1966) theorized that microbiota such as coliforms become encapsulated within copepods and are afforded protection from chlorination. They further theorized that upon reaching the consumers tap, many of these microcrustaceans have broken apart or have disintegrated, releasing the encapsulated microbiota. From minimal research data and experience, MacKenthun and Keup (1970) stated that "perhaps none of these macroinvertebrates found in waters used for domestic purposes are injurious to health". Smerda et al. (1971) fed *Salmonella sp.* to nematodes, surface sterilized the animals, and later recovered viable *Salmonella* in the excreta. Such data established a possible route for bacteria and viruses to bypass or breach the disinfection barrier and put consumers at risk. Haney (1978) reported that macroinvertebrates (in this case nematodes) could pose a threat to the public health, but only if they were exposed to potential pathogens. Gerardi and Grimm (1982) reported the view-point held by the majority of water purveyors and researchers when they wrote that, for lack of any epidemiological or microbiological data to the contrary, "these animals are not known to present a serious public health risk".

Most recently Huq et al. (1983) suggested that the attachment of *Vibrio cholerae* to live copepods might be an important factor in the ecology of this species as well as in the epidemiology of cholera. These studies suggest that macroinvertebrates could provide a possible mode of transmission and protection for naturally associated or recently adsorbed bacteria and viruses. Levy et al. (1984a, 1984b) demonstrated that bacteria were associated with invertebrates, but that these bacteria were probably nonpathogenic.

These data suggest that the occurrence and persistence of invertebrates in potable waters should not be tolerated for a variety of reasons, several of which relate directly to the capacity of the water authority to maintain high quality drinking water as required by the U.S. Safe Drinking Water Act. These reasons may be summarized as follows:

1. The presence of adult invertebrates in drinking water indicates inadequate physical or chemical treatment of the raw water source, or contamination of filtration beds in the water treatment plant.

2. The presence of certain invertebrates, e.g. amphipods, isopods, and certain worms, indicates the presence of an active infestation in the system, most likely in deadend areas. The water quality in these areas of the distribution system should be suspect, and steps outlined above should be taken to eliminate the infestation. This would concurrently improve the overall water quality in those areas of the distribution system.

3. The occurrence of invertebrates in potable water may present a health hazard if the flora of the invertebrates include potential pathogens. The risk to water consumers is however problematic. If bacterial-laden invertebrates or their decaying parts are ingested directly by the water user, a dose sufficient to elicit a pathological response may occur if the bacteria are pathogenic. If invertebrates are allowed to collect and disintegrate in point-of-use filters or other such devices, a large population of bacteria may become established. If the bacteria break through the device, or if endotoxin is released, their continued presence may constitute a health hazard.

4. Fecal matter produced by amphipods, and probably other invertebrates, contains viable bacteria and therefore contributes to the bacterial load entering into the drinking water system. These fecal-associated bacteria may be protected from disinfection by virtue of this

association, and can continue to survive in spite of residual disinfectant.

5. Once past the disinfection barrier, invertebrates and their associated bacteria and invertebrate fecal matter may serve as a stimulus and/ or inoculum for bacterial regrowth within the distribution system.

6. The ability of disinfectants to inactivate invertebrate-associated bacteria is reduced. This necessitates increasing the disinfectant dose, contact time, or both, to insure complete inactivation of potential pathogens.

7. Theoretically, invertebrates could enter the distribution, adsorb bacteria from the water within the system, and transport as well as protect those bacteria from subsequent disinfection. The chance of these events taking place will be mediated by the concentration of ambient bacteria and the length of time the invertebrates are exposed to those bacteria.

8. The probable public health risk these invertebrates pose may be gauged by the frequency of their occurrence (intact bodies, body parts, or frasse) within the distribution system waters. Subsequent microbiological analysis of the bacteria associated with these invertebrates may indicate the nature of the health risk associated with their continued presence within the system.

11.8 Conclusions and Recommendations

The public health significance of invertebrates in drinking water is now more clear. Invertebrates entering and persisting within the distribution system are very likely to harbor associated bacteria, some of which may be potentially pathogenic to humans under certain circumstances. These bacteria are afforded protection from chlorine and chloramines at levels commonly encountered in the drinking water, even when the invertebrates are completely inactivated. It should be emphasized that these organisms, with the possible exception of copepods, were infrequently sampled from drinking water or point-of-use filters and hydrants, and then in low numbers. It is clear that additional studies on the disinfection of bacteria associated with particulates is needed to insure adequate disinfection of potable waters. Filtration alone may not be sufficient to prevent invertebrate occurrences and infestations, therefore new approaches to the disinfection of drinking water should be investigated.

Control of invertebrates within the reservoir and distribution system should be accomplished by selective use of the methods summarized in Table 11.2. Frequent system maintenance (e.g., hydrant flushing, elimination of low flow and deadends and the installation and proper operation of filtration facilities) can minimize the occurrence of invertebrates. Water authorities, treat-

Table 11.2 Summary of methods for locating and controlling invertebrates in potable water distribution systems

Factor	Methodology
Site evaluation:	Hydrant, faucet and in-line filter sampling
	Routine examination of point-of-use filters
	Fiber optic examination of water distribution pipes
Chemical Control:	Copper sulfate
	Pyrethrins (pesticide used in England)
	Disinfectants: chlorine, chloramines and cuprichloramines
Physical Control:	Filtration of source waters
	Hydrant flushing program
	Cleaning and relining of pipes
	Elimination of dead ends and areas of low flow

ment plant operators, engineers, and public health authorities should be cognizant of the potential microbiological significance of the presence of invertebrates in drinking water. Monitoring turbidity and appearance of finished water should be a prerequisite for attacking the problem. Since larger invertebrates such as isopods and amphipods and fly larvae can colonize distribution systems, they should be located using hydrant sampling or by using fiber optic examination of pipelines (for example, use of the Olympus Industrial Fibroscope plus hydrant adaptor), removed physically (water or air scouring of pipelines), or killed within the system by hydraulic isolation of infestation areas followed by superchlorination or other chemical treatment of suspected areas.

Today, the best approach to invertebrate control is the systematic elimination of these invertebrates before they pass into the distribution system and to eliminate them within the system by conscientious monitoring of drinking water and routine maintenance and upgrading of the distribution system.

References

Allen, M., Jr., Taylor, R.H., and Geldreich, E.E. 1980. The occurrence of microorganisms in water main encrustations. *Journal American Water Works Association*, 72:614–625.

American Public Health Association. 1985. *Standard Methods for the Examination of Water and Wastewater*, 16th ed. American Public Health Association, N.Y.,

Amsbary, F.C., Jr. 1928. Iron removal plant at Champaign Illinois. *Journal American Water Works Association* 19:522–525.

Arnold, G.E. 1936. Plankton and insect larvae control in California waters. *Journal American Water Works Association* 28:1469–1479.

Arthur, J.W. and Leonard, E.N. 1970. Effects of copper on *Gammarus pseudolimnaeus*, *Phya integra* and *Campeloma decisum* in soft water. *Journal Fisheries Research Board of Canada* 27:1277–1283.

Atlas, R.M., Busdosh, M., Krichevski, E.J., and Kaneko, T. 1982. Populations associated with the artic amphipod *Boeckosimus affinis. Canadian Journal of Microbiology*, 28:92–99.

Bahlman, C. 1932. Larval contamination of a clear water reservoir. *Journal American Water Works Association* 24:660–664.

Barnes, R.D. *Invertebrate Zoology* Holt, Rinehart, and Winston, New York, N.Y. 1089pp.

Brown, K.W. 1933. Experience with well water in an uncovered reservoir. *Journal American Water Works Association* 25:337–342.

Boyle, P.J. and Mitchell, R. 1981. External microflora of a marine wood-boring isopod. *Applied and Environmental Microbiology* 42:720–729.

Bousfield, E.L. 1973. *The Shallow-water Gammaridean Amphipoda of New England.* Cornell University Press, Ithaca, N.Y.

Buchmann, W. 1933. Chironomos control in bathing establishments, swimming pools and water supplied by means of chlorine and copper. *Z. Gesundheitstech. Statehyg.* (Ger.) 24:B:6: 235; Abstracted in *Journal American Water Works Association* 25:1317.

Butterfield, C.T., Wattie, Megregian, S., and Chambers, C.W. 1943. Influence of pH and temperature on the survival of coliforms and enteric pathogens when exposed to free chlorine. *Public Health Reports* 58:1837–1866.

Chang, S.L., Austin, J.H., Poston, H.W., and Woodward, R.L. 1959. Occurrence of a nematode worm in a city water supply. *Journal American Water Works Association* 51:671–676.

Chang, S.L., R.L. Woodward, and Kabler, P.W. 1960. Survey of free-living nematodes and amoebas in municipal supplies. *Journal American Water Works Association* 52:613.

Chang, S.L., Berg, G., Clarke, N.A., and Kabler, P.W. 1960. Survival and protection against chlorination of human enteric pathogens in free-living nematodes isolated from water supplies. *American Journal Tropical Medicine and Hygiene* 9:136–142.

Chang, S.L. and Kabler, P.W. 1962. Free living nematodes in aerobic treatment plant effluents. *Journal Water Pollution Control Federation* 34:1256–1261.

Cipolla, J. 1976. Amphipoda Study Final Report, Summer 1976. Department of Public Health, City of Worcester, Worcester, Massachusetts, September 29, 1976.

Cobb, N.A. 1918. Filter-bed nemas: nematodes of the slow sand filter-beds of American cities. *Contributions Science Nematology* (Cobb) 7:189–212.

Cooper, W.E. 1965. Dynamics and production of a natural population of a fresh-water amphipoda, *Hyalella azteca. Ecological Monographs* 35:377–394.

Crabill, M.P. 1956. Biological infestation at Indianapolis. *Journal American Water Works Association* 48:269–274.

Crabill, M.P., Becker, R.J., Derby, R.L., Ingram, W.M., Olsen, T.A., Ongerth, H.J., Palmer, C.M., Renn, C.E., and Silvey, J.K.G. 1960. Questions and answers on biological infestations. *Journal American Water Works Association* 52:1081–1084.

Croll, N.A. and Matthews, B.E. 1970. *Biology of Nematodes*, Halsted Press, New York, N.Y. 201pp.

Ducklow, H.W., Boyle, P.J., Maugel, P.W., Strong, C., and Mitchell, R. 1979. Bacterial flora of the schistosome vector snail *Biomphalaria glabrata. Applied and Environmental Microbiology* 38:667–672.

Ducklow, H.W., Clausen, K., and Mitchell, R. 1981. Ecology of the schistosomiasis vector snail *Biomphalaria glabrata. Microbial Ecology* 7:253–274.

Edmondson, W.T., (ed.). 1973. *Ward and Whipple's Freshwater Biology*, 2nd ed. John Wiley and Sons, Inc., New York, N.Y.

English, E. 1958. Biological problems in distribution systems. Infestation of water mains. *Proceedings Society Water Treatment and Examination* 7:127–143.

Flentje, M.E. 1945a. Twenty years of trouble shooting. Unpublished as cited in Flentje, M.E., *Journal American Water Works Association* 37:1194–1203.

Flentje, M.E. 1945b. Control and elimination of pest infestation in public water supplies. *Journal American Water Works Association* 37:1194–1203.

Flentje, M.E. 1945c. Elimination of midge fly larvae with DDT. *Journal American Water Works Association* 37:1053.

Floris, R.B. 1935, Metropolitan Water Board (London), 28th chemical and bacteriological report for year ending December 31, 1933. Abstracted in, *Journal American Water Works Association* 27:1194–1203.

George, M.G. 1966. Further studies on the nematode infestation of surface water supplies. *Environmental Health* 8:93–102.

Gerardi, M.H. and Grimm, J.K. 1982. Aquatic invaders. *Water/Engineering and Management,* 10: 22–23.

Haney, P.D. 1978. Evaluation of microbiological standards for drinking water. *Water and Sewage Works* 125:R126–134.

Hargrave, B.T. 1970a. The utilization of benthic microflora by *Hyalella azteca. Journal of Animal Ecology* 39:427–437.

Hargrave, B.T. 1970b. distribution, growth and seasonal abundance of *Hyalella azteca* (Amphipoda) in relation to sediment microflora. *Journal Fisheries Research Board of Canada* 27:685–699.

Hechmer, C.A. 1932. Chironomos in water supply. *Journal American Water Works Association* 24(5):665–668.

Hudgins, B. 1931. Turbidity, plankton, and mineral content of the Detroit water supply. *Journal American Water Works Association* 23:435–444.

Huq, A., Small, E.B., West, P.A., Huq, M., Rahman, R., and Colwell, R.R. 1983. Ecological relationships between *Vibrio cholerae* and planktonic crustacean copepods. *Applied and Environmental Microbiology,* 45:275–283.

Kelley, S.N. 1955. Infestation of the Norwich, England water system. *Journal American Water Works Association* 47:330–334.

King, C.H., Schotts, Jr., E.B., Wooley, R.E., and Porter, K.G. 1988. Survival of coliforms and bacterial pathogens within protozoa during chlorination. *Applied and Environmental Microbiology,* Vol 54:3023–3033.

Koslucher, D.G. and Minshall, G.W. 1973. Food habits of some benthic invertebrates in a northern cool-desert stream. *Transactions of the American Microscopical Society* 92:441–452.

Klug, M.J. and Kotarski, S. 1980. Bacteria associated with the gut tract of larval stages of the aquatic cranefly *Tipula abdominalis* (Diptera; Tipulidae). *Applied and Environmental Microbiology* 40:408–416.

LeChevallier, M.W., Hassenauer, T.S., Camper, A.K., and McFeters, G.A. 1985. Disinfection of bacteria attached to granular activated carbon. *Applied and Environmental Microbiology* 48:918–923.

Levy, R.V., Cheetham, R.D., Davis, J., Winer, G., and Hart, F.L. 1984a. Novel method for studying the public health significance of macroinvertebrates occurring in potable water. *Applied and Environmental Microbiology* 47:889–894.

Levy, R.V., Hart, F.L., and Cheetham, R.D. 1986. Occurrence and public health significance of invertebrates in drinking water systems. *Journal American Water Works Association* 78:105–110.

Levy, R.V. 1985. *The occurrence and disinfection of invertebrates and associated bacteria in water supplies and distribution systems.* Ph.D. Thesis, Worcester Polytechnic Institute, Worcester, MA., pp. 112.

Lorenz, R.C., McGill, N.R., and Beer, T.A. 1986. The effects of free chlorine on nematode mortality. pp. 787–793 in *Advances in water analysis and treatment.* Technology Conference Proceedings, American Water Works Association, Portland, Oregon.

Luczak, J., Rybak, M., and Ranke-Rybicka, B. 1980. Wystepobwanie organizmov wodnych w wodzie wodociagowej (Aquatic organisms in tap water). *Rocz. Panstw Zakl. Hig. T.* XXXI, NR3:319.

MacKenthun, K.M., and Keup, L.E. 1970. Biological problems encountered in water supplies. *Journal American Water Works Association,* 62:520–526.

Malloch, J.R. 1915. *The Chironomidae, or midges of Illinois, with particular reference to the species occurring in the Illinois River.* Illinois State Laboratory of Natural History Bulletin No. 10:6.

Mathias, J. 1971. Energy flow and secondary production of the amphipods *Hyalella azteca* and *Crangonyx richmondensis occidentalis* in Marion lake, British Columbia. *Journal Fisheries Research Board Canada* 28:711–726.

Mitcham, R.P., Shelley, M.W., and Wheadon, C.M. 1983. Free chlorine versus ammonia-chlorine: disinfection, trihalomethane formation, and zooplankton removal. *Journal American Water Works Association* 75:196–198.

Mitcham, R.P. and Shelley, M.W. 1980. The control of animals in water mains using permethrin, a synthetic pyrethroid. *Institute of Water Engineers and Scientists* 34(5):474–483.

Mott, J.B. and Harrison, A.D. 1983. Nematodes from river drift and surface drinking water supplies in southern Ontario. *Hydrobiologia* 102:27–38.

Muchmore, C.B. 1978. Algae control in water-supply reservoirs. *Journal American Water Works Association* 70:273–279.

Muller, H.G. 1980. Experiences with test systems using *Daphnia magna*. *Ecotoxicology and Environmental Safety* 4:21–25.

Nicholas, W.L. 1975. *The biology of free living nematodes*. Oxford University Press, London. 216pp.

Odum, E.P. *Fundamentals of Ecology*, 3rd ed., W.B. Saunders Company, Philadelphia, Pa. 574pp.

Organization Conference of Virginia Water-Sewage Works Association, Proceedings. 1929. Abstracted in, *Journal American Water Works Association* 21:1603.

Palmer, L.E. and Fowler, H.S. *Field Book of Natural History*, 2nd ed., McGraw Hill Book Company, New York, N.Y. 779 pp.

Pennak, R.W. *Freshwater invertebrates of the United States*. Ronald Press Co., New York, N.Y.

Reasoner, D.J. and Geldreich, E.E. 1985. A new medium for the enumeration and sub-culture of bacteria from potable water. *Applied and Environmental Microbiology* 49:1–7.

Remmer, S. 1977. Amphipoda presence in the Worcester water supply. Rationale for proposed study procedure. Unpublished reports. Report 1977a, May 13, 1977; 1977b, May 19, 1977; 1977c, June 3, 1977; 1977d, June 27, 1977.

Ridgeway, H.F. and Olson, B.H. 1982. Chlorine resistance patterns of bacteria from two drinking water distribution systems. *Applied and Environmental Microbiology* 44:972–987.

Safe Drinking Water Act. 1974. Public Law 93-523. SWDA, 42 USC 300f, et seq.

Sands, J.R. 1968. The control of animals in water mains. Medmenham, *Water Research Association*.

Sarai, D.S. 1976. Total and fecal coliform bacteria in some aquatic and other insects. *Environmental Entomology* 5:365–367.

Schaible, G.J., Jr. and Lahrman, J.R. 1969. *"Report to the Department of Public Health, City of New York on the Larvae in the Croton Water Supply"*. Unpublished Report. December 10, 1969.

Small, I.C. and Greaves, G.F. 1986. A survey of animals in distribution systems. *Journal of Water Treatment and Examination* 19:150–183.

Smerda, S.M., Jensen, H.J., and Anderson, A.W. 1971. Escape of *Salmonellae* from chlorination during ingestion by *Pristionchus lheritieri* (Nematoda: Diplogasterinae). *Journal of Nematology* 3:201–204.

Sochard, M.R., Wilson, D.F., Austin, B., and Colwell, R.R. 1979. Bacteria associated with the surface and gut of marine copepods. *Applied and Environmental Microbiology* 37:750–759.

Strong, D.R., Jr. 1972. Life history variation among populations of an amphipod (*Hyalella azteca*). *Ecology* 53:1103–1111.

Strong, D.R., Jr. 1973. Amphipod amplexus, the significance of ecotypic variation. *Ecology* 54:1383–1388.

Tombes, A.S., Abernathy, A.R., Welch, D.M., and Lewis, S.A. 1979. The relationship between rainfall and nematode density in drinking water. *Water Research* 13:619–622.

Touvinen, O.H., Button, K.S., Vuorinen, Carlson, L., Mair, D.M., and Yut. L. 1980. Bacteriological, chemical and mineralogical characteristics for tubercles in distribution pipelines. *Journal American Water Works Association* 72:626–635.

Tracy, H.W., Camarena, V.M., and Wing, F. 1966. Coliform persistence in highly chlorinated waters. *Journal American Water Works Association* 58:1151–1159.

United States Environmental Protection Agency. 1977. National Primary Drinking Water Regulations, Washington, D.C. 159pp.

United States Public Health Service. 1970. Public Health Service Drinking Water Standards, 1962. Publication No. 956, U.S. Government Printing Office, Washington, D.C. 61pp.

Walker, A.P. 1983. The microscopy of consumer complaints. *Journal of the Institution of Water Engineers and Scientists* 37:200–214.

Ward, R.W. and DeGraeve, G.M. 1978. Residual toxicity of several disinfectants in domestic and industrial wastewaters. *Journal Water Pollution Control Federation* 50:2703–2722.

Ward, N.R., Wolfe, R.L., and Olson, B.H. 1984. Effect of pH, application technique, and chlorine-to-nitrogen ratio on disinfectant activity of inorganic chloramines with pure culture bacteria. *Applied and Environmental Microbiology* 48:508–514.

Weber, C.I. 1973. Biological field and laboratory methods for measuring the quality of surface waters and effluents. National Environmental Research Center, U.S. Environmental Protection Agency, Cincinnati, Ohio. EPA-670/4-73-001.

Weiss, C.M. 1971. Biological control by water-intake location. *Journal American Water Works Association* 63:185–188.

Wilbur, C.C. 1961. Microstrainers to Remove Insect Larvae. *Public Works* 92:118.

Williams, D.N. 1974. An infestation by a parthenogenic chironomid. *Water Treatment and Examination* 23:215–231.

Willoughby, L.G. and Earnshaw, R. 1982. Gut passage times in *Gammarus pulex* (Crustacea, Amphipoda) and aspects of summer feeding in a stony stream. *Hydrobiologia* 97:105–117.

Wolfe, R.L., Ward, N.R., and Olson, B.H. 1984. Inorganic chloramines as drinking water disinfectants: A review. *Journal American Water Works Association* 76:74–78.

12

Biofilms in Potable Water Distribution Systems

E. van der Wende and W. G. Characklis

Institute for Biological and Chemical Process Analysis, Montana State University, Bozeman, Montana

12.1 Introduction

High bacterial populations in potable water distribution systems, sometimes referred to as events or blooms, have troubled utilities because of their possible implications for the hygienic safety and taste and odor of their product. Before considering the contribution of biofilm accumulation to these high bacterial populations in distribution systems, some terminology must be clarified with regard to drinking water bacteriology.

The water utility industry uses the terms "regrowth" and "aftergrowth" synonymously to describe the processes contributing to the increase in number of organisms in distribution systems with distance away from the treatment plant. However, Brazos and O'Connor (1985) proposed the following more specific definitions: **regrowth** is the recovery of disinfectant-injured cells which have entered the distribution system from the water source or treatment plant and **aftergrowth** is growth of microorganisms native to a water distribution system. Both terms implicate the microbial growth process in the phenomenon of increased microbial cell numbers. These terms do not clearly discriminate between the two primary mechanisms by which the microorganisms appear in the distribution system, i.e., breakthrough in the treatment plant and growth within the distribution system.

Breakthrough is the increase in bacterial numbers in the distribution system resulting from viable bacteria passing through the disinfection process. The surviving cells inoculate the biofilms in the distribution system and/or reproduce in the bulk water. Accordingly, injured organisms entering the distribution system recover and contribute to the process of breakthrough. Transport of viable bacterial cells through the disinfection process is very common in drinking water treatment (McFeters et al. 1986). Disinfection only means ridding the water of viable pathogenic organisms.

Growth is the increase in viable bacterial numbers in the distribution system resulting from bacterial growth downstream of the disinfection process. Growth may occur in biofilms or as planktonic growth in the distribution system water. (*Planktonic cells* are viable bacterial cells suspended in the water phase). An **episode** is an occurrence of "excessive" viable bacteria in the distribution system. Episodes result from breakthrough, growth, and, sometimes, from cross-connections and back-siphoning. Thus, an episode refers to the occurrence while breakthrough and growth refer to its causes.

Excessive viable bacteria are numbers of viable bacteria exceeding drinking water standards. Since present U.S. standards only refer to coliforms, excessive bacteria only refers to these bacteria. Coliforms frequently enter distribution systems via breakthrough without an episode occurring. The standard for the membrane filtration method permits a maximum of one coliform per 100 milliliters based on a monthly average of all samples.

Insufficient disinfectant concentration is a disinfectant residual which permits excessive viable bacteria to exist at some location in the distribution system. Since the disinfectant residual varies spatially and temporally in the distribution system, an insufficient disinfectant concentration may exist in one part of the system (e.g., "dead ends") and not in another (e.g., constantly flowing regions). Disinfectant concentration may also vary radially in the pipe at any given axial position.

12.2 Biofilms

A biofilm is a layer of microorganisms in an aquatic environment held together in a polymeric matrix attached to a substratum (Figure 12.1). The matrix consists of organic polymers that are produced and excreted by the biofilm microorganisms and are referred to as extracellular polymeric substances (EPS). The chemical structure of the EPS varies among different types of organisms and is also dependent on environmental conditions. Biofilms sometimes form continuous, evenly distributed layers but are often quite patchy in appearance. Biofilms in water distribution systems are thin, reaching maximum thicknesses of perhaps a few hundred micrometers. Biofilms from natural environments are generally heterogeneous, frequently containing more than one distinct microenvironment. For example, biofilms with aerobic as well as anaerobic strata are common. This means that different microenvironments inhabited by different microorganisms exist in the direction perpendicular to the substratum. Consequently, the term biofilms does certainly not only refer to a uniform surface community. In addition, the film often contains organic or inorganic debris from external sources. Inorganic particles may result from the adsorption of silt and sediment, precipitation of inorganic salts, or corrosion products.

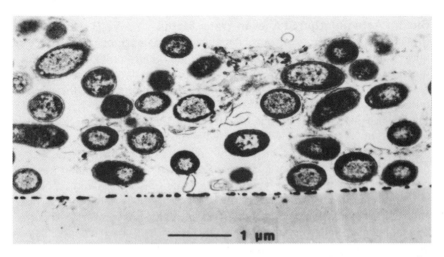

Figure 12.1   Transmission Electron Microscopy (TEM) picture of a cross section of a very thin biofilm.

12.3   Observations of Biofilms in Water Distribution Systems

Limitations in Water Distribution System Research   A thorough study of biofilm accumulation and biofilm composition throughout a water distribution system has not been reported, probably because access to these systems is difficult. Most observations consist of analysis of samples obtained during flushing and pigging of water mains, physical main cleaning procedures used to remove biofilm, corrosion products, and sediment from the distribution system. Samples obtained this way consist of biofilm as well as sediments and corrosion products dislodged by the pigging and flushing processes. Results of bacterial enumeration procedures of such samples can not be accurately related to a biofilm surface area because the relation between sample volume and the corresponding pipe wall surface area is unknown. In addition, the extent of biofilm removal from the pipe wall surface is unknown and limits the interpretation of the data.

The biofilm on the often heavily corroded pipe walls is hard to sample and sample handling causes difficulties. For example, instead of water, the pipe surface with the biofilm is often exposed to air for long periods of time preceding viable cell enumerations. This significantly influences these analyses. Direct cell counts on distribution system samples are complicated by the presence of inorganic particles (e.g., corrosion products). Electron micrographs of in situ biofilm of a single pipe surface sample are fragmentary and frequently neglect the spatial heterogeneity of the biofilm.

Biofilm Accumulation Studies in Water Mains   Allen (1980) examined the occurrence of microorganisms on main encrustations (tubercles) by electron microscopy. Water utilities throughout the United States provided water main tubercles obtained during water main servicing, repair, or replacement. Tubercles from seven different water utilities were examined and the electron micrographs of these tubercles showed a number of common characteristics. The surfaces had a clear porous texture, while beneath the surface veneer a multitude of crystal arrays was found with several tubercles having laminar deposits near the surface. On most tubercles, microorganisms were mainly observed at or near the surface. However, active bacterial colonization was also detected beneath the surface veneer while some cells were found embedded in the granular matrix deep in the tubercle. Allen (1980) stated that bacterial enumeration by conventional methods (scraping of the tubercle surface followed by viable cell counts) greatly underestimates the true bacterial populations present in the tubercles. Certainly, viable counting techniques based on colony growth on solid media only recover a fraction of the living cells, especially when mixed populations are enumerated. Furthermore, electron microscopy cannot distinguish between intact dead cells and living cells.

Ridgway and Olson (1981) used scanning electron microscopy to observe the bacterial colonization of a distribution main consisting of galvanized iron pipe with a 0.5 cm thick hard, black lining, presumably cement or porcelain. The main had been in service 25 years and appeared to be in good condition since iron tubercles were not observed. The main carried unchlorinated ground water for most of the year but it was blended with 10 to 15% chlorinated, fully treated surface water during the hot summer months. A low magnification scanning electron micrograph of a representative area of the surface of the liner showed that it was covered with a thin (10 to 100 μm) amorphous mineral encrustation. The crust contained a network of fissures and cavities, increasing the total available surface area of the pipe, thus providing a variety of potential microhabitats for microorganisms. Aluminum, silicon, phosphorus, sulfur, calcium, and iron were the predominant elements at the pipe surface. This was quite different from the situation in the water phase where both iron and phosphorus were found to be minor soluble constituents. Similarly, manganese and zinc were present in very low concentrations in the water supply but readily detected on the pipe surface. Thus, cells attached to the pipe surface were exposed to concentrations of dissolved compounds that were significantly different from the concentrations in the bulk water. Such altered concentrations of minerals and nutrients could have a profound effect on the growth of attached organisms and their susceptibility towards disinfectants. The surface was examined for microorganisms at high magnification. Many of the observed cells had a coccoid shape and the cells were sometimes linked together in chains like streptococci. Filamentous bacteria that bore a clear morphological similarity to the streptomycetes were the most frequently observed microorganisms. Iron-oxidizing bacteria of the genus *Gallionella* were also frequently observed. *Streptomycetes* and *Gallionella* were also often found as planktonic cells in water samples. Although the water main had been in service for 25 years, no continuous biofilm was found on its surface. Only sparsely and randomly distributed microcolonies were present on the pipe surface, frequently associated with crevices in the mineral layer. The small amount of biomass accumulation in the system may have been the result of low organic carbon concentrations in the ground water (no data available).

Groundwater often contains smaller amounts of organic carbon than surface water. O'Connor and Banerji (1984) found that the microbial growth on pipe walls was related to total organic carbon (TOC) levels in the water and that TOC levels of 5.0 mg/l or less limited the amount of microbial growth in their laboratory experimental system. Very little growth was observed at an organic-carbon level of 0.5 mg/l. A typical TOC level for treated groundwater is 0.5 to 3.0 mg/l while the TOC level in treated surface water is in the range of 2.0 to 5.0 mg/l.

Characklis (1988) studied the biofilm accumulation on the surface of PVC pipe that was exposed to chlorine-free drinking water prepared from surface water. A biofilm that covered the entire pipe surface and was easily visible had developed after an exposure time of several weeks. Viable cell counts of the biofilm cells after an exposure time of 85 days revealed cell densities of approximately 5×10^{10} cells/m^2. O'Connor and Banerji (1984) found no significant differences in the accumulation of cells on copper, PVC, or iron surfaces. Thus, this relatively high level of biofilm accumulation was probably not a result of different pipe materials. In a somewhat similar situation, Haudidier et al. (1988) studied the accumulation of biomass in a plug-flow system comprised of six serially disposed loops. The influent water was fully-treated river water without any disinfectant. At steady state, the viable biofilm cell density at the beginning of the plug-flow system was almost 10^{11} cells/m^2 while a constant decline was observed with increasing residence time.

Physicochemical and Bacterial Analyses of Water Main Tubercles

Tuovinen (1980) examined tubercles from uncoated or poorly coated iron pipes for their bacterial, physicochemical, and mineral properties. Using selective growth media, Tuovinen isolated a variety of different bacterial groups including sulfate reducers, nitrate reducers, nitrite oxidizers, ammonia oxidizers, sulfur oxidizers, and various unidentified heterotrophic microorganisms. The relative importance of the various groups in the bacterial community could not be estimated since cell numbers were not determined. Oxidation-reduction (redox) potential measurements throughout the tubercles indicated different microenvironments that could explain the presence of microorganisms from various bacterial groups. High redox potentials, reflecting aerobic conditions, were found at the tubercle surface (the oxygen concentration in the bulk water was approximately 8 mg/l). Low redox potentials, reflecting highly reduced conditions, were found at 1 or 2 mm depth in the tubercles. Thus, bacterial niches were available for aerobic, facultative anaerobic, and anaerobic bacteria. The reducing conditions of the tubercle interior may have been a suitable habitat for the anaerobic, sulfate-reducing bacteria. The obligate aerobic organisms, including those that oxidize inorganic compounds for energy, apparently remain in the exterior layer which is exposed to oxygen where reduced nitrogen and sulfur compounds are accessible from the tubercle interior. The presence of large amounts of reduced iron in the tubercles (an $Fe(II)/Fe(III)$ ratio between 1.2 and 2.0) may contribute to the chlorine demand of the pipe wall and therefore may hinder efforts to control bacteria in the distribution system.

LeChevallier et al. (1987) examined and characterized pipe scrapings, tubercles, biofilm flocs, and sediments from a water distribution system in New Jersey for bacterial and mineral properties. The samples were taken approximately 1.7 km from the water treatment plant from a 15 cm ID, 28 year old, bituminous-coated main. Pipe scrapings were taken just before a flushing and pigging event. At the onset of the flushing, a dark brown floc suspension and a small amount of sediment was collected. More sediment and tubercles were collected during the pigging process. The flox contained 21.4% zinc, probably resulting from accumulation of zinc orthophosphate used for corrosion control. It also contained 2.4×10^6 heterotrophic plate count (HPC) bacteria/ml, mainly *Flavobacterium sp.* and *Pseudomomas vesicularis*. These same species were the predominant planktonic species found in the bulk water. The flushing sediment contained 8.4×10^6 HPC bacteria/gram, mainly *Arthrobacter sp.* The pipe scraping contained 1.0×10^7 HPC bacteria/m^2, again mainly *Flavobacterium sp.* and *P. vesicularis*. The dislodged iron tubercles consisted of 98.7% iron and were the only particulate samples that contained significant numbers of coliforms (>160 colony-forming units (CFU)/gram). The number of HPC bacteria in tubercle material was 1.3×10^7 cells/gram. The number of cells accumulated on the pipe wall surface could not be determined from these data. The chlorine concentration was probably much higher in the bulk water than in the iron tubercles, which may have contributed to the coliform survival in the tubercles. The tubercle environment somehow provided a suitable microenvironment for the coliform bacteria. The most superficial layers (flocs + scrapings) contained essentially the same bacterial species that were found in the water phase, suggesting that the planktonic cells present in the bulk water were mainly cells detached from the biofilms.

Camper (personal communication, 1988) analyzed tubercles and flush sediments from a Connecticut water distribution system for the presence of HPC, iron oxidizing, and sulfate-reducing bacteria. A free-chlorine residual of approximately 1 mg/l was maintained in this system. Tubercle and sediment material contained high levels of HPC bacteria. Iron bacteria and low numbers of sulfate-reducing bacteria were only associated with iron tubercles.

Conclusions on Biofilm Distribution   Observations of biofilms in water distribution systems indicate little qualitative or quantitative uniformity. Although some bacterial species were more frequently detected than others, a fairly large variation in population composition was seen. Also, cell numbers and distribution on and in the pipe wall lining or corrosion product layer varied widely. Considering the large differences in the water distribution system environments, this is not surprising. Differences exist in the quality of the treatment plant effluent (the finished water) and in the structure and chemical composition of the pipe wall. The flow rates may also affect the biomass accumulation at the pipe wall surface. Important water quality parameters appear to include the nutrient level, the presence of disinfectants, and other chemical and physical parameters like pH, hardness, and temperature.

12.4   Relevance of Biofilms in the Water Distribution System

Water Quality Problems   The depletion of dissolved oxygen, reduction of sulfate to hydrogen sulfide, occurrence of bad taste and odor, and the occurrence of red water in relation to corrosion of cast-iron mains may result from microbial activity. Lee et al. (1977) conducted a survey to determine the extent and nature of water quality problems in the distribution systems in the United States. The results indicated that 60% of the responding utilities report taste and odor to be their most common water quality problems. Red water caused by ferric iron ranked second, with 48% of the utilities having this problem. Complaints of cloudy and black water (manganese oxides) were the next two most frequently cited water quality problems. In addition, virtually every utility periodically experienced some water quality problem originating in the distribution system. Evaluations of distribution system water quality in five Missouri communities indicated that higher bacterial plate counts were generally observed at locations where consumers reported water quality problems (Banerji, 1978).

Biofilms and Water Quality Problems   Waterborne diseases, especially those caused by bacteria, have become rare in the United States and other developed parts of the world, although the number of cases is increasing.

Legionella pneumophila is one of the very few bacterial waterborne diseases that occasionally still causes problems in the United States. *L. pneumophila* sometimes grows in water systems, especially hot-water systems and causes legionellosis, a disease which is often fatal. Colbourne et al. (1988) studied the presence of *L. pneumophila* in public water supplies in England and found culturable *L. pneumophila* to be associated with deposits or biofilms in mains although this organism only appeared in water when these materials were dislodged from the pipe wall.

Serious events of black water are sometimes caused by bacteria of the genus *Hyphomicrobium*. A clear example of this is the biofilm sloughing that occurred repeatedly in a drinking water distribution system in Queensland, Australia (Hamilton, personal communication, 1986). The biofilm that had accumulated on the pipe wall of the distribution system contained high numbers of *Hyphomicrobium*. These bacteria form hypha-like structures that are readily covered with a black precipitate of manganese oxide arising from low concentrations of reduced manganese present in a dissolved form in the water. The cell wall of *Hyphomicrobium* plays a catalytic role in the otherwise spontaneous oxidative precipitation process. The hypha-like structures yielded a biofilm that was relatively thick and heavy from the manganese oxide precipitate. When a sloughing event occurred in the system, biofilm downstream of the point of initiation would also slough in an avalanche-like event. The result was black water that caused problems such as irreversible staining of laundry.

Controlled Biofilm Experiments in Distribution Systems   The role that biofilms play in the chemical and microbiological deterioration of drinking water is clear in cases like the manganese oxide precipitation caused by *Hyphomicrobium* and to a lesser extent in contamination with *Legionella*. However, generally, we do not know the extent to which biofilm cell growth and detachment contribute to the increase in planktonic cell numbers in the distribution water. Experiments designed to examine the contribution of biofilms to elevated planktonic cell numbers in the distribution system were reported by Van der Wende et al. (1989). The purpose of the experiments was to quantify the relative contribution of detached biofilm cells and growth of planktonic cells to the total planktonic cell concentrations in a drinking water distribution system.

The experiments were conducted with a RotoTorque System (RTS; Figure 12.2) which consists of four (4) RotoTorque (RT) reactors (Figure 12.3) in series. Each RT is a continuous flow stirred tank reactor (CFSTR) with a 15 cm (6 in) ID water pipe section forming the outer wall of the reactor. A solid PVC drum spins inside the pipe section. The reactor can be easily dismantled so that pipe sections of different material or varying age can be conveniently substituted in the reactor. The liquid volume of each RT is 1.575 l and the wetted surface area is 0.207 m².

The RTS simulates the plug flow nature of a water main in that concentration gradients exist in the direction of flow. Mains of variable length carrying water at variable flow rates can be modelled with this system. The volumetric flow rate determines the hydraulic residence time of the RTS. The rotational speed of the drum creates the shear stress imposed on the inside pipe wall surface of a water main. Therefore, combinations of main length and flow velocity can be modelled by selecting the appropriate combination of rotational speed and hydraulic residence time and shear stress, their effects on biofilm processes can be studied independently. The four RTs serve as "windows" at fixed positions along the modelled main. Water and biofilm samples can be collected from these "windows."

Two RTs were located at a water treatment plant that purified lake water using the "direct filtration" process. Water quality parameters and other physical and chemical properties of the treatment plant effluent are presented in Table 12.1.

When the hydraulic residence time in the RTS is significantly shorter than

Figure 12.2   A RotoTorque system as used at the water treatment plant.

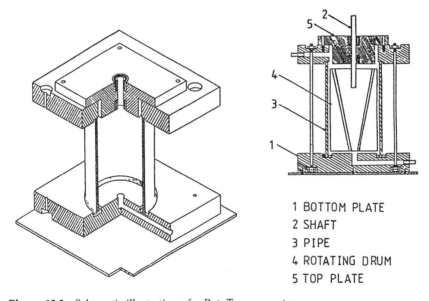

1 BOTTOM PLATE
2 SHAFT
3 PIPE
4 ROTATING DRUM
5 TOP PLATE

Figure 12.3   Schematic illustration of a RotoTorque reactor.

the generation time of the planktonic cells, replication of planktonic cells in the RTS is negligible. As a consequence, the measurement of biofilm (X_b) and planktonic cell numbers (X) in the RTS for two suitable hydraulic residence times (V/Q), one short and one long, permits the calculation of the specific growth rate in the water (μ) and in the biofilm (μ_b) by means of material

Table 12.1 Physical and chemical properties of the water treatment plant effluent (RTS influent) during the period of biofilm accumulation (average values)

Parameter	Value
pH	7.2
Alkalinity	15 mg/l (as $CaCO_3$)
TOC	2 mg/l (as carbon)
Color	1 Color Unit
Turbidity	0.3 nepholometric turbidity unit (NTU)
Ammonia	0.1 mg/l (as nitrogen)
Nitrate	0.2 mg/l (as nitrogen)
Temperature	21.1 °C

balances. The following mathematical expressions describe the balance for cells in RT1:

Planktonic cell balance in RT1:

$$V \frac{dX_1}{dt} = Q[X_0 - X_1] + \mu X_1 V + r_d X_{bl} A \qquad 1$$

rate of accumulation	net rate of transport out of reactor	net rate of growth in the bulk liquid	net rate of detachment from the biofilm

Biofilm cell balance in RT1:

$$A \frac{dX_{bl}}{dt} = \mu_b X_{bl} A - r_d X_{bl} A \qquad 2$$

rate of accumulation	net rate of growth in the biofilm	net rate of detachment from the biofilm

At steady state, equation 1 and 2 simplify to:

$$\frac{Q}{V} [X_1 - X_0] = \mu X_1 + r_d X_{bl} \frac{A}{V} \qquad 3$$

$$\mu_b = r_d \qquad 4$$

The equations contain three unknowns: μ, μ_b and r_d. However, at high flow rates (short hydraulic residence times), dilution rates are much larger than the specific planktonic cell growth rate (μ) and the planktonic cell growth rate (μX_1) is negligible.

$$\frac{Q}{V}[X_1 - X_0] = r_d X_{bl} \frac{A}{V} \qquad\qquad 5$$

Thus, the set of equations is determinant and r_d can be calculated from equation 5 while μ_b can be determined from equation 4. The values for μ_b and r_d determined for the short hydraulic residence time can be used to determine μ in the long hydraulic residence experiments (equation 3).

The RTS hydraulic residence times used for the experiments were 7.3 h (Q = 0.87 1 h^{-1}) and 0.22 h (Q = 28.2 1 h^{-1}). Dilution rate (D) is the reciprocal of hydraulic residence time (V/Q). Thus, the corresponding dilution rates are 0.14 h^{-1} and 4.5 h^{-1}, respectively. The dilution rate of 0.14 h^{-1} is low enough to expect cellular reproduction in suspension in addition to growth in the biofilm. A dilution rate of 4.5 h^{-1} is too high for significant microbial cell replication in suspension (van der Kooij and Hijnen, 1982; van der Kooij et al., 1982). In order to improve the accuracy of the experiment, RTS influent was filtered during the conditions of the high dilution rate so that the planktonic cell concentration in the influent (X_0) was zero, thus eliminating background noise in the measurement of cell numbers. The rotational speed in all of the experiments provided a shear stress at the reactor surface simulating 0.92 m s^{-1} (3 ft s^{-1}) flow velocity in a pipe of 0.15 m (6 in) ID. The pipe sections in the RT were made of new cement-lined iron pipe with a bitumastic coating.

Most microbial activity occurred near the inlet to the system as indicated by the relatively high numbers of biofilm cells and the large increase in planktonic cell numbers (Figure 12.4). Thus, cell growth-rate calculations were made for RT1 using equations 3, 4, and 5. The calculated specific growth rate for biofilm cells was 0.0025 h^{-1}. This means that the cell turnover time or cell generation time in the biofilm (the reciprocal of the specific growth rate) was approximately 17 days. The specific growth rate of the planktonic cells was -0.12 h^{-1}. The negative value may be explained by analytical error (a small error in the count of detached biofilm cells at the high flow rate results in a large error in the calculation of the specific planktonic cell growth rate) or by decreasing cell viability or cell dieoff in RT1 water. The latter is quite possible as suggested by the declining viable bacterial numbers in the consecutive RT operating at the low flow rate (Figure 12.4). The decrease in cell numbers with distance (time) through the reactor may reflect a nutrient limitation.

The specific growth rate of planktonic cells in the finished water was very low. Thus, the significant increase of cell numbers in the bulk water was primarily due to detachment of biofilm cells. Net cell production rate is the product of specific growth rate and population cell numbers. In the biofilm, the low specific cell growth rate was countered by a high density of cells resulting in a net biofilm cell production rate that was relatively high compared to the planktonic cell production rate and the input rate of cells with the influent. Another determination also suggests that planktonic growth can be neglected

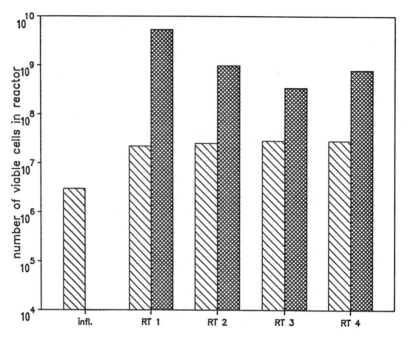

Figure 12.4 Number of viable cells (Heterotrophic Plate Count) in the bulk water (= planktonic cells) and biofilm in a RotoTorque system after an exposure time of 85 days to chlorine-free water. The overall hydraulic residence time was 7.3 hours. The data represent cell numbers per RotoTorque or per 1.575 1 (= RT volume) influent.

⧄ = planktonic cells.   ▨ = biofilm cells.

relative to the number of cells detaching from the biofilm. If all microbial growth was occurring in suspension, μ would have to be approximately 0.5 h^{-1} to account for the growth observed in RT1 for the low flow rate. Such a high growth rate may be possible for bacteria living under rather ideal conditions but not for organisms in unsupplemented tap water at low temperature (van der Kooij and Hijnen, 1982; van der Kooij et al., 1982).

LeChevallier (1987) observed a large increase in coliforms between a water treatment plant effluent and water samples taken 1.1 km down stream of that plant. The hydraulic residence time in this 1.1 km stretch moved from 91 to 102 min. The large increase in coliform numbers could be explained by the planktonic growth of these bacteria only if they had a specific growth rate of more than 2 h^{-1}. This is much too high for the low nutrient conditions, low temperature, and high chlorine residuals found in the water main. High pipe line pressures and other operating parameters ruled out cross connections and backsiphonage as causes of the coliform episode. The only remaining possible explanation was that these coliform bacteria came from biofilms in the water main.

12.5 Disinfectants and Biofilm in the Water Distribution System

Drinking Water Disinfection   Water utilities usually apply one or more disinfectants during or at the end of the water treatment process. The most often applied disinfectant in the United States is chlorine, although the use of chloramines is on the increase. In some cases, chlorine dioxide is applied. After a disinfectant has been applied to the water, the disinfectant residual concentration begins to decline due to the reaction with various dissolved and suspended compounds in the water phase and with biofilm and pipe wall materials. The disinfectants are reactive compounds which oxidize substances like Fe^{2+}, Mn^{2+}, sulfide, and organic compounds (Characklis et al., 1980). Disinfectant booster stations throughout the distribution system are sometimes used to maintain appropriate disinfectant residuals in the extremities of the system.

Effect of Chlorine on Bacteria in the Water Distribution System
Van der Wende et al. (1988) studied the effect of different chlorine levels on biofilm accumulation and bacteriological water quality in the RTS system described earlier. RTS influent was prepared by mixing chlorine-free filter effluent with chlorinated treatment plant effluent in two feed tanks equipped with overflows. One RTS received influent with a chlorine concentration of 0.8 mg/l free chlorine while another RTS received influent with a free chlorine concentrations of 0.2 mg/l. Chlorine reacted with various components of the system resulting in declining chlorine concentrations through the RTS. The flow rate through both systems was 1.8 l/h and the rotational speed of the drums was 150 rpm, simulating a flow velocity of 0.92 m/s (3 ft/s).

Biomass accumulation in the system with the low chlorine concentration (0.2 mg/l) was determined for exposure times of 17 and 45 days. Biofilm cell numbers increased through the RTS (Figure 12.5a and b). The accumulation of planktonic cells was correlated with biofilm accumulation, i.e., as biofilm accumulation increased, planktonic cell numbers increased. The maximum number of planktonic cells observed at low chlorine concentrations was approximately the same as for the chlorine-free system. However, the free-chlorine residual in the bulk water never decreased below 0.05 mg/l at any point in the system.

Biomass accumulation in the system with the high chlorine concentrations (0.8 mg/l) was determined for exposure times of 17 and 38 days (Figure 12.6a and b). Not surprisingly, viable cell numbers were lower than in other experiments due to higher chlorine levels. For the first 17 days, the RTS remained essentially free of cells. However, after approximately 38 days, a "patchy" biofilm had accumulated throughout most of the RTS. At the same time, low numbers of planktonic cells were found throughout the RTS. Free-chlorine residuals never decreased below 0.5 mg/l in the bulk water at any point of the system.

The presence of chlorine caused reduced accumulation of biofilm and in-

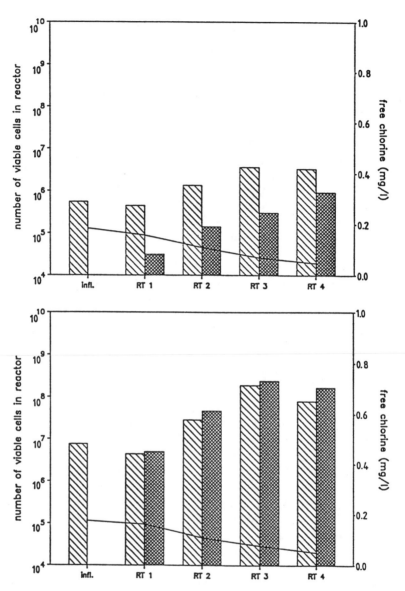

Figure 12.5 a. Numbers of viable cells (Heterotrophic Plate Count) in the bulk water (= planktonic cells) and biofilm of a RotoTorque System after an exposure time of 17 days to water with a low chlorine concentration. The overall hydraulic residence time was 3.5 hours. The data represent cell numbers per RotoTorque or per 1.575 1(= RT volume) influent. b. Numbers of viable cells (Heterotrophic Plate Count) in the bulk water (= planktonic cells) and biofilm of a RotoTorque System after an exposure time of 45 days to water with a low chlorine concentration. The overall hydraulic residence time was 3.5 hours. The data represent cell numbers per RotoTorque or per 1.575 1 (= RT volume) influent.

= planktonic cells.   = biofilm cells.   —— = chlorine.

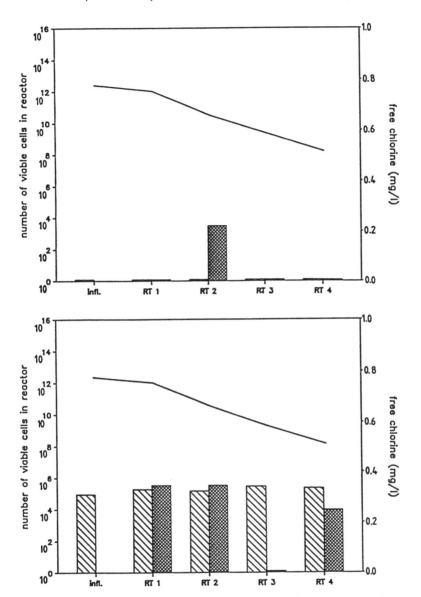

Figure 12.6   a. Numbers of viable cells (Heterotrophic Plate Count) in the bulk water (= planktonic cells) and biofilm of a RotoTorque System after an exposure time of 17 days to water with a high chlorine concentration. The overall hydraulic residence time was 3.5 hours. The data represent cell numbers per RotoTorque or per 1.575 1 (= RT volume) influent. b. Numbers of viable cells (Heterotrophic Plate Count) in the bulk water (= planktonic cells) and biofilm of a RotoTorque System after an exposure time of 38 days to water with a high chlorine concentration. The overall hydraulic residence time was 3.5 hours. The data represent cell numbers per RotoTorque or per 1.575 1 (= RT volume) influent.

⬚ = planktonic cells.   ▨ = biofilm cells.   —— = chlorine.

fluenced its distribution in the RTS. For the chlorine-free system, most biofilm accumulated at the entrance to the reactor system, probably through the rapid consumption of nutrients. Less biofilm accumulation was observed in the remaining RTS (Figure 12.4). A low chlorine concentration (0.2 mg/l) in the RTS influent reduced the biofilm in the first reactor, thus leaving a larger potential for biofilm accumulation in the subsequent reactors where the free-chlorine residual was lower (Figure 12.5). Planktonic cell numbers increased to 1 to 2 orders of magnitude in conjunction with increased biofilm accumulation between the exposure times of 17 and 45 days.

At the higher chlorine concentration (0.8 mg/l), planktonic cells were observed only in the presence of biofilm (Figure 12.6). Planktonic cells appeared only after an initial biofilm had been established. These results are consistent with the results from the growth rate experiment which showed the importance of biofilm cell growth in the increase of planktonic cell numbers.

Biofilm cell growth and detachment may be even more dominant in water distribution systems with low chlorine concentrations than in chlorine-free systems. The biofilm environment is believed to protect cells against the activity of chlorine by diffusional resistance and neutralization of chlorine as a result of the reaction with biofilm and pipe wall materials. Thus, microbial growth in the biofilm will be less inhibited than the growth of planktonic cells which experience a higher chlorine concentration.

Bacterial species less susceptible to chlorine may accumulate selectively in the distribution system. Biofilm and planktonic cells isolated from the RTS at high chlorine concentration after 38 days represented only a few species. An estimated 75 to 90% belonged to one species of the genus *Pseudomonas*. Wolfe et al. (1985) found that these bacteria can be highly chlorine tolerant. The colonies of this species, on R_2A-agar, were clearly of the "smooth type," indicating the production of significant amounts of extracellular polymer substances (EPS). Rate and extent of EPS formation in situ is influenced by many environmental conditions but it is clear that this species has the potential of producing significant amounts of EPS. The large difference in species diversity between chlorinated and unchlorinated water was also observed by other experimenters (Brazos and O'Connor, 1985; Maki et al., 1986). Ridgway and Olson (1982) described large differences in chlorine and chloramine susceptibility between bacteria obtained from chlorinated drinking water and those from a chlorine-free system. Wolfe et al. (1985) found that a red pigmented *Flavobacterium sp.* was highly chlorine-tolerant. Flavobacteria have been frequently isolated from disinfected drinking water. Thus, chlorine may lead to the selection of a limited number of microbial species in the early stages of biofilm accumulation. For example, in chlorine-free water, different species may randomly adsorb but subsequent cell growth will be dominated by the "fast growers." In chlorinated water, organisms less susceptible to the damaging effects of chlorine may adsorb to the pipe surface and can selectively accumulate in the system. Further enhancement of chlorine tolerance may be achieved by different processes such as continued selection, mutation, and/or recombination of genetic material (Kelly et al., 1983). In addition, a film of organisms containing high amounts of EPS or accumulating various reduced species characterized by a high chlorine demand may further insulate other organisms from the effects of chlorine. Although this protective mechanism is often suggested by researchers, little, if any, evidence is available that EPS protects planktonic cells against disinfection. LeChevallier et al. (1987) studied the effect of cell encapsulation with EPS on the cell susceptibility towards chlorine and monochloramine. They used two strains of *Klebsiella pneumoniae*, an encapsulated strain and a mutant which was incapable of EPS production. Dilute suspensions of the cells in low nutrient water were exposed to the disinfectants. Despite large differences in the amount of EPS, no change in the susceptibility to free chlorine or monochloramine was observed between the two strains. Obviously, cell encapsulation was not a sufficient penetration barrier to cause a measurable difference in the level of disinfection at the disinfection concentration examined. However, EPS exists in many chemical forms and some may be more protective than others depending, on the reaction kinetics with chlorine. Furthermore, a protective function of EPS may be more important in biofilms where the large quantity of EPS protects the cells deeper in the biofilm.

Disinfectant Chemistry and Biofilm Control   LeChevallier et al. (1987) conducted experiments examining the disinfection of biofilms with chlorine, monochloramine, and chlorine dioxide. They first determined the disinfectant doses (CT = concentration $\times$ time) that caused 99% kill in dispersed biofilm cell populations. These doses (different for the various disinfectants) were then applied to the initial biofilm to determine the percentage kill in these sessile cell populations. The results indicated the relatively high biofilm disinfection efficacy of monochloramine as compared to free chlorine (HOCl and OCl$-$). Apparently, monochloramine is more successful in penetrating the biofilm than free chlorine. The extent of penetration is dependent on two competing rate processes: diffusion and reaction. Siegrist and Gujer (1985) studied the diffusion rate of several different molecules and ions in biofilm accumulated on a permeable membrane. The biofilm had a total solids content of 1.8 to 2.6% and reduced the molecular diffusion rate of the test compounds to 50 to 80% of the value in water. The biofilm diffusivities of the species (glucose, sodium, and bromide) were linearly related to their diffusivities in water. The diffusivity of monochloramine will probably be similar to that of free chlorine. However, there is a large difference in the reactivity of chlorine and monochloramine with biological matter.

   In the last decade, researchers have determined that chlorine reacts much faster and with a larger variety of biological organic compounds than the chloramines. Chlorine causes oxidation, hydrolysis, or deamination of virtually every component of the bacterial cell. In contrast, monochloramine reacts rather specifically with nucleic acids, tryptophan, and sulfur-containing amino acids (Jacangelo and Olivieri, 1985) but does not react with EPS or sugars such as ribose. The reaction of chloramines with sensitive amino acids has been reported by several researchers but not the reaction with nucleic acids. Olivieri et al. (1980) reported that chlorine reacted much faster with RNA than mono- and dichloramine did. These chemical properties of chloramines do not make them particularly good primary disinfectants. Wolfe et al. (1985) studied the inactivation of heterotrophic bacterial populations in finished drinking water. They found that prereacted chloramines were less effective disinfectants than chlorine. Members of the genera *Pseudomonas, Enterobacter, Acinetobacter, Moraxella*, and *Alcaligenes* which were inactivated after 1 min exposure to 1.0 mg/l free chlorine were recovered in low levels after 30 and 60 min of exposure to 1.0 mg/l prereacted chloramines. Water consumers that are situated close to a water treatment plant often receive water with a very short disinfectant contact time. Approximately 25% of the U.S. water utilities report contact times of less than 10 min between disinfectant application and first consumer. However, the same chemical properties that negatively influence the efficacy of this compound as a primary disinfectant increase the biofilm penetration capabilities of this disinfectant. Thus, cells deeper in the biofilm are not as well protected against the activity of this disinfectant. The chemical properties of chloramines also make them more effective than chlorine in inhibiting biofilm accumulation in the extremities of a water distribution network or in parts of the system with low flow rates. A highly reactive compound like chlorine will not even reach such areas that are far from the point of chlorine addition.

12.6   Summary

Biofilms accumulate in drinking water distribution systems primarily by growth at the expense of nutrients in the water. The biofilms are continuously seeded with planktonic organisms entering the system via breakthrough. Biofilm cells are generally less susceptible to disinfectants than planktonic cells. As a consequence, excessive viable cell numbers may accumulate in the bulk water from detachment of cells from the biofilm in areas where insufficient disinfectant residual is maintained.

   Certain disinfectants may have properties more conducive to controlling biofilm populations. Less reactive, more persistent compounds, such as chloramines, maintain a higher disinfectant residual throughout the distribution system and may penetrate the biofilm more effectively and, thus, control biofilm organisms better than free chlorine. However at present, little quantitative information is available regarding disinfection kinetics in biofilms.

The authors gratefully acknowledge the financial support and in-kind services from the South Central Connecticut Regional Water Authority and the American Water Works Service Company. The National Science Foundation, the Office of Naval Research, and the IPA Industrial Associates are also acknowledged for their partial support.

References

Allen, M.J., Taylor, H., and Geldreich, E.E. 1980. The occurrence of microorganisms in water main encrustations. *American Water Works Association Journal* 72:614–625.

Banerji, S.K. 1978. Quality deterioration in mains. *Journal of the Missouri Water & Sewage Conference* 1:46–56.

Brazos, B.J. and O'Connor, J.T. 1985. A transmission electron micrograph survey of the planktonic bacteria population in chlorinated and nonchlorinated drinking water. *Proceedings, Water Quality Technology Conference: Advances in Water Analysis and Treatment.* Houston, TX, 275–305.

Characklis, W.G. 1988. *Bacterial Regrowth in Distribution Systems.* Research Report. American Water Works Association Research Foundation. Denver, CO, 107–114.

Characklis, W.G., Trulear, M.G., Stathopoulos, N., and Chang, L.C. 1980. Oxidation and destruction of microbial films. pp. 349–368 *in* Jolley, R.L., W.A. Brungs, R.B. Cumming, and V.A. Jacobs (eds.), *Water Chlorination: Environmental Impact and Health Effects,* Ann Arbor Science Publishers, Inc., Ann Arbor, MI.

Colbourne, J.S., Dennis, P.J., Trew, R.M., Berry, C., and Vesey, G. 1988. *Legionella* and public water supplies. *Proceedings, International Conference on Water and Wastewater Microbiology.* Newport Beach, CA.

Haudidier, K., Paquin, J.L., Francais, T., Hartemann, P., Grapin, G., Colin, F., Jourdain, M.J., Block, J.C., Cheron, J., Pascal, O., Levi, Y., and Miazga, J. 1988. Biofilm growth in drinking water network: a preliminary industrial pilot plant experiment. *Proceedings, International Conference on Water and Wastewater Microbiology.* Newport Beach, CA.

Jacangelo, J.G. and Olivieri, V.P. 1985. Aspects of the mode of action of monochloramine. pp. 575–586 *in* R.L. Jolley, R.J. Bull, W.P. Davis, S. Katz, M.H. Roberts, Jr., and V.A. Jacobs (eds.), *Water Chlorination, Chemistry, Environmental Impact and Health Effects,* Lewis Publishers Inc., Chelsea, MI.

Kelly, A.J., Justice, C.A., and Nagy, L.A. 1983. Predominance of chlorine tolerant bacteria in drinking water systems. *Abstracts Annual Meeting American Society for Microbiology* 0, Q122.

LeChevallier, M.W., Babcock, T.M., and Lee, R.G. 1987. Examination and characterization of distribution system biofilms. *Applied and Environmental Microbiology* 53:2714–2724.

Lee, S.H., O'Connor, J.T., and Banerji, S.K. 1977. Biologically mediated deterioration of water quality in distribution systems. *Proceedings, American Water Works Association 5th Annual Water Quality Technology Conference.* Kansas City, MO.

Maki, J.S., LaCroix, S.J., Hopkins, B.S., and Staley, J.T. 1986. Recovery and diversity of heterotrophic bacteria from chlorinated drinking water. *Applied and Environmental Microbiology* 51:1047–1055.

McFeters, G.A., Kippin, J.S., and LeChevallier, M.W. 1986. Injured coliforms in drinking water. *Applied and Environmental Microbiology* 51:1–5.

O'Connor, J.T. and Banerji, S.K. 1984. *Biologically Mediated Corrosion and Water Quality Deterioration in Distribution Systems.* EPA-600/52-84-056. U.S. Environmental Protecting Agency, Cincinnati, OH.

Olivieri, V.P., Dennis, W.H., Snead, M.C., Richfield, D.T., and Kruse, C.W. 1980. Reaction of chlorine and chloramines with nucleic acids under disinfection conditions. pp. 651–653 *in* Jolley, R.L., W.A. Brungs, R.B. Cumming, and V.A. Jacobs (eds.), *Water Chlorination: Environmental Impact and Health Effects,* Ann Arbor Science Publishers, Inc., Ann Arbor, MI.

Ridgway, H.F. and Olson, B.H. 1982. Chlorine resistance pattern of bacteria from 2 drinking water systems. *Applied and Environmental Microbiology* 44:972–987.

Ridgway, H.F. and Olson, B.H. 1981. Scanning electron microscope evidence for bacterial colonization of a drinking-water distribution system. *Applied and Environmental Microbiology* 41:274–287.

Siegrist, H. and Guyer, W. 1985. Mass transfer mechanisms in a heterotrophic biofilm. *Water Research* 19:1369–1378.

Tuovinen, O.H., Button, K.S., Vuorinen, A., Carlson, L., Mair, D.M., and Yut, L.A. 1980. Bacterial, chemical and mineralogical characterization of tubercles in distribution pipelines. *Journal of the American Water Works Association.* 72:626–635.

van der Kooij, D. and Hijnen, W.A.M. 1982. Nutritional versatility of a starch-utilizing flavobacterium at low substrate concentrations. *Applied and Environmental Microbiology* 45:804–810.

van der Kooij, D., Visser, A., and Oranje, J.P. 1982. Multiplication of fluorescent pseudomonas at low substrate concentrations in tap water. *Antonie van Leeuwenhoek* 48:229–243.

van der Wende, E., Characklis, W.G., and Smith, D.B. 1989. Biofilms and bacterial drinking water quality. *Water Research,* in press.

Wolfe, R.L., Ward, N.R., and Olson, B.H. 1985. Inactivation of heterotrophic bacterial populations in finished drinking water by chlorine and chloramines. *Water Research* 19:1393–1404.

Part 4

Pathogenic Organisms and Drinking Water

13

Waterborne Giardiasis

Charles P. Hibler and Carrie M. Hancock

CH Diagnostic & Consulting Service, Ft. Collins, Colorado

13.1 Introduction

Giardia duodenalis is a protozoan parasite that infects the upper portion of the small intestine of man and many different species of mammals. In an early monograph on *Giardia*, Filice (1952) used morphology as the criterion to lump most of the mammalian "species" into *G. duodenalis*, the type species found in European hares. *Giardia duodenalis* has been found in humans, beaver, muskrat, mule deer, domestic sheep, cattle, elk, coyotes, dogs, cats, horses, moose, and a number of small wild and laboratory rodents. Some investigators considered the parasite found in humans (*G. lamblia*) to be host-specific, but the majority of the research performed to date questions this assumption. The parasite has been successfully transmitted from humans to beaver (Davies and Hibler, 1979; Erlandsen et al., 1987), beaver to humans (Davies and Hibler, 1979), humans to dogs and cats (Davies and Hibler, 1979), humans to muskrat and from muskrat to dogs and cats (Wegrzyn, 1988), from mule deer to humans (Davies and Hibler, 1979), and from humans to a number of laboratory rodents (Davies and Hibler, 1979; Swabby et al., 1987). While the parasite has been successfully cross-transmitted, thus proving it is not host-specific, not every source from any given animal will cross-transmit every time the effort is made, indicating there may be "strains" that do not readily adapt to a new host (Hibler et al., 1987). Filice (1952) also recognized that some wild and laboratory mice were infected with another species, *Giardia muris*, a parasite that can be distinguished from *G. duodenalis* by the internal morphology of the organisms. *Giardia muris* is not known to be infectious for humans. Hibler (1987a) and S.L. Erlandsen (personal communication) have found species of *Giardia* morphologically similar to the *G. duodenalis* type in waterfowl and shore-birds. While *Giardia* found in birds probably is not infectious for humans, their presence in a sample of surface water may cause diagnostic problems for the analyst and, potentially, the municipality using this water. Parasitologists also recognize *Giardia agilis*, a parasite found in frogs, salamanders, and toads, as a distinct species. The morphology of this particular species is sufficiently distinctive that experienced parasitologists can easily recognize the difference.

271

Fortunately, none of the microscopists who regularly analyze surface water samples have encountered *G. agilis*; therefore, we assume it is a rare infection of amphibians.

Species of *Giardia* possess eight flagellae to facilitate motility and a ventral sucking disc to maintain their position attached to the cells of the small intestinal villi. The trophozoites possess a pair of nuclei, situated anteriorly, a pair of axonemes or axostyles traversing the length of the organism, and a pair of median bodies (Figure 13.1). In *G. duodenalis*, these median bodies are claw-shaped while in *G. muris* they are spherical. The organisms attach to the surfaces of cells in the small intestine and secure nourishment either from the cells or from the ingesta present in the small intestine. They multiply by simple binary fission. When the upper portion of the small intestine is heavily populated and/or an immune response from the host stimulates the organisms, some detach and excrete a cyst wall about their body as they pass through the digestive tract. Generally the cyst is mature and infectious when passed in the excreta from the host. A mature cyst of *G. duodenalis* possesses some very important morphological features that enables an experienced microscopist to recognize it as a cyst of *G. duodenalis* in a stool specimen or in a water sample. The mature cyst (Figure 13.1) has four nuclei, a pair of axonemes, and a pair of claw-shaped median bodies. The axonemes and median bodies are actually doubled (two of each) but are positioned so close together that they appear as a pair. These are extremely important features for the microscopist: no other organism or cyst found in water has these internal structures and

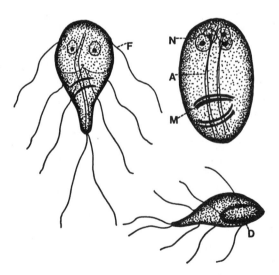

Figure 13.1   Semi-diagrammatic illustration of *Giardia duodenalis*. Left: dorsal view of trophozoite; upper right: dorsal view of mature cyst; lower right: ventro-lateral view of trophozoite showing suction disk. F, flagella; N, nucleus; A, axostyle; M, median body; D, suction disk. (Original by C.P. Hibler).

before *Giardia* can be reported as present in water, at least two of the three internal structures must be identified. Usually these structures are the nuclei and median bodies.

The cyst wall protects the organism from adverse environmental conditions (drying and desiccation) for a short period of time, but unless it is transmitted directly by fecal-oral contact within a few hours to a few days, depending upon environmental conditions, it will die. However, if the cysts are discharged into water their life span may be increased to about two months, especially in cold water.

When the intestine is heavily populated with trophozoites, symptoms of giardiasis will appear. This usually takes 5 to 6 days post-exposure but may be as long as 10 days, depending on the initial dose and the resistance of the individual. The most common symptoms are diarrhea, abdominal distension, flatulence, and malaise. The disease may be asymptomatic in some individuals; symptomatic but with a short-term duration in other individuals; or symptomatic and long-term, often debilitating, in a few individuals. In most infected individuals, some symptoms develop. Recovery from the disease, either by medication or through the individual's immune system, may result in complete immunity or the individual may become a carrier; this is not completely known. Neither is it known if individuals, once recovered, are again susceptible.

13.2 Waterborne Giardiasis

Giardia is recognized as the most common animal parasite of humans in the United States and it has been reported to be the most common animal parasite of humans in the entire world. The usual mode of transmission is fecal-oral, from person to person, and this accounts for the majority of cases. The parasite is commonly found among children in day-care centers, kindergartens and nurseries, etc. where it spreads from child to child and then within the families. It is also a common infection of homosexuals and among those who are immunologically compromised. The *least common mode of transmission is waterborne*; yet this mode of transmission usually receives the most notoriety. Regrettably, most of the cases in individuals or within families are blamed on the water; physicians seem reluctant to accept the efficient and highly effective fecal-oral mode of transmission from person to person.

Giardiasis has become increasingly important as a waterborne disease of humans over the past 20 to 25 years and currently is the most common cause of waterborne disease. Between 1965 and 1980, 42 outbreaks of waterborne giardiasis were reported and several outbreaks have occurred since 1980. Craun (1986) compiled valuable statistical data on these outbreaks, showing that 67% of the waterborne outbreaks and 52% of the cases resulted from consumption of untreated surface water or surface water in which disinfection was the only treatment. Ineffective filtration was responsible for five (12%) of the outbreaks from 1965 to 1980. Craun (1988) recently published an updated synopsis of this material.

There is not sufficient space or need to detail each of the outbreaks reported to date, but it is necessary to provide some examples to point out how and why they occur. The first U.S. outbreak suspected to be caused by a waterborne source of *Giardia* was in Aspen, Colorado during the 1965 to 66 winter tourist season. A retrospective epidemiologic survey indicated sewage contamination of wells, although the creek also may have been contaminated (Moore et al., 1969). In 1974 to 75, an outbreak of giardiasis occurred in Rome, New York (Shaw et al., 1977). During this outbreak, investigators from the Centers for Disease Control (U.S.) detected cysts in the water. This report provided the first direct proof that *Giardia* was present in the source. In addition, material concentrated from the source was introduced into specific-pathogen-free beagle dogs maintained at Colorado State University (Hibler et al., 1975). Some dogs became infected with *Giardia* from this source; however, this proved only that the cysts were alive and infectious for dogs.

In the spring of 1976 an outbreak of giardiasis occurred in a filtered water source at Camas, Washington (Kirner et al., 1978; Dykes et al., 1980). *Giardia* cysts were found in the water using an improved sampling device developed by the Environmental Protection Agency. Moreover, infected beaver were found near the intakes (Dykes et al., 1980). Since the creek used as a source originated in a remote, isolated area with little human activity, and there was no evidence of sewage contamination, beaver were promptly incriminated as the probable source of the cysts. This was the first time a wild animal source had been implicated; the previous two outbreaks were considered to be the result of sewage contamination. While beaver, some other wild animal (muskrat, voles, etc.), or domestic animals (cattle, etc.) may have been responsible for this outbreak, incrimination of beaver often resulted in limited epidemiologic investigation of subsequent outbreaks. The second outbreak of waterborne giardiasis to occur in a filtered water supply was in the spring of 1977 at Berlin, New Hampshire (Lippy, 1978). A source was not determined, but beaver were implicated as the probable cause. In both outbreaks involving filtered water problems with filtration and chlorination had occurred. In the fall of 1979 another outbreak of waterborne giardiasis occurred in Bradford, Pennsylvania which used an unfiltered source of water (Lippy, 1981). Between 1980 and 1983 the frequency of outbreaks of waterborne disease increased to the highest level since 1942 (Craun, 1986). The pathogen most frequently identified was *Giardia*, accounting for 41 of the 126 outbreaks in the 35 states reporting an outbreak of waterborne disease. Most of these outbreaks were small, involving only a few individuals. Colorado reported about half of the outbreaks of giardiasis recorded, but this is primarily the result of an efficient epidemiologic surveillance program. For example, less than 50 individuals were involved in 19 of the 21 outbreaks reported in Colorado; likely many other states have had similar problems but did not have a surveillance program sufficient to detect these small outbreaks. Several large outbreaks occurred in communities in Pennsylvania between 1982 and 1986. An outbreak involving over a thousand individuals occurred in Utah. Massachusetts had an outbreak in 1985 that reported over 500 individuals infected, and Nevada had an outbreak including over 300 individuals in 1982. New Mexico had two outbreaks, each involving over 100 individuals, but only one was reported; the other occurred on a private ranch using ineffective filtration and inadequate disinfection (Hibler, unpublished). Most of the outbreaks between 1980 and 1983 were in communities using unfiltered, disinfected water, but some were in communities with filtration systems that were bypassed or malfunctioning and five occurred in communities with filtration systems that had inadequate treatment (Craun, 1986).

13.3   Effect of Water Treatment on Presence of *Giardia* Cysts

Over the past 12 years we have analyzed over 7500 water samples from 361 municipalities in 30 of the 48 contiguous United States, three provinces in Western Canada, and throughout the Commonwealth of Puerto Rico. A total of 4423 samples from 301 municipalities (sites) in the United States were analyzed at the request of the U.S. Environmental Protection Agency (USEPA), Office of Drinking Water, Washington, D.C. for use as part of the justification to establish new drinking water regulations (Hibler, 1987a). Data acquired from 1979 through 1986 have revealed that *Giardia* cysts are extremely common contaminants of surface water sources in North America (Table 13.1). A total of 798/4423 (18%) samples had *Giardia* cysts present (Figures 13.2 and 13.3). Figure 13.3 suggests that cysts may be more numerous in the winter months, but this probably is a reflection of the lower turbidity, facilitating recovery of the cysts without interference from other organisms or from detritus. Likely the number of sites contaminated with *Giardia* cysts is higher than the data indicate (102/301; 34%) because less than 4 samples were acquired from 199

Table 13.1 Detection of *Giardia* cysts in water samples: classification by type of water

Water Type	Number of samples examined	Percent of total samples	Number of samples positive	Percent of samples positive
Raw	1968	44.5	512	26
Finished	2372	53.6	267	11
Unknown	83	1.9	15	18
Totals	4423	100	794	18

Samples originated from 301 municipal sites in 28 states; 102 (34%) sites were positive for *Giardia* cysts. Negative samples were obtained from 199 sites, but only 651 samples (average 3.27 sample/site) were examined.

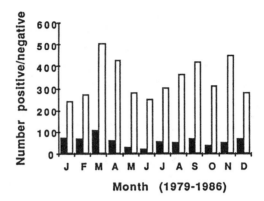

Figure 13.2 Seasonal variation in content of *Giardia* cysts in surface water in North America. Composite of samples from all sources: 794/4423 (18%) positive, 1979–1986. Solid bars, positive; open bars, negative.

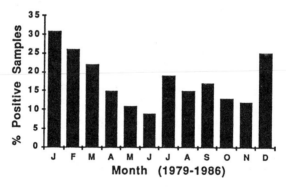

Figure 13.3 Seasonal variations in content of *Giardia* cysts in positive water samples. Composite samples from all sources: 794/4423 (18%) positive, 1979–1986.

(66%) of the sites examined. *Giardia* cysts are not evenly dispersed in water and frequently are few in number. Repeated sampling of negative sources has invariably produced positive results. These data were further analyzed by source and type of water to determine if some sources were more highly contaminated (Table 13.2). Municipalities using creeks, rivers, and open (unprotected) springs were more likely to have *Giardia* cysts present than municipalities using lakes. The "settling effect" in lakes (reservoirs) resulted in fewer positive samples and (generally) fewer cysts. Analysis of the finished water data by type of filtration (if used) provided some revealing information regarding the presence of *Giardia* cysts in the final product (Table 13.3). Most of the unfiltered but chlorinated finished water originated from lakes or reservoirs where the "settling effect" had already reduced the number of cysts. Chlorine does, of course, inactivate the cysts and the time delay between sam-

Table 13.2 Detection of *Giardia* cysts in water samples: classification by source and type of water

Source*	Number of sites	Number of sites positive	Percentage of sites positive	Number of samples	Number of samples positive	Percent of samples positive
Creeks (total)	75	38	51	1218	346	28
Raw	–	–	–	444	181	41
Finished	–	–	–	774	165	21
Rivers (total)	74	38	51	828	212	26
Raw	–	–	–	449	163	36
Finished	–	–	–	379	49	13
Lakes (total)	49	19	39	1983	193	10
Raw	–	–	–	829	138	17
Finished	–	–	–	1154	55	5
Springs (total)**	36	5	14	84	16	19
Wells (total)**	40	2	5	63	2	3

* Municipalities using a combination of sources (i.e. creek and spring) have been excluded from totals.
** Most water from springs and wells is unfiltered and may/may not be disinfected before consumption.

Table 13.3 Detections of *Giardia* cysts in finished drinking water supplies

Classification	Number of sites	Number of sites positive	Percent of sites positive	Number of samples	Number of samples positive	Percent of samples positive
Unfiltered, chlorinated	94	16	17	1214	80	6.6
Direct Filtration*	92	17	18.5	615	148	24.0
Conventional treatment	86	5	3.4	357	12	5.8
Commercial and/or pressure filters	12	2	16.7	33	4	12.1
Slow sand and diatomaceous earth	3	0	0	18	0	0
Mechanical or cartridge filters	13	7	53.8	51	11	21.6
Infiltration gallery	16	5	31.3	37	7	18.9
Filter type unknown**	24	6	25.0	132	15	11.4

* Most of the positive samples originated from systems that were not using coagulation prior to filtration.
** Many of these samples are also included in the 83 samples (Table 13.1) where the type of sample (raw or finished) is also unknown.

pling, shipping, and processing of a sample further reduced the percentage of positive samples because (in time) chlorine will render a *Giardia* cyst unrecognizable to the microscopist. The highest percentage of samples with *Giardia* cysts present originated from municipalities using direct filtration without chemical pretreatment; a simple screening process used (generally in the winter months) when water supplies were cold ($<5°C$) and the turbidity was at its lowest for the year (<1 nephelometric turbidity unit, NTU). A scattergraph analysis of 20 systems using direct filtration (without chemical pretreatment) is given in Figure 13.4 to illustrate that simple screening does not remove the cysts. However, when chemical coagulants were added (Figure 13.5), the cysts were effectively removed. Engineering results from pilot filtration plants had already revealed (under experimental conditions) that removal of 75 to 80% of the *incoming* turbidity was necessary to insure removal of the cysts. Actual plant conditions have verified these experimental results. Conventional treatment is an effective procedure for removing *Giardia* from a source of water (Table 13.3). While 5/86 sites sometimes had cysts present in the finished product, it was discovered that 3 of these 5 sites were by-passing the conventional treatment procedure and simply screening the water. Data from 20 conventional treatment plant sites are presented in a scattergraph (Figure 13.6) to illustrate the efficiency of this type of filtration. Diatomaceous earth and slow sand filtration are both extremely effective for removal of *Giardia*. However, we had only 11 samples of slow sand (1 site) and 7 samples of diatomaceous earth (2 sites); an inadequate number to be useful for any conclusions.

The number of samples obtained from commercially manufactured filters also is an inadequate number to be useful for any conclusions; moreover, many of the municipalities using these systems did not have any *Giardia* cysts in the source samples. Since the purpose of the evaluation was removal of the cysts,

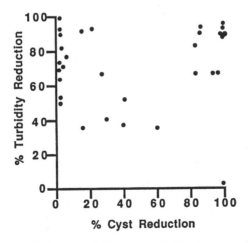

Figure 13.4   Removal of *Giardia* cysts by direct filtration (no coagulation); composite data from 20 water systems, 1979–1986.

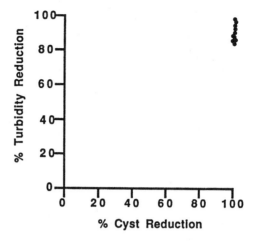

Figure 13.5   Removal of *Giardia* cysts by direct filtration after coagulation; composite data, 1979–1986.

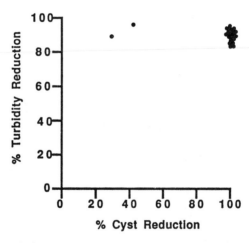

Figure 13.6   Removal of *Giardia* cysts from 20 conventional treatment plants; composite of samples from 1979–1986.

the 2/12 (12.1%) only reflects the two sites that had cysts, and these were not removed. Most of the 12 sites using these types of filters were passing particulates the size of or larger than *Giardia*.

A number of small sites (guest lodges, camps, etc.) used mechanical filters and 7/13 (53.8%) sites using these devices did not always remove the cysts. Generally the device was not used as recommended by the manufacturer or was not properly maintained.

Most of the 16 infiltration galleries examined were the only filtration system used by the municipally, but a few were used prior to in-line direct filtration or conventional treatment. These generally functioned as designed; a simple screening process. Five of the 16 (31.3%) did not remove *Giardia*. These were the only five galleries that had cysts in the source water; most of them were passing particulates the size of or larger than *Giardia*.

While *Giardia* cysts have been found in raw and/or finished (filtered as well as unfiltered, disinfected) water at all times of the year, analysis of these data suggests a seasonal distribution with more cysts from fall to spring. Indeed, 35 to 45% of the raw water samples from rivers and creeks often were contaminated with cysts during these months (Hibler, 1987a). However, during these months water supplies were often low, turbidities stabilized to a minimum, and cyst longevity increased by the lower water temperatures. Analytical procedures are more effective when the turbidity is low. These factors may function to increase recovery and/or visualization of cysts during the winter and give a false indication of seasonal distribution.

Although data indicating that *Giardia* cysts have a seasonal distribution may be complicated by the above factors, analysis of water samples has shown that more filtration plants are passing cysts in the winter months. Generally this is due to the mode of operation. Municipalities often switch from conventional treatment to direct filtration and often use filtration without coagulants during these months, primarily because the source of water meets and frequently exceeds existing government regulations and no need is perceived to perform any better. Coagulation and flocculation of cold water often is complicated by turbidities less than 0.6 NTU since chemical reactions are slower and there are fewer particles to form an effective floc. Needless to say, *Giardia* cysts usually are present in this cold, low-turbidity water and without coagulation there was seldom more than 60% removal (Hibler, 1987a). The risk for waterborne giardiasis increases considerably as water temperatures decrease. Not only is longevity of the cyst increased by cold water, but most disinfectants in general use need a longer contact time in cold water to inactivate the cyst. This is because chemical reaction rates decrease by two or three fold for every 10 degree decrease in temperature (Weber, 1972; White, 1972).

As might be expected, 83 of the samples in this compilation of data are "unknowns," and no data were provided as to type of water or filtration system used. The majority of these samples were secured by EPA, state, county, or local health authorities and sent for analysis. These represent less than 2% of the total, but it is unfortunate that those individuals responsible for enforcing regulations were lax in providing the information necessary to effectively evaluate the sample.

13.4 Techniques for Analysis of *Giardia* in Water

Analysis of a sample of surface water for cysts of *Giardia* is an extremely difficult and time-consuming technique, necessitating considerable experience with *Giardia* as well as knowledge of the numerous organisms and organic and inorganic particulates present in surface water. Currently there are no reliable procedures available to effectively isolate and culture the few *Giardia* cysts present in a sample. The cyst must be selectively isolated from the other organisms and detritus present in water and visually identified by direct bright-field and/or phase-contrast microscopy or with a monoclonal antibody using an epifluorescent microscope. Cysts are undoubtedly much more prevalent in surface water than our data indicate because the techniques available are extremely inefficient (Hibler, 1987b).

The presently accepted procedure for capturing *Giardia* cysts is to filter 100 to 1000 gallons of water through a 1 micrometer (nominal rating) polypropylene, cotton, or orlon filter cartridge. The cartridge is cut to the core and the fibers hand-washed in distilled water, usually several times, to release the particulates trapped by the cartridge. The suspended material is concentrated to a small volume (usually 100 to 200 ml) and then 20 to 25 ml aliquots are processed for examination. One processing technique is generally referred to as the "Reference Method" (*Standard Methods for the Examination of Water and Wastewater*, 1985). This often is modified to remove unwanted detritus by an "overlay" or "underlay" procedure. In the Reference Method, zinc sulfate, sucrose, potassium citrate or percoll is used to separate the organisms from the other particulates present by gradient centrifugation. A *Giardia* cyst has a specific gravity (density) of about 1.05. When mixed with one of the above chemicals whose specific gravity (density) has been adjusted from about 1.13 to as high as 1.3, the cysts "float" and the other particulates "settle." Centrifugation for about five minutes will speed this action. Material "floating" to the top of the centrifuge tube is then processed for direct or fluorescent microscopy. In the "overlay" technique, a 25 ml aliquot of the concentrated suspension is carefully pipetted over the chemical and centrifuged. The end result is that the cysts are trapped at the interface between the suspension and the chemical while the dense particulates pass through the interface into the chemical. In the "underlay" technique an aliquot of the concentrated suspension is placed into a centrifuge tube and the chemical inserted beneath the concentrate. As with the "overlay" technique, the cysts are trapped at the interface following centrifugation. The particulates trapped at the interface are selectively removed by passage through a 1, 3 or 5 micrometer Nuclepore® membrane and further processed for direct, phase, or fluorescent microscopy.

These techniques are fraught with problems (Hibler, 1987b). The first loss of cysts occurs during the washing technique. Moreover, if the concentrate is heavily laden with organic detritus, cysts become trapped and do not "float," or the trapped cysts are pulled through the chemical interface in the "overlay" or "underlay" technique. While cysts have a specific gravity (density) of about 1.05, so do algae, diatoms, flagellates, ciliates, crustaceans, etc. and if many such particles that are the same size and shape as a *Giardia* cyst are present, the microscopist quickly develops fatigue and may miss the cysts present. The monoclonal antibody technique offers some promise to visualize the cysts, but it does not prevent the losses that occur during processing the sample, nor does it prevent fatigue. In summary, the techniques available probably fail to recover more than 50% of the cysts present, and this may be giving the analyst the benefit of doubt.

13.5   Sources of *Giardia* in Drinking Water

The source of contamination with *Giardia* cysts often is difficult to determine, although some times it is clearly human sewage, or wild or domestic animals. In our experience, the cysts in raw sewage are frequently too numerous to count and the numbers present in treated sewage effluent are almost equivalent to the numbers in the raw sewage. As indicated earlier, *Giardia* is found in many wild and domestic animals. Monzingo (1985) has found 40 to 45% of the beaver infected in some areas of Colorado and each animal was shedding up to 1×10^8 cysts/animal/day. Wegrzyn (1988) found all of 386 muskrat infected in Colorado and 48% of the animals were shedding as many as 3×10^6 cysts/animal/day. Up to 20% of the cattle examined and about 5% of the domestic sheep (not accurate because sheep fecal pellets dry quickly, making the results conservative) were shedding cysts. The environment of municipal source water often is the habitat for many wild animals, especially beaver, muskrat and the small water voles. Unfortunately, in several outbreaks wildlife (beaver) have been incriminated on the basis of insufficient evidence. Investigators seem reluctant to place any blame on sewage.

Records from the analysis of water samples done in this laboratory over the last 12 years and statistics provided by Craun (1986) have resulted in considerable information about the water supplied to consumers in communities across the United States. Surface water often is consumed raw. However, if the source supplies a community (a community is defined as an installation with at least 15 service connections; Craun, 1986), there is generally some form of treatment. High quality water originating in a pristine environment and/or from a protected watershed frequently is unfiltered, but is disinfected to protect consumers from the risk of a waterborne disease. We have found cysts of *Giardia* as frequently in water originating in pristine and/or protected sources as in unprotected sources; protection from animal contamination generally is not possible. If the source traverses public and/or private property, the multiple-use concept applied by the U.S Forest Service (Forestry, Agriculture and Recreation), as well as impact of grazing by domestic animals effectively prevent protection of that source.

The potential for outbreaks of waterborne giardiasis in unfiltered surface water sources in the United States is a primary concern to the USEPA. Because of the history of epidemics, the USEPA has established criteria to determine when surface water sources providing water to communities must filter this water (National Primary Drinking Water Regulations, 1987). Some communities oppose the requirement for construction of filtration plants specifically to prevent the possibility of waterborne giardiasis, arguing that cost would be prohibitive to use *Giardia* as the sole reason for construction. Most of those opposed to mandatory filtration provide a high quality water originating in a pristine environment and/or a protected area. Undoubtedly variances will be granted for communities able to demonstrate that the only barrier needed is disinfection of the water. However, many high quality surface water sources are often subject to temperatures less than 5°C in winter months; therefore, before regulations could be enacted, information was needed describing the inactivation of *Giardia* cysts at low temperatures.

13.6   Disinfection for *Giardia* Removal

Recently, Hoff (1986) undertook the monumental task of assembling and interpreting the published information on the use of the CT values (disinfectant concentration, C, in milligrams/liter or ppm multiplied by contact time, T, in minutes) required to inactivate different types of pathogens (viruses, bacteria and protozoans) to certain levels under specified conditions as established by laboratory experimentation. The CT concept is based on an empirical equation developed by Watson (1908) to examine the effects of changing disinfectant concentration on rates of microbial inactivation. Watson's law is $K = C^nT$. The terms are the same in the $K = CT$ except for n, the coefficient of dilution, an important addition which determines the order of the chemical reaction. If $n = 1$, the CT value remains constant regardless of the disinfection concentration used. Therefore, if $n = 1$, concentration of disinfectant and exposure time are of equal importance. If $n > 1$, disinfectant concentration is the dominant factor and if $n < 1$, exposure time is more important than disinfection concentration. As Hoff (1986) stated, n is very important in determining the degree to which extrapolation of data from disinfection experiments is valid. For example if $n = 1$, CT values can be used to predict efficiency over a broad range of disinfection concentration levels and exposure time periods.

Baumann and Ludwig (1962) used Watson's Law to illustrate its use for making disinfection recommendations for small nonpublic water supplies relying only on disinfection for inactivation of pathogens (Hoff, 1986) but the CT concept received little further attention until the Safe Drinking Water Committee (1980) used CT as the method of comparing biocidal efficiency. Thereafter the CT concept for interpretation of disinfection data, both for assessing comparative efficiencies of different disinfectants and expressing comparative resistance of the different pathogens, became more widely used and now is the accepted approach (Hoff, 1986).

The USEPA was directed by Congress to prepare new criteria for treatment of surface water sources (National Primary Drinking Water Regulations, 1987). As a minimum, the treatment must include disinfection and provide a 1000-fold removal and/or inactivation (99.9%) of *Giardia duodenalis* cysts. The data currently available for inactivation of *Giardia* cysts in the 5 to 25°C range is based on 99% inactivation and some of the information was generated using

Giardia muris, a parasite of mice that is not infectious for humans. Data accumulated by Hoff (1986) indicate that the CT values necessary for inactivation of pathogens frequently are specific for that pathogen. Since the proposed regulations for inactivation of *Giardia* cysts specify *G. duodenalis* of human origin, data generated for *G. muris* may not be applicable. Meyer and Schaefer (1984) pointed out that interpretation of results using *G. muris* must be done with caution.

Hoff (1986) pointed out that the validity of extrapolating from CT values required for 99% inactivation to CT values required for other levels of inactivation (in this case, the 99.9% specified for *Giardia* in the proposed regulations) is dependent upon the nature of the inactivation curves from which the 99% inactivation CT value was determined. These two factors together placed the USEPA in the unenviable position of establishing new regulations based on an inadequate and incomplete database for inactivation of *Giardia* cysts in the 5 to 25°C range.

Chick (1908) characterized the inactivation of microorganisms as a first order chemical reaction using the disinfectant and the microorganisms as the two chemical reactants. Essentially, Chick's Law proposes that the number of organisms remaining (alive and infectious) after a period of time is a proportional constant. However, microorganisms do not behave as chemical reactants (Hoff, 1986). There may be an initial lag period before inactivation begins for some organisms (e.g., those protected by a cyst wall), and a "tailing off" of inactivation for those organisms that have a more resistant segment of the population. Whether these are real problems or the result of experimental conditions is unknown but the possibility they are real problems must be accepted until proven otherwise.

The failure of microorganisms to conform to Chick's Law has important implications. For instance, it would be hazardous to attempt extrapolation of the data generated for 99% inactivation of pathogens to 99.9% or 99.99% inactivation. The requirement for 1000-fold (99.9%) inactivation of *Giardia* cysts specified in the proposed new regulations is a prime example. Most of the data generated for *Giardia* in the 5 to 25°C range has been for 99% inactivation using artificial excystation as the technique to determine inactivation by disinfectants. Jarroll et al. (1980) developed and improved the excystation procedure sufficiently to evaluate the effect of different disinfectants on *Giardia* cysts and then evaluated the effects of chlorine on these protozoans (Jarroll et al., 1981). Similar research, comparing the effect of chlorine on *Giardia* cysts obtained from symptomatic and asymptomatic individuals, was performed by Rice et al. (1982). Hoff et al. (1985) compared in vitro excystation with animal infectivity to determine the effects of disinfection on *G. muris* cyst viability and found that excystation was an adequate indication of viability. However, this has not been done for *G. duodenalis,* the species infectious for humans. Although excystation techniques have improved over the years, use of *G. duodenalis* cysts directly from a human source is fraught with problems: clinical patients seldom are available when needed, the history of the patient seldom is known, and cysts from different human sources seldom respond the same to the excystation procedure. Generally, less than 90% of the cysts will excyst, necessitating a correction factor and creating the problem of interpreting results with confidence. Moreover, it is not currently possible to obtain accurate CT values beyond 99% inactivation with the excystation technique because it is logistically difficult to accurately count the number of excysted organisms necessary to have confidence in the results for 99.9% inactivation.

The only current alternative to develop CT values for 99.9% inactivation is to use a sensitive biological model involving a laboratory animal highly susceptible to infection with *G. duodenalis* of human origin. The most sensitive biological model that has been evaluated for susceptibility to *Giardia* cysts of human origin is the Mongolian gerbil (*Meriones unguiculatus*). This model was first used by Belosevic et al. (1983) after Davies and Hibler (1979) showed that another species of gerbil (*Gerbillus gerbillus*) was a reasonably good model for human-source *Giardia*. After Belosevic et al. (1983) showed that the gerbil was a potentially excellent biological model, we began using this animal in our laboratory for production of cysts from human sources as well as from other animals (cattle, horses, cats, etc.). The colony has been specific-pathogen-free (for *Giardia*) and the animals have been sensitive to about 60% of the human sources of *Giardia* cysts to which they have been exposed (Swabby et al., 1987). Experimentation has shown that as few as five cysts will infect 100% of the animals if the human source was well adapted to the gerbil. Probably they are susceptible to a single cyst, but determining if a single cyst is alive or dead is difficult and we have succeeded in infecting only 10 to 20% of the animals when a single cyst was used.

Use of a biological model should not be considered a panacea for the problems associated with development of a 99.9% CT value for inactivation of *Giardia* cysts. While use of an animal model is not subject to the limitations of the excystation technique, the sensitivity of the animal to five or fewer cysts probably will result in a CT value closer to 99.99% inactivation, a 10,000-fold reduction.

Prior to our research (Hibler et al., 1987), the experimental database for inactivation of *Giardia* cysts with the disinfectants in common use was extremely limited and most of the data were obtained in the range of 5 to 25°C. DeWalle and Jansson (1983) were the only investigators to provide any free chlorine data for temperatures lower than 5°C. They used the *G. muris* model and applied the excystation technique to generate information on the CT values necessary to inactivate cysts with chlorine in an unbuffered raw water source. Most of their trials were at 1°C and 5°C, varying the concentration of chlorine and evaluating inactivation through a pH range of 5 to 8.7. Unfortunately, since they did not use chlorine demand-free water, reporting chlorine as the initial levels, interpretation of their results for comparative purposes is difficult because there is no indication of the chlorine demand or the time span of this demand. Use of final concentrations would have been a better approach.

From the time that we began analyzing water samples for *Giardia* we have been asked "will chlorine kill *Giardia*?" Initially we had to respond with "Yes, but we do not know how much is needed, or how much time is necessary." After Jarroll et al. (1981) and Rice et al. (1982) developed CT values for inactivation of cysts in the temperature range from 5 to 25°C, we could respond to these questions with more confidence. Following outbreaks in several large communities over the last few years, requests for analyses of surface water increased considerably, accompanied by a tremendous number of questions regarding the effect of chlorine on these cysts. Municipal authorities, public health authorities, engineers, and water treatment plant operators needed to know how much chlorine and contact time was necessary to inactivate the cysts. Unfortunately, rarely were we asked these questions until cysts had been found in the distribution system of a community (often during the peak of the winter tourist season) when the temperature of the water was near freezing.

These questions and problems prompted us to initiate research on the amount of chlorine and time necessary to inactivate the cysts of *Giardia* between 0.5°C and 5.0°C, over a pH range of 6 to 8 (Hibler et al., 1987). We opted to use the animal model as an indicator for inactivation because the model is susceptible to five or fewer viable *G. duodenalis* cysts if a well-adapted human source is used. This would provide considerable confidence that inactivation could be detected between 99.9 and 99.99%, at least equal to and probably greater than the level specified by the proposed regulations. Since these regulations specify *G. lamblia* (*G. duodenalis* of human origin) we did not feel that development of CT values with *G. muris*, a species infectious for a limited number of rodents, would be considered acceptable.

We used five different sources of *Giardia* cysts during the research, primarily to include possible variations in susceptibility of different sources to chlorine. This was fortunate because during the course of the research some human sources proved to be more resistant to chlorine than others.

The data used to calculate CT values for inactivation of *Giardia* cysts with chlorine at temperatures 0.5°C, 2.5°C, and 5.0°C, pH 6, 7, and 8 were generated in 48 experimental trials using 744 groups of 5 animals/group (3720 gerbils). Each animal in the chlorine exposure groups was inoculated with a calculated dose of 5×10^4 *Giardia* cysts that had been exposed to a specified concentration of chlorine over a period of time. An equal number of positive control animals was inoculated with a calculated dose of 50 *Giardia* cysts that had not been exposed to chlorine. All positive control animals used in these trials became infected. Five negative control animals were used in each trial. These animals were not exposed to *Giardia* cysts and none became infected. All five human source isolates were used in these trials, and two to three sources were used to generate CT values at each pH and temperature. Data generated with all five of these sources were used to calculate the predicted CT values at each temperature and pH (Table 13.4).

Table 13.4 includes the coefficient of correlation (R), slope, probability, and standard error. CT values are expressed as the predicted range, and predicted mean CT for animal groups in which 1 to 4 animals were infected and for

Table 13.4 CT values for 99.99% inactivation of *Giardia* cysts at temperatures of 0.5°C to 5.0°C at pH ranges of 6 to 8

pH	Temp (°C)	N	Animals infected	R	Slope	Probability	Standard error	Predicted CT range	Predicted mean CT
6	0.5	14	1–4	0.979	1.068	<.001	0.051	176–197	185
		20	None	0.948	0.917	<.001	0.075	202–233	220
6	2.5	7	1–4	0.998	0.997	<.001	0.018	141–142	142
		16	None	0.859	0.881	<.001	0.124	156–190	175
6	5.0	12	1–4	0.946	1.004	<.001	0.089	146–147	146
		10	None	0.937	0.876	<.001	0.106	139–171	157
7	0.5	15	1–4	0.997	1.020	<.001	0.028	285–295	289
		20	None	0.995	0.975	<.001	0.025	302–315	310
7	2.5	14	1–4	0.991	0.977	<.001	0.027	246–256	252
		15	None	0.998	0.980	<.001	0.017	260–268	265
7	5.0	14	1–4	0.993	0.984	<.001	0.027	159–163	161
		15	None	0.992	0.998	<.001	0.034	165–166	166
8	0.5	10	1–4	0.888	1.159	<.003	0.140	312–389	342
		10	None	0.962	1.137	<.001	0.083	392–475	425
8	2.5	15	1–4	0.950	1.121	<.001	0.083	250–295	268
		17	None	0.912	1.037	<.001	0.108	336–353	343
8	5.0	17	1–4	0.854	0.993	<.001	0.133	278–281	280
		13	None	0.930	1.299	<.001	0.090	243–367	290

N (number of data points), R (correlation coefficient), slope, probability, standard error (all in logs), predicted CT ranges, and predicted mean CT (all in antilog) for results in which 1 to 4 animals/group and no animals/group were infected at 0.5°C, 2.5°C, 5.0°C, pH 6.0, 7.0 and 8.0.
CT, product of concentration (C) and time (T), at which 99.99% of the cysts are inactivated.

animal groups in which none were infected. Data generated for CT values where 1 to 4 animals/group were infected represent the "break-point," a chlorine concentration and time value where 99.9% to 99.99% of the cysts were inactivated by chlorine. Data generated for CT values where none of the animals/group were infected include the range and means for greater than 99.99% inactivation of the cysts.

Chlorine concentrations used in these trials ranged from 0.3 to greater than 4.0 mg/l for all temperatures examined at pH 6, 7, and 8. However, the only chlorine concentrations used to calculate predicted CT values were those in the range of 0.3 to 2.5 mg/l. Throughout the course of these trials we observed that use of chlorine concentrations above 2.5 mg/l often produced erratic and unpredictable results, suggesting a "lag period" before complete inactivation was achieved at temperatures in the 0.5 to 5.0°C range. This prompted a comparison using regression analysis of: 1) data generated for all chlorine concentrations; 2) data generated with chlorine concentrations of 0.3 to 2.5 mg/l; and 3) data generated for chlorine concentrations of 2.5 mg/l or greater. For most of the temperatures at any pH, use of data generated with chlorine concentrations of 0.3 to 2.5 mg/l resulted in a higher coefficient of correlation and a slope very close to 1, indicating that chlorine concentration and time were of equal importance for concentrations up to 2.5 mg/l (Table 13.4). When all of the chlorine concentration values were used for analysis, the coefficient of correlation was lower, and the slope varied from 0.5 to 0.9, indicating that time was somewhat more important than chlorine concentrations. When the data using chlorine concentrations of 2.5 mg/l or greater were analyzed, the slope varied from 0.15 to 0.64 for all temperatures at pH 6 and pH 8, definitely indicating that time was more important than chlorine concentration. These same results were obtained with all temperatures at pH 7, but changes in the slope were not as significant as the changes at pH 6 and pH 8.

Since these trials were not specifically designed to evaluate the possibility of a "lag period" before inactivation, it would be inappropriate to make this interpretation; the database for chlorine concentrations greater than 2.5 mg/l is too limited for a valid statistical comparison. Moreover, the predicted CT values were not appreciably different when data using all chlorine concentrations were compared with chlorine concentrations of 0.3 mg/l to 2.5 mg/l. However, the function of a cyst wall on *Giardia* or any other protozoan is to protect the organism from adverse environmental conditions, at least for a short period of time, and a "lag period" before complete inactivation is achieved should not be unexpected for the chlorine concentrations used in these trials. Until more information becomes available, we must caution public health agencies, engineers, and municipal water treatment operators that increasing chlorine concentrations to reduce contact time may not be a prudent solution. Use of predicted CT values obtained in these trials should be restricted to final chlorine concentrations of 0.3 to 2.5 mg/l.

As was expected, higher CT values are necessary for the inactivation of

Giardia cysts at temperatures between 0.5°C and 5.0°C than at temperatures above 5.0°C. If municipalities do not have a filtration system capable of removing 99.9% of the cysts, and the source is either contaminated with *Giardia* cysts or the source is at risk for contamination with *Giardia* cysts, these municipalities must either increase the concentration of chlorine or increase contact time (e.g., prechlorination or storage) to arrive at a CT necessary for inactivation of the cyst when temperatures are less than 5.0°C.

The chemical species primarily responsible for biocidal activity is HOCl (Weber, 1972; White, 1972). At pH 6 about 96% of the chlorine is in this form. This is reduced to about 75% at pH 7. Above pH 7.5 the chemical equilibrium shifts very quickly to the OCl− form. At pH 8 about 23% of the chlorine is in the HOCl form and at pH 9 less than 4% is in this form. The reduced rate of chemical reaction at lower temperatures, when combined with considerable loss of biocidal efficiency at a pH above 7.5, resulted in very high CT values predicted for inactivation of cysts at pH 8. Indeed, we experienced considerable difficulty to even establish predicted CT values for various temperatures at pH 8. Our confidence in these results is best exemplified by the range of CT values predicted for inactivation. The decrease in biocidal efficiency of chlorine above pH 8 is so great that we cannot recommend extrapolation of the current CT values for water sources with a higher pH.

Many municipalities using surface water as a source have developed facilities designed to accommodate the physical, chemical, and biological prop-

Table 13.5 Interpolation of CT values for temperatures 0.5 to 5.0 °C, pH 6.0 to 8.0*

Temperature °C	pH				
	6.0	6.5	7.0	7.5	8.0
0.5	185	237	289	316	342
1.0	174	227	280	302	324
1.5	164	217	271	288	305
2.0	153	207	261	274	287
2.5	142	197	252	260	268
3.0	143	188	234	252	270
3.5	144	180	216	244	273
4.0	144	171	197	237	275
4.5	145	163	179	229	278
5.0	146	154	161	221	280

* If the temperature or pH of the water is between the interpolated values given, use the value for the lower temperature and the higher pH. Individuals responsible for establishing CT values should not attempt to increase chlorine concentrations above 2.5 mg/l to arrive at the CT necessary for inactivation. The *Giardia* cyst wall may be capable of resisting external stimulus for a short period of time, especially at lower temperatures; increasing chlorine to decrease time of contact would be risky. Municipalities with short contact times and/or with a pH above 7.5 must consider other options, such as storage, to obtain a CT necessary to insure inactivation.

erties unique to that source and provide effective treatment of the water. However, if the source is contaminated with the cysts of *Giardia*, the treatment may not be adequate to prevent the risk of waterborne giardiasis, even though water quality meets or even exceeds existing regulations. If filtration is not adequate to remove 99.9% of the *Giardia* cysts or treatment is considered inadequate to prevent a risk of waterborne giardiasis, municipalities using chlorine as the biocidal barrier will need to introduce sufficient chlorine and/or increase the time of contact with chlorine to inactivate at least 99.9% of the *Giardia* cysts present in that water. The method by which this is to be accomplished is contingent upon the properties unique to that source and the physical layout of the facility.

Public health agencies, engineers, and municipal water treatment operators can use Table 13.4 to arrive at a CT value necessary for inactivation of *Giardia* cysts if temperatures and pH values are close to those used in this study. Likely the temperature and pH of the water will be between the exact levels studied, necessitating interpolation. To facilitate this application, the CT values have been calculated and presented in matrix format (Table 13.5). If the temperature or pH of the water is between the interpolated values given, it would be best to use the value for the lower temperature and the higher pH. Individuals using these interpolated values should be cognizant that these are mathematical interpolations and not experimental values.

References

Baumann, E.R. and Ludwig, D.D. 1962. Free available chlorine residuals for small non-public water supplies. *Journal of the American Water Works Association* 54:1379–1388.

Belosevic, M., Faubert, G.M., MacLean, J.D., Law, D., and Croll, N.A. 1983. *Giardia lamblia* infections in mongolian gerbils: an animal model. *Journal of Infectious Disease* 147:222–226.

Chick, H. 1908. An investigation of the laws of disinfection. *Journal of Hygiene* 8:92–158.

Craun, G.F. 1986. *Waterborne Diseases in the United States*. CRC Press, Inc., Boca Raton, Florida. 295 pp.

Craun, G.F. 1988. Surface water supplies and health. *Journal of the American Water Works Association* 80:40–52.

Davies, R.B. and Hibler, C.P. 1979. Animal reservoirs and cross-species transmission of *Giardia*. pp. 104–126 in Jakubowski, W. and Hoff, J.C. (editors), *Waterborne Transmission of Giardiasis*, Proceedings of Symposium, United States Environmental Protection Agency, Cincinnati, Ohio.

DeWalle, F.B. and Jansson, C.R.E. 1983. Chlorine and ultraviolet treatment to ensure inactivation of the *Giardia lamblia* cyst. *Proceedings Seminar Giardiasis and Public Water Supplies*. British Columbia Water and Waste Association. Richmond, British Columbia.

Dykes, A.C., Juranek, D.D., Lorenz, R.A., Sinclair, S.P., Jakubowski, W., and Davies, R.B. 1980. Municipal waterborne giardiasis: an epidemiologic investigation. Beavers implicated as a possible reservoir. *Annals of Internal Medicine* 92:165–170.

Erlandsen, S.L., Bemrick, W.J., Kamp, L.E., Sherlock, L.F., and Schupp, D.G. 1987. Cross-species transmission of giardiasis: infection of beavers with human *Giardia lamblia*. *Calgary Conference on Giardiasis*, February 1987. Calgary, Canada.

Filice, F.P. 1952. Studies on the cytology and life history of a *Giardia* from the laboratory rat. *University of California Publications in Zoology* 57:53–143.

Hibler, C.P., MacLeod, K.I., and Lyman, D.O. 1975. Giardiasis in residents of Rome, New York. *Morbidity and Mortality Weekly Report* 24:366.

Hibler, C.P. 1988. Advances in *Giardia* Research, edited by P.M. Wallis and B.R. Hammond. Univ. of Calgary Press. Analysis of Municipal Water Samples for *Giardia*. pp. 237–245.

Hibler, C.P. 1988. Advances in *Giardia* Research, edited by P.M. Wallis and B.R. Hammond. Univ. of Calgary Press. An Overview of the Techniques for Diagnosis of Waterborne *Giardia*. pp. 197–204.

Hibler, C.P., Hancock, C.M., Perger, L.M., Wegrzyn, J.G. and Swabby, K.D. 1987. Inactivation of *Giardia* cysts with chlorine at 0.5°C to 5.0°C. *American Water Works Association Research Foundation Research Report*, Denver, Colorado.

Hoff, J.C,. Rice, E.W., and Schaefer, F.W. III. 1985. Comparison of animal infectivity and excystation as measures of *Giardia muris* cyst inactivation by chlorine. *Applied Environmental Microbiology* 50:1115–1117.

Hoff, J.C. 1986. Strengths and weaknesses of using CT values to evaluate disinfection practice. *American Water Works Association Water Quality Technology Conference XIV*, Portland, Oregon.

Jarroll, E.L., Bingham, A.K., and Meyer, E.A. 1980. *Giardia* cyst destruction: effectiveness of six small quantity water disinfection methods. *American Journal of Tropical Medicine and Hygiene* 29:8–11.

Jarroll, E.L., Bingham, A.K., and Meyer, E.A. 1981. Effect of chlorine on *Giardia lamblia* cyst viability. *Applied Environmental Microbiology* 41:483–487.

Kirner, J.C., Littler, J.D., and Angelo, L.A. 1978. A waterborne outbreak of giardiasis in Camas, Washington. *Journal of the American Water Works Association* 70:35–40.

Lippy, E.C. 1978. Tracing a giardiasis outbreak at Berlin, New Hampshire. *Journal of the American Water Works Association* 70:512–520.

Lippy, E.C. 1981. Waterborne disease: occurrence is on the upswing. *Journal of the American Water Works Association* 72:57–62.

Meyer, E.A. and Schaefer, F.W. III. 1984. Models for excystation. pp. 131–146 in Erlandsen, S.L. and Meyer, E.A. (editors), *Giardia and Giardiasis*, Plenum Press, New York.

Monzingo Jr., D.L. 1985. The Prevalence of *Giardia* in a Beaver Colony and the Resulting Environmental Contamination. *Ph.D. Dissertation, Colorado State University*, Fort Collins.

Moore, G.T., Cross, W.M., McGuire, C.D., Mollohan, C.S., Gleason, N.N., Healy, G.R., and Newton, L.H. 1969. Epidemic giardiasis at a ski resort. *New England Journal of Medicine* 281:402–407.

National Primary Drinking Water Regulations: Filtration, Disinfection, Turbidity, *Giardia lamblia*, Viruses, *Legionella*, and Heterotrophic Bacteria. 1987. Proposed Rule, 40 CFR parts 141 and 142. *Federal Register* 52:212:42718.

Rice, E.W., Hoff, J.C., and Schaefer, F.W. III. 1982. Inactivation of *Giardia* cysts by chlorine. *Applied Environmental Microbiology* 43:250–251.

Safe Drinking Water Committee. 1980. The Disinfection of Drinking Water. pp. 5–138 in *Drinking Water and Health*, Vol. 2, National Academy Press, Washington, D.C.

Shaw, P.K., Brodsky, R.E., Lyman, D.O., Wood, B.T., Hibler, C.P., Healy, G.R., MacLeod, K.I., Stahl, W., and Schultz, M.G. 1977. A community-wide outbreak of giardiasis with evidence of transmission by municipal water supply. *Annals of Internal Medicine* 87:426–432.

Standard Methods for the Examination of Water and Wastewater. 1985. American Public

Health Association, American Water Works Association, Water Pollution Control Federation, 16th Edition, American Public Health Association, Washington, D.C. p. 1268.

Swabby, K.D., Hibler, C.P., and Wegrzyn, J.C. 1988. Advances in *Giardia* Research. pp. 75–77 in Wallis, P.M. and Hammond, B.R. (editors), University of Calgary Press.

Watson, H.E. 1908. A note in the variation of the rate of disinfection with change in the concentration of the disinfectant. *Journal of Hygiene* 8:536–592.

Weber, W.J. 1972. *Physiochemical Processes for Water Quality Control*. Wiley-Interscience, New York.

Wegrzyn, J.G. 1988. *Giardia in Colorado muskrats: prevalence, host-specificity and cross-species transmission*. Ph.D. Dissertation, Colorado State University, Fort Collins.

White, G.C. 1972. *Handbook of Chlorination*. Van Nostrand Reinhold Co., New York.

14

Occurrence and Control of *Cryptosporidium* in Drinking Water

Joan B. Rose

Department of Environmental and Occupational Health, University of South Florida, Tampa, Florida

14.1 Introduction

Although *Cryptosporidium*, an enteric coccidian protozoan, was described in 1907, it has only been recognized as a cause of disease in humans since 1980. Within five years of its acceptance as a human pathogen, the first waterborne outbreak associated with *Cryptosporidium* was described in the United States. *Cryptosporidium* must now be included in a list of pathogens which may occur and be transmitted by contaminated water.

The importance of *Cryptosporidium* in water may be better defined by understanding the nature of the organism, the presence of *Cryptosporidium* disease in the population, and the occurrence of the organism in the environment. *Cryptosporidium* is newly recognized as a waterborne agent, but without an historical record of outbreaks one can only guess at the potential of waterborne transmission and the significance of this route in initiating disease. Despite this deficiency in knowledge, large waterborne outbreaks such as that which occurred in Carollton, Georgia, in 1987 should be prevented. Approaches for the control of *Cryptosporidium* in water will need to be assessed.

14.2 History of Studies on *Cryptosporidium*

The first formal description of the genus *Cryptosporidium* came in 1907 by E.E. Tyzzer (1907). The organism was isolated from the gastric mucosa within the stomach of the laboratory mouse. This isolate was named *C. muris*; and four years later *C. parvum*, a second species which inhabited the small intestine of the mouse, was identified (Tyzzer, 1910; Tyzzer, 1912). During the initial investigations the life cycle was described and it was recognized that the oocyst stage could be found in the feces of infected mice.

Cryptosporidium was not recognized as a pathogen until 1955 (Slavin, 1955). A new species, *C. meleagridis*, was described during an outbreak of diarrhea in a domesticated turkey flock. The organism again was found in the intestinal tract. By 1971 *Cryptosporidium* was first associated with diarrhea in calves (Panciera et al., 1971). After numerous reports and studies on experimental infections over the next ten years, it was recognized that *Cryptosporidium* was a cause of illness and, in some cases, significant mortality in calves and lambs (Anderson, 1982; Anderson, 1982a; Anderson and Bulgin, 1981; Angus et al., 1982; Current, 1987).

Two cases of *Cryptosporidium* in humans were reported in 1976 (Meisel et al., 1976; Nime et al., 1976). One infection was noted in an immunosuppressed patient (Meisel et al., 1976), while a second infection occurred in an apparently healthy 3-year-old child who lived in a farm house in a rural area (Nime et al., 1976). During 1982 reports of cryptosporidiosis in humans increased, with the majority of the infections being found in patients who had impaired immune functions (Navin and Jaranek, 1984). The onslaught of Acquired Immunodeficiency Syndrome (AIDS) brought with it a recognition of the role *Cryptosporidium* played as a causative agent of diarrhea in immunosuppressed individuals. Because the infection was thought to be a major factor in the mortality of these patients, there was an urgency to find a rapid diagnostic test and a treatment.

No effective treatment was provided by any of the therapeutic agents tested (Fayer and Ungar, 1986). However, the diagnosis of infection by oocyst detection in feces in lieu of intestinal biopsies was a major advancement (Current et al., 1983; Garcia et al., 1983; Ma and Soave, 1983).

The availability of relatively quick and easy techniques for the diagnosis of *Cryptosporidium* infections augmented studies on immunocompetent individuals presenting with diarrhea. Hence, by 1985 it was well recognized that *Cryptosporidium* was a cause of diarrhea in humans worldwide (Fayer and Ungar, 1986).

Studies on the environmental occurrence of *Cryptosporidium* began in 1985 (Musial et al., 1987), and concurrently the first waterborne outbreak occurred in Texas (D'Antonio et al., 1985). In January, 1987, a major waterborne outbreak attributed to *Cryptosporidium* took place in Georgia (Hayes et al., 1988; Mason, 1988). The Environmental Protection Agency had proposed including *Cryptosporidium* on the Drinking Water Priority List (DWPL) (Federal Register, 1987). The agency must provide regulations for this list of contaminantes within three years. Finalized in January of 1988, *Cryptosporidium* was added to the DWPL (Federal Register, 1988).

14.3   Taxonomy of *Cryptosporidium*

Cryptosporidium is a protozoan parasite. All protozoa are comprised of a single cell and are differentiated by morphology and life cycle. Although the protozoa had been considered a single phylum of animals, there are currently seven recognized phyla (Schmidt and Roberts, 1985). Figure 14.1 diagrams *Cryptosporidium's* taxonomic position within the protozoa.

Cryptosporidium belongs to the phylum *Apicomplexa*, subclass *coccidia*. The majority of coccidia are small protozoa which complete their life cycles intracellularly within the digestive tract, epithelium, liver, kidneys, blood cells, or other tissues of animals (Schmidt and Roberts, 1985). The suborder *Eimeriina* contains thousands of species predominately parasitizing domestic animals, with very few human parasites. The three families under the suborder include *Sarcocystidae, Eimeriidae* and *Cryptosporidiidae*.

The genera *Sarcocystis* and *Toxoplasma* within *Sarcocystidae* are heterogenous, and their life cycles are completed in two hosts with asexual development in the intermediate host while the sexual life cycle is complete in the carnivorous definitive host (Long, 1982). Both genera may invade muscle cells. *Toxoplasma* may invade the intestinal mucosa but the oocysts are not immediately infective upon excretion in feces and must first mature in the environment.

Eimeria and *Isospora* are enteric coccodia and inhabit the intestinal tract of animals. One species, *I. belli*, causes rare infections in man; however the relationship to disease is ill defined (Dubey, 1977). *Eimeria* is found only in animals and most species are host specific (Long, 1982). With few exceptions,

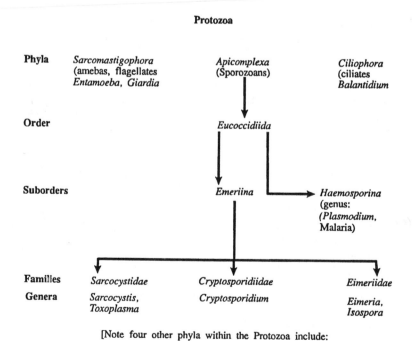

Protozoa

| Phyla | *Sarcomastigophora* (amebas, flagellates *Entamoeba, Giardia* | *Apicomplexa* (Sporozoans) | *Ciliophora* (ciliates *Balantidium* |

Order — *Eucoccidiida*

Suborders — *Emeriina* → *Haemosporina* (genus: (*Plasmodium*, Malaria))

| Families | *Sarcocystidae* | *Cryptosporidiidae* | *Eimeriidae* |
| Genera | *Sarcocystis, Toxoplasma* | *Cryptosporidium* | *Eimeria, Isospora* |

[Note four other phyla within the Protozoa include: *Labyrinthomorpha, Microspora, Ascetospora, and Myxozoa*].

Figure 14.1   Status of *Cryptosporidium* within the Protozoa (Navin and Juranek, 1984).

newly shed oocysts in feces must sporulate outside the body before becoming infective (Long, 1982).

The family *Cryptosporidiidae* contains the single genus *Cryptosporidium.* This genus differs from the genera in the *Sarcocystidae* family as it is monoxenous, completing its life cycle within a single host. Like members of the *Eimeriidae, Cryptosporidium* is an enteric coccidian and develops within the gastric or intestinal mucosal epithelium in mammals (Levine, 1984).

It is interesting that the oocyst of *Cryptosporidium* is morphologically different from all the other genera in the suborder *Eimerina.* Most of the coccidian oocysts are large (10 to 40 μm) and contain varying numbers of sporocysts which in turn contain sporozoites (Long, 1982). The oocyst of *Cryptosporidium* contains no sporocysts but house four sporozoites and is much smaller (anywhere from 4 to 8 μm) (Levine, 1984).

After Tyzzer's first description of *Cryptosporidium* from mice (Tyzzer, 1907), many species were named based on isolation from a specific host. It was thought that, like *Eimeria, Cryptosporidium* species were host specific. This assumption turned out to be erroneous. Cross-transmission studies have shown that mammalian isolates demonstrate a wide range of infectivity for other animals (Current, 1987; Fayer and Ungar, 1986). The most extensive studies

Table 14.1 Cross-transmission of *Cryptosporidium* between humans and animals

	Oocyst Source		
	Human	Cattle	Chicken
Animals Infected	Cat	Human	Turkey
	Cattle	Cat	Duck
	Dog	Dog	
	Goat	Goat	
	Mouse	Guinea Pig	
	Pig	Mouse	
	Rat	Pig	
	Sheep	Rabbit	
		Rat	
		Sheep	
		Chicken	
Animals Not Infected	Chicken	Chicken	Cotton Rat
			Gerbil
			Guinea Pig
			Hamster
			Mouse
			Pig
			Quail
			Rat

Fayer and Ungar (1986).

indicated oocysts from humans and cattle would infect a variety of animals (Table 14.1), while avian isolates were not infective for mammals. Two species have been described as infecting mammals (Upton and Current, 1985). *Cryptosporidium parvum* appears to be the predominant species responsible for disease in man and animals (Current, 1987). This species has oocysts measuring 4 to 6 μm in diameter and inhabits the intestinal tract. *Cryptosporidium muris* is found in the stomach and has larger oocysts (5.6 × 7.4 μm) and a narrower host range (Upton and Current, 1985). Two other species, *C. meleagridis* isolated from turkeys (Slavin, 1955) and *C. baileyi* isolated from chickens, have been described (Current et al., 1986). Isolates from snakes and fish may also constitute separate species, yet this classification awaits further validation (Levine, 1984).

The life cycle of *Cryptosporidium* is like that of most coccidia and has been described in detail elsewhere (Current, 1987; Fayer and Ungar, 1986; Long, 1982). Briefly, the oocyst, the environmentally stable stage, after ingestion undergoes excystation releasing the sporozoites which then initiate the intracellular infection within the epithelial cells of the gastrointestinal tract. The sporozoite differentiates into the trophozoite which undergoes asexual multiplication to form type I meronts and then merozoites which may infect new host cells. Merozoites from type II meronts produce microgametocytes and macrogametocytes which undergo sexual reproduction to form the oocyst. The oocysts are immediately infective upon excretion in the feces.

14.4   Epidemiology of the Disease, Cryptosporidiosis

Knowledge of the clinical manifestations and epidemiology of *Cryptosporidium* infections has been increasing. As more cases of cryptosporidiosis have been identified and documented, more insight has been gained into the illness, infectivity potential, and prevalence of the disease.

Cryptosporidium in humans is associated with profuse watery diarrhea; fluid losses may average three liters per day (Navin and Juranek, 1984). Abdominal pain, nausea, vomiting, and fever may also be present (Fayer and Ungar, 1986). The symptoms may begin on an average of three to six days after exposure (Jokipii and Jokipii, 1986). Oocyst excretion coincides with the symptoms which have been reported from less than 3 to 30 days with an average duration of 12 days (Fayer and Ungar, 1986; Jokipii and Jokipii, 1986). Thus, individuals apparently "well" may remain infectious. The disease is self-limiting in those individuals with normal immune functions.

The dose necessary to cause an infection in humans is unknown. In mice inoculated with 100, 500, and 1000 oocysts, 22%, 66%, and 78% became infected, respectively (Ernest et al., 1987). In a study using primates, two of two subjects became infected after exposure to 10 oocysts (Miller et al., 1986). This preliminary work suggests that once ingested, low numbers of oocysts may be capable of causing disease. The size of the inoculum was not shown to be related to the severity of the disease. However, illness patterns may vary, which may be a factor of host age, immune status, and possibly variations in the organism's virulence (Tzipori, 1983).

Infections occur more commonly in young mammals (Current, 1987). In animals, the majority of calves infected were under 3 weeks of age (Pavlasek, 1982); and lambs, young goats, and piglets were also found to be primarily affected. In humans the documentation of outbreaks in day care centers has demonstrated the susceptibility of young children in that setting to cryptosporidiosis (Alpert et al., 1984; Alpert et al., 1986; Anonymous, 1984; Taylor et al, 1985). Numerous other studies on the prevalence of *Cryptosporidium* infections have reported a greater frequency in children (Addy et al., 1986; Højlyng et al. 1986; Pape et al., 1987), particularly those under two years of age. *Cryptosporidium* was the predominant enteric parasite infection in children less than 1.5 years old but decreased in older children (3 to 5 years old) (Addy et al., 1986; Højlyng et al., 1986). It has been suggested that *Cryptosporidium* is an important cause of chronic childhood diarrhea, and in one study *Cryptosporidium* was the most prevalent pathogen isolated in children ranging in age from 2 to 15 months (Sallon et al., 1988). Infections were more prevalent in bottle-fed infants. Other surveys have supported this finding, with increased prevalence in metropolitan areas where many infants were not breast-fed (Mata et al., 1984) or by demonstrating rare *Cryptosporidium* infection in breast-fed infants (Pape et al., 1987). This may be due to some protection afforded through breast feeding but studies in calves and mice have not demonstrated a protective effect (Current, 1987). Another possible explanation is that the water used for bottle or formula preparation was contaminated.

Fayer and Ungar (1986) have reviewed prevalence data for human cryptosporidiosis in all age groups. Prevalence ranged from a low of 0.12% in an Australian study to a high of 23% in a study in Spain (excluding outbreak data). Many of the large surveys included 2000 to 7000 patients and prevalence ranged from 0.5 to 1.4% (Fayer and Ungar, 1986; Marshall et al., 1987). In some surveys for agents causing gastroenteritis, *Cryptosporidium* has been found to be the most predominant intestinal protozoan, (in contrast to the protozoa *Giardia* and *Entamoeba*) ranking sixth behind the enteric bacteria (Marshall et al., 1987), and the most predominant parasite infection (Holley and Dover, 1986). In addition, little infection was detected in control patients without diarrhea, suggesting the majority of infections are symptomatic (Marshall et al., 1987). The excretion of oocysts in the absence of symptoms, however, has been reported (Pohjola et al., 1986; Stehr-Green et al., 1987).

Analysis of IgG and IgM antibodies to *Cryptosporidium* in populations in Peru and Venezuela suggested 19.8% and 15.5% recent or active infections, respectively (Ungar et al., 1988). However, 64% of these populations had antibodies indicating infection at some time during their lives. The protective nature of the antibodies remains uncertain.

It was recognized early on that *Cryptosporidium*, because of its fecal-oral route of transmission, was similar to other enteric organisms. Person-to-person,

Table 14.2 Reported *Cryptosporidium* infections in animals

Fishes
Naso tang (*Naso lituratus*)
Carp (*Cyprinus carpio*)

Reptiles
Red-bellied black snake (*Pseudechis porphyriacus*)
Corn snake (*Elaphae guttata*)
Trans-pacos ratsnake (*E. subocularis*)
Madagascar boa (*Sansinia madagascarensis*)
Timber rattlesnake (*Crotalus horridus*)

Birds
Chicken (*Gallus gallus*)
Turkey (*Meleagris gallopavo*)
Bobwhite quail (*Colinus virginianus*)
Peafowl (*Pavo cristatus*)
Black-throated finch (*Poephila cincta*)
Domestic goose (*Anser anser*)
Red-lored parrot (*Amazona autumnalis*)

Mammals
Macaques (*Macaca mulatta*)
  (*M. radiata*)
  (*M. fascincularis*)
Calf (*Bos taurus*)
Lamb (*Ovis aries*)
Goat (*Capra hircus*)
Horse (*Equus caballus*)
Swine (*Sus scrofa*)
Roe deer (*Capreolus capreolus*)
Red deer (*Cervus elaphus*)
White tail deer (*Odocoileus virginianus*)
Fox squirrel (*Sciurus niger*)
Gray squirrel (*S. carolinensis*)
Pocket gopher (*Geomys bursarius*)
Chipmunk (*Tamias striatus*)
Flying squirrel (*Glaucomys volans*)
13-Lined ground squirrel (*Spermophilus tridecemlineatus*)
Beaver (*Castor canadensis*)
Muskrat (*Ondatra zibethicus*)
Woodchuck (*Marmota monax*)
Nutrina (*Myocastor coypus*)
Domestic rabbit (*Oryctolagus cuniculus*)
Cottontail rabbit (*Sylvilagus floridanus*)
Exotic undulates (~14 species)

(Continued next page)

Table 14.2 (continued)

Domestic dog (*Canis familiaris*)
Coyote (*C. latrans*)
Red fox (*Vulpes vulpes*)
Grey fox (*Orocyon cinereoargenteus*)
Domestic cat (*Felis catus*)
Striped skunk (*Mephitis mephitis*)
Raccoon (*Procyon lotor*)
Black bear (*Ursus americanus*)

Current (1987)

water, and foodborne routes may all play a role in the transmission of pathogen. Zoonotic transmission is also a possibility. Person-to-person transmission has been suggested as a major route of transmission documented in day care settings and hospitals (Anonymous, 1984; Baxby et al., 1983; Koch et al., 1984; Taylor et al., 1985). Foodborne and particularly waterborne transmission has been suggested through association with traveller's diarrhea (Elsser et al., 1986; Jokipii et al., 1985; Ma et al., 1985). Zoonotic transmission may be of less importance (Fayer and Ungar, 1986), but many animals have been shown to be infected with *Cryptosporidium* (Table 14.2). Both wild and domesticated animals may be important reservoirs of human infection, contributing also to the contamination of the environment. It has been suggested that contaminated water plays a major role in transmission of giardiasis (Jakubowski, 1988). *Cryptosporidium* with its similar epidemiology (Current, 1987) may also be significantly associated with waterborne disease.

14.5 Occurrence of *Cryptosporidium* in the Environment

The presence of *Cryptosporidium* in the environment is demonstrated by the occurrence of oocysts. Infected humans and animals may contribute to this contamination through deposition of fecal material and sewage. Since 1985, methods have been developed to specifically recover and detect *Cryptosporidium* oocysts from water. As more data are accumulated, oocyst contamination may be evaluated to assess levels of pollution, sources of contamination, and potential exposure.

Oocyst Occurrence   Many enteric pathogens may be present in sewage, and wastewater may contain varying numbers of *Cryptosporidium* oocysts. This may be influenced by the size of the community as well as the rate of infection within the population. During a survey of raw wastewater from a community undergoing a waterborne outbreak of *Cryptosporidium*, concentrations decreased from 34 oocysts/l to 1.2 oocysts/l after the outbreak was over (Rose et al., 1987). Madore et al., (1987) examined both raw and treated wastewater. They reported levels as high as 13,700 oocysts/l in raw sewage, averaging 5180 oocysts/l in four samples. Wastes from a cattle slaughter house may have been responsible for the high concentrations in at least three of the four samples which came from a single sewer system. Other studies with limited sampling have reported 521 oocysts/l and 4 oocysts/l in untreated wastewater (DeLeon et al., 1988; Rose, 1988). After sewage treatment by activated sludge, oocyst levels were found to range from 4 oocysts/l to 1297 oocysts/l (Madore et al., 1987; Rose, 1988). Following examination of 14 samples, DeLeon et al. (1988) reported an average of 39.7 oocysts/l in treated effluents. Table 14.3 summarizes the results of these studies.

Although data on *Cryptosporidium* in wastewater are preliminary, it seems evident that sewage discharge may be a significant source of contamination of oocysts in the environment. This may not only be a potential risk for humans but also for animals. Many surface waters destined for other uses (potable supplies, irrigation) are receiving wastewater discharges. As populations increase and more water is continually recycled, risk of waterborne cryptosporidiosis needs to be taken into consideration.

Cryptosporidium oocysts may be ubiquitous in the water environment as is the infection in animal and human populations. In a survey of waters in the western U.S. (Rose, 1988), 91% of the sewage samples were found to contain oocysts at varying levels while 77% and 75% of the rivers and lakes, respectively, were also found to contain *Cryptosporidium* (Figure 14.2). Interestingly, even 83% of pristine water supplies with no human activity in the watershed contained *Cryptosporidium* oocysts. Limited sampling of treated drinking water reported 28% of the samples positive for oocysts.

Table 14.3 Concentrations of *Cryptosporidium* oocysts reported in wastewater

State	Type of sample	Number of samples	Oocysts/l Range	Oocysts/l Mean	Reference
Arizona	Raw sewage	4	850–13,700	5180[a]	(Madore et al., 1987)
Arizona	Raw sewage	5	NR[c]	521[b]	(DeLeon et al., 1988)
Texas	Raw sewage	6	NR	4[b]	(Rose, 1988)
Arizona	Activated sludge effluent	9	140–3960	1297[a]	(Madore et al., 1987)
Arizona	Activated sludge effluent	14	NR	39.7[b]	(DeLeon et al., 1988)
Colorado	Activated sludge effluent	2	NR	4[b]	(Rose, 1988)

[a] Arithmetic means
[b] Geometric means
[c] Not reported

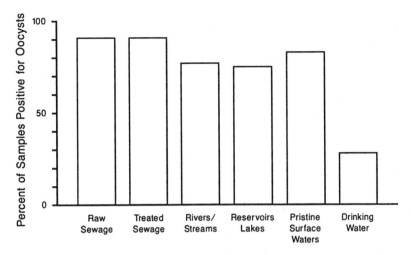

Figure 14.2   Percentage of positive isolations for *Cryptosporidium* from various water types (Rose, 1988).

The percentage of positive isolations may be misleading as to the true significance of the contamination, and concentrations or levels of contamination should be examined as well. Average levels of oocysts reported in surface waters, including lakes, reservoirs, streams and rivers, ranged from 0.02 oocysts/l to 1.3 oocysts/l (Table 14.4) (Ongerth and Stibbs, 1987; Rose, 1988; Rose et al., 1988; Rose et al., 1988a). Pristine areas contained levels of 0.02 to 0.08 oocysts/l; while in reservoirs and lakes, oocysts were reported at 0.91 and 0.58/l. Streams and rivers seemed to contain slightly greater concentrations of 0.94, 1.09, and 1.2 oocyst/l.

The work of Madore et al. (1987) has suggested that agricultural sources of water pollution may be as much a concern as human wastewater. This may include runoff from dairies and grazing lands. In three separate surveys of surface waters, samples could be separated by probable agricultural sources of pollution in contrast to sewage discharges (Madore et al., 1987; Ongerth and Stibbs, 1987; Rose et al., 1988). The concentrations of oocysts were 1.5 to 1.9

Table 14.4 *Cryptosporidium* levels in surface waters

Type of water	Average oocysts/l (geometric means)
Pristine rivers	0.02, 0.08
Reservoirs/lakes*	0.91, 0.58
Streams/rivers*	0.94, 1.3, 1.09

* Receiving wastewater discharges or agricultural pollution.
Data of Ongerth and Stibbs (1987), Rose (1988), Rose et al. (1988), Rose et al. (1988a).

times greater in waters where agricultural pollution seemed to be the major contaminating source (Table 14.5). This was found to be true in each survey.

Other evidence has also suggested that animals were an important source of pollution in one of these studies (Rose et al., 1988). In an extensive survey of a single watershed, ratios of fecal coliforms to fecal streptococci implied animals as the source of the bacterial contamination. In addition, human enteric viruses which would have originated from sewage were not recovered from this water system. Peaks of contamination were observed for *Cryptosporidium* and also for *Giardia*. After correlation analysis, it was found that *Cryptosporidium* oocyst levels were significantly correlated with *Giardia* cyst levels. Thus, both of these protozoa may have been originating from a common animal source. Neither protozoan was statistically associated with total or fecal coliforms or turbidity. This lack of relationship between *Giardia* and the indicator bacteria has been reported previously (Akin and Jakubowski, 1986). However, this is the first report indicating that the same may be true for *Cryptosporidium*. It should be kept in mind that routine coliform methodologies may be underestimating indicator concentrations by as much as 90% (McFeters et al., 1986; see also Chapter 23). In addition, while large volumes of water are collected for parasite analysis (up to 400 l), routinely 100 ml samples are analyzed for coliform bacteria. Further studies will be needed to determine whether fecal streptococci bacteria may be a better indicator of protozoan contamination in water.

Other studies have also included concurrent analyses of water samples for *Cryptosporidium* and *Giardia* (DeLeon et al., 1988; Rose et al., 1988; Rose et al., 1988a). In both wastewater and surface water, levels of *Cryptosporidium* oocysts were 10 to 100 times greater than *Giardia* cyst levels (Table 14.6). It is possible that the methods used to recover *Cryptosporidium* oocysts were more efficient than that used for *Giardia* cysts. Previous studies have reported *Giardia* levels in sewage between 7 and 1242 cysts/l, which in some cases is closer to the published levels of oocysts (Akin and Jakubowski, 1986). It may also be possible that more animals were infected with *Cryptosporidium* than

Table 14.5 Concentrations of *Cryptosporidium* oocysts in surface waters of different contaminating sources

Study	Probable source of contamination	
	Agricultural runoff (dairy/ranch)	Human wastewater (treated)
	——— Average numbers of oocyst/l ———	
Madore et al. (1987)	2904	1864
Ongerth et al. (1987)	1.53	1.0*
Rose et al. (1988)	1.09	0.58

* Possible agricultural impact.

Table 14.6 Comparison of *Cryptosporidium* oocysts and *Giardia* cysts in contaminated water

Comparison	Wastewater		Lakes/reservoirs		Rivers	
	Crypto-sporidium oocysts/l	*Giardia* cysts/l	*Crypto-sporidium* oocysts/l	*Giardia* cysts/l	*Crypto-sporidium* oocysts/l	*Giardia* cysts/l
1	521	3.7	0.58	0.08	1.09	0.22
2	39.7	1.2	7.1	0.35	28.5	1.2
3	5191	51			0.08	0.009
4	1374	1.3				

Data of DeLeon et al. (1988), Rose et al. (1988), Rose et al. (1988a).

Table 14.7 Levels of *Cryptosporidium* oocysts in drinking water

Type of treatment	Number of samples collected	Number of samples positive	Average oocysts/l
Conventional*	4	1	0.002
Direct filtration	3	1	0.006
Filtration, no coagulation	3	1	0.005
Disinfection only	3	2	0.009
Conventional, Carrollton waterborne outbreak	7	6	0.460

* Coagulation, sedimentation, rapid sand filtration, and disinfection.
Data of Rose et al. (1987a), Rose (1988), Rose et al. (1987).

Giardia. Considering Cryptosporidial excretion patterns differ significantly from those of *Giardia*, *Cryptosporidium* oocysts are shed in three liters of fluid or more a day during illness, whereas *Giardia* cysts may be shed intermittently (Wok, 1984). Thus, greater numbers of *Cryptosporidium* oocysts than *Giardia* cysts may be passed into the environment. A combination of factors no doubt may be influencing the occurrence of these protozoa, which may change depending on season, prevalence of infection, as well as a number of other as yet unidentified conditions.

It appears that *Cryptosporidium* oocysts may be found in waters quite commonly; however, it may be unacceptable for any pathogen to be present in potable waters. Twenty-eight percent of the treated drinking water samples examined contained *Cryptosporidium* oocysts (Figure 14.2), although levels were low, ranging between 0.002 and 0.009 oocysts/l. Types of treatment included disinfection alone, filtration, direct filtration, and conventional: coagulation, sedimentation, filtration, and disinfection (Table 14.7). This contamination may be contributing to an endemic level of disease in the exposed population. During a waterborne outbreak in Carrollton, Georgia, levels of oocysts were approximately 100 times greater in the filtered water (0.46 oocysts/l) (Rose et al., 1987) than concentrations reported in other drinking waters. This resulted in approximately a 40% attack rate in the exposed population (Hayes et al., 1988). Perhaps future use of risk assessment models can aid in assessing the potential hazard of low level contamination.

Data Interpretation   A number of considerations must be taken into account in the interpretation and significance of the occurrence data. These are directly related to the limitations of the methods used to recover and detect *Cryptosporidium* oocysts from environmental samples.

In order to detect low levels of oocysts in large volumes of water, filtration techniques have been developed. Once the organisms have been concentrated on a filter, the filter can be processed to recover the debris and organisms which have accumulated. Both centrifugation and density gradients (the most common being gradients of sucrose and potassium citrate) have been utilized to clarify the sample. Musial et al. (1987) reported that sucrose (1.29 g/ml specific gravity) could be used in combination with 1% Tween-80 and sodium dodecyl sulfate to efficiently separate the oocysts from the other debris found in the sample. Finally, oocysts are detected microscopically on membrane filters (1.2 μm pore size) using antibodies specific to the oocyst wall. Both polyclonal and monoclonal antibodies have been used in direct or indirect immunofluorescent procedures (Garcia et al., 1987; Sterling and Arrowood, 1986; Stibbs and Ongerth, 1986). Fluorescein isothiocyanate (FITC) is conjugated to the antibodies and epifluorescent microscopy used for final detection. A characteristic apple-green fluorescence around the oocyst wall is produced. The procedures are summarized in Table 14.8.

The overall recovery efficiency of the methods range from 9 to 59% and is highly influenced by water quality. Thus, the oocyst levels reported in water samples are no doubt underestimating the true levels of contamination. Ongerth and Stibbs (1987) have reported that levels of 0.44 to 5.6 oocysts/l may be mathematically adjusted to 2 to 112 oocysts/l based on recoveries that the different methods exhibited in various river waters. However, *Cryptosporidium* may not be detected at all in the majority of the samples. The interpretation of negative samples may be difficult and does not necessarily signify the absence of contaminating oocysts.

In addition to potentially false negatives or underestimation of levels, false positives may occur. The presence of oocysts in negative controls has been reported (Ongerth and Stibbs, 1987). This may be a result of contamination from one sample to another due to inadequate cleaning of equipment where common equipment is shared. Great caution should be exercised, particularly when processing of both seeded and environmental samples, and separate equipment is recommended. Negative controls should always be run during routine laboratory processing and in conjunction with any drinking water samples to ensure accurate reporting of data. A good quality control program for

Table 14.8 Methods for the recovery and detection of *Cryptosporidium* oocysts from water

Sample step	Procedures used	Comments
Collection	Ten to 1000 liters of water passed through yarn-wound cartridge or polycarbonate disc filters, 1.0 μm pore size.	Water quality will influence volume processed. 88–99% efficient.
Elution	Distilled water (500–3000 ml) used to rinse or wash filters to recover the organisms. Backwashing and use of Tween-80 may enhance recoveries.	Water quality influences recovery efficiency. 16–78%.
Concentration and clarification	Centrifugation used to concentrate the sample. Density gradients (1.18–1.29 g/ml) used to clarify the sample (sucrose with Tween-80 sodium dodecyl sulfate and potassium citrate commonly used).	Balance recovery of oocysts with removal of extraneous debris. Algal contamination a common problem. 66–77% recoveries.
Detection	Filter sample through membrane filter, strain with oocyst antibodies in a direct or indirect fluorescent procedure. Examine using epifluorescent microscopy.	Interfering debris may decrease detection, various antibodies may be used. Overall method recovery 9–59%.

Data of Musial et al. (1987), Ongerth and Stibbs (1987), Rose et al. (1988a).

checking equipment and reagents in concert with negative controls will aid in validation of the data.

The use of antibodies has greatly enhanced the ability to detect *Cryptosporidium*. Fluorescent antibody detection has been previously utilized with great success for *Giardia* cysts in environmental samples (Sauch, 1985; see also Chapter 13). Fluorescence alone, however, may not be sufficient for identification of the organism of interest. Background fluorescence due to naturally fluorescing organisms and nonspecific binding of the antibody may decrease the accurate identification of *Cryptosporidium* oocysts.

A number of antibodies have been used for *Cryptosporidium*. Ongerth and Stibbs (1987) have used polyclonal antibody (PAb) specific for the oocyst wall. Used in an indirect immunoassay, background fluorescence is reduced. A polyclonal antibody may detect a variety of isolates (including *Cryptosporidium* species from birds; however, no information is available regarding the specificity of this particular antibody to bird isolates (Stibbs and Ongerth, 1986). The PAb was shown to react with human, monkey, and bovine oocysts in fecal samples and showed no cross reactivity with yeast (i.e., *Candida*) and other enteric organisms (i.e., *Giardia*, *Entamoeba*, and *Chilomastix*).

Two monoclonal antibodies (MAb) have also been developed for detection of *Cryptosporidium* oocysts (Sterling and Arrowood, 1986). The first MAb (IgG) directly conjugated to FITC was shown to cross-react with yeasts (Sterling and Arrowood, 1986) as well as bird isolates of *Cryptosporidium*. This MAb has been used for numerous water studies (Madore et al., 1987; Musial et al., 1987; Rose, 1988; Rose et al., 1988). The second MAb (IgM), used in an indirect procedure, was tested extensively within the clinical setting (Garcia et al., 1987), and did not demonstrate a cross reactivity with a variety of yeasts, enteric organisms, or with *C. baileyi*, a bird species (Darlington and Blagburn, 1988; Sterling et al., 1988). Recent studies have shown some affinity of this MAb (IgM) for *C. meleagridis*, another *Cryptosporidium* species isolated from turkeys (Darlington and Blagburn, 1988). The MAb (IgM) has also been used in conjunction with the first MAb (IgG) for *Cryptosporidium* detection in water samples (Rose et al., 1987; Rose, 1988; Rose et al., 1988).

Monoclonal antibodies may have a greater specificity for the organism, however any antibody must be evaluated under the conditions in which it will be used. Because MAb are generated against a specific antigenic epitope, it must be demonstrated that the antibody-antigen complex will be stable, particularly under a variety of environmental conditions.

For detection of *Cryptosporidium* oocysts in environmental samples using antibodies, a number of criteria have been suggested for identification (Rose et al., 1988a): 1) characteristic fluorescence specifically around the oocyst wall; 2) shape (spherical); 3) size (4 to 6 μm in diameter); and 4) characteristic folding in the oocyst wall. A training period for operators, as well as various seeded studies with any new antibody, will enhance the users familiarity with the reagent and accuracy of the identification.

During the examination of environmental samples, one may be underestimating *Cryptosporidium* contamination as stated previously, or one may be detecting bird isolates which perhaps have no significance to human health. Perhaps the greatest limitation with the methodologies is the inability to determine the viability of the oocysts. To take a conservative approach one must assume that any oocyst detected in potable water is potentially infectious.

14.6   Waterborne Outbreaks of *Cryptosporidium*

Many enteric pathogens which are transmitted by the fecal-oral route have been shown to cause waterborne disease (Craun, 1986). The enteric protozoa associated with waterborne transmission include *Cryptosporidium*, *Entamoeba*, and *Giardia* (see Figure 14.1). *Entamoeba* was first recognized early in the 1920s for its potential to cause waterborne disease, with seven outbreaks due to sewage contamination, yet its transmission by this route has largely been curtailed in the United States through wastewater treatment practices. Study of waterborne giardiasis has an extensive history since 1965, and 95 documented outbreaks through 1985 have been mainly associated with unfiltered water supplies (Craun, 1988; see also Chapter 13). *Cryptosporidium* has only been recently recognized as a cause of waterborne disease since 1985 with two waterborne outbreaks (Table 14.9). In over 50% of the documented waterborne outbreaks, however, no etiological agent was identified (Craun, 1986), and *Cryptosporidium* may have been responsible for some of these previous outbreaks.

Cryptosporidium shares many characteristics with *Giardia* (the most predominant waterborne etiological agent) which enhances its potential for transmission through water. Both animals as well as humans may serve as sources of environmental contamination and human infection. Both protozoa produce an environmentally stable stage (cyst) which may be resistant to water treatment processes. There is no simple or routine test which can be used to evaluate the occurrence of these protozoa in water, and the bacterial indicator system used to assess microbial water quality is inadequate for the determination of parasitological water quality. The occurrence of these two parasites in water may be related. At least in one study, *Cryptosporidium* oocyst contamination in water was correlated with *Giardia* cyst levels (Rose et al., 1988), yet oocysts were found in greater concentrations (see Section 14.5). A similar risk of waterborne disease could therefore be implied.

The role of water transmission in *Cryptosporidium* infections was initially implicated from studies on traveler's diarrhea (Jokipii et al., 1985; Ma et al., 1985). Again, there was an association with other waterborne agents such as *Giardia*. In 1985 the first waterborne outbreak of *Cryptosporidium* was documented in the United States (D'Antonio et al., 1985). The outbreak occurred in Texas in a community of 1,000 persons in July, 1984 (Table 14.10). The rate of diarrhea was 12 times that of the neighboring communities, and oocysts were detected in 47 of 79 individuals with diarrhea. In a control group not exposed to the community water supply, 12 of 194 were found to be excreting oocysts (Fayer and Ungar, 1986). The water supply was an artesian well, and

Table 14.9 Comparison of protozoan waterborne disease in the United States

Genera	Cryptosporidium	Entamoeba	Giardia
Sources of environmental contamination	Infected humans and animals	Infected humans and chronic carriers	Infected humans and animals, chronic infections
First documentation of waterborne disease	1985	1920	1965
Documented waterborne outbreaks	2	7	90
Major cause of waterborne disease	?	Sewage contamination of potable waters	Use of unfiltered surface waters as sources

Data of Craun (1986), D'Antonio et al. (1985), Hayes et al. (1988).

Table 14.10 Waterborne disease outbreaks of *Cryptosporidium*

	Outbreak 1	Outbreak 2
Date	July, 1984	January, 1987
Location	Braun Station, Texas	Carrollton, GA
Water source	Artesian well	River
Treatment	Chlorination	Conventional*
Infection rate	34%	40%
Population exposed	5,900	32,400
Probable contamination	Sewage	Unknown
Water quality standards**	Met U.S. coliform and turbidity standards	Met U.S. coliform and turbidity standards
Potential treatment deficiency	Disinfection only treatment	Inadequate coagulation, restarting of unwashed filters

* Coagulation, sedimentation, rapid sand filtration, and chlorination.
** Total coliforms <1/100 ml, turbidity of ≤1 nephelometric turbidity unit.
Data of D'Antonio et al. (1985), Hayes et al. (1988).

chlorination was the only treatment. Sewage contamination was suggested as the cause of the outbreak, since a second pathogen, the Norwalk virus, was also detected among the infected population. Although the disinfection was adequate in the control of coliform bacteria, it was apparently insufficient for the inactivation of *Cryptosporidium* and the Norwalk virus.

The second outbreak occurred in Carrollton, Georgia, in January, 1987 (Table 14.10) (Hayes et al., 1988; Mason, 1988). Illness was widespread throughout the county of 64,000 serviced by a common water supply. *Cryptosporidium* oocysts were identified in 39% of the stool specimens and no other pathogen was detected. It was also determined that illness was significantly greater in the population using city water in comparison to those individuals using well water. The overall attack rate was 40% for the county with an estimated 13,000 individuals affected. Diarrheal illness peaked the last two weeks in January and the first two weeks in February .

In Carrollton, Georgia, the drinking water underwent conventional treatment, including coagulation, sedimentation, rapid sand filtration, and disinfection. *Cryptosporidium* oocysts were detected in the treated water at the water treatment plant on January 28 by the laboratory at the Center for Disease Control. Subsequently, seven of nine samples collected were positive for *Cryptosporidium* oocysts on February 4 and 5 (Rose et al., 1987). Concentrations averaged 0.63 oocysts per liter within the distribution system. The highest level was found in a 24-hour sample taken post-filtration (2.2 oocysts/l). No unusually high levels were observed in samples from the lake, which entered the stream feeding the drinking water treatment plant, or in the stream itself (0.08 oocysts/l). Samples collected from treated water, collected February 10–13, were negative for oocysts. Raw sewage was found to contain 34 oocysts/l on February 5, while levels dropped to 1.2 oocysts/l by March 20. Table 14.11 summarizes these results.

Investigations into the characteristics of the treatment plant revealed no violations for coliforms or turbidity levels. Chlorine concentrations were 1.5 mg/l after disinfection and 0.5 mg/l within the distribution system. It was found that some individual filters were processing water with as much as 3.0 nephelometive turbidity units (NTU) of turbidity. However, once blending had occurred the water was ≤1.0 NTU, thus satisfying the standard. Other operational irregularities included start-up of dirty filters without backwashing and inadequate chemical mixing for coagulation (Hayes et al., 1988; Logsdon, 1987).

The source of the contamination was never determined. Through serological studies of the population, it has been speculated that there may have been low level oocyst contamination of the drinking water prior to the outbreak. It is plausible that at some point a peak of contamination entered the lake or river and passed through the treatment plant. Residual levels of oocysts were all that was detected within the water distribution system during the peak of the outbreak. Secondary spread from person-to-person probably contributed to heightening the outbreak in addition to primary infection.

Waterborne cryptosporidiosis has also been tentatively identified in New Mexico. Laboratories had been reporting on cases of cryptosporidiosis since 1984 to the New Mexico Office of Epidemiology (Anonymous, 1987). From July to October 1986, 76 cases of cryptosporidiosis were found through laboratory identification of oocysts in diarrheal specimens. An investigation was initiated to determine the source of infection and risk factor associated with illness. By use of questionnaires, matching controls (46) to case controls (24), it was found that there was a statistically significant association between drinking untreated surface water and illness. An increased risk of infection was also related to swimming in surface water, attending a day-care center, camping, and having a pet which was ill or young.

It is important to realize that because investigators in New Mexico initiated

Table 14.11 Summary of oocyst concentrations in water samples collected during the 1987 Carrollton outbreak

Date	Site of collection	Positive samples/ number collected	Average oocysts/l
2/4–2/6	Post-filtration	7/9	0.63
2/10–2/13	Post-filtration	0/5	
2/4–2/11	Carrollton Lake	2/2	0.08
2/5	Raw sewage	1/1	34
3/20	Raw sewage	1/1	1.2

Data of Rose et al. (1987).

a more active role in detection of disease, clusters of cases were recognized and followed up with a more thorough investigation. Among the many possible routes of transmission for *Cryptosporidium* in this study, drinking untreated surface water was the most significant suspected transmission route.

The risk of waterborne disease may be related first to the occurrence of the pathogen. The risk may be further defined by evaluating the concentration of the organism in water, its ability to survive in the environment, and its viability or infectivity potential. Secondly, risk will be dependent on the exposure (i.e., drinking water or recreational exposure) and susceptibility of the population. *Cryptosporidium* is commonly detected in surface waters; however, it is unknown what factors are associated with peaks of contamination which may be responsible for waterborne outbreaks.

14.7   Control of *Cryptosporidium* in Potable Water Supplies

Two major treatment processes are responsible for the control of microbial contamination of drinking water, filtration and disinfection. A thorough evaluation of drinking water treatment processes for removal or inactivation of *Cryptosporidium* oocysts will help to ensure adequate control of waterborne cryptosporidiosis.

Filtration Efficacy   No detailed information is available on the efficiency of filtration for the removal of *Cryptosporidium* oocysts, as these types of investigations have just begun. The oocysts are similar in shape and size to some yeasts, averaging 4 to 5 μm in diameter, much smaller than *Giardia* cysts. Apparently the oocysts are somewhat flexible, as 57% were able to pass through a 3.0 μm membrane filter (Rose et al., 1988a). Water filtration should be able to achieve some level of oocyst reduction; however, the real questions are: 1) What are the expected removal rates? 2) What are the optimal operating conditions to achieve optimal oocyst reductions? and 3) What variables may be monitored to maintain reliable treatment?

Data on the occurrence of indigenous *Cryptosporidium* oocysts indicate high concentrations (up to 24,306 oocysts/l) may be detected in backwash waters from rapid sand filters (Rose et al., 1986). It is reasonable to assume that oocysts are being concentrated on these filters. Comparison of levels of oocysts in source waters to levels in treated waters may be utilized to tenuously estimate removals (Table 14.12). Filtration without the use of coagulation achieved estimated oocyst removals of 54.6% and 91.4% from effluent and river water, respectively. An 85% reduction was seen with direct filtration; however, this plant was new, just initiating operation, and the rapid mixers as well as the ozonator were operating poorly or not operating at all. In one case, oocysts were detected in water undergoing conventional treatment (coagulation, sedimentation, rapid sand filtration, and chlorination), and removal was esti-

Table 14.12 Comparisons of *Cryptosporidium* oocyst concentrations in source water and treated water*

| Type water | Treatment | Oocysts/l | | |
		Source water	Treated water	Estimated removal
Secondary effluent[a]	Filtration, no coagulation	39.7	18	54.6%
River	Filtration, no coagulation	0.07	0.006	91.4
River	Direct filtration with ozone[b]	0.04	0.006	85
River	Conventional[c]	0.12	0.008	93.3
Spring	Chlorination	0.053	0.0054	89.8

[a] Averages based on six samples, two facilities.
[b] Rapid mixer and ozonator not operating optimally.
[c] Coagulation, sedimentation, rapid sand filtration, and chlorination.
Data of DeLeon et al. (1988), Rose et al. (1986), and Rose (1988).

mated at 93.3%. No operational problems were identifiable. Chlorination (0.45 mg/l) as the sole treatment of spring water achieved an 89.8% reduction of oocysts. It has been stated that to maximize filtration efficiency, operational conditions must be optimized, including coagulant dosage, mixing, adjustment of filter runs, and maintenance of filters (Logsdon, 1987; see also Chapter 6). The outbreak in Carrollton also demonstrates that simply meeting standards for coliforms and turbidity levels may not be a sufficient measure of treatment reliability for removal of *Cryptosporidium*.

Disinfection Studies   Drinking water treatment should rely on a multi-barrier concept. Therefore, microorganisms which may not be removed by filtration may then be inactivated (reduced further) using a final disinfection step. In order to assess inactivation, a measure of viability needs to be used. Parasite viability may be determined by using a bioassay in an animal model or by excystation.

Oral inoculation of newborn mice has been used to evaluate the effect of disinfectants on *Cryptosporidium* oocysts (Campbell et al., 1982). Shedding of oocysts and histological evidence was used to determine infectivity. Gut homogenates were mixed with an equal volume of disinfectant and after incubation periods of 18 hours at 4°C were orally inoculated into the mice. Formaldehyde (10%), cresylic acid (2.5%), and ammonia (5%) were found to be effective disinfectants. A three percent solution of hypochlorite was not able to destroy the infectivity of the suspension under these conditions. In addition, iodophor (4%), benzalkonium chloride (10%), and sodium hydroxide (0.02 M) were also ineffective. Although this study did not simulate water disinfection conditions, it did demonstrate that oocysts remained infectious even after 18 hours exposure to chlorine concentrations approaching undiluted bleach.

Although animal models may be advantageous in that infectivity is evaluated, quantitation of inactivation rates is difficult. Variation in animal susceptibility and the assessment of the infectious dose are disadvantages using this system.

Excystation has also been used to evaluate viability of *Cryptosporidium baileyi* oocysts after exposure to a variety of disinfectants (Sundermann et al., 1987). Bleach (sodium hypochlorite) was tested in addition to other surface disinfectants (formaldehyde, ammonia, and chlorhexidine). One million purified oocysts were added to 10 ml of bleach solutions and exposed for 30 minutes at 25°C. Percent excystation was determined by enumerating the number of empty oocysts and full oocysts. The greater the percent excystation the greater the viability. A 25% and 89% inactivation was reported after exposure to 4.7% and 50% bleach, respectively. The extreme resistance of *C. baileyi* oocysts to greater than 2000 mg/l of sodium hypochlorite suggests that water disinfection practices may be of little value in controlling *Cryptosporidium* in water. Yet the interpretation of excystation studies and extrapolation to the in situ environment needs to be further examined. It must be kept in mind that oocyst pretreatment, storage conditions, age, and counting procedures may all influence the results.

Excystation in *Cryptosporidium* is the process wherein the oocyst opens up and releases the stage which initiates intracellular infection, the sporozoites. The sporozoites have been shown to be readily inactivated if released into the water environment (Current, 1987). Thus, if excystation occurs outside the host, the organism will lose its infectivity.

Excystation may be induced in the laboratory. For *Cryptosporidium parvum*, this can be accomplished by incubation at 37°C in an excystation solution (bile salts [0.75% sodium taurocholate], trypsin [0.25%], and Hanks balanced salt solution (HBSS) [calcium and magnesium free]). Empty and full oocysts as well as sporozoites (an average of 4 sporozoites/oocyst) may be enumerated using phase or differential interference contrast microscopy to determine percent excystation and sporozoite to oocyst ratios (Reducker and Speer, 1985). Maximum excystation (>70%) was achieved after 60 minutes incubation. Interestingly, oocyst pretreated with 1.05 or 1.75% sodium hypochlorite excysted readily in HBSS after incubation at 37°C without exposure to the excystation media. Very little excystation was observed at room temperature or 4°C regardless of the pretreatment or exposure to the excystation solution.

This is not the only report where hypochlorite enhanced *Cryptosporidium* excystation (Reducker et al., 1985; Woodmansee, 1987). Apparently the sodium hypochlorite alters the permeability of the oocyst wall, perhaps by removing an outer layer whereby at 37°C the sporozoites may escape in the absence of excysting fluid (Reducker and Speer, 1985; Reducker et al., 1985). Excystation, therefore, can occur without host stimuli after exposure to bleach. This phenomenon has direct relevance to water disinfection practices. This may suggest that disinfection enhances the potential infectivity of *Cryptosporidium* by perhaps enhancing in vivo excystation. If, however, the oocyst undergoes excystation in tap water after exposure to a disinfectant, or begins to disintegrate, this may be regarded as a mechanism of inactivation.

Table 14.13 summarizes the excystation studies on *Cryptosporidium*. It has been demonstrated that oocysts will apparently excyst in water (0.8–26%), but this may be related to incubation temperature (37°C), optimal age of the oocysts, and storage conditions (Current, 1987; Fayer and Leek, 1984). Current (1987) has suggested that the structural and functional integrity of the oocyst wall may be degraded by bacterial and fungal enzymes, and this may influence the rate of excystation over time of improperly stored oocysts (Current, 1987). In water a true excystation may not be occurring but rather a disintegration of the oocyst. Little change was observed in excystation rates for purified oocysts stored in $K_2Cr_2O_7$ (2.5%), tap water with antibiotics, and tap water alone at 4°C for up to 140 days. Excystation ranged form 0.8 to 15% with tap water incubation (37°C, 4 hours) and 78 to 94% in excystation media (37°C, 1 hour) (Current, 1987). However, the number of sporozoites observed decreased after 140 days of storage from 136 to 81 when stored in $K_2Cr_2O_7$, 122 to 51 in water plus antibiotics and 117 to 21 in water alone (Current, 1987). Speer and Reducker (1986) reported excystation rates of 30.8, 57.4, and 20.9% after 98, 224, and 301 days, respectively, of storage of $K_2Cr_2O_7$ (2.5%). However, the ratio of sporozoites to empty oocysts decreased dramatically, reported as 224, 21.5, and 2.8, respectively (number of sporozoites/number of empty oocyst $\times$ 100). Infectivity of *Cryptosporidium* oocysts in animals has also been shown to decrease after 60 to 180 days storage at 4°C (Moon and Bemrick, 1981; Sherwood et al., 1982).

Excystation in *Cryptosporidium* is a complex process. The relationship between extent of excystation, numbers of sporozoites, and infectivity needs to be determined; therefore, preliminary data should be interpreted cautiously. To evaluate disinfection data and oocyst inactivation, a thorough understanding of possible mechanisms involved in these processes will be necessary.

14.8   Conclusions

The significance of waterborne cryptosporidiosis is unknown. There is not an extensive record of *Cryptosporidium* outbreaks associated with water, and the potential risk of waterborne transmission is speculative. The data so far indicate that *Cryptosporidium* oocysts may be commonly recovered from water at various concentrations. Peaks of contamination have been observed. The waterborne outbreak in Carrollton, Georgia, has certainly demonstrated that *Cryptosporidium* has the potential to cause major waterborne disease and that water treatment may not be reliably assessed based on the common indicators. Routine filtration and disinfection practices may not adequately control *Cryptosporidium* in water.

More information will be needed on *Cryptosporidium* occurrence in water, survival in the environment, potential infectivity, removal, and inactivation by

Table 14.13 Summary of excystation studies on *Cryptosporidium* oocysts

Storage conditions	Pretreatment with sodium hypochlorite	Conditions to achieve excystation	Amount of excystation*	Reference
$K_2Cr_2O_7$	−	Sodium Taurocholate and Trypsin 37°C	20–57%	Reducker and Speer (1985)
$K_2Cr_2O_7$	+	HBSS 37°C	~48–70%	Reducker and Speer (1985)
Tap Water	−	Tap Water 37°C	12–26%	Fayer and Leek (1984)
Tap Water	−	Sodium Taurocholate and Trypsin 37°C	56–91%	Fayer and Leek (1984)
Tap Water	−	Sodium Taurocholate and Trypsin 20°C	8–12%	Fayer and Leek (1984)
$K_2Cr_2O_7$	+	Taurocholic Acid 37°C	~68%	Woodmansee (1987)
$K_2Cr_2O_7$	−	Taurocholic Acid 37°C	~50%	Woodmansee (1987)
$K_2Cr_2O_7$	+	PBS 37°C	~40%	Woodmansee (1987)
$K_2Cr_2O_7$	−	Tap Water 37°C**	0.8–13%	Current (1987)
$K_2Cr_2O_7$	−	Sodium Taurocholate and Trypsin 37°C	87–94%	Current (1987)
H_2O + Antibiotic	−	Tap Water 37°C**	2.3–9.5%	Current (1987)
H_2O + Antibiotic	−	Sodium Taurocholate and Trypsin 37°C	78–92%	Current (1987)
H_2O	−	Tap Water 37 °C**	2–15%	Current (1987)
H_2O	−	Sodium Taurocholate and Trypsin 37°C	79–91%	Current (1987)
$K_2Cr_2O_7$	−	Sodium Taurocholate and Trypsin 37 °C	64–73%	Current (1987)
$K_2Cr_2O_7$	−	Sodium Taurocholate 37°C	29%	Current (1987)
$K_2Cr_2O_7$	−	Trypsin 37°C	34%	Current (1987)
$K_2Cr_2O_7$	−	PBS 37°C	2.5%	Current (1987)
$K_2Cr_2O_7$	−	Tap Water 22°C**	1.4%	Speer and Reducker, 1986

(Continued next page)

Table 14.13 (continued)

Storage conditions	Pretreatment with sodium hypochlorite	Conditions to achieve excystation	Amount of excystation*	Reference
$K_2Cr_2O_7$	—	Tap Water 37°C**	0.1%	Speer and Reducker (1986)
$K_2Cr_2O_7$	—	Sodium Taurocholate and Trypsin 22 °C	0.5%	Speer and Reducker (1986)
$K_2Cr_2O_7$	—	Sodium Taurocholate and Trypsin 37 °C	30.8%	Speer and Reducker (1986)

* Based on ratio of empty to total oocysts, 60 min. incubation
** 4-hour incubation

water treatment processes. In addition, more information on the significance of the disease in the population will aid in evaluation of the risk of waterborne disease due to *Cryptosporidium*.

References

Addy, P.A.K. and Aikens-Bekoe, P. 1986. Cryptosporidiosis in diarrheal children in Kumasi Ghana. *Lancet* i:735.

Akin, E.W. and Jakubowski, W. 1986. Drinking water transmission of giardiasis in the United States. *Water Science & Technology* 18:219–226.

Alpert, G., Bell, L.M., Kirkpatrick, C.E., Budnick, L.D., Campos, J.M., Friedman, H.M., and Plotkin, S.A. 1984. Cryptosporidiosis in a day-care center. *New England Journal Medicine* 311:860–861.

Alpert, G., Bell, L.M., Kirkpatrick, C.E., Budnick, L.D., Campos, J.M., Friedman, H.M., and Plotkin, S.A. 1986. Outbreak of cryptosporidiosis in a day-care center. *Pediatrics* 77:152–157.

Anderson, B.C. 1982. Cryptosporidiosis in Idaho lambs: natural and experimental infections. *Journal of the American Veterinary Medical Association* 181:151–153.

Anderson, B.C. 1982. Is cryptosporidial infection responsible for diarrhea? *California Veterinary* 36:9–10.

Anderson, B.C. and Bulgin, M.S. 1981. Enteritis caused by *Cryptosporidium* in calves. *Veterinary Medicine Small Clinic* 76:865–868.

Angus, K.W., Appleyard, W.T., Menzies, J.D., Campbell, I., and Sherwood, D. 1982. An outbreak of diarrhea associated with cryptosporidiosis in naturally reared lambs. *Veterinary Record* 110:129.

Anonymous. 1984. Cryptosporidiosis among children attending day-care centers–Georgia, Pennsylvania, Michigan, California, New Mexico. *Morbidity Mortality Weekly Reports* 33:599–601.

Anonymous. 1987. Cryptosporidiosis New Mexico, 1986. *Morbidity Mortality Weekly Report* 36:561–563.

Baxby, D., Hart, C.A., and Taylor, C. 1983. Human cryptosporidiosis: a possible case of hospital cross infection. *British Medical Journal* 187:1760–1761.

Campbell, I., Tzipori, S., Hutchison, G., and Angus, K.W. 1982. Effect of disinfectants on survival of *Cryptosporidium* oocysts. *Veterinary Record* 111:414–415.

Craun, G.F. 1986. *Waterborne Diseases in the United States*. CRC Press, Boca Raton, FL.

Craun, G.F. 1988. Waterborne outbreaks of giardiasis: why they happen, how to prevent them. *Health and Environmental Digest* 2:3–4.

Current, W.L., Reese, N.C., Ernst, J.V., Bailey, W.S., Heyman, M.B., and Weinstein, W.M. 1983. Human cryptosporidiosis in immunocompetent and immunodeficient persons. Studies of an outbreak and experimental transmission. *New England Journal Medicine* 308:1252–1257.

Current, W.L., Upton, S.J., and Haynes, T.B. 1986. The life cycle of *Cryptosporidium baileyi* n. sp. (Apicomplexa, Cryptosporidiidae) infecting chickens. *Journal Protozoology* 33:289–296.

Current, W.L. 1987. *Cryptosporidium*: its biology and potential for environmental transmission. *CRC Critical Reviews Environmental Control* 17:21–51.

D'Antonio, R.G., Winn, R.E., Taylor, J.P., Gustafson, T.L., Current, W.L., Rhodes, M.M., Gary, G.W., and Zajac, R.A. 1985. A waterborne outbreak of cryptosporidiosis in normal hosts. *Annals Internal Medicine* 103:886–888.

Darlington, M.V. and Blagburn, B.L. 1988. The Merifluor™ Immunofluorescent detection procedure is nonspecific for *Cryptosporidium* species. *Annual Meeting Southeastern Society of Parasitologists Abstract 34*, Clemson University, March 30–April 1.

DeLeon, R., Naranjo, J.E., Rose, J.B., and Gerba, C.P. 1988. Enterovirus, *Cryptosporidium* and *Giardia* monitoring of wastewater reuse effluent in Arizona. pp. 833–846 *in Implementing Water Reuse, Symposium IV*, American Water Works Assoc., Denver, CO.

Dubey, J.P. 1977. *Toxoplasma, Hammondia, Besnoitia, Sarcocystis* and other tissue cyst-forming coccidia of man and animals. pp. 101–237 *in Parasitic Protozoa*, Vol. III. Gregarines, Haemogregar-ines, Coccidia, Plasmodia and Haemoproteids. J.P. Kreiber (ed.), Academic Press, NY.

Elsser, K.A., Moricz, M., and Proctor, E.M. 1986. *Cryptosporidium* infections: a laboratory survey. *Canadian Medical Association Journal* 135:211–213.

Ernest, J.A., Blagburn, B.L., Lindsay, D.S, and Current, W.L. 1987. Infection dynamics of *Cryptosporidium parvum* (Apicomplexa: Cryptosporidiidae) in neonatal mice (*Mus musculus*). *Journal of Parasitology* 75:796–798.

Fayer, R. and Leek, R.G. 1984. The effects of reducing conditions, medium, pH, temperature and time on in vitro excystation of *Cryptosporidium. Journal of Protozoology* 31:567–569.

Fayer, R. and Ungar, B.L.P. 1986. *Cryptosporidium* and cryptosporidiosis. *Microbiological Reviews* 50:458–483.

Federal Register. 1987. Part III 40 CFR Part 141. Wednesday, July 8. 52:25732.

Federal Register. 1988. January 22. 53:141892.

Garcia, L.S., Brewer, T.C., and Bouckner, D.A. 1987. Fluorescent detection of *Cryptosporidium* oocysts in human fecal specimens by using monoclonal antibodies. *Journal Clinical Microbiology* 25:119–121.

Garcia, L.S., Bruckner, D.A., Brewer, T.C., and Shimizu, R.Y. 1983. Techniques for the recovery and identification of *Cryptosporidium* oocysts from stool specimens. *Journal Clinical Microbiology* 18:185–190.

Hayes, E.B., Matte, T.D., O'Brien, T.R., McKinley, T.W., Logsdon, G.S., Rose, J.B., Ungar, B.L.P., Word, D.M., Pinsky, P.F., Cummings, M.L., Wilson, M.A., Long, E.G., Hurwitz, E.S., and Juranek, D.D. Contamination of a conventionally treated filtered public water supply by *Cryptosporidium* associated with a large community outbreak of cryptosporidiosis. *New England Journal of Medicine* 320:1372–1376.

Højlyng, N., Molbak, K., and Jepsen, S. 1986. *Cryptosporidium* spp., a frequent cause of diarrhea in Liberian children. *Journal Clinical Microbiology* 23:1109–1113.

Holly, H.P. Jr. and Dover, C. 1986. *Cryptosporidium*: a common cause of parasitic diarrhea in otherwise healthy individuals. *Journal of Infectious Disease* 153:365–368.

Jakubowski, W. 1984. *Giardia* and Giardiasis. pp. 263–285 *in Detection of Giardia Cysts in Drinking Water; State of the Art*. Erlandsen, S.L. and E.A. Meyer (eds.), Plenum Press, New York.

Jakubowski, W. 1988. Purple burps and the filtration of drinking water supplies. *American Journal Public Health* 78:123–125.

Jokipii, L., Pohjola, S., and Jokipii, A. 1985. Cryptosporidiosis associated with traveling and giardiasis. *Gastroenterology* 89:838–842.

Jokipii, L. and Jokipii, A.M.M. 1986. Timing of symptoms and oocyst excretion in human cryptosporidiosis. *New England Journal of Medicine* 315:1643–1647.

Koch, K.L., Phillips, D.J., and Current, W.J. 1984. Cryptosporidiosis in hospital personnel: evidence for person-to-person transmission. *Annals Internal Medicine* 102:593–596.

Levine, N.D. 1984. Taxonomy and review of the coccidian genus *Cryptosporidium* (Protozoa, Apicomplexa). *Journal of Protozoology* 31:94–98.

Long, P.L. (ed.). 1982. *The Biology of the Coccidia*. University Park Press, Baltimore, MD.

Logsdon, G.S. 1987. Evaluating treatment plants for particulate contaminant removal. *Journal American Water Works Association* 79(9):82–92.

Ma, P., Kaufman, D.L., Helmick, C.G., D'Souza, A.J., and Navin, T.R. 1985. Cryptosporidiosis in tourists from the Caribbean. *New England Journal Medicine* 312:647–648.

Ma, P. and Soave, R. 1983. Three-step stool examination for cryptosporidiosis in 10 homosexual men with protracted watery diarrhea. *Journal Infectious Diseases* 147:824–828.

Madore, M.S., Rose, J.B., Gerba, C.P., Arrowood, M.J., and Sterling, C.R. 1987. Occurrence of *Cryptosporidium* oocysts in sewage effluents and select surface waters. *Journal Parasitology* 73:702–705.

Marshall, A.R., Al-Jumaili, I.J., Fenwick, G.A., Bint, A.J., and Record, C.O. 1987. Cryptosporidiosis in patients at a large teaching hospital. *Journal Clinical Microbiology* 25:172–173.

Mason, L. 1988. Experience with *Cryptosporidium* at Carrollton, Georgia. pp. 889–898 *in Water Quality Technology Conference*. Nov. 15–20, 1987, Baltimore, MD, American Water Works Association, Denver, CO.

Mata, L., Bolanos, H., Pezarro, D., and Vives, M. 1984. Cryptosporidiosis in children from some highland Costa Rican rural and urban areas. *American Journal of Tropical Medicine and Hygiene* 33:24–29.

McFeters, G.A., Kippin, J.S, and LeChevallier, M.W. 1986. Injured coliforms in drinking water. *Applied and Environmental Microbiology* 51:1–5.

Meisel, J.L., Perera, D.R, and Rubin, C.E. 1976. Overwhelming watery diarrhea associated with *Cryptosporidium* in an immunosuppressed patient. *Gastroenterology* 70:1156–1160.

Miller, R.A., Bronsdon, M.A. and Morton, W.R. 1986. Determination of the infectious dose of *Cryptosporidium* and the influence of inoculum size on disease severity in a primate model. *Abstract Annual Meeting of American Society of Microbiologists*, Washington, D.C., 23–28 March. p. 49.

Moon, H.W. and Bemrick, W.J. 1981. Fecal transmission of calf cryptosporidia between calves and pigs. *Veterinary Pathology* 18:248–255.

Musial, C.E., Arrowood, M.J., Sterling, C.R., and Gerba, C.P. 1987. Detection of *Cryptosporidium* in water using polypropylene cartridge filters. *Applied Environmental Microbiology* 53:687–692.

Navin, T.R. and Juranek, D.D. 1984. Cryptosporidiosis: clinical epidemiologic and parasitologic review. *Reviews on Infectious Disease* 6(B):313–327.

Nime, F.A., Burek, J.D., Page, D.L., Holscher, M.A., and Yardley, J.H. 1976. Acute enterocolitis in a human being infected with the protozoan *Cryptosporidium*. *Gastroenterology* 70:592–598.

Ongerth, J.E. and Stibbs, H.H. 1987. Identification of *Cryptosporidium* oocysts in river water. *Applied Environmental Microbiology* 53:672–676.

Panciera, R.J., Thomassen, R.W., and Garner, F.M. 1971. Cryptosporidial infection in a calf. *Veterinary Pathology* 8:479–484.

Pape, J.W., Levine, E., Beaulieu, M.E., Marshall, F., Verdier, R., and Johnson, W.D. Jr. 1987. Cryptosporidiosis in Haitian children. *American Journal Tropical Medicine and Hygiene* 36:333–337.

Pavlasek, I. 1982. Dynamics of the release of oocysts of *Cryptosporidium* sp. in spontaneously infected calves. *Folia Parasitology* 29:295–296.

Pohjola, S., Jokipii, A.M., and Jokipii, L. 1986. Sporadic cryptosporidiosis in a rural population is asymptomatic and associated with contact to cattle. *Acta Vet. Scand.* 27:91–102.

Reducker, D.W. and Speer, C.A. 1985. Factors influencing excystation in *Cryptosporidium* oocysts from cattle. *Journal Parasitology* 71:112–115.

Reducker, D.W., Speer, C.A., and Blixt, J.A. 1985. Ultrastructural changes in the oocyst wall during excystation of *Cryptosporidium parvum* (Apicomplexa; Eucoccidiordia). *Canadian Journal Zoology* 63:1892–1896.

Rose, J.B., Cifrino, A., Madore, M.S., Gerba, C.P., Sterling, C.R., and Arrowood, M.J. 1986. Detection of *Cryptosporidium* from wastewater and freshwater environments. *Water Science Technology* 18:233–239.

Rose, J.B., Madore, M.S., Riggs, J.L., and Gerba, C.P. 1987a. Detection of *Cryptosporidium* and *Giardia* in environmental waters. pp. 417–424 in: *Advances in Water Analysis and Treatments*, Water Technology Conference, Portland, OR, Nov. 1986. American Water Works Association, Denver, CO.

Rose, J.B., Gerba, C.P., Logsdon, G.A., Hayes, E., Matti, T., and Juranek, D.D. 1987. Occurrence of *Cryptosporidium* oocysts in a distribution system after a waterborne outbreak. *Water Quality Technology Conference*. Baltimore, MD. Nov. 15-19. Journal American Water Works Association 79:47.

Rose, J.B. 1988. Occurrence and significance of *Cryptosporidium* in water. *Journal American Water Works Association* 80:53–58.

Rose, J.B., Darbin, H., and Gerba, C.P. 1988. Correlations of the protozoa, *Cryptosporidium* and *Giardia* with water quality variables in a watershed. *Proceedings International Conference on Water and Wastewater Microbiology*, Newport Beach, CA, Feb 8-11, 2:43-1–43-6.

Rose, J.B., Kayed, D., Madore, M.S., Gerba, C.P., Arrowood, M.J., and Sterling, C.R. 1988a. Methods for the recovery of *Giardia* and *Cryptosporidium* from environmental waters and their comparative occurrence. pp. 205–209 in: *Advances in Giardia Research*, Wallis, P. and Hammond, B. (eds.), Univ. Calgary Press.

Sallon, S., Deckelbaum, R.J., Schmid, I.I., Harlap, S., Baras, M., and Spira, D.T. 1988. *Cryptosporidium*, malnutrition and chronic diarrhea in children. *American Journal of Diseases of Children* 142:312–315.

Sauch, J.F. 1985. Use of immunofluorescence and phase-contrast microscopy for detection and identification of *Giardia* cysts in water samples. *Applied and Environmental Microbiology* 50:1434–1438.

Schmidt, G.D. and Roberts, L.S. 1985. *Foundation of Parasitology* 3rd Ed. Bowen, D.L. (ed.), Times Mirrow/Mosby College Publishing, St. Louis, MO.

Sherwood, D., Angus, K.W., Snodgrass, D.R., and Tzipori, S. 1982. Experimental cryptosporidiosis in laboratory mice. *Infection and Immunity* 38:471–475.

Slavin, D. 1955. *Cryptosporidium meleagridis* (sp. nov.). *Journal of Comparative Pathology* 65:262–266.

Speer, C.A. and Reducker, D.W. 1986. Oocyst age and excystation of *Cryptosporidium parvum*. *Canadian Journal Zoology* 64:1254–1255.

Stehr-Green, J.K., McCaig, L., Remsen, H.M., Rains, C.S., Fox, M., and Juranek, D.D.

1987. Shedding of oocysts in immunocompetent individuals infected with *Cryptosporidium. American Journal of Tropical Medicine and Hygiene* 36:338–342.

Sterling, C.R. and Arrowood, M.J. 1986. Detection of *Cryptosporidium* sp. infections using a direct immunofluorescent assay. *Pediatric Infectious Disease* 5:5139–5142.

Sterling, C.R., Arrowood, M.J., Marshall, M.M., and Stetzenbach, L.D. 1988. The detection of *Giardia* and *Cryptosporidium* from water sources using monoclonal antibodies. pp. 271–279 in *Water Quality Technology Conference*. Nov. 15-20, 1987, Baltimore, MD. *American Water Works Association*, Denver, CO.

Stibbs, H.H. and Ongerth, J.E. 1986. Immunofluorescence detection of *Cryptosporidium* oocysts in fecal smears. *Journal of Clinical Microbiology* 24:517–521.

Sundermann, C.A., Lindsay, D.S., and Blagburn, B.L. 1987. Evaluation of disinfectants for ability to kill avian *Cryptosporidium* oocysts. *Companion Animal Practice* Nov. 1:36–39.

Taylor, J.P., Perdue, J.W., Dingley, D., Gustafson, T.L., Patterson, M., and Reed, L.A. 1985. Cryptosporidiosis outbreak in a day-care center. *American Journal of Diseases of Childhood* 139:1023–1025.

Tyzzer, E.E. 1907. A sporozoan found in the peptic glands of the common mouse. *Proceedings of Society for Experimental Biology and Medicine* 5:12–13.

Tyzzer, E.E. 1910. An extracellular coccidium, *Cryptosporidium muris* (gen. et sp. nov.), of the gastric glands of the common mouse. *Journal of Medical Research* 23:487–516.

Tyzzer, E.E. 1912. *Cryptosporidium parvum* (sp. nov.), a coccidium found in the small intestine of the common mouse. *Arch. Protistenkd.* 26:394–412.

Tzipori, S. 1983. Cryptosporidiosis in animals and humans. *Microbiological Reviews* 47:84–96.

Ungar, B.L., Gilman, R.H., Lanata, C.F., and Perez-Schael, I. 1988. Seroepidemiology of *Cryptosporidium* infection in two Latin American populations. *Journal of Infectious Diseases* 57:551–556.

Upton, S.J. and Current, W.L. 1985. The species of *Cryptosporidium* (Apicomplexa: Cryptosporidiidae) infecting mammals. *Journal Parasitology* 71:625–629.

Woodmansee, D.B. 1987. Studies of in vitro excystation of *Cryptosporidium parvum* from calves. *Journal Protozoology* 34:398–402.

15

Yersinia enterocolitica in Drinking Water

Donald A. Schiemann

Department of Microbiology, Montana State University, Bozeman, Montana

Beginning in the 1930s, the New York State Health Department reported sporadic isolations over nearly twenty years of unusual bacteria associated most often with enteritis in children (Schleifstein and Coleman, 1939). The organisms were gram-negative fermentive bacilli that had several identical biochemical characteristics (sucrose and indole positive, rhamnose negative, motile at 20°C but not at 37°C), demonstrated common antigens, and were highly virulent for mice. They were tentatively named *Bacterium enterocoliticum* (Schleifstein and Coleman, 1943).

During the late 1940s, reports first appeared in Europe on the isolation of bacteria that resembled *Pasteurella pseudotuberculosis* from humans, hares, pigs, and from chinchillas which were highly susceptible to fatal infections. The primary differences between these isolates and *P. pseudotuberculosis*, which had long been recognized as an animal pathogen, were in sucrose fermentation, the inability to ferment esculin, rhamnose, and adonitol, and decarboxylate ornithine, in antigenicity, and in the absence of pathogenicity in mice, guinea pigs and rabbits. Frederiksen (1964) examined 55 of these strains and concluded that they were sufficiently different from *P. pseudotuberculosis* to be placed in a new genus and species, *Yersinia enterocolitica*. The new genus *Yersinia* encompassed three species from the genus *Pasteurella*: *Y. enterocolitica*, *Y. pseudotuberculosis*, and *Y. pestis* and the genus was placed in the family *Enterobacteriaceae*.

Until the late 1960s, the number of reported human isolates of *Y. enterocolitica* was very small, but within a few years it increased to thousands, reports coming first from European and Scandinavian countries, and then worldwide. The organism has been associated most often with gastrointestinal disease including enteritis, terminal ileitis, and mesenteric lymphadenitis.

In the United States, where the story began, the incidence of human yersiniosis has been much lower than in many other countries, including neighboring Canada. However, two events during the 1970s in this country stimulated new interest in this organism. The first involved a single case of septicemia caused by a bacterium identified as *Y. enterocolitica* but which differed from European strains biochemically, antigenically (serotype 0:8), and in being highly pathogenic in mice (Carter et al., 1973). The same organism was isolated from a mountain stream which the patient had used as a source of drinking water (Keet, 1974). The second episode involved an outbreak in school children in the same state, New York, with contaminated chocolate milk implicated as the source (Black et al., 1978). The serotype was also 0:8. This outbreak received considerable public attention through reports of unnecessary appendectomies, the consequence of lower right-quadrant abdominal pain associated commonly with yersiniosis. Foodborne outbreaks have characterized the epidemiology of yersiniosis in the United States, unlike many other countries where the disease appears to be endemic.

15.1 Description and Classification

The original biochemical criteria set forth by Frederiksen for *Y. enterocolitica* proved later to circumscribe a large heterogeneous group of bacteria. Several biotyping schemes were proposed, with biotype correlating to some degree with host adaptation. Biotype 4, for example, became nearly synonymous with serotype 0:3 and with human infections and was isolated frequently from porcine throats and tongues.

The application of DNA homology techniques led to division in 1980 of the single species into four: *Y. enterocolitica* (Bercovier et al., 1980a), *Yersinia frederiksenii* (Ursing et al., 1980), *Yersinia kristensenii* (Bercovier et al., 1980b) and *Yersinia intermedia* (Brenner et al., 1980). An unusual biotype of *Y. enterocolitica* (biotype 5, Table 15.1) includes strains that are negative for trehalose and nitrate reductase. Some workers have separated another biochemical group presently referred to only as group X1 (Bercovier et al., 1980). A fifth species, *Yersinia aldovae*, was proposed later to define a biochemically-distinguishable group that appears to be confined to water (Bercovier et al., 1984). Another new species has also been proposed, *Yersinia rohdei*, for strains that are Voges-Proskauer negative, positive on Simmons citrate, and positive for trehalose and nitrate reductase (Aleksic et al., 1987). Many workers feel that the species *Yersinia ruckeri*, a fish pathogen, should not be included in the genus *Yersinia* (Farmer et al., 1985). Some key biochemical tests for separating species of *Yersinia* (other than *Yersinia pseudotuberculosis*, *Yersinia pestis* and *Y. ruckeri*) are identified in Table 15.1

Serotyping   Fifty-four heat-stable somatic antigens have been identified for *Y. enterocolitica* which have been used to describe 57 different serotypes (Wauters, 1981). The complete antigens are best expressed during growth at 25°C, a temperature that should be used both for preparing the antigens for making antisera and for growing the bacteria for typing by slide agglutination.

Table 15.1 Key biochemical tests for differentiation of *Y. enterocolitica* and related species

Biochemical test (25°C)	*Y. enterocolitica* Biotypes 1–4	*Y. enterocolitica* Biotype 5	*Y. kristensenii*	*Y. frederiksenii*	*Y. intermedia*	*Y. aldovae*	*Y. rohdei* Biotype 1	*Y. rohdei* Biotype 2
Voges-Proskauer	+	+	–	+	+	+	–	–
Sucrose	+	v	–	+	+	–	+	+
L-rhamnose	–	–	–	+	+	+	–	–
D-melibiose	–	–	–	–	+	–	+	–
α-methyl-D-glucoside	–	–	–	–	+	–	–	–
Indole	v	–	v	+	+	–	–	–
D-raffinose	–	–	–	–	+	–	+	–
Simmons citrate	–	–	–	v	+	+	v	+
D-sorbitol	+	–	+	+	+	+	+	+

v = variable.

Cross-reactions are common and can be removed by adsorption or, in the case of weak reactions, by dilution of the antiserum.

Several antigenic factors described originally for *Y. enterocolitica* appear to be confined to species that were defined afterwards. It has been suggested, therefore, that the serotyping scheme for the species *Y. enterocolitica* as now defined be revised to include only 20 antigenic factors (Aleksic and Bockemuhl, 1984). Serotype 0:13 has been revised to include two factors: 13a and 13b (Toma et al., 1984).

Serotyping based on flagellar antigens has not been widely used. More recent studies on the flagellar antigens of *Y. enterocolitica* suggest that they may be useful for separating the species of *Yersinia* (Aleksic et al., 1986). It may be of some interest that the predominant North American serotypes 0:8 and 0:21 share common flagellar antigens.

Bacteriophage Typing   Three different bacteriophage typing systems have been described for *Y. enterocolitica* and related species (Baker and Farmer, 1982; Bergan, 1978; Nilehn, 1973). However, typing with lytic bacteriophages has not been applied as yet for epidemiological purposes.

15.2   Recognition of Pathogenic *Y. enterocolitica*

Bacteria identified as the species *Y. enterocolitica* have been isolated commonly from the environment, including water supplies of various types. Because these environmental isolates often differ biochemically from those associated with human disease, they have sometimes been referred to as "*Y. enterocolitica*-like" bacteria. An urgency for judging the health significance of environmental strains of *Yersinia* motivated intensive studies on pathogenicity using a variety of animal and laboratory models. A summary is presented here of various models and markers that have been described for identifying pathogenic strains of *Y. enterocolitica*.

Yersinia **Species**   At least two of the species that have been separated from *Y. enterocolitica* as defined originally, *Y. kristensenii* and *Y. frederiksenii*, are not infrequently isolated from humans with some disease condition. Isolation is often accomplished only after enrichment, indicating low numbers of the bacteria in the specimen which is not indicative of acute infection. These two species, and also *Y. intermedia*, have never demonstrated any laboratory-determined attributes and markers of pathogenicity. Furthermore, there has never been evidence presented of a serological response associated with the isolation of these species from the body. It is possible, of course, that bacteria of minimal or low virulence may be pathogenic in the immunocompromised host. However, the preponderance of evidence indicates that virulence under most conditions is associated only with the species *Y. enterocolitica*.

Serotype   If one isolates a strain of *Y. enterocolitica* that is either serotype 0:3, 0:9, 0:5,27 or 0:8, it will nearly always present a biochemical profile unique for that serotype. It will also demonstrate all the recognized markers of path-ogenicity, so long as the 42–48 megadalton virulence plasmid has not been lost. On occasion, strains of serotypes 0:3 and 0:8 have been isolated from the environment that could not be recognized as pathogenic but they were always biochemically different. Some serotypes such as 0:21 and 0:4,32 are isolated frequently from the environment, are not pathogenic, and differ biochemically from strains of the same serotype that are clearly associated with disease conditions. Strains of serotypes 0:5 and 0:6,30 are especially common in human stool specimens and the environment, but always fail to demonstrate any marker of pathogenicity. Table 15.2 identifies the serotypes of *Y. enterocolitica* that are known to include pathogenic strains.

Plasmids   All strains of the species *Y. enterocolitica* that are recognizable as pathogenic by several other characteristics carry a 42–48 megadalton (Mdal) plasmid (identified henceforth as pYV) that mediates most of the properties that have been associated with pathogenicity (Portnoy et al., 1981; Skurnik, 1985; Zink et al., 1980). There is not complete homology among the pYV plasmids taken from different serotypes of *Y. enterocolitica*. However, a useful DNA probe prepared from the virulence plasmid permits rapid identification of plasmid-bearing *Y. enterocolitica* (Hill et al., 1983). The use of plasmid anal-ysis for identifying pathogenic strains has been questioned because of the occasional presence of plasmids of similar size in strains that do not demon-strate other markers of pathogenicity (Wachsmuth et al., 1984).

Table 15.2 Serotypes that include pathogenic strains of *Y. enterocolitica*

Serotype O:	Occurrence			Distribution	
	Common	Occasional	Rare	Worldwide	North America
3	X			X	
9	X			X[a]	
5,27 (5B)	X			X	
8	X				X
21 ('O:Tacoma)		X			X
13a,13b		X			X
4,32		X			X[b]
1			X	X	
1,2,3			X	X	
2,3			X	X	
20			X		X
40			X		X

[a] Extremely rare in North America.
[b] Primarily in western region.

Animal Models   Virulence of *Y. enterocolitica* in any animal model requires the presence of the pYV plasmid. However, production of a heat-stable
enterotoxin *in vitro*, which is measured with the suckling mouse assay (Table
15.3), is chromosomally mediated. *Y. enterocolitica* can be divided into two
groups based on pathogenicity in animal models: 1) lethal for adult mice in
low doses and producing conjunctivitis in the guinea pig (serotypes 0:8 and
0:21) (Zink et al., 1980), and 2) nonlethal for mice and producing only weak
and erratic conjunctivitis (serotypes 0:3, 0:9, and 0:5,27) (Schiemann and Devenish, 1980). Strains from both groups will produce lethal infections in iron-
overloaded mice (Robins-Browne and Prpic, 1985) and in suckling mice (Aulisio
et al., 1983), will invade the spleen in cold-stressed mice (Bakour et al., 1985),
cause diarrhea in infant rabbits (Pai et al., 1980), and colonize the mouse
intestine producing signs of diarrhea in some strains of mice and fecal excretion
in high numbers for prolonged periods in all strains of mice (Ricciardi et al.,
1978; Schiemann and Devenish, 1982; Ueno et al., 1981). Production of a heat-
stable enterotoxin measured by the suckling mouse assay is characteristic of
pathogenic strains but is also common among environmental strains (Pai et
al., 1978). Enterotoxin production has never been demonstrated in vivo, and
the heat-stable toxin has no recognized role in pathogenesis (Schiemann,
1981).

Table 15.3 Animal models used to demonstrate pathogenic *Y. enterocolitica*

Animal model	Plasmid-mediated	Reaction for serotype		
		0:8	0:3	0:5,27
Guinea pig (or mouse) conjunctivitis	Yes	+	±/−	±/−
with iron-chelator[a]	Yes		+	
Lethal infections in mice and gerbils (ip or oral)	Yes	+	−	−
with iron-chelator[a]	Yes		+	
Lethal infections in suckling mice (ip)[b]	Yes	+	+	+
Mouse spleen culture positive (oral)[a]	Yes		+	+
Intestinal colonization, diarrhea and long-term fecal excretion in mice (oral)	Yes	+	+	+
Diarrhea in infant rabbit (oral)	Yes		+	
Heat-stable enterotoxin produced *in vitro* and measured with suckling mouse assay	No	+	+	+

[a] Also demonstrated for serotype 0:9.
[b] Also demonstrated for plasmid-positive serotypes 0:9, 0:4,32, 0:21 and 0:13.

In vitro Virulence Markers   Calcium dependence, measured by growth restriction on magnesium oxalate agar (Gemski et al., 1980), and autoagglutination at 35–37°C (Laird and Cavanaugh, 1980), have been the most popular indirect markers for identifying pathogenic strains of *Y. enterocolitica* (Table 15.4). Both tests have been questioned because of a lack of correlation among all recognized attributes of virulence in some strains. The test for calcium dependence can give equivocal results because some strains do not grow well at high temperatures, the control medium has not been supplemented with calcium to demonstrate this temperature effect, and because some dependent strains will form small colonies on magnesium oxalate agar especially with extended incubation. There are some strains of nonpathogenic *Yersinia* that will produce a type of autoagglutination in broth medium at 35–37°C. Nevertheless, some workers believe autoagglutination to be the most reliable indirect marker of the presence of the virulence plasmid and pathogenicity in *Y. enterocolitica* (Prpic et al., 1985).

The other markers listed in Table 15.4 have been used less often for identifying pathogenic strains of *Y. enterocolitica*. There has been some controversy regarding the role of the plasmid and growth temperature on resistance to the action of natural bactericidal factors present in serum (Chiesa and Bottone, 1983). Uptake of Congo red has not been widely confirmed (Prpic et al., 1983). The V and W antigens, first described in *Yersinia pestis*, are common to all three pathogenic species of *Yersinia*, and can only be demonstrated by special separation techniques and specific antiserum (Une and Brubaker, 1984). Pyrazinamidase activity is one of the most recent virulence markers described and merits additional investigation (Kandolo and Wauters, 1985). Negative reactions for salicin utilization (35°C) and esculin hydrolysis (25°C) have been

Table 15.4 *In vitro* virulence markers for *Y. enterocolitica*

Marker	Plasmid-mediated
Calcium requirement for growth at 35–37°C	yes
Autoagglutination at 35–37°C	yes
Small colony size at 37 °C (or 25°C)	yes
Unique outer membrane proteins at 37°C	yes
Production of V and W antigens at 37°C	yes
Serum resistance* at 37°C	?
Congo red uptake at 25 or 37°C	yes
Hydrophobicity at 37°C	yes
Hydrophobicity at 25°C	no
Pyrazinamidase	no
Mannose-resistant hemagglutination	yes
Salicin (35°C) and esculin (25°C) negative (≤4 days)	no

* Resistance to the action of natural bactericidal factors present in serum.

used reliably to differentiate pathogenic and nonpathogenic strains within the species *Y. enterocolitica* (Schiemann and Devenish, 1982). The salicin reaction can be confusing if read beyond 4 days of incubation, and both tests are not useful for this purpose when applied to species other than *Y. enterocolitica*.

Outer Membrane Proteins   At least 20 additional polypeptides are synthesized by plasmid-positive strains of *Y. enterocolitica* grown in broth medium at 37°C (Skurnik, 1985). Three to five of these polypeptides are located in the outer membrane. These unique proteins have provided the antigens for preparation of a pathogen-specific antiserum (Doyle et al., 1982). They are also responsible for increased hydrophobicity (Lachica and Zink, 1984) and possibly serum resistance (Balligand et al., 1985). At least one large protein can be seen as a fibril structure that appears to be responsible for autoagglutination (Skurnik et al., 1984).

Epithelial Cell Interactions In Vitro   Pathogenic strains of *Y. enterocolitica* demonstrate greater adhesion to epithelial cells in tissue cultures than do nonpathogenic strains (Schiemann and Devenish, 1982). This adhesion is not mediated by fimbriae but is related to greater hydrophobicity, a property further increased in plasmid-positive strains grown at 37°C. Penetration of the epithelial cells is not plasmid-mediated and occurs more readily with 25- than 37°C-grown bacteria. There is considerable evidence that this invasion is mediated by an antigenic heat-labile outer membrane protein (Isberg et al., 1987). Unlike other invasive bacteria such as *Shigella*, intracellular multiplication in continuous cell lines does not occur (Devenish and Schiemann, 1981a). Plasmid-positive strains of *Yersinia enterocolitica* that are invasive also produce a cytotoxin that causes detachment of cell monolayers (Portnoy et al., 1981).

15.3 Epidemiology

The complex epidemiology of yersiniosis as we now understand it is summarized in Figure 15.1. Transmission by pork from a swine reservoir is probably quite common in countries where eating of raw pork or ground beef contaminated with pork is a cultural practice. Direct transmission from swine to humans is supported by circumstantial evidence only (Schiemann and Fleming, 1981). Transmission from feral rodents either directly or through fleas is strictly a hypothesis. It is supported by just two reports on the isolation of *Y. enterocolitica* from rodent fleas, and the idea that North American serotypes 0:8 and 0:21 that are unusually virulent in mice, are closely related to *Y. pestis* which is transmitted in this manner.

Animal Reservoirs   Isolations of *Y. enterocolitica* have been reported from a large number of animals but in many early cases these were organisms not now recognizable as pathogenic strains (Hurvell, 1981). Occasional reports

Epidemiology of *Yersinia enterocolitica*

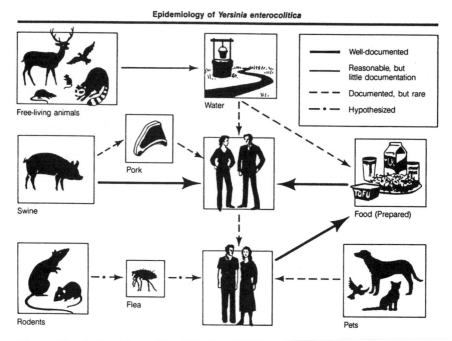

Figure 15.1   Epidemiology of *Yersinia enterocolitica*. (Reprinted with permission of Marcel Dekker, Inc.)

have described pet animals such as dogs as likely sources of human infections (Gutman et al., 1973). However, there is far more evidence to indicate that the major animal reservoir and source of human infections is swine. Swine appear to carry *Y. enterocolitica* in the throat as part of the normal flora, suggesting that the organism has a special tropism for porcine pharyngeal epithelial cells. The predominant serotypes associated with human illness have been isolated at rates of 50% and more from the throats of slaughter-age swine and fresh porcine tongues (Schiemann and Fleming, 1981; Wauters, 1979). The exceptions are common North American serotypes 0:8, which has only once been reported in swine, and 0:21, which has never been reported in swine. However, there have been only a few small surveys completed in the United States for *Y. enterocolitica* in swine, and, therefore, complete information on swine carriage in this country is unavailable. Evidence for an association between presence in swine and human infection is primarily circumstantial (Schiemann and Fleming, 1981), and no outbreak or case of yersiniosis has been described with bacteriological evidence of a link between patient and pork. This may be, however, due to inhibition of *Y. enterocolitica* growth by antagonistic flora in pork (Fukushima and Gomyoda, 1986), and the inability of available laboratory methods to isolate low numbers of *Y. enterocolitica* from mixed populations of related bacteria. Convincing epidemiological evidence has been presented recently of a link between yersiniosis in humans and consumption of raw pork in Belgium, a country with a very high incidence of this disease (Tauxe et al., 1987).

The wastes from pig farms can be contaminated with *Y. enterocolitica* which quickly colonizes new animals introduced into common pens. It has been suggested that runoff water from pig farms may contaminate surface waters (Walker and Grimes, 1985). This type of epidemiological link was suggested in a large multi-state outbreak in the United States involving pasteurized milk. The milk was apparently inoculated during opening and storage by organisms present on the outside of consumer cartons which had been transported in mud-contaminated crates returned to the dairy from a pig farm. Infectious doses were no doubt the consequence of psychrotrophic growth in the milk.

Foodborne Yersiniosis   Food was often suggested as the major vehicle for transmission of human yersiniosis even before any epidemiological evidence was available. Subsequently, several large outbreaks were reported in Japan (Zen-Yoji, 1981) and Czechoslovakia (Olsovsky et al., 1975) in which the epidemiology indicated a common vehicle such as food or water was involved but it was never identified. Four major foodborne outbreaks have been documented in the United States (Black et al., 1978; Morse et al., 1984; Tacket et al., 1984; Tacket et al., 1985). Three of these involved serotype 0:8 and the fourth serotype 0:13a,13b. Human carriers were probably the source in two outbreaks where the food vehicles were chocolate milk and chow mein.

An outbreak of yersiniosis in the state of Washington emphasized the potential of water used in food manufacture and processing for transmission of bacterial pathogens. The vehicle in this outbreak was identified as Tofu manufactured with contaminated water. The same serotype of *Y. enterocolitica* was recovered from water and several patients (Tacket et al., 1985).

The fourth and largest foodborne outbreak of yersiniosis in the United States also indirectly involved water. This outbreak was spread over several states that received pasteurized milk from a Memphis, Tennessee, dairy (Tacket et al., 1984). The milk cartons had been distributed in crates used to transport outdated milk to a pig farm. Water runoff from the animal pens apparently contaminated the crates placed on the ground. The same serotype, 0:13a,13b, was isolated from mud taken from a milk crate. This unusual serotype had been associated only with illness in monkeys prior to this outbreak. Subsequent studies established that the epidemiological strain was able to survive for at least 21 days on the exterior surface of milk cartons held at 4°C (Stanfield et al., 1985). Contamination of the milk likely occurred during opening and handling by consumers, and psychrotrophic *Y. enterocolitica* multiplied to infectious doses during refrigerated storage.

Waterborne Yersiniosis   There have been only a few episodes, each involving only a small number of individuals, in which drinking water was identified as the source of infection with *Y. enterocolitica* (Table 15.5). A large outbreak of intestinal illness at a Montana ski resort in 1977 was suggested to be waterborne yersiniosis (Eden et al., 1977). However, as new knowledge on pathogenicity emerged, the varied strains of *Y. enterocolitica* isolated from water sources were later identified as nonpathogenic. In addition, no clinical evidence of yersiniosis was presented in this outbreak.

Certain characteristics of *Y. enterocolitica* suggest that waterborne yersiniosis should be found more commonly than the available evidence suggests. Animal reservoirs are common, especially pigs, and their wastes represent a ready source for introduction of the organism into contaminated water supplies. *Y. enterocolitica* is highly adaptable to aquatic environments, having minimal nutrient requirements (Highsmith et al., 1977), tolerating low temperatures well, and being capable of surviving in water for long periods (Schillinger and McFeters, 1978). Standard water treatment and disinfection practices do provide an effective safeguard; however, these barriers have not always been effective for other enteropathogenic bacteria. It may be that the infectious doses for *Y. enterocolitica* by the oral route is higher than the numbers commonly achieved in water. Waterborne outbreaks of infectious diseases are more likely to be recognized when a large number of cases is involved. The available epidemiological data on waterborne yersiniosis indicate that limited episodes are more common, and these are more likely to go unrecognized. Furthermore, health authorities are probably not equally vigilant regarding waterborne *Y. enterocolitica* as they are with other pathogens, and public health laboratories do not normally include the special isolation and diagnostic examinations re-

Table 15.5 Reported waterborne yersiniosis

Place	Number ill	Serotype	Epidemiology notes	Reference
Norway	2	0:7,13	Same serotype isolated from feces and well water. Patient serum titers to isolates 10,320 and 640.	Lassen, 1972
New York	1	0:8 ("WA")	Same serotype isolated from mountain stream used for drinking water. Patient serum titers 8,192.	Keet, 1974
Montana	750	various	Isolates from water only. No clinical evidence of yersiniosis.	Eden et al., 1977
Denmark	1	0:3	Same serotype isolated from well water used to prepare baby food. Patient serum titer 800.	Christensen, 1979
Ontario, Canada	2	0:3	Same serotype isolated from polluted well.	Thompson and Gravel, 1986

quired for recognition of *Y. enterocolitica*. An accurate description of waterborne yersiniosis is also frustrated by the lack of a laboratory method that will selectively recover pathogenic *Y. enterocolitica* from the mixed population of related bacteria commonly present in water.

15.4   Isolation of *Y. enterocolitica*

Most of the laboratory methods described for isolation of *Y. enterocolitica* from water have used "cold enrichment." Low-temperature incubation inhibits or decreases the growth rate of competitive bacteria and thereby allows *Y. enterocolitica*, which is capable of psychrotrophic growth, to reach numbers that facilitate isolation (Schiemann and Olson, 1984). Rich media have proven to be better for nonselective pre-enrichment than simple buffers that were used earlier.

No selective enrichment procedures have been described specifically for isolation of *Y. enterocolitica* from water. A much greater research effort on isolation methods has been exerted relative to foods. Some of these procedures should be equally effective for isolation of *Y. enterocolitica* from water.

Some early reports described the isolation of *Y. enterocolitica* bacteria from water by the use of a membrane filter and standard coliform isolation medium (Highsmith et al., 1977; Schiemann, 1978). In no case could any of the isolates obtained by this method by recognized as pathogens. Later a membrane filter method for quantitative recovery of *Y. enterocolitica* from water with sequential transfer of the membrane for biochemical confirmation was described (Bartley et al. 1982). There are no published studies confirming the reliability of this procedure, nor is it likely that the growth medium is of sufficient selectivity to inhibit the background bacteria that will be present in many types of water. Furthermore, a quantitative technique as provided by the membrane filter is not necessary for determining the presence of an enteropathogenic bacterium in water. As is true for *Salmonella*, presence alone usually provides all the information desired. The biggest problem associated with isolation of *Y. enterocolitica* from water is the failure of available methods to selectively recover pathogenic strains of *Y. enterocolitica* while excluding nonpathogenic related bacteria that are so common in water (Saari and Jansen, 1979; Schiemann, 1978).

Y. enterocolitica is more tolerant of high pH than other gram-negative bacteria. Treatment of food enrichments and fecal specimens with alkali has been used to destroy background organisms and thereby increase the probability of finding *Y. enterocolitica* among the colonies growing on the plating agar medium (Aulisio et al., 1980). Mixed results have been reported with this technique (Ratnam et al., 1983), and one study shows it to be difficult to standardize and control (Schiemann, 1983). The use of a selective enrichment medium such as bile-oxalate-sorbose (BOS) broth (Schiemann, 1982), which does however lengthen the time, and a plating agar with high selectivity are considered more reliable methods (Schiemann, 1979).

MacConkey agar has been one of the most successful of the traditional enteric media available for isolating *Y. enterocolitica*. However, MacConkey agar is not highly selective or differential for *Y. enterocolitica*, and requires two days incubation at room temperature for optimum recovery. Cefsulodin-irgasan-novobiocin (CIN) agar (available from Oxoid and Difco as "*Yersinia* Selective Agar") was designed specifically for isolation of *Y. enterocolitica* (Schiemann, 1979). It is a highly selective agar medium, provides a very characteristic colonial morphology with *Y. enterocolitica*, and requires just 18 hours incubation at 32°C for recovery of all strains. The superiority of this medium with foods and fecal specimens has been demonstrated in several studies (Harmon et al., 1983; Head et al., 1982). CIN agar is of sufficient selectivity to require just a few biochemical tests for confirming the identity of presumptive colonies as *Y. enterocolitica* (see below) (Devenish and Schiemann, 1981b).

A Suggested Method for Isolating *Y. enterocolitica* from Water   The following method was originally developed and subsequently evaluated for the isolation of *Y. enterocolitica* from foods. This procedure has never been evaluated with water. However, after concentration of bacteria from water samples, the method should be directly applicable and provide equivalent success.

1. **Concentration of bacteria from water samples:** This is a dilemma in bacteriological examinations of water, especially surface or wastewaters that contain large amounts of suspended material. Filtration is a desirable concentration technique, but surface filters clog easily and depth filters may allow unacceptable bacterial passage. The Whatman GF/F glass fiber filter provides two desired properties, a high "dirt-loading" capacity, and a high degree of bacterial retention (Schiemann et al., 1978). Not all glass fiber filters are equivalent in these respects.

2. **Pre-enrichment:** The GF/F filter (or replicates if more than one is required for filtering the volume of sample desired for examination) should be immersed in about 50 ml of pre-enrichment medium (PEM) which contains (concentrations are final): Na_2HPO_4, 0.05 M; NaCl, 0.1%; KCl, 0.1%; Special Peptone (Oxoid), 1.0%; yeast extract (Difco), 2.0%. The pH of this basal medium is adjusted to 8.3 and the solution is autoclaved. The following filter-sterilized solutions are added to the basal medium (concentrations are final): $MgSO_4 \cdot 7H_2O$, 0.001%; $CaCl_2$, 0.001%. The medium should be incubated at 4°C for 7–9 days, or, with nearly equal effectiveness, at 15°C for 2 days (Schiemann and Olson, 1984).

3. **Selective enrichment:** One ml of PEM is transferred to 100 ml of bile-oxalate-sorbose (BOS) broth which contains (per liter): Na_2HPO_4, 9.14 g; Na oxalate, 5.0 g; bile salts (Difco), 2.0 g; NaCl, 1.0 g; $MgSO_4$

· 7H$_2$O, 10.0 ml × 0.1%; distilled water, 649 ml. The pH of this basal medium is adjusted to 7.6 and the mixture sterilized by autoclaving. The following filter-sterilized solutions are added to the cooled basal broth; Sorbose, 100 ml × 10%; asparagine, 100 ml × 1.0%; methionine, 100 ml × 1.0%; yeast extract, 10 ml × 2.5 mg/ml; Na pyruvate, 10 ml × 0.5%; metanil yellow, 10 ml × 2.5 mg/ml; Na nitrofurantoin, 10 ml × 1.0 gm/ml; irgasan, 1 ml × 0.4% in 95% ethanol. The inoculated BOS broth is incubated at 25°C and streaked to the plating agar after three and five days (Schiemann, 1982.

4. **Plating agar:** The isolation medium is CIN agar (*"Yersinia"* Selective Agar, YSA) (Schiemann, 1979). It should be incubated at 32°C since higher temperatures are inhibitory for occasional strains. Colonies about 1.0–1.5 mm in diameter will develop within 18 hours, and show slightly different morphologies depending upon serotype. Generally, the colony will show a dark red "bullseye" with an outer transparent zone that can be identified by placing the colony over a black wax pencil mark on the stage of a stereo-microscope with transmitted light. Two typical colonies should be transferred with an inoculation needle to slants of Kligler iron agar. After overnight incubation at 35°C, *Yersinia* will give an alkaline (surface)/acid (butt) reaction with no hydrogen sulfide and little or no gas. Tubes with this typical reaction should be transferred to a slant of Christensen's urea agar and incubated at 35°C. A positive reaction will appear usually in 2–4 hours, but negative tubes should be held overnight. The species *Y. enterocolitica* can be identified by inoculating the following biochemicals: sucrose, rhamnose, raffinose, melibiose, alpha-methyl-glucoside and Simmons citrate. All are incubated at 25°C. *Y. enterocolitica* is sucrose-positive but negative in the other tests. Pathogenic strains of *Y. enterocolitica* can also be identified by negative reactions for fermentation of salicin (35°C) and esculin hydrolysis on bile-esculin medium (25°C).

References

Aleksic, S. and Bockemuhl, J. 1984. Proposed revision of the Wauters et al. antigenic scheme for serotyping of *Yersinia enterocolitica. Journal of Clinical Microbiology* 20:99–102.

Aleksic, S., Bockemuhl, J., and Lange, F. 1986. Studies on the serology of flagellar antigens of *Yersinia enterocolitica* and related *Yersinia* species. *Zentrablatt Bakteriologie, Parasitenkunde, Infektionskrankheiten und Hygiene. Abt 1 Originale Reihe A* 261:299–310.

Aleksic, S., Steigerwalt, A.G., Bockemuhl, J., Huntley-Carter, G.P., and Brenner, D.J. 1987. *Yersinia rohdei* sp. nov. isolated from human and dog feces and surface water. *International Journal of Systematic Bacteriology* 37:327–332.

Aulisio, C.C.G., Hill, W.E., Stanfield, J.T., and Morris, J.A. 1983. Pathogenicity of *Yersinia enterocolitica* demonstrated in the suckling mouse. *Journal of Food Protection* 46:856–860.

Aulisio, C.C.G., Mehlman, I.J., and Sanders, A.C. 1980. Alkali method for rapid recovery of *Yersinia enterocolitica* and *Yersinia pseudotuberculosis* from foods. *Applied and Environmental Microbiology* 39:135–140.

Baker, P.M. and Farmer, J.J. III. 1982. New bacteriophage typing system for *Yersinia enterocolitica*, *Yersinia kristensenii*, *Yersinia frederiksenii*, and *Yersinia intermedia*: correlation with serotyping, biotyping, and antibiotic susceptibility. *Journal of Clinical Microbiology* 15:491–502.

Bakour, R., Balligand, G., Laroche, Y., Cornelis, G., and Wauters, G. 1985. A simple adult-mouse test for tissue invasiveness in *Yersinia enterocolitica* strains of low experimental virulence. *Journal of Medical Microbiology* 19:237–246.

Balligand, G., Laroche, Y., and Cornelis, G. 1985. Genetic analysis of virulence plasmid from a serogroup 9 *Yersinia enterocolitica* strain: role of outer membrane protein P1 in resistance to human serum and autoagglutination. *Infection and Immunity* 48:782–786.

Bartley, T.D., Quan, T.J., Collins, M.T., and Morrison, S.M. 1982. Membrane filter technique for the isolation of *Yersinia enterocolitica*. *Applied and Environmental Microbiology* 43:829–834.

Bercovier, H., Brenner, D.J., Ursing, J., Steigerwalt, A.G., Fanning, G.R., Alonso, J.M., Carter, G.P., and Mollaret, H.H. 1980a. Characterization of *Yersinia enterocolitica sensu stricto*. *Current Microbiology* 4:201–206.

Bercovier, H., Steigerwalt, A.G., Guiyoule, A., Huntley-Carter, G., and Brenner, D.J. 1984. *Yersinia aldovae* (formerly *Yersinia enterocolitica*-like group X2): a new species of *Enterobacteriaceae* isolated from aquatic ecosystems. *International Journal of Systematic Bacteriology* 34:166–172.

Bercovier, H., Ursing, J., Brenner, D.J., Steigerwalt, A.G., Fanning, G.R., Carter, G.P., and Mollaret, H.H. 1980b. *Yersinia kristensenii*: a new species of *Enterobacteriaceae* composed of sucrose-negative strains (formerly called atypical *Yersinia enterocolitica* or *Yersinia enterocolitica*-like). *Current Microbiology* 4:219–224.

Bergan, T. 1978. Bacteriophage typing of *Yersinia enterocolitica*. *Methods in Microbiology* 12:25–36.

Black, R.E., Jackson, R.J., Tsai, T., Medvesky, M., Shayegani, M., Feeley, J.C., MacLeod, K.I.E., and Wakelee, A.M. 1978. Epidemic *Yersinia enterocolitica* infection due to contaminated chocolate milk. *New England Journal of Medicine* 298:76–79.

Brenner, D.J., Bercovier, H., Ursing, J., Alonso, J.M., Steigerwalt, A.G., Fanning, G.R., Carter, G.P., and Mollaret, H.H. 1980. *Yersinia intermedia*: a new species of *Enterobacteriaceae* composed of rhamnose-positive, melibiose-positive, raffinose-positive strains (formerly called *Yersinia enterocolitica* or *Yersinia enterocolitica*-like). *Current Microbiology* 4:207–212.

Carter, P.B., Varga, C.F., and Keet, E.E. 1973. A new strain of *Yersinia enterocolitica* pathogenic for rodents. *Applied Microbiology* 26:1016–1018.

Chiesa, C. and Bottone, E.J. 1983. Serum resistance of *Yersinia enterocolitica* expressed in absence of other virulence markers. *Infection and Immunity* 39:469–472.

Christensen, S.G. 1979. Isolation of *Yersinia enterocolitica* 0:3 from a well suspected as the source of yersiniosis in a baby. *Acta Veterinaria Scandinavica* 20:154–156.

Devenish, J.A. and Schiemann, D.A. 1981a. HeLa cell infection by *Yersinia enterocolitica*: evidence for lack of intracellular multiplication and development of a new procedure for quantitative expression of infectivity. *Infection and Immunity* 32:48–55.

Devenish, J.A. and Schiemann, D.A. 1981b. An abbreviated scheme for identification of *Yersinia enterocolitica* isolated from food enrichments on CIN (cefsulodin-irgasan-novobiocin) agar. *Canadian Journal of Microbiology* 27:937–941.

Doyle, M.P., Hugdahl, M.B., Chang, M.T., and Beery, J.T. 1982. Serological relatedness of mouse-virulent *Yersinia enterocolitica*. *Infection and Immunity* 37:1234–1240.

Eden, K.V., Rosenberg, M.L., Stoopler, M., Wood, B.T., Highsmith, A.K., Skaliy, P., Wells, J.G., and Feeley, J.C. 1977. Waterborne gastrointestinal illness at a ski resort. *Public Health Reports* 92:245–250.

Farmer, J.J., III, Davis, B.R., Hickman-Brenner, F.W., McWhorter, A., Huntley-Carter, G.G., Asbury, M.A., Riddle, C., Wathan-Grady, H.G., Elias, C., Fanning, G.R., Steigerwalt, A.G., O'Hara, C.M., Morris, G.K., Smith, P.B., and Brenner, D.J. 1985. Biochemical identification of new species and biogroups of *Enterobacteriaceae* isolated from clinical specimens. *Journal of Clinical Microbiology* 21:46–76.

Frederiksen, W. 1964. A study of some *Yersinia pseudotuberculosis*-like bacteria ("*Bacterium enterocoliticum*" and "*Pasteurella X*"). *Scandinavian Congress of Pathology and Microbiology* 14:103–104.

Fukushima, H. and Gomyoda, M. 1986. Inhibition of *Yersinia enterocolitica* serotype 0:3 by natural microflora of pork. *Applied and Environmental Microbiology* 51:990–994.

Gemski, P., Lazere, J.R., and Casey, T. 1980. Plasmid associated with pathogenicity and calcium dependency of *Yersinia enterocolitica*. *Infection and Immunity* 27:682–685.

Gutman, L.T., Ottesen, E.A., Quan, T.J., Noce, P.S., and Katz, S.L. 1973. An inter-familial outbreak of *Yersinia enterocolitica* enteritis. *New England Journal of Medicine* 288:1372–1377.

Harmon, M.C., Yu, C.L., and Swaminathan, B. 1983. An evaluation of selective differential plating media for the isolation of *Yersinia enterocolitica* from experimentally inoculated fresh ground pork homogenates. *Journal of Food Science* 48:6–9.

Head, C.B., Whitty, D.A., and Ratnam, S. 1982. Comparative study of selective media for recovery of *Yersinia enterocolitica*. *Journal of Clinical Microbiology* 16:615–621.

Highsmith, A.K., Feeley, J.C., Skaliy, P., Wells, J.G., and Wood, B.T. 1977. The isolation and enumeration of *Yersinia enterocolitica* from well water and growth in distilled water. *Applied and Environmental Microbiology* 34:745–750.

Hill, W.E., Payne, W.L., and Aulisio, C.C.G. 1983. Detection and enumeration of virulent *Yersinia enterocolitica* in food by DNA colony hybridization. *Applied and Environmental Microbiology* 46:636–641.

Hurvell, B. 1981. Zoonotic *Yersinia enterocolitica* infection: host range, clinical manifestations, and transmission between animals and man. pp. 145–159. *in* E.J. Bottone (ed.), *Yersinia enterocolitica*. CRC Press, Boca Raton, Florida.

Isberg, R.R., Voorhis, D.L., and Falkow, S. 1987. Identification of invasion: a protein that allows enteric bacteria to penetrate cultured mammalian cells. *Cell* 50:769–778.

Kandolo, K. and Wauters, G. 1985. Pyrazinamidase activity in *Yersinia enterocolitica* and related organisms. *Journal of Clinical Microbiology* 21:980–982.

Keet, E.E. 1974. *Yersinia enterocolitica* septicemia. Source of infection and incubation period identified. *New York State Journal of Medicine* 74:2226–2229.

Lachica, R.V. and Zink, D.L. 1984. Plasmid-associated cell surface charge and hydrophobicity of *Yersinia enterocolitica*. *Infection and Immunity* 44:540–543.

Laird, W.J. and Cavanaugh, D.G. 1980. Correlation of autoagglutination and virulence in yersiniae. *Journal of Clinical Microbiology* 11:430–432.

Lassen, J. 1972. *Yersinia enterocolitica* in drinking water. *Scandinavian Journal of Infectious Diseases* 4:125–127.

Morse, D.L., Shayegani, M., and Gallo, R.J. 1984. Epidemiologic investigation of a *Yersinia* camp outbreak linked to a food handler. *American Journal of Public Health* 74:589–592.

Nilehn, B. 1973. Host range, temperature characteristics and serologic relationships among *Yersinia* phages. *Contributions to Microbiology and Immunology* 2:59–67.

Olsovsky, Z., Olsakova, V., Chobot, S., and Sviridov, V. 1975. Mass occurrence of *Yersinia enterocolitica* in two establishments of collective care of children. *Journal of Hygiene, Epidemiology, Microbiology and Immunology* 19:22–29.

Pai, C.H., Mors, V., and Seemayer, T.A. 1980. Experimental *Yersinia enterocolitica* enteritis in rabbits. *Infection and Immunity* 28:238–244.

Pai, C.H., Mors, V,. and Toma, S. 1978. Prevalence of enterotoxigenicity in human and nonhuman isolates of *Yersinia enterocolitica*. *Infection and Immunity* 22:334–338.

Portnoy, D.A., Moseley, S.L., and Falkow, S. 1981. Characterization of plasmids and plasmid-associated determinants of *Yersinia enterocolitica* pathogenesis. *Infection and Immunity* 31:775–782.

Prpic, J.K., Robins-Browne, R.M., and Davey, R.B. 1983. Differentiation between virulent and avirulent *Yersinia enterocolitica* isolates by using Congo red agar. *Journal of Clinical Microbiology* 18:486–490.

Prpic, J.K., Robins-Browne, R.M., and Davey, R.B. 1985. In vitro assessment of virulence in *Yersinia enterocolitica* and related species. *Journal of Clinical Microbiology* 22:105–110.

Ratnam, S., Looi, C.L., and Patel, T.R. 1983. Lack of efficacy of alkali treatment for isolation of *Yersinia enterocolitica* from feces. *Journal of Clinical Microbiology* 18:1092–1097.

Ricciardi, I.D., Pearson, A.D., Suckling, W.G., and Klein, C. 1978. Long-term fecal excretion and resistance induced in mice infected with *Yersinia enterocolitica*. *Infection and Immunity* 21:342–344.

Robins-Browne, R.M. and Prpic, J.K. 1985. Effects of iron and desferrioxamine on infections with *Yersinia enterocolitica*. *Infection and Immunity* 47:774–779.

Saari, T.N. and Jansen, G.P. 1979. Waterborne *Yersinia enterocolitica* in the midwestern United States. *Contributions to Microbiology and Immunology* 5:183–196.

Schiemann, D.A. 1978. Isolation of *Yersinia enterocolitica* from surface and well waters in Ontario. *Canadian Journal of Microbiology* 24:1049–1052.

Schiemann, D.A. 1979. Synthesis of a selective agar medium for *Yersinia enterocolitica*. *Canadian Journal of Microbiology* 25:1298–1304.

Schiemann, D.A. 1981. An enterotoxin-negative strain of *Yersinia enterocolitica* serotype 0:3 is capable of producing diarrhea in mice. *Infection and Immunity* 32:571–574.

Schiemann, D.A. 1982. Development of a two-step enrichment procedure for recovery of *Yersinia enterocolitica* from food. *Applied and Environmental Microbiology* 43:14–27.

Schiemann, D.A. 1983. Alkalotolerance of *Yersinia enterocolitica* as a basis for selective isolation from food enrichments. *Applied and Environmental Microbiology* 46:22–27.

Schiemann, D.A., Brodsky, M.H., and Ciebin, B.W. 1978. Salmonella and bacterial indicators in ozonated and chlorine dioxide-disinfected effluent. *Journal Water Pollution Control Federation* 50:158–162.

Schiemann, D.A. and Devenish, J.A. 1980. Virulence of *Yersinia enterocolitica* determined by lethality in Mongolian gerbils and by the Sereny test. *Infection and Immunity* 29:500–506.

Schiemann, D.A. and Devenish, J.A. 1982. Relationship of HeLa cell infectivity to biochemical, serological, and virulence characteristics of *Yersinia enterocolitica*. *Infection and Immunity* 35:497–506.

Schiemann, D.A. and Fleming, C.A. 1981. *Yersinia enterocolitica* isolated from throats of swine in eastern and western Canada. *Canadian Journal of Microbiology* 27:1326–1333.

Schiemann, D.A. and Olson, S.A. 1984. Antagonism by gram-negative bacteria to growth of *Yersinia enterocolitica* in mixed cultures. *Applied and Environmental Microbiology* 48:539–544.

Schillinger, J.E. and McFeters, G. 1978. Survival of *Escherichia coli* and *Yersinia enterocolitica* in stream and tap waters. *Abstracts of the Annual Meeting, American Society of Microbiology* N79, p. 175.

Schleifstein, J. and Coleman, M.B. 1939. An unidentified microorganism resembling *B.*

lignieresi and *Pasteurella pseudotuberculosis* pathogenic for man. *New York State Journal of Medicine* 39:1749–1753.

Schleifstein, J. and Coleman, M.B. 1943. *Bacterium enterocoliticum. Annual Report New York State Department of Health, Albany*, p. 56.

Skurnik, M. 1985. Studies on the virulence plasmids of *Yersinia* species. *Acta Universitatis Ouluensis Series A. Scientiae Rerum Naturalium No. 169*, Biochemica No. 53.

Skurnik, M., Bolin, I., Heikkinen, H., Piha, S., and Wolfe-Watz, H. Virulence plasmid-associated autoagglutination in *Yersinia* spp. *Journal of Bacteriology* 158:1033–1036.

Stanfield, J.T., Jackson, G.J., and Aulisio, C.C.G. 1985. *Yersinia enterocolitica*: survival of a pathogenic strain on milk containers. *Journal of Food Protection* 49:947–948.

Tacket, C.O., Ballard, J., Harris, N., Allard, J., Nolan, C., Quan, T., and Cohen, M.L. 1985. An outbreak of *Yersinia enterocolitica* infections caused by contaminated tofu (soybean curd). *American Journal of Epidemiology* 121:705–711.

Tacket, C.O., Narain, J.P., Sattin, R. ,Lofgren, J.P., Konigsberg, C., Jr., Rendtorff, R.C., Rausa, A., Davis, B.R., and Cohen, M.L. 1984. A multistate outbreak of infections caused by *Yersinia enterocolitica* transmitted by pasteurized milk. *Journal of American Medical Association* 251:483–486.

Tauxe, R.V., Vandepitte, J., Wauters, G., Martin, S.M., Goossens, V., de Mol, P., van Noyen, R., and Thiers, G. 1987. *Yersinia enterocolitica* infections and pork: the missing link. *Lancet* 1:1129–1132.

Thompson, J.S. and Gravel, M.J. 1986. Family outbreak of gastroenteritis due to *Yersinia enterocolitica* serotype 0:3 from well water. *Canadian Journal of Microbiology* 32:700–701.

Toma, S., Wauters, G., McClure, H.M., Morris, G.K., and Weissfeld, A.S. 1984. 0:13a, 13b, a new pathogenic serotype of *Yersinia enterocolitica. Journal of Clinical Microbiology* 20:843–845.

Ueno, H. Kaneko, K.-I., and Hashimoto, N. 1981. Fecal excretion of *Yersinia enterocolitica* in mice and rats inoculated intragastrically. *Japanese Journal of Veterinary Research* 29:67–72.

Une, T. and Brubaker, R.R. 1984. In vivo comparison of avirulent Vwa− and Pgm− or Pst− phenotypes of yersiniae. *Infection and Immunity* 43:895–900.

Ursing, J., Brenner, D.J., Bercovier, H., Fanning, G.R., Steigerwalt, A.G., Brault, J., and Mollaret, H.H. 1980. *Yersinia frederiksenii*: a new species of *Enterobacteriaceae* composed of rhamnose-positive strains (formerly called atypical *Yersinia enterocolitica* of *Yersinia enterocolitica*-like). *Current Microbiology* 4:213–217.

Wachsmuth, K., Kay, B.A., and Birkness, K.A. 1984. Diagnostic value of plasmid analyses and assays for virulence in *Yersinia enterocolitica. Diagnostic Microbiology and Infectious Disease* 2:219–228.

Walker, P.J. and Grimes, D.J. 1985. A note on *Yersinia enterocolitica* in a swine farm watershed. *Journal of Applied Bacteriology* 58:139–143.

Wauters, G. 1979. Carriage of *Yersinia enterocolitica* serotype 3 by pigs as a source of human infection. *Contributions to Microbiology and Immunology* 5:249–252.

Wauters, G. 1981. Antigens of *Yersinia enterocolitica*. pp. 41–53 *in* E. Bottone (ed.), *Yersinia enterocolitica*. CRC Press, Boca Raton, Florida.

Zen-Yoji, H. 1981. Epidemiologic aspects of yersiniosis in Japan. pp. 205–216 *in* E.J. Bottone (ed.), *Yersinia enterocolitica*. CRC Press, Boca Raton, Florida.

Zink, D.L., Felley, J.C., Wells, J.G., Vanderzant, C., Vickery, J.C., Roof, W.D., and O'Donovan, G.A. 1980. Plasmid-mediated tissue invasiveness in *Yersinia enterocolitica. Nature* 283:224–226.

16

Legionella in Drinking Water

Stanley J. States[1], Robert M. Wadowsky[2], John M. Kuchta[1],
Randy S. Wolford[1], Louis F. Conley[1], and Robert B. Yee[3]

[1]*City of Pittsburgh Water Department,* [2]*School of Medicine, University of Pittsburgh,
and* [3]*Graduate School of Public Health, University of Pittsburgh, Pittsburgh,
Pennsylvania*

In July 1976, an outbreak of acute respiratory illness occurred during an American Legion convention in Philadelphia, Pennsylvania. Of 4,400 attendees, and some other individuals not directly associated with the convention, 221 became ill and 34 of these cases died (Fraser et al., 1977). The cause of the epidemic was unknown until later that year, when investigators at the Centers for Disease Control isolated the responsible bacterium, subsequently termed *Legionella pneumophila* (lung-loving) (McDade et al., 1977).

Reexamination of preserved sera and bacterial specimens from earlier explosive outbreaks of pneumonia indicated that *Legionella* pneumonia was not a new syndrome but had occurred uncharacterized for some years (McDade et al., 1979). Additionally, this bacterium was determined to be responsible for another illness, labeled Pontiac fever after an epidemic of non pneumonic respiratory illness occurring in Pontiac, Michigan in 1968.

While both legionellosis syndromes appear to be caused by the same group of bacteria, the illnesses occur in two distinct clinicoepidemiologic forms. *Legionella* pneumonia (Legionnaires' disease) is a severe, acute pneumonia with multisystem involvement and an incubation period ranging from two to ten days. Attack rates, which range between 0.1 and 4.0% of those exposed, are highest in immunocompromised individuals. Other risk factors predisposing to *Legionella* pneumonia include being male, being over 50 years of age, smoking, heavy alcohol consumption, and underlying chronic disease. Erythromycin is used in treatment, but case-fatality ratios range from 10 to 20%. The other form of legionellosis, Pontiac fever, is a nonpneumonic, self-limited, nonfatal influenza-like syndrome consisting of fever, myalgia, and headache (Fleming and Reingold, 1986). The incubation period, 5 to 66 hours, is much shorter than that of *Legionella* pneumonia. Pontiac fever is also much less dependent on host factors with an attack rate approaching 100% of those exposed.

Legionellosis occurs in epidemic clusters, and in even higher frequency, as sporadic community acquired cases. Many cases occur as nosocomial infections. While the illness has been documented in nearly every country, the actual incidence of legionellosis is not known. Based on a retrospective study of pneumonia among people enrolled in a prepaid medical care group in Seattle (Foy et al., 1979), it has been estimated that more than 25,000 cases occur annually in the United States. However, other estimates suggest that as many as 50,000 to 100,000 cases may occur within this country each year (U.S. Environmental Protection Agency, 1985).

Legionellae are aerobic, nonspore-forming, gram-negative bacilli that are typically flagellated. When cultured on charcoal yeast-extract agar, the colonies are circular, grey to white, and present a characteristic cut glass appearance with green to purple iridescence. Studies of the biochemical characteristics, molecular weight of the genome, guanine-cytosine content, and DNA homology have not demonstrated relatedness between these and any other known family of bacteria (Band and Fraser,1986). The legionellae are not a single species but, based on DNA homology, are a collection of at least 34 species grouped under the genus *Legionella* and the family *Legionellaceae*.

At least three of the species, *L. pneumophila*, *L. longbeachae*, and *L. bozemaneae*, contain serogroups which differ in serogroup-specific antigens. *L. pneumophila*, the species most frequently causing illness, currently includes fourteen serogroups of which serogroup 1 is the most commonly encountered.

The method by which legionellae are transmitted to humans, both in sporadic cases and in most outbreaks of legionellosis, is not well understood. Person-to-person spread has not been documented. Rather, the mechanism of infection appears to be direct transmission from the environment by inhalation of the bacterium in aerosolized contaminated water. Surveys of lakes, ponds, and streams indicate that *Legionella* is a common inhabitant of natural waters (Fliermans et al, 1979, 1981; Politi et al., 1979). The organism has also been isolated from a variety of artificial aquatic habitats including cooling towers and evaporative condensers (Glick et al., 1978; Cordes et al., 1980; Dondero et al., 1980; Gorman et al., 1985), and the plumbing systems of hospitals, hotels, gymnasiums, and homes (Cores, 1981; Tobin, 1981; Wadowsky et al., 1982; Stout et al., 1987).

Characterization of the illness and isolation of a responsible waterborne agent pose two questions to those working with drinking water. What role does potable water, both in the municipal system and within internal plumbing systems, play in the spread of *Legionella*? And, which factors within potable water systems influence survival and growth of the organism? This chapter describes research conducted jointly by the City of Pittsburgh Water Department and the University of Pittsburgh Graduate School of Public Health aimed at studying these questions. It also summarizes the work of a number of other investigators interested in the same problems.

16.1 Development of a Selective Medium and Water Stock Culture

Legionella was first cultured from tissue specimens obtained from victims of the pneumonia outbreak at the 1976 Legionnaires' convention. The bacterium was isolated using a rickettsiologic technique involving intraperitoneal inoculation of guinea pigs and culturing of infected tissue in embryonated hens' eggs (McDade et al., 1977). While successful, the method is time-consuming and expensive. Quite specific growth requirements, involving high levels of cysteine and iron, made isolation on commonly used bacteriologic media difficult. Supplementation of Mueller-Hinton agar with 1% hemoglobin and 1% Isovitale X permitted the initial isolation on artificial medium. An improvement to this medium was the development of Feeley-Gorman agar, in which ferric pyrophosphate and L-cysteine hydrochloride were incorporated as additives (Feeley et al., 1978). Supplementation of charcoal yeast-extract agar with these substances (Feeley et al., 1979) and buffering with ACES buffer (BCYE agar) (Pasculle et al., 1980) resulted in a yet improved means for cultivation of these bacteria.

Even with these advances, recovery of *Legionella* from environmental samples was difficult. Legionellae often comprise only a small minority of the total bacterial flora in environmental specimens and may exhibit a lag period of several days for growth. As a result, the Legionellaceae may be inhibited or masked by other bacteria when direct culturing is attempted.

To overcome this problem, selective media have been developed consisting of a buffered charcoal yeast extract base supplemented with iron, cysteine, and a variety of bacterial inhibitors. Wadowsky and Yee (1981) prepared a selective medium, designated differential glycine-vancomycin-polymyxin B (DGVP) agar, that incorporates glycine and the antibiotics vancomycin and polymyxin B as selected agents, and the dyes bromothymol blue and bromocresol purple (Vickers, 1981). Bopp et al. (1981) produced a medium containing cephalothin, colistin, cycloheximide, and vancomycin. Additionally, Edelstein (1982) designed a medium containing glycine, polymyxin B, vancomycin, and anisomycin as well as bromothymol blue and bromocresol purple. These selective media inhibit other bacteria and facilitate recovery of Legionellaceae.

Isolation of legionellae has also been improved by the development of pretreatment techniques. Exposure of water samples to acid, pH 2.2 for five minutes (Bopp, 1981), heating to 50°C for 30 minutes (Dennis et al., 1984), or both, prior to plating on selective media, further enhances recovery. A membrane filtration technique, utilizing a polyvinylidene fluoride membrane, has also been used to process large volumes of water in order to provide a more sensitive method for isolation of low numbers of legionellae (R.S. Wolford, J.M. Kuchta, R.M. Wadowsky, and R.B. Yee, Abstr. Annu. Meet. Am. Soc. Microbiol. 1988, Q25). However, while all of these methods are effective, studies on the sensitivity of *Legionella* spp. to selective isolation procedures still indicate a need for care when interpreting quantitative culture results from heavily contaminated environmental samples (Calderon and Dufour, 1984; Roberts et al., 1987).

In addition to requiring a selective medium for isolation of *Legionella*, a method was needed to maintain naturally occurring legionellae in the laboratory. This would facilitate study of the organism under conditions more similar to those in the environment. Earlier work had shown that some medium-growth bacteria behave differently than water-grown strains. Experiments with *Pseudomonas aeruginosa* indicated that growing a single subculture of this bacterium on artificial media reduces its resistance to disinfectants (Carson et al., 1972) and lessens its ability to grow in distilled water (Favero et al., 1971). Similarly, studies with *L. pneumophila* showed that prolonged cultivation on artificial media reduces its virulence in guinea pigs (Ormsbee et al., 1981).

To alleviate these problems, a system was developed to maintain a water stock culture of naturally occurring *L. pneumophila* and associated microbiota (Yee and Wadowsky, 1982). The initial source of the stock culture is a water sample collected from a location, such as a hospital hot water tank, known to be contaminated with *Legionella*. Rather than culturing the sample on media, it is transferred, at a 1:100 dilution, into sterilized tap water. The water culture is then incubated at 35°C and transferred again into sterile tap water when the growth of *L. pneumophila* is in the late-exponential to early-stationary phase. This generally occurs 18 to 22 days following the previous transfer (Figure 16.1). Water cultures have been perpetuated in this way for over fifty transfers into sterile tap water.

While cumbersome to work with, this system for maintaining water stock cultures of *L. pneumophila* allows investigation of the bacterium in the laboratory under conditions closer to those which exist in the natural environment. This has permitted more realistic examination of the ecological relationships of *Legionella* with biotic and abiotic components of the environment, and a more accurate appraisal of methods used for disinfection and control.

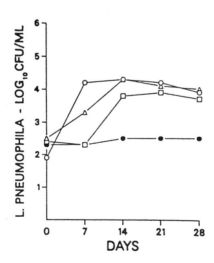

Figure 16.1 Multiplication of *L. pneumophila* in three serial passages in membrane filter-sterilized tap water incubated at 35°C. Symbols: O, Passage 4; △, Passage 5; □, Passage 6; ●, Distilled water.

16.2   Occurrence of *Legionella* in Potable Water Systems

With the availability of selective media, it has been possible to analyze environmental samples to determine the occurrence, and potential for survival and growth, of *Legionella* sp. in potable water systems.

Yee and Wadowsky (1982) examined the survival and multiplication in tap water of *L. pneumophila* environmental isolates which had never been passaged on artificial medium. Laboratory growth experiments were conducted in which a legionellae-contaminated showerhead, bacteria sedimented from hospital tap water, and contaminated water and sediment from the bottom of a hot water tank were individually immersed or inoculated into tap water. In all three experiments, *Legionella* and non-Legionellaceae bacteria multiplied freely. *L. pneumophila* increased from two to three log colony-forming units (CFU) at somewhat higher temperatures (32–42°C) than non-*Legionella* bacteria (25–37°C). This demonstrated that *Legionella* may not merely contaminate plumbing fixtures, hot water tanks, and cooling towers but can survive and multiply in these tap water environments, especially at warm temperatures.

Investigations have been conducted to determine the extent of *Legionella* sp. contamination of internal plumbing systems in hospitals, nonhospital buildings, and homes. The initial isolations from plumbing systems were accomplished by concentrating the bacteria in the water samples and using guinea pig inoculation for detection. With this technique Tobin et al. (1980) and Cordes et al. (1981) isolated *L. pneumophila* from showerheads and associated fixtures, and demonstrated that this bacterium can inhabit the plumbing systems of hospitals. Subsequently, Tobin et al (1981a), sampled fixtures and water in hotels as well as hospitals in Great Britain and showed that *L. pneumophila* was present in the potable water of many of these buildings, both with and without associated disease.

Selective media made it possible to conduct still more extensive surveys of internal potable water systems. Wadowsky et al. (1982) examined plumbing systems in hospitals, institutions, and homes in Pittsburgh. In this study, swab samples from the inner surfaces of faucets and showers, and water and sediment from the bottom of hot water tanks were cultured onto DGVP medium. The results showed that tap water contained only small or undetectable numbers of legionellae. However, aerators, faucet spouts, hot water valve seats, and shower assemblies in the hospitals, non-hospital institutions, and one of the five houses surveyed were more commonly contaminated. Additionally, water and sediment from hot water tanks maintained at 30 degrees to 54 degrees C, but not at 70 degrees C, also contained *Legionella*. In one case, the density in the tank bottom reached 13,850 CFU/ml. The findings suggested that hot water tanks support the growth of legionellae and are a major source of contamination and amplification of legionellae for plumbing systems. This is especially true in hospital and domestic tanks in which the heating elements are not located at the bottom of the tank, and in hospital recirculating hot water systems in which lower temperatures are maintained to prevent scalding of patients. These conditions permit the temperature at the bottom of the tanks to fall within the range 32 to 42°C, m which is ideal for *Legionella* multiplication.

Arnow and Weil (1985) evaluated the prevalence and significance of *Legionella* contamination in apartment buildings in Chicago by sampling tap water and plumbing scrapings. They isolated *L. pneumophila* from 37% of the apartments surveyed, but only from those in buildings containing large institutional type hot water tanks having water temperatures lower than 55°C. While health questionnaires and blood serum analysis did not indicate that the bacterial contamination represented a significant hazard to the immunocompetant population, the authors considered this reservoir a potential source for sporadic cases of Legionnaires disease.

Lee and coworkers (1988) studied *Legionella pneumophila* colonization in the water systems of single family homes in the Pittsburgh area. Of 55 homes sampled, *L. pneumophila* was found in the hot water tanks in two and was isolated from faucets or showerheads in four. The presence of *L. pneumophila* was associated with lower water temperatures (less than 48.8°C) within the hot water system and with the use of electric hot water heaters. Water temperatures in the electrically heated tanks, especially at the bottoms of the tanks, were significantly lower than in gas heated tanks. None of the residents of the homes colonized with legionellae had a history of pneumonia or recent hospitalization, demonstrating the importance of host susceptibility and other undefined factors in addition to presence of the organism. In a similar study, Joly et al. (1985) found that 11 of 54 hot water tanks in Quebec residences showed *L. pneumophila* colonization. There was again a significant association between colonization and use of electric heaters.

The detection of *L. pneumophila* in internal plumbing systems has raised the suspicion that legionellae inhabit municipal drinking water supplies, and that these supplies serve as a pathway for contamination of buildings (Fliermans, 1984; Hsu et al., 1984; Stout et al., 1982; Wadowsky et al., 1982). The possibility has been explored by a number of investigators who, having detected legionellae within internal plumbing systems, cultured water from mains entering these buildings (Fisher-Hoch) et al., 1981; Tobin et al., 1981a, 1981b; Desplaces et al., 1984). However, legionellae were not isolated from any of the samples. Tison and Seidler (1983) examined the incidence of *Legionella* sp. in raw water, at various stages of treatment, and in the distribution system of several water supplies. While they obtained positive direct fluorescent antibody (DFA) results, suggesting the presence of the bacterium, they were unable to isolate legionellae by animal inoculation or culture procedures.

The question of *Legionella* contamination of municipal drinking water systems was pursued further in a study of the City of Pittsburgh system (States et al., 1987a). In this study, surveys to detect the occurrence of *Legionella* were conducted over a period of several years. Sampling sites included the river water supply, locations within the treatment plant, finished water reservoirs, mains, and distribution taps. Despite the use of several isolation techniques,

Legionella sp. could not be cultured in any of the samples other than those from the river. The suitability of the public water supply environment for *Legionella* growth was then investigated by collecting additional water samples from throughout the system and dechlorinating, pasteurizing, and inoculating them with an *L. pneumophila* water stock culture. The growth study indicated that samples from several locations within the treatment plant and distribution network can support legionellae multiplication. These included water from the top of treatment plant filters, the bottom of sedimentation basins and finished water reservoirs, and some distribution taps. Chemical analysis suggested that turbidity, organic carbon, and certain metals are of particular importance to growth in these locations. Additionally, these are locations in which water is pooled and nutrients tend to accumulate. More recent studies also indicate that these sites contain populations of free living amoebae (States et al., 1988).

Colbourne et al (1986, 1988) conducted a 15 month legionellae survey of nine public water supplies in England and also initially failed to culture the organism from source works and most distribution samples. However, employing an indirect immunofluorescent assay (IFA), based on monoclonal antibodies specific to *L. pneumophila* serogroup 1, they detected legionellae in underground and surface water sources and distribution systems. Suspecting that culturability is related to water temperature, they heat treated some of the samples (45°C for 10 to 20 minutes). Following this treatment, five of seven IFA positive, culture negative distribution samples yielded low numbers of culturable *L. pneumophila*. The authors hypothesized that the nonculturable legionellae forms may be induced at low temperatures as a survival mechanism and that legionellae may possess heat shock genes similar to those occurring in *Escherichia coli*. However, another possibility, discussed later in this chapter, is that legionellae culturability in this habitat is reduced because the bacteria become sequestered within amoebae or other protozoal cysts. Excystation induced by heat, or mechanical opening of cysts by sonication or other artificial means, may be required for true bacterial levels to be determined.

These investigations indicate that while legionellae are widely present in raw source waters, and have been detected in numerous internal plumbing systems, the organism probably occurs in public supplies only in low numbers or sporadically. Factors that would tend to keep legionellae numbers low in municipal supplies include:

Removal of raw water legionellae from and reduction of growth-supporting nutrients by coagulation-sedimentation, carbon absorption, and filtration processes during water treatment.

Existence of chlorine residuals at long contact times thereby producing high CT values (concentration times time).

Absence of amplifier conditions that provide warm temperatures and elevated levels of nutrients and growth supporting metals.

A much lower environmental surface-to-volume ration than in internal plumbing systems. This reduces the prime solid-liquid interface sites suitable for *Legionella* growth and increases the dilution factor for any legionellae that are present.

While these factors make the occurrence of large, or even detectable, levels of *Legionella* sp. unlikely, it does not eliminate the possibility that low numbers of legionellae pass from municipal supplies into cooling towers and internal plumbing systems. This emphasizes the need for sound conventional public water treatment practices, including maintenance of a disinfectant residual, and, where indicated, the application of additional control measures at the point of use.

16.3 Biotic Factors Affecting *Legionella*

All of the carbon and energy requirements of *Legionella* sp. can be met with nine amino acids (Tesh and Miller, 1981). *L. pneumophila* can apparently synthesize all other necessary constituents de novo and seems to have no vitamin requirements (Ristroph et al., 1981). In aquatic environments, *Legionella* must obtain these amino acids from the environment either from other living organisms which produce them in excess or from the decomposition of organic matter, or both. Since legionellae typically exist in association with other microorganisms, this suggests that these other microorganisms may support *Legionella* growth. The importance of associations between *Legionella* and other microorganisms for its survival, growth, and pathogenicity is still not completely understood. However, it has been the subject of a number of studies.

The microbial group initially considered for their influence on legionellae were the algae. Tison et al., (1980) isolated *L. pneumophila* from an algal-bacterial mat community growing in an artificial thermal effluent. When they subsequently inoculated the *L. pneumophila* into cultures of *Fischerella* sp., a blue-green algae, they observed bacterial multiplication. Multiplication occurred, however, only when the test system was exposed to light. Little *Legionella* growth was observed under conditions of darkness or in the absence of *Fischerella*.

Pope et al. (1982) expanded these findings to include additional species of green algae. They inoculated environmental and clinical isolates of *L. pneumophila* into minimal salts media with various algal strains. The media, containing no organic nutrients other than those provided by algae, supported legionellae multiplication. Growth reoccurred with successive transfers to new green algal or cyanobacterial cultures.

In addition to supporting growth, algae may promote transmission of legionellae. Brendt (1981) showed that survival of *L. pneumophila* in aerosols was improved when the bacterium was associated with *Fischerella* sp. The enhancement may result from physical protection from desiccation provided by the algal mucilaginous matrix.

Legionella growth is also enhanced by other bacteria found associated with it in plumbing systems. In one set of experiments, a satellite assay was used to test the ability of non-Legionellaceae bacteria to produce L-cysteine, needed for legionellae multiplication (Wadowsky and Yee, 1983). Portions of a suspension of *L. pneumophila* originally cultured from a water stock culture, were spread onto the surface of buffered charcoal-yeast extract (BCYE) agar which had been supplemented with ferric pyrophosphate but not with L-cysteine. L-cysteine-deficient BCYE agar does not normally support legionellae growth. Non-Legionellaceae bacteria, also originally cultured from the water stock culture, were inoculated onto the center of these same plates. After incubation at 35°C, zones of satellite growth were observed around colonies of three of the four strains of non-Legionellaceae bacteria. Since the critical substance eliminated from the BCYE agar was L-cysteine, it was inferred that the growth factor produced by the three strains is most likely L-cysteine or a related compound.

Similar findings were reported by Stout et al. (1985). These investigators demonstrated that the sediment in hospital water systems provide nutrients which stimulate the growth of non-*Legionella* bacteria including, among others, *Flavobacterium* sp., *Pseudomonas* sp., *Alcaligenes* sp., and *Acinetobacter* sp. These, in turn, stimulate multiplication of *L. pneumophila*.

The hypothesis that non-Legionellaceae bacteria support legionellae growth was further investigated by cocultivation experiments in tap water, of non-Legionellaceae bacteria and *L. pneumophila* recovered on artificial medium from water stock cultures (Wadowsky and Yee, 1985). In these studies, isolates of the non-Legionellaceae bacteria failed to support multiplication of agar-grown *L. pneumophila* which had been seeded into the suspension. However, naturally occurring *L. pneumophila* which had not been passed on agar multiplied in the presence of the associated microbiota. This indicates that the non-*Legionella* bacteria are not growth-supporting but merely growth-enhancing and suggests that still other microorganisms, present in the water stock culture, are responsible for supporting the multiplication of *L. pneumophila* in tap water.

Free-living amoebae are strong candidates for this primary role in supporting legionellae multiplication in water. The importance of these organisms is discussed in detail later in this chapter.

16.4 Abiotic Factors Affecting *Legionella*

In addition to biotic factors, a number of physical and chemical parameters have been evaluated for their influence on *Legionella* sp. Several studies have shown that legionellae can exist in habitats having a wide range of temperatures, and are particularly well adapted to warmer conditions. Fliermans et al. (1981) sampled 67 rivers and lakes in the United States and, using guinea pig inoculation, recovered legionellae in waters varying from 5.7°C to 63°C. While these data did not determine whether *Legionella* multiplies throughout this range, they demonstrated that the bacterium can survive and remain viable. The authors suggested that the relationship between *L. pneumophila* and thermal environments is also indicated by its cellular fatty acid composition, which is similar to that of known thermophilic bacteria (Moss et al., 1977). Dennis et al. (1984) compared the effects of higher temperatures on several types of bacteria and found that *L. pneumophila* exhibited little loss in viability at 50°C relative to a *Pseudomonas* sp., a *Micrococcus* sp., and coliforms. Wadowsky et al. (1985b) investigated the effects of temperature on multiplication of naturally occurring legionellae seeded into membrane-filtered tap water. They observed that *L. pneumophila* multiplied between 25 and 37°C (Figure 16.2), with a maximum increase occurring in test systems adjusted to 32°C and 35°C.

Legionella sp. also tolerate a wide pH range. While developing one of the original culture media for legionellae, Feeley et al. (1978) reported a narrow pH limit of 6.5 to 7.5 for growth of *L. pneumophila* stock cultures on artificial media. However, Fliermans et al. (1981) recovered *L. pneumophila* from natural bodies of fresh water in the United States having pH values from 5.4 to 8.1 and Wadowsky and co-workers (1985b), studying the multiplication of water stock cultures of *L. pneumophila*, found the acceptable pH to extend from 5.5 to 9.2. The multiplication of *L. pneumophila* over such a wide pH range is believed to be a reflection of its natural habitat being the outdoor environment (Fliermans et al., 1981).

Legionella appears to be positively affected by certain nonmetallic plumbing materials. During an investigation of a nosocomial legionellosis outbreak

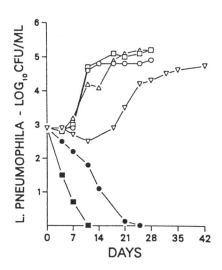

Figure 16.2 Effects of temperature on the multiplication of naturally occurring *L. pneumophila* in membrane filter-sterilized tap water. Symbols: ▽, 25°C; ○, 32°C; □, 35°C; △, 37°C; ●, 42°C; ■, 45°C.

in a London hospital, Colbourne et al. (1984) isolated legionellae from rubber faucet washers. They found that the fittings are an excellent substrate for growth and provide an ecological niche within the plumbing system. After replacement of the washers with a National Water Council (UK)-approved rubber material, known to give a negative microbial growth response, legionellae could no longer be cultured. Similar observations on rubber fittings were made by Schofield and Locci (1985) using a model hot water plumbing system in the laboratory.

Legionellae have also been shown to be sensitive to certain metals and to have some special metal requirements. Media used for isolation of *Legionella* sp. are routinely supplemented with iron to enhance recovery (Feeley et al., 1978). Furthermore, investigators developing chemically defined media for growth of *L. pneumophila* determined that magnesium and potassium stimulate multiplication in the laboratory (Reeves et al., 1981; Tesh and Miller, 1982). Given the metallic nature of plumbing systems, a question arises concerning the effects on *Legionella* of metals leached from hot water tanks and pipes. States et al. (1985) studied the chemical environment in plumbing systems and the influence of this environment on *L. pneumophila* growth. The chemical environment was surveyed by analyzing water and sediment samples from the bottom of hospital hot water tanks known to have supported *L. pneumophila* populations. The influence of the chemical environment on growth was studied by observing legionellae multiplication in tap water supplemented with metals and *L. pneumophila*, and the multiplication of legionellae inoculated into hot water tank samples collected as part of the chemical survey. Analysis of the chemical environment indicated that certain metals, such as iron and zinc, reach high concentrations due to corrosion or leaching of the tanks, the associated plumbing, and solders. The growth studies showed that while elevated levels of a number of metals are toxic to *Legionella*, low concentrations of iron, zinc, and potassium substantially enhance legionellae multiplication. Figure 16.3 illustrates the relationship between *L. pneumophila* multiplication and iron in the chemically analyzed water tank samples. Enhancement of legionellae growth by low concentrations of Zn was also documented by Colbourne et al. (1986, 1988) who observed legionellae colonization of taps fed by water passing through galvanized steel pipe. These data indicate that metal components and corrosion products are important factors in the survival and growth of *Legionella* sp.

The observations on the influence of temperature, pH, nonmetallic materials, and metals help to explain the successful adaptation of *Legionella* to plumbing systems, cooling towers, and other artificial habitats.

16.5 Control of *Legionella* in the Environment

A reason for obtaining a better understanding of the occurrence and ecology of *Legionella* in the environment is to assist in the development of improved methods for controlling the bacterium in artificial aquatic habitats. Studies have

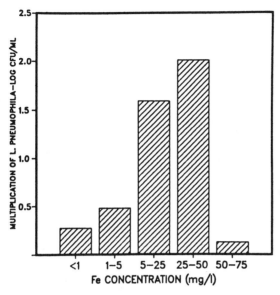

Figure 16.3 *L. pneumophila* multiplication and iron (Fe) concentration in hot-water tank samples.

been conducted to determine the suitability of conventional biocides and the effectiveness of operational decontamination procedures.

Several disinfectants have been tested in the laboratory and in contaminated aquatic systems. However, the results have been mixed. Since chlorine is the most widely used water disinfectant, it has naturally received the most attention.

In the laboratory, Kuchta et al. (1983, 1985), using chlorine residuals typical of those found in public water supplies, found agar-passaged legionellae to be more resistant than coliforms to chlorine. This was true for coliforms maintained on artificial media as well as for a natural population of coliforms in a river water sample. At 21°C, pH 7.6, and 0.1 mg of free chlorine residual per liter, a 99% kill of *L. pneumophila* was achieved within 40 minutes, compared with less than one minute for coliforms. Further testing indicated that more natural, water stock cultures of *L. pneumophila* have even greater resistance than their agar medium-passaged counterpart (Figure 16.4). Additionally, samples from a previously contaminated sink faucet in a hospital having a hyperchlorinator and maintaining a 2 ppm residual for the previous three weeks, yielded legionellae that were still more chlorine resistant. These findings indicate that legionellae can survive for relatively long periods of time at low concentrations of chlorine, and imply that small numbers of legionellae may occasionally survive in waters judged to be microbiologically acceptable. The reason is unclear for the chlorine resistance of *Legionella*, and for the further enhanced resistance of tap water maintained *L. pneumophila* and legionellae

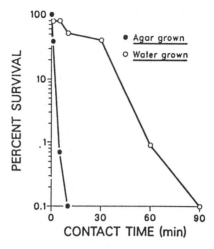

Figure 16.4 Comparison of bactericidal effect of 0.25 mg. free chlorine per liter on agar-passaged and water-grown L. pneumophila in tap water at pH 7.8 and 20°C.

previously exposed to chlorine. Protection may be associated with variations in growth rate, degree of nutrient limitation, or changes in the cellular envelope. It may also result from interactions between legionellae and microbiota in the same habitat that could include becoming sheltered within protozoa. Regardless of the explanation, the data emphasized the need to use natural, non agar-passaged legionellae when evaluating disinfectants.

In field studies of chlorine, a number of investigators have demonstrated the effectiveness of continuous hyperchlorination. At levels reaching 6.0 mg/l free chlorine, *Legionella* contamination in hospital hot water systems has been controlled or eliminated, and the incidence of legionellosis decreased (Fisher-Hoch et al., 1981; Helms et al., 1983; Shands et al., 1985). In whirlpool spas, Witherall et al. (1984) observed that *L. pneumophila* numbers were reduced to below detection limits within the first week of continuous chlorination at 1-3 mg/l free chlorine. In cooling systems, Fliermans et al. (1982) concluded that elevated *L. pneumophila* densities could be reduced and maintained near levels found in the source water by either continuous or shock chlorination. However, Band et al. (1981) found that while continuous chlorination of a contaminated cooling tower at 3 mg/l eliminated *L. pneumophila*, a seven-day time lag occurred between the initiation of treatment and elimination of the organism. The dosage of chlorine needed to control legionellae in artificial aquatic habitats is unclear and may require additional controlled studies (Mead et al, 1988).

The effectiveness of chlorine dioxide for inactivating *L. pneumophila* was examined by Berg et al. (1984, 1985). Their test system involved an agar-passaged *L. pneumophila* strain subsequently maintained in a chemostat containing a liquid growth medium. Under these conditions, both *L. pneumophila* and *E. coli* were inactivated at a comparable rate by 0.75 mg/l chlorine dioxide.

In another experiment, conducted to compare the effectiveness of chlorine dioxide and free chlorine in the disinfection of batch grown legionellae cultures, chlorine dioxide was superior to chlorine when both were applied at equal mass doses.

Edelstein and colleagues (1982) studied ozonation of supply water in an unoccupied hospital building containing water fixtures contaminated with *L. pneumophila*. Although experimental conditions were not optimal, ozone concentrations of 0.79 mg/l appeared to eradicate *L. pneumophila* from the fixtures.

Several studies have dealt with the susceptibility of *Legionella* to ultraviolet radiation. Antopol and Ellner (1979) examined UV disinfection of *L. pneumophila* grown on chocolate agar plates and suspended in distilled water. They found legionellae suspensions to be sensitive to low doses of germicidal UV radiation with 99% kill occurring at a dosage of 1840 $\mu W/s/cm^2$. Gilpin (1984) investigated the UV sensitivity of *Legionella* suspended in phosphate buffer and found dosages required for 99% kill to be in the moderate range compared to those reported for other bacteria. Dutka (1984) tested the sensitivity of agar-passaged *L. pneumophila* to sunlight in water and demonstrated that both UV and visible light play a role in killing the microorganisms. Knudson (1985) exposed lawns of seven *Legionella* species on charcoal yeast extract agar to low doses (240 $\mu W/s/cm^2$) of UV light. He found that while legionellae were sensitive to short exposures, the germicidal effect of the UV light was reduced for all species when subsequently exposed to photoreactivating light. This suggests a need for caution in assessing the effectiveness of UV radiation, particularly when the water will later be exposed to visible light.

Some investigations have compared the effectiveness of more than one disinfectant. Muraca et al. (1987) evaluated the individual efficiency of chlorine (4 to 6 mg/l), heat (50°C to 60°C), ozone (1 to 2 mg/1), and UV light (30,000 $\mu W/s/cm^2$) and showed that all four methods successfully inactivated agar-passaged *L. pneumophila* in a model plumbing system. Similarly, Dominque et al. (1988) compared the effects of ozone, hydrogen peroxide, and free chlorine on agar-maintained *L. pneumophila* and found that ozone was the most potent, with 99% kill occurring during a five minute exposure to 0.10 to 0.30 mg/l of O_3, and hydrogen peroxide the least effective, with a 30 minute exposure to 1,000 mg/l H_2O_2 being required for a 99% reduction.

In addition to controlling *Legionella* sp. with bactericides, mechanical or operational techniques have also been utilized for potable water system disinfection. Increasing the temperature of the hot water plumbing system, either permanently or periodically, has been successfully used to control the bacterium. Having failed to eradicate legionellae from the hot water system in Kingston hospital by chlorination alone, Fisher-Hoch et al. (1981) succeeded by also increasing the ambient hot water temperature from 45°C to 55°C–60°C.

Similarly, heat treatment can be accompanied by mechanical flushing of hot water tanks and plumbing fixtures. Best et al. (1983) designed a schedule for control of *L. pneumophila* and *L. micdadei* in a hospital plumbing system that included intermittent boosting of hot water tank temperature to 60 to

77°C, and two 30-minute fixture flushes (24 hours apart) in patient areas. After 72 hours, the water temperature was returned to less than 54°C. The frequency of heating and flushing was then adjusted according to the results of surveillance culturing.

Still another operational disinfection technique has been proposed for cooling towers based upon a study by States et al. (1987b). In this investigation, water samples were collected over a period of months from the condenser and cooling tower basin of a hospital cooling system. The samples were analyzed for a variety of chemical parameters, inoculated with naturally occurring *L. pneumophila*, and evaluated for their ability to support *Legionella* growth. The results indicated that legionellae growth is inversely correlated with water pH and alkalinity. (Figure 16.5, 16.6) Since the operation of cooling towers at elevated alkalinity and pH had previously been recommended for corrosion control and chlorine persistence (Kemmer and McCallion, 1979; Characklis et al., 1980), these experimental findings suggest that increasing pH and alkalinity, out of the tolerance range of *Legionella* sp., may be a useful approach for controlling this organism. Elevated pH may also be a factor in reduced legionellae numbers in some hospital potable hot water systems (States et al., 1989).

Although disinfectants and operational modifications can control environmental legionellae, there are serious drawbacks to both approaches. High chlorine concentrations are corrosive to plumbing systems and can increase ambient tap water levels of potentially carcinogenic trihalomethanes (Helms et al., 1988). Elevated hot water temperatures pose a scalding risk for patients and staff. Furthermore, continuous disinfection and manual flushing routines are

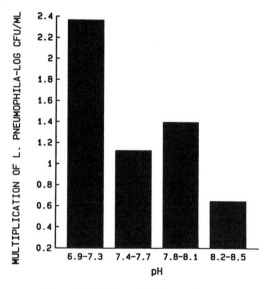

Figure 16.5 *L. pneumophila* multiplication and pH in cooling tower condenser samples.

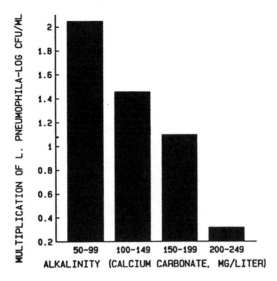

Figure 16.6 *L. pneumophila* multiplication and alkalinity in cooling tower condenser samples.

expensive and require a continual investment of time and attention. While some larger hospitals have adopted these approaches, most smaller institutions, with high risk populations, have not committed the required resources.

A solution to these problems may lie in the development of passive control measures which reduce the risk of plumbing contamination but require little system manipulation or ongoing effort. Such measures primarily involve the redesign of institutional plumbing systems to discourage bacterial amplification and colonization. For example, installation of additional internal heating elements near the bottom of centrally heated storage or heating tanks would increase temperature in the critical amplification zone without elevating overall water temperature. Construction of plumbing systems to avoid stagnation, dead ends, and materials conducive to legionellae multiplication would curtail the problem of recurring growth. Contouring the bottoms of tanks to direct settling solids to a central location and incorporating a continuous slow discharge at that point would eliminate growth-supporting sediment. Utilization of single pass instantaneous steam heating systems in place of traditional steam heated hot water storage tanks would heat water to temperatures exceeding 88°C and then reduce temperature by blending with cool water (Bartlett et al., 1986). Additionally, as suggested by Witherall (1988), extending the cold water inlet pipe to near the bottom of the hot water tank would increase mixing and prevent thermal stratification and accumulation of solids. Several guidance manuals have been published by government agencies and trade associations outlining changes in the design, operation, and maintenance of health care institution building services to reduce contamination by *Legionella* sp. (Public

Works Canada, 1986a, b; Chartered Institution of Building Services Engineers, 1987; Department of Health and Social Security, 1989). Although passive control measures would probably not entirely eliminate legionellae contamination, bacterial levels would be reduced. More definitive active steps could still be taken in the event of serious contamination.

16.6   Amoebae and *Legionella*

Evidence has been presented indicating that amoebae and other protozoa act as natural hosts and amplifiers for legionellae in the environment (Rowbotham, 1980, 1983; Anand et al., 1983; Tyndall and Dominique, 1982; Skinner et al., 1983; Fields et al., 1984; Holden et al., 1984; Barbaree et al., 1986). It has also been suggested that the host relationship affects the virulence of *Legionella* sp. (Rowbotham, 1980; Fields et al., 1986). The interaction in the environment parallels the situation in humans in that legionellae in the body multiply within host macrophages, an ability apparently important in *Legionella*'s pathogenic mechanism (Eisenstein, 1987). In amoebae, it is believed that legionellae are phagocytosed by trophozoites, multiply within vesicles, and are either released when the vesicles and amoebae rupture or, under certain conditions, remain encapsulated when the amoebae encyst (Rowbotham, 1986).

The legionellae-amoebae relationship may be an important factor in the ecology of *Legionella* and the epidemiology of legionellosis. A review of previous studies suggests that an interaction with amoebae could explain many of the observations that have already been made on the behavior of *Legionella* in the laboratory and in potable water. Of particular importance is the potential protection granted by amoebic hosts of *Legionella* to the effects of disinfectants, low pH, and heat. Also interesting is the possible role of amoebic hosts in explaining observed differences between cultures and natural legionellae.

Earlier disinfection experiments showed legionellae to be relatively resistant to inactivation by chlorine (Kuchta et al., 1983, 1985). They also demonstrated enhanced chlorine resistance among *Legionella* populations which had survived exposure to elevated chlorine concentrations in a hospital plumbing system. DeJonckheere and van de Voorde (1976) examined inactivation of amoebic cysts by free chlorine. They found that cysts of *Acanthamoeba* sp., including a strain isolated from tap water, were highly resistant, surviving a free chlorine concentration of 4 mg/l (ppm) for more than three hours of contact time. Additionally, Chang (1978) observed that at a free Cl_2 concentration of 1.4 mg/l, a 25 minute contact time was required to inactivate 99.9% of pathogenic *Naegleria* cysts. Skinner et al. (1983) hypothesized that protection by an amoebic cyst wall may allow legionellae to survive in chlorinated waters. Furthermore, Rowbotham (1986), explaining the failure of periodic dosing with biocide to control *L. pneumophila* in a cooling tower, suggested that this practice may select for chlorine-resistant amoebae infected with legionellae. It is possible that much of the chlorine resistance observed in legionellae can be attributed to the presence of chlorine-resistant protozoa. Further, it may be that chlorination of a hospital plumbing system selects for legionellae protected by cysts of particularly resistant amoebic strains. King et al. (1988) have demonstrated in the laboratory that *Legionella gormanii*, as well as a variety of coliforms and other bacterial pathogens, can survive ingestion by amoebae and ciliate protozoa and consequently tolerate levels of free chlorine that kill free-living bacteria.

Survival and growth of *Legionella* has also been shown to be substantially affected by pH. Laboratory and field studies indicated that naturally occurring *L. pneumophila* multiplies at pH levels of 5.5 to 9.2 (Fliermans et al., 1981; Wadowsky et al., 1985). Work with cooling tower water suggested that elevating pH inhibits legionellae multiplication (States et al., 1987). Additionally, samples being cultured for *Legionella* sp. are often pretreated with a hydrochloric acid-potassium chloride solution, pH 2.2, to eliminate contaminating organisms (Bopp, 1981). Several in vitro studies have shown that the amoeba *Naegleria* sp. tolerates a relatively wide pH range (4.6 to 9.5) (Carter, 1970; Kyle and Noblet, 1985). Sykora et al. (1983) observed 100% survival of *Naegleria fowleri* cysts, in vitro, at pH levels, as low as 2.1, but reduced survival at pH 8.7 and no survival at pH 10.0. These findings indicate that the susceptibility of *L. pneumophila* to high pH levels in laboratory stock cultures, natural waters, and cooling tower waters may be related to the susceptibility of amoebae to elevated pH. They also suggest that the resistance of amoebae to low pH may be a factor in the effectiveness of the low-pH selective technique for isolating *Legionella*.

As in the case of chlorine and pH, the observed resistance of legionellae to higher temperatures may also be associated with a corresponding resistance of amoebae to heat. *Legionella* sp. tolerate elevated temperatures well, as demonstrated by their occurrence in hot water tanks and hot water plumbing systems. Furthermore, preheating samples is another selective procedure commonly used to eliminate interfering microorganisms when culturing legionellae from environmental samples. The method involves heating at 60°C for two minutes, or 50°C for 30 minutes (Dennis et al., 1984). In the case of amoebae, cysts of some pathogenic *Naegleria* strains have been shown to survive 51°C for 145 minutes and 58°C for 45 minutes (Chang, 1978). Similarly, preheating samples at 60°C for 30 minutes has been successfully used for the selective isolation of acanthamoebae from soil (personal communication with T.J. Rowbotham). The sequestering of legionellae within amoebae would be expected to confer some resistance to heat.

Differences have been detected between the characteristics of agar-passaged and naturally occurring legionellae. Kuchta et al. (1985) showed that *L. pneumophila* maintained in tap water is more resistant to chlorine than strains passaged on artificial media. Roberts et al. (1987), studying the effects of selective techniques on legionellae themselves, found that natural water cultures of legionellae are significantly more resistant to the combinations of acid treatment, antibiotics, and heat treatment than are BCYE-passaged cultures. Additionally, Haggerty and Duma (P.G. Haggerty and R.J. Duma, Program Abstr. 24th Intersci. Conf. Antimicro. Agents Chemother., abstr. no. 165, 1984) observed that *L. pneumophila* from natural waters is more resistant to acid (pH 2.2) pretreatment than are laboratory stock cultures. While variations in physiological state may cause the observed differences, a loss of an amoebic host upon culturing on nutrient-rich artificial media serves as another explanation. If this is true, the parallel observations that wild strains of *Pseudomonas aeruginosa* passaged on artificial media lose their ability to multiply in distilled water (Favero et al., 1971) and become less resistant to disinfectants (Carson et al., 1972) suggest the possibility that protozoal hosts may also be involved in the environmental survival of pseudomonads and other bacteria. The findings of King et al. (1988) indicate that the ability to form a protozoa-bacteria host relationship may exist for: *Escherichia coli, Enterobacter* sp., *Klebsiella* sp., *Salmonella typhimurium, Yersinia entercolitica, Shigella sonnei,* and *Campylobacter jejuni,* as well as for *Legionella* spp.

Finally, some additional laboratory and field studies provide further evidence that free-living amoebae are important determinants for survival and multiplication of *Legionella* in drinking water systems. Fields et al. (1989) isolated hartmannellid amoebae from a tap water stock culture used for the laboratory study of *L. pneumophila* multiplication. Subsequent experiments indicated that a growth-supporting factor which could be removed from the culture by filtration with a 1 μm filter, could be reintroduced by inoculating hartmannellid amoebae into the culture (Wadowsky et al., 1988). Since these amoebae are widely present in internal potable water systems, they may contribute significantly to amplification of *L. pneumophila* in this habitat. Concerning public water systems, a survey conducted for the occurrence of amoebae within the City of Pittsburgh supply show amoebae present in areas of slow moving water including the bottom and corners of reservoirs and the surface of treatment plant filters (States et al., 1988). These locations correlate well with sites previously demonstrated to be capable of supporting legionellae growth (States et al., 1987a), suggesting that amoebae could also be an important factor in potential *Legionella* contamination of public drinking water systems. This hypothesis may be consistent with the finding of Colbourne et al. (1988) that heat increases the culturability of legionellae colonizing municipal water systems. While heat shock genes may be responsible for the observation, cyst formation, induced by low temperature, and excystation with release of sequestered legionellae, initiated by heat, provides a possible alternate explanation.

While there are still many questions concerning the extent of the association between amoebae and legionellae, the *Legionella*-amoeba model is consistent with many of the earlier findings. This suggests that the best way to understand the ecology of environmental *Legionella* sp. may be through an improved understanding of the ecology of amoebae. It also implies that the optimal approach to controlling legionellae in potable water systems may be to control protozoa in these habitats.

16.7 Conclusions

Research over the past decade has shown that *Legionella* is a common inhabitant, usually in low numbers, of natural aquatic habitats and is also able to survive and multiply in water treated to meet drinking water standards. Legionellae are capable of multiplying within certain niches in public drinking water treatment and distribution systems, and can be cultured from these sites, in low densities, following heat treatment. *Legionella* sp. within municipal supplies could then enter buildings, sporadically or in low numbers, and seed plumbing and cooling systems. Legionellae proliferate within certain artificial structures that serve as bacterial amplifiers. Following amplification in cooling towers, the organism is presumably able to directly infect individuals, particularly those at higher risk due to immunosuppression, underlying illness, advanced age, heavy smoking, or alcoholism. Following amplification in hot water tanks, *Legionella* sp. are available to colonize other parts of the internal plumbing systems of hospitals, hotels, and even some homes, from which they may also potentially infect susceptible persons.

A number of abiotic factors significantly influence *Legionella* survival and growth, including temperature, pH, and the availability of key metals. *Legionella* sp. are also affected by other microorganisms, with their survival and multiplication being enhanced by associated bacteria, algae, and protozoa. In fact, much of the ecological interaction of *Legionella* with biotic and abiotic components in natural and artificial environments may be explained by relationships with amoebae.

Disinfection of *Legionella* is complicated by a demonstrated resistance to chlorine and has led to investigation of alternate means to control the contamination of artificial aquatic systems.

While a substantial amount of information has been accumulated concerning *Legionella* in the environment, several key issues remain unresolved. For example, an explanation is unavailable for the observation that while some outbreaks of legionellosis can be associated with reservoirs of legionellae in the environment, the mere presence of large numbers of these bacteria does not necessarily result in human illness. Possibilities include host relationships with protozoa (Rowbotham, 1980), species differences within the genus *Legionella* (Bernstein et al., 1986), and subspecific differences within legionellae species (Joly and Winn, 1984; Watkins et al., 1985; Tobin et al., 1986; Brindle et al., 1987; Dourman et al., 1988). Additionally, it is unclear why two different clinicopathological responses, *Legionella* pneumonia and Pontiac fever, apparently result from the same bacterial group.

For those involved in protecting public health and in providing safe drinking water, an important consideration is the extent to which attempts should be made to monitor for and control legionellae in the environment. Effective filtration and disinfection at the treatment plant, maintenance of a chlorine residual throughout the distribution system, covering of open reservoirs, cross-connection and back siphonage control, and disinfection of newly-repaired or constructed components of the water distribution systems are all standard measures utilized to ensure the sanitary quality of drinking water. These measures should also be helpful in discouraging survival and growth of legionellae in public supplies. Other than these techniques, no exceptional steps for the treatment of drinking water have been recommended. The United States Environmental Protection Agency (USEPA) (1985) decided that due to the ubiquity of the organism, its resistance to chlorine, its ability to colonize plumbing systems, and uncertainty concerning the health hazard posed by its mere presence in drinking water, control of *Legionella* is probably more appropriate at locations where susceptible human populations reside, than at the waterworks.

Currently, there is no consensus for measures to be taken at point-of-use locations such as internal plumbing systems, hot water tanks, cooling systems, and cooling towers. Edelstein (1985) concluded that disinfection should occur only when the system is implicated as a source of a legionellosis outbreak, is present in a hospital ward housing especially high risk patients, or is found in a building which has not been used for some time. Barbaree et al. (1987) suggested that since legionellae not related to disease may be found in hospital cooling towers and plumbing systems, an epidemiologic association with the probable source of an outbreak should be established before intervention methods are undertaken. Broome (1984) advised that due to the difficulty in eradicating legionellae and uncertainties associated with altering established treatment protocols, cooling towers should only receive routine maintenance and should not normally be cultured for legionellae except during an outbreak. However, USEPA (1987) has recommended that, based on the high incidence and mortality rate of legionellosis, health care institutions should consider preventive measures for the control of legionellae in their plumbing systems. EPA stated that these measures would be additionally beneficial in controlling other opportunistic pathogens in the system that might also cause nosocomial infections.

Health care institutions may need to favor EPA's recommendation. Since these institutions have the responsibility for providing a safe environment for susceptible patients, would they otherwise be ignoring infection-control guidelines? Furthermore, is it ethically or legally responsible to wait until an outbreak occurs before monitoring and treating environmental sources? This point is supported by a study in which Muder et al. (1983), surveying a hospital in which legionellosis had never been documented, discovered that 14.3% of the nosocomial pneumonias were caused by legionellae and 64% of test sites in the water system were positive for *Legionella* sp.

To the extent that monitoring and control are deemed necessary, better methods are still needed to detect legionellae among large numbers of other environmental microorganisms. Additional passive techniques are also needed to more effectively control proliferation of legionellae, and colonization by supporting microorganisms, in bacterial amplifiers, in particular, and in plumbing and cooling systems in general. A better understanding of the ecology of *Le-*

gionella and supporting microbiota, in both natural and artificial aquatic ecosystems, would enhance all of the detection and control efforts.

This work was supported in part by the City of Pittsburgh Water Department. It was also sponsored by the Environmental Epidemiology Center of the Graduate School of Public Health of the University of Pittsburgh under the support of cooperative agreement CR80681-01-2 with the U.S. Environmental Protection Agency.

References

Abrams, A.P., Garbett, M., Rees, H.B., and Lewis, B.G. 1980. Bacterial aerosols produced from a cooling tower using wastewater effluent as makeup water. *Journal of Water Pollution Control Federation* 52:498–501.

Anand, C.M., Skinner, A.R., Malic, A., and Kurta, J.B. 1983. Interaction of L. *pneumophila* and a free-living amoeba (*Acanthamoeba palestinensis*). *Journal of Hygiene* 91:167–178.

Antopol, S.C. and Ellner, P.D. 1979. Susceptibility of *Legionella pneumophila* to ultraviolet radiation. *Applied and Environmental Microbiology* 38:347–348.

Arnow, P.M., Weil, D., and Para, M.F. 1985. Prevalence and significance of *Legionella pneumophila* contamination of residential hot-tap water systems. *Journal of Infectious Diseases* 152:145–151.

Band, J.D., LaVenture, M., David, J.P., Mallison, G.F., Skaliy, P., Hayes, P.S., Schell, W.L., Weiss, H., Greenberg, D.J., and Fraser, D.W. 1981. Endemic Legionnaires Disease: Airborne transmission down a chimney. *Journal of the American Medical Association* 245:2404–2407.

Barbaree, J.M., Fields, B.S., Feeley, J.C., Gorman, G.W., and Martin, W.T. 1986. Isolation of protozoa from water associated with a legionellosis outbreak and demonstration of intracellular multiplication of *Legionella pneumophila*. *Applied and Environmental Microbiology* 51:422–424.

Barbaree, J.M., Gorman, G.W., Martin, W.T., Fields, B.S., and Morrill, W.E. 1987. Protocol for sampling environmental sites for legionellae. *Applied and Environmental Microbiology* 53:1454–1458.

Bartlett, C.L.R. Macrae, A.D., and MacFarlane, J.T. 1986. *Legionella Infections*, Edward Arnold, Ltd. London, U.K.

Berendt, R.F. 1981. Influence of blue-green algae (cyanobacteria) on survival of *Legionella pneumophila* in aerosols. *Infection and Immunity* 32:690–692.

Berg, J.D., Hoff, J.C., Roberts, P.V., and Matin, A. 1984. Growth of *Legionella pneumophila* in continuous culture and its sensitivity to inactivation by chlorine dioxide. pp. 68–70 in Thornsberry, C., Balows, A., Feeley, J.C., and Jakubowski, W. (editors), *Legionella: Proceedings of the Second International Symposium*, American Society for Microbiology, Washington, DC.

Berg, J.D., Hoff, J.C., Roberts, P.V., and Matin, A. 1985. Disinfection resistance of *Legionella pneumophila* and *Escherichia coli* grown in continuous and batch culture, pp. 603–613 in Jolly, R.L., Bull, R.L., Davis, W.P., Katz, S., Roberts, M.H., and Jacobs, V.A. (editors), *Water Chlorination: Environmental Impact and Health Effects*, Vol. 5. Lewis Publishers.

Best, M., Stout, J., Muder, R.R., Yu, V.L., Goetz, A., and Taylor, F. 1983. Legionellosis in the hospital water supply. *Lancet* ii:307–310.

Bopp, C.A., Sumner, J.W., Morris, G.K., and Wells, J.G. 1981. Isolation of *Legionella* spp.

from environmental water samples of low pH treatment and use of a selective medium. *Journal of Clinical Microbiology* 13:714–719.

Bornstein, N., Veilly, C., Nowicki, M., Paucod, J.C., and Fleurett, J. 1986. Epidemiological evidence of legionellosis transmission through domestic hot water supply systems and possibilities of control. *Israel Journal of Medical Sciences* 22:655–661.

Broome, C.V. 1984. Current issues in epidemiology of legionellosis. pp. 205–209 in Thornsberry, C., Balows, A., Feeley, J.C., and Jakubowski, W. (editors), *Legionella: Proceedings of The Second International Symposium*, American Society for Microbiology, Washington, DC.

Calderon, R.L. and Dufour, A.P. 1984. Media for detection of *Legionella* spp. in environmental water supplies. pp. 290–292 in Thornsberry, C., Balows, A., Feeley, J. and Jakubowski, W. (editors), *Legionella: Proceedings of the Second International Symposium*, American Society for Microbiology, Washington, D.C.

Carson, L.A., Favero, M.S., Bond, W.W., and Peterson, N.J. 1972. Factors affecting comparative resistance of naturally occurring and subcultured *Pseudomonas aeruginosa* to disinfectants. *Applied and Environmental Microbiology* 23:863–869.

Carter, R.F. 1970. Description of a *Naegleria* species isolated from two cases of primary amoebic meningoencephalitis and of the experimental pathological changes induced by it. *Journal of Pathology* 100:217–244.

Chang, S.L. 1978. Resistance of pathogenic *Naegleria* to some common physical and chemical agents. *Applied and Environmental Microbiology* 35:368–375.

Characklis, W.G., Trulear, M.G., Stathopoulos, N., and Chang, L.C. 1980. Oxidation and destruction of microbial films. pp. 349–368 in Jolly, R.L. (editor), *Water Chlorination: Environmental Impact and Health Effects*, Butterworth Publishers, Stoneham, Massachusetts.

Chartered Institution of Building Services Engineers. 1987. *Minimizing the Risk of Legionnaires Disease*, Technical Memorandum (TM) 13, London.

Colbourne, J.S., Pratt, P.J., Smith, M.G., Fisher-Hoch, S.P., and Harper, D. 1984. Water fittings as sources of *Legionella pneumophila* in a hospital plumbing system. *Lancet* i:210–213.

Colbourne, J.S. and Trew, R.M. 1986. Presence of *Legionella* in London's water supplies. *Israel Journal of Medical Sciences* 22:633–639.

Colbourne, J.S., Dennis, P.J., Trew, R.M., Berry, C., and Vesey, G. 1988. *Legionella* and public water supplies. pp. 31–36 in *Proceedings of the International Conference on Water and Wastewater Microbiology*, Newport Beach, California.

Cordes, L.G., Wisenthal, A.M., Gorman, G.W., Phair, J.P., Sommers, H.M., Brown, A., Yu, V.L., Magnussen, M.H., Meyer, R.D., Wolf, J.S., Shands, K.N., and Fraser, D.W. 1981. Isolation of *Legionella pneumophila* from shower heads. *Annals of Internal Medicine* 94:195–197.

Cordes, L.G., Fraser, D.W., Skaliy, P., Perlino, C.A., Elsea, W.R., Mallison, G.F., and Hayes, P.S. 1980. Legionnaires disease outbreak at an Atlanta, Georgia, country club: Evidence for spread from an evaporative condenser. *American Journal of Epidemiology* 111:425–431.

DeJonckheere, J. and van de Voorde, H. 1975. Differences in destruction of cysts of pathogenic and nonpathogenic *Naegleria* and *Acanthamoeba* by chlorine. *Applied and Environmental Microbiology* 31:294–297.

Dennis, P.J., Bartlett, C.L.R., and Wright, A.E. 1984. Comparison of isolation methods for *Legionella* spp. pp. 294–296 in Thornsberry, C., Balows, A., Feeley, J. and Jakubowski, W. (editors), *Legionella: Proceedings of the Second International Symposium*, American Society for Microbiology, Washington, DC.

Dennis, P.J., Green, D., and Jones, B.P. 1984. A note on the temperature tolerance of *Legionella*. *Journal of Applied Bacteriology* 56:349–350.

Department of Health and Social Security and the Welsh Office. 1989. *The Control of*

Legionellae in Health Care Premises: A Code of Practice, Her Majesty's Stationery Office, London.

Desplaces, N., Bure, A., and Dournon, E. 1984. *Legionella* spp. in environmental water samples in Paris. pp. 320–321 in Thornsberry, C., Balows, A., Feeley, J.C. and Jakubowski, W. (editors), *Legionella: Proceedings of the Second International Symposium*, American Society for Microbiology, Washington, D.C.

Dominque, E.L., Tyndall, R.L., Mayberry, W.R., and Pancorbo, O.C. 1988. Effects of three oxidizing biocides on *Legionella pneumophila* serogroup 1. *Applied and Environmental Microbiology* 54:741–747.

Dondero, T.J., Rendtorff, R.C., Mallison, G.F., Weeks, R.M., Levy, J.S., Wong, E.W., and Schaffner, W. 1980. An outbreak of Legionnaires disease associated with a contaminated air conditioning cooling tower. *New England Journal of Medicine* 302:362–370.

Dournan, E. Bibb, W.F., Rajagopalan, P., Desplaces, N., and McKinney, R.M. 1988. Monoclonal antibody reactivity as a virulence marker for *Legionella pneumophila* serogroup 1 strains. *Journal of Infectious Diseases* 157:496–501.

Dutka, B.J. 1984. Sensitivity of *Legionella pneumophila* to sunlight in fresh and marine waters. *Applied and Environmental Microbiology* 48:970–974.

Edelstein, P.H., Whittaker, R.E., Kreiling, R.L., and Howell, C.L. 1982. Efficacy of ozone in eradication of *Legionella pneumophila* from hospital plumbing fixtures. *Applied and Environmental Microbiology* 44:1330–1334.

Edelstein, P.H. 1982. Comparative study of selective media for isolation of *Legionella pneumophila* from potable water. *Journal of Clinical Microbiology* 16:697–699.

Edelstein, P.H. 1985. Environmental aspects of *Legionella*. *American Society for Microbiology News* 51:460–467.

Eisenstein, B.J. 1987. Pathogenic mechanisms of *Legionella pneumophila* and *Escherichia coli*. *American Society for Microbiology News* 53:621–624.

Favero, M.S., Carson, L.A., Bond, W.W., and Peterson, N.J. 1971. *Pseudomonas aeruginosa*: growth in distilled water from hospitals. *Science* 173:836–838.

Feeley, J.C., Gorman, G.W., Weaver, R.E., Mackel, D.C., and Smith, H.W. 1978. Primary isolation media for Legionnaires' disease bacterium. *Journal of Clinical Microbiology* 8:320–325.

Feeley, J.C., Gibson, R.J., and Gorman, G.W. 1979. Charcoal-yeast extract agar: primary isolation medium for *Legionella pneumophila Journal of Clinical Microbiology* 10:437–441.

Fields, B.S., Shotts, E.B., Feeley, J.C., Gorman, G.W., and Martin, W.J. 1984. Proliferation of *Legionella pneumophila* as an intracellular parasite of the ciliated protozoan *Tetrahymena pyriformis*. *Applied and Environmental Microbiology* 47:467–471.

Fields, B.S., Barbaree, J.M., Shotts, E.B., Feeley, J.C., Morrill, W.E., Sanden, G.N., and Dykstra, M.J. 1986. Comparison of guinea pig and protozoan models for determining virulence of *Legionella* species. *Infection and Immunity* 53:553–559.

Fields, B.S., Sanden, G.N., Barbaree, J.M., Morrill, W.E., Wadowsky, R.M., White, E.H., and Feeley, J.C. 1989. Intracellular multiplication of *Legionella pneumophila* in amoebae isolated from hospital hot water tanks. *Current Microbiology* 18:131–137.

Fisher-Hoch, S.P., Tobin, J.O'H., Nelson, A.M., Smith, M.G., Talbot, J.M., Bartlett, C.L.R., Gillet, M.B., Pritchard, J.E., Swann, R.A., and Thomas, J.A. 1981. Investigation and control of an outbreak of Legionnaires' Disease in a district general hospital. *Lancet* i:932–936.

Fleming, D.W., and Reingold, A.L. 1986. *Legionella*. pp. 352–358 in Braude, A.I., Davis, C.E. and Fierer, J. (editors), Infectious Diseases and Medical Microbiology, W.B. Saunders Company, Philadelphia.

Fliermans, C.B., Cherry, W.B., Orrison, L.H., and Thacker, L. 1979. Isolation of *Legionella pneumophila* from nonepidemic related aquatic habitats. *Applied and Environmental Microbiology* 37:1239–1242.

Fliermans, C.B., Cherry, W.B., Orrison, L.H., Smith, S.J., Tison, D.L., and Pope, D.H.

1981. Ecological distribution of *Legionella pneumophila*. *Applied and Environmental Microbiology* 41:9–16.

Fliermans, C.B., Bettinger, G.E., and Fynsk, A.W. 1982. Treatment of cooling systems containing high levels of *Legionella pneumophila*. *Water Research* 16:903–909.

Fliermans, C.B. 1984. Philosophical ecology: *Legionella* in historical perspective. pp. 285–289 in Thornsberry, C., Balows, A., Feeley, J.C. and Jakubowski, W. (editors), *Legionella: Proceedings of the Second International Symposium*, American Society for Microbiology, Washington, DC.

Foy, H.M., Hayes, P.S., Cooney, M.K., Broome, C.V., Allan, I., and Tobe, R. 1979. Legionnaires' disease in a prepaid medical-care group in Seattle, 1963–75. *Lancet* i: 767–770.

Fraser, D.W., Tsai, T.F., Orenstein, W., Parkin, W.E., Beecham, H.J., Sharrar, R.G., Harris, J., Mallison, G.F., Martin, S.M., McDade, J.E., Shepard, C.C., Brachman, P.S., and The Field Investigation Team. 1977. Legionnaire's disease: A description of an epidemic of pneumonia. *New England Journal of Medicine* 297:1189–1197.

Gilpin, R.W. 1984. Laboratory and field applications of U.V. light disinfection on six species of *Legionella* and other bacteria in water. pp. 337–339 in Thornsberry, C., Balows, A., Feeley, J.C., and Jakubowski, W. (editors), *Legionella: Proceedings of the Second International Symposium*. American Society for Microbiology, Washington, DC.

Glick, T.H., Gregg, M.B., Berman, B., Mallison, G.W., Rhodes, W.W., and Kassanoff. 1978. Pontiac fever: An epidemic of unknown etiology in a Health Department. I. clinical and epidemiological aspects. *American Journal of Epidemiology* 107:149–160.

Gorman, G.W., Feeley, J.C., Steigerwalt, A., Edelstein, P.H., Moss, C.W., and Brenner, D.J. 1985. *Legionella anisa* : A new species of *Legionella* isolated from potable waters and a cooling tower. *Applied and Environmental Microbiology* 49:305–309.

Helms, C.M., Massonari, R.M., Zeitler, R., Streed, S., Gilchrist, M.J.R., Hall N., Hausler, W.J., Sywassin, J., Johnson, W., Wintermeyer, L., and Hierholzer, W.V. 1983. Legionnaires' disease associated with a hospital water system: a cluster of 24 nosocomial cases. *Annals of Internal Medicine* 99:172–178.

Helms, C.M., Massanari, M., Wenzel, R.P., Pfaller, M.A., Moyer, N.P., and Hall, M.T. 1988. Legionnaires' disease associated with a hospital water system: A five-year progress report on continuous hyperchlorination. *Journal of the American Medical Association* 259:2423–2427.

Holden, E.P., Winkler, H.H., Wood, D.O., and Leinbach, E.D. 1984. Intracellular growth of *Legionella pneumophila* within *Acanthamoeba castellanii* Neff. *Infection and Immunity* 45:18–24.

Hsu, S.C., Martin, R., and Wentworth, B.B. 1984. Isolation of *Legionella* species from drinking water. *Applied and Environmental Microbiology* 48:830–832.

Joly, J.R. and Winn, W.C. 1984. Correlation of sub-types of *Legionella pneumophila* defined by monoclonal antibodies with epidemiological classification of cases and environmental sources. *Journal of Infectious Diseases* 150:667–671.

Joly, J.R., Dewaily, E., Bernard, L., Ramsey, D., and Brisson, J. 1985. *Legionella* and domestic water heaters in Quebec City area. *Canadian Medical Association Journal* 132:160.

Kemmer, F.N. and McCallion J. (editors). 1979. *The Nalco Water Handbook*, McGraw-Hill Book Co., New York.

King, C.H., Shotts, E.B., Wooley, R.E., and Porter, K.G. 1988. Survival of coliforms and bacterial pathogens within protozoa during chlorination. *Applied and Environmental Microbiology* 54:3023–3033.

Knudson, G.B. 1985. Photoreactivation of U.V. irradiated *Legionella pneumophila* and other *Legionella* species. *Applied and Environmental Microbiology* 49:975–980.

Kuchta, J.M., States, S.J., McNamara, A.M., Wadowsky, R.M., and Yee, R.B. 1983. Susceptibility of *Legionella pneumophila* to chlorine in tap water. *Applied and Environmental Microbiology* 46:1134–1139.

Kuuchta, J.M., States, S.J., McGaughlin, J.E., Overmeyer, J.H., Wadowsky, R.M., Mc-Namara, A.M., Wolford, R.S., and Yee, R.B. 1985. Enhanced chlorine resistance of tap water adapted *Legionella pneumophila* as compared with agar-medium passaged strains. *Applied and Environmental Microbiology* 50:21–26.

Kurtz, J.B., Bartlett, C.J.R., Newton, V.A., White, R.A., and Jones, N.L. 1982. *Legionella pneumophila* in cooling water systems: Report on a survey of cooling towers in London and a pilot trial of selected biocides. *Journal of Hygiene* 88:369–381.

Kyle, D.E. and Noblett, G.P. 1985. Vertical distribution of potentially pathogenic free-living amoebae in freshwater lakes. *Journal of Protozoology* 32:99–105.

Lee. T.C., Stout, J.E., and Yu, V.L. 1988. Factors predisposing to *Legionella pneumophila* colonization in residential water systems. *Archives of Environmental Health* 43:59–62.

McDade, J.E., Shepard, C.C., Fraser, D.W., Tsai, T.F., Redus, M.A., Dowdle, W.R., and The Laboratory Investigation Team. 1977. Legionnaire's disease: Isolation of a bacterium and demonstration of its role in other respiratory disease. *New England Journal of Medicine* 297:1197–1205.

McDade, J.E., Brenner, D.J., and Bozeman, F.M. 1979. Legionnaires disease bacterium isolated in 1947. *Annals of Internal Medicine* 90:659–661.

Mead, P.B., Lawson, J.M., and Patterson, J.W. 1988. Chlorination of water supplies to control *Legionella* may corrode the pipes. *Journal of the American Medical Association* 260:2216.

Moss, C.W., Weaver, R.E., Dees, S.B., and Cherry, W. 1977. Cellular fatty acid composition of isolates from Legionnaires disease. *Journal of Clinical Microbiology* 6:140–143.

Muder, R.R., Yu, V.L., McClure, J.K., Kroboth, F.J., Kominos, S.D., and Lumish, R.M. 1983. Nosocomial legionnaires' disease uncovered in a prospective pneumonia study. *Journal of the American Medical Association* 249:3184–3188.

Muraca, P., Stout, J.E., and Yu, V.L. 1987. Comparative assessment of chlorine, heat, ozone and U.V. light for killing *Legionella pneumophila* within a model plumbing system. *Applied and Environmental Microbiology* 53:447–453.

Ormsbee, R.A., Peacock, M.G., Bickel, W.D., and Fiset, P. 1981. A comparison of some biologic characteristics of isolates of the Legionnaire's disease bacterium. *Annals of Clinical Laboratory Science* 11:53–62.

Pasculle, A.W., Feeley, J.C. Gibson, R.J., Cordes, L.G., Myerowitz, R.L., Patton, C.M., Gorman, G.W., Carmack, C.L., Ezzell, J.W., and Dowling, J.N. 1980. Pittsburgh pneumonia agent: Direct isolation from human lung tissue. *Journal of Infectious Disease* 141:727–732.

Politi, B.D., Fraser, D.W., Mallison, G.F., Mohatt, J.V., Morris, G.K., Patton, C.M., Feeley, J.C., Telle, R.D., and Bennett, J.V. 1979. A major focus of Legionnaire's disease in Bloomington, Indiana. *Annals of Internal Medicine* 90:587–591.

Pope, D.H., Soracco, R.J., Gill, H.K., and Fliermans, C.B. 1982. Growth of *Legionella pneumophila* in two-membered cultures with green algae and cyanobacteria. *Current Microbiology* 7:319–322.

Public Works Canada. 1986a. *Legionella Control in Hospitals—Standards and Guidelines (Supplement)*, MD15161. Engineering Technology Directorate, Canada.

Public Works Canada. 1986b. *Legionella Control in Hospitals—Standards and Guidelines*, MD15161S. Engineering Technology Directorate, Canada.

Reeves, M.W., Pine, L., Hutner, S.H., George, J.R., and Harrell, W.K. 1981. Metal requirements of *Legionella pneumophila*. *Journal of Clinical Microbiology* 13:688–695.

Ristroph, J.D., Hedlund, K.W., and Gowda, S. 1981. Chemically defined medium for *Legionella pneumophila* growth. *Journal of Clinical Microbiology* 13:115–119.

Roberts, K.P., August, C.M., and Nelson, J.D. 1987. Relative sensitivities of environmental legionellae to selective isolation procedures. *Applied and Environmental Microbiology* 12:2704–2707.

Rowbotham, T.J. 1980. Preliminary report on the pathogenicity of *Legionella pneumophila* for freshwater and soil amoebae. *Journal of Clinical Pathology* 33:1179–1183.

Rowbotham, T.J. 1983. Isolation of *Legionella pneumophila* from clinical specimens via amoebae and the interaction of those and other isolates with amoebae. *Journal of clinical Pathology* 36:978–986.

Rowbotham, T.J. 1986. Current views on the relationship between amoeba, Legionella, and Man. *Israel Journal of Medical Sciences* 22:679–689.

Schofield, G.M. and Locci, R. 1985. Colonization of components of a model hot water system by *Legionella pneumophila*. *Journal of Applied Bacteriology* 58:151–162.

Shands, K.N., Ho, J.L., Meyer, R.D., Gorman, G.W., Edelstein, P.H., Mallison, G.F., Finegold, S.M., and Fraser, D.W. 1985. Potable water as a source of Legionnaires' disease. *Journal of the American Medical Association* 253:1412–1416.

Skinner, A.R., Anand, C.M., Malic, A., and Kurtz, J.B. 1983. Acanthamoebae and environmental spread of *Legionella pneumophila*. *Lancet* ii:289–290.

States, S.J., Conley, L.F., Ceraso, M. Stephenson, T.E., Wolford, R.S., Wadowsky, R.M., McNamara, A.M., and Yee, R.B. 1985. Effects of metals on *Legionella pneumophila* growth in drinking water plumbing systems. *Applied and Environmental Microbiology* 50:1149–1154.

States, S.J., Conley, L.F., Kuchta, J.M., Oleck, B.M., Lipovich, M.J., Wolford, R.S., Wadowsky, R.M., McNamara, A.M., Sykora, J.L., Keleti, G., and Yee, R.B. 1987a. Survival and multiplication of *Legionella pneumophila* in municipal drinking water systems. *Applied and Environmental Microbiology* 53:979–986.

States, S.J., Conley, L.F., Towner, S.G., Wolford, R.S., Stephenson, T.E., McNamara, A.M., Wadowsky, R.M., and Yee, R.B. 1987b. An alkaline approach to treating cooling towers for control of *Legionella pneumophila*. *Applied and Environmental Microbiology* 53:1775–1779.

States, S.J., Conley, L.F., Knezivich, C.R., Keleti, G., Sykora, J.L., Wadowsky, R.M., and Yee, R.B. 1988. Free-living amoebae in public water supplies: Implications for *Legionella, Giardia,* and *Cryptosporidia*, in *Proceedings of the American Water Works Association Water Quality Technology Conference*, St. Louis, Missouri.

States, S.J., Conley, L.F., Kuchta, J.M., Wolford, R.S., Wadowsky, R.M. and Yee, R.B. 1989. Chlorine, pH and control of *Legionella* in hospital plumbing systems. *Journal of the American Medical Association* 261:1882–1883.

Stout, J., Yu, V.L., Vickers, R.M., Zuravleff, J., Best, M., Brown, A., Yee, R.B., and Wadowsky, R. 1982. Ubiquitousness of *Legionella pneumophila* in the water supply of a hospital with endemic Legionnaires' Disease. New England Journal of Medicine 306: 466–468

Stout, J.E., Yu, V.L., and Best, M.G. 1985. Ecology of *Legionella pneumophila* within water distribution systems. *Applied and Environmental Microbiology* 49:221–228.

Stout, J.E., Yu, V.L., and Muraca, P. 1987. Legionnaires' disease acquired within the homes of two patients. *Journal of the American Medical Society* 257:1215–1217.

Sykora, J.L., Keleti, G., and Martinez, J. 1985. Occurrence and pathogenicity of *Naegleria fowleri* in artificially heated waters. *Applied and Environmental Microbiology* 45:974–979.

Tesh, M.J. and Miller, R.D. 1981. Amino acid requirements for *Legionella pneumophila* growth. *Journal of Clinical Microbiology* 13:865–869.

Tesh, M.J. and Miller, R.D. 1982. Growth of *Legionella pneumophila* in defined media: Requirement for magnesium and potassium. *Canadian Journal of Microbiology* 28: 1055–1058.

Tison, D.L., Pope, D.H. Cherry, W.B., and Fliermans, C.B. 1980. Growth of *Legionella pneumophila* in association with blue-green algae (cyanobacteria). *Applied and Environmental Microbiology* 39:456–459.

Tison, D.L. and Seidler, R.J. 1983. *Legionella* incidence and density in potable drinking water supplies. *Applied and Environmental Microbiology* 45:337–339.

Tobin, J.O'H., Beare, J., Dunnill, M.S., Fisher-Hoch, S., French, M., Mitchell, R.G.,

Morris, P.J., and Muers, M.F. 1980. Legionnaires' disease in a transplant unit: Isolation of the causative agent from shower baths. *Lancet* ii:118–121.

Tobin, J.O'H., Swann, R.A., and Bartlett, C.L.R. 1981a. Isolation of *Legionella pneumophila* from water systems: Methods and preliminary results. *British Medical Journal* 282: 515–517.

Tobin, J.O'H., Bartlett, C.L.R., and Waitkins, S.A. 1981b. Legionnaires' disease: Further evidence to implicate water storage and distribution systems as sources. *British Medical Journal* 282:573.

Tobin, J.O'H., Watkins, I.D., Woodhead, S., and Mitchell, R.G. 1986. Epidemiological studies using monoclonal antibodies to *Legionella pneumophila*. *Israel Journal of Medical Sciences* 22:711–714.

Tyndall, R.L. and Domingue, E.L. 1982. Cocultivation of *Legionella pneumophila* and free-living amoebae. *Applied and Environmental Microbiology* 44:954–959.

U.S. Environmental Protection Agency. 1985. National primary drinking water regulations; volatile synthetic organic chemicals; final rule and proposed rule. *Federal Register* 50:46880–47025.

U.S. Environmental Protection Agency. 1987. Control of *Legionella* in plumbing systems. In *Health Advisories for Legionella and Seven Inorganics*, Office of Drinking Water, Washington, DC.

Vickers, R.M., Brown, A., and Garrity, C.L.R. 1981. Dye-containing buffered charcoal-yeast extract medium for differentiation of the family *Legionellaceae*. *Journal of Clinical Microbiology* 13:380–382.

Wadowsky, R.M. and Yee, R.B. 1981. Glycine-containing selective medium for isolation of Legionellaceae from environmental specimens. *Applied and Environmental Microbiology* 42:768–772.

Wadowsky, R.M., Yee, R.B., Mezmar, L., Wing, E.J., and Dowling, J.N. 1982. Hot water systems as sources of *Legionella pneumophila* in hospital and nonhospital plumbing fixtures. *Applied and Environmental Microbiology* 43:1104–1110.

Wadowsky, R.M. and Yee, R.B. 1983. Satellite growth of *Legionella pneumophila* with an environmental isolate of *Flavobacterium breve*. *Applied and Environmental Microbiology* 46:1447–1449.

Wadowsky, R.M. and Yee, R.B. 1985a. Effect of non-Legionellaceae bacteria on the multiplication of *Legionella pneumophila* in potable water. *Applied and Environmental Microbiology* 49:1206–1210.

Wadowsky, R.M., Wolford, R., McNamara, A.M., and Yee, R.B. 1985b. Effect of temperature, pH, and oxygen level on the multiplication of naturally occurring *Legionella pneumophila* in potable water. *Applied and Environmental Microbiology* 49:1197–1205.

Wadowsky, R.M., Butler, L.J., Cook, M.K., Verma, S.M., Paul, M.A., Fields, B.S., Keleti, G., Sykora, J.L., and Yee, R.B. 1988. growth-supporting activity of *Legionella pneumophila* in tap water cultures and implication of harmannellid amoebae as growth factors. *Applied and Environmental Microbiology* 54:2677–2682.

Watkins, I.D., Tobin, J.O'H., Dennis, P.J., Brown, W., Newnham, R., and Kurtz, J.B. 1985. *Legionella pneumophila* serogroup 1 subgrouping by monoclonal antibodies. *Journal of Hygiene* 95:211–216.

Witherall, L.E., Orciari, L.A., Spitalny, K.C., Pelletier, R.A., Stone, K.M., and Vogt, R.L. 1984. Disinfection of *Legionella pneumophila*—contaminated whirlpool spas. pp. 339–340 in Thornsberry, C., Balows, A., Feeley, J.C., and Jakubowski, W. (editors), *Legionella: Proceedings of the Second International Symposium*, American Society for Microbiology, Washington, D.C.

Witherell, L.E., Duncan,R.W., Stone, K.M., Stralton, L.J., Orciari, L., Kappel, S., and Jillson, D.A. 1988. Investigation of *Legionella pneumophila* ion drinking water. *American Water Works Association Journal* 80:87–93.

Yee, R.B. and Wadowsky, R.M. 1982. Multiplication of *Legionella pneumophila* in unsterilized tap water. *Applied and Environmental Microbiology* 43:1330–1334.

17

Injury of Enteropathogenic Bacteria in Drinking Water

Ajaib Singh and Gordon A. McFeters

Department of Microbiology, Montana State University, Bozeman, Montana

Outbreaks of gastroenteritis due to contaminated drinking water are frequently documented occurrences. A variety of etiological agents have been implicated in this disease. However, the causative agent varies with the geographical area, climate, general level of sanitation, endemic persistance of the disease, as well as cultural and socioeconomic characteristics of the population. In less developed countries, the prevalence of one of a combination of these factors is, quite often, associated with frequent recurrences of gastroenteritis caused by a typical enteropathogen. In the United States, contaminated ground water is the major source of waterborne disease outbreaks (Craun, 1984). Contamination generally occurs in local areas by seepage of sewage into aquifers and improperly developed wells or is due to inadequately treated drinking water. It can also occur after the entry of sewage-contaminated surface water into poorly protected wells.

Potability of drinking water is assessed by enumerating coliform bacteria. However, the effectiveness of a monitoring program is largely dependent on the efficiency of the procedure employed. It is well documented that a sizable portion of the bacterial population in drinking water and sewage may be injured as a result of various stress factors (LeChevallier and McFeters, 1984). The injured cells are unable to grow in the presence of selective agents incorporated in the media commonly employed for their detection and enumeration. This can lead to a significant underestimation of the organisms and result in an inaccurate evaluation of the potential health risks (McFeters et al., 1986). It is now recognized that an efficient procedure for monitoring the potability of drinking water should also allow the enumeration of injured coliform bacteria (*Standard Methods for the Examination of Water and Waste Water*, Section 920. APHA, 1985).

Pathogenic bacteria, like index organisms, can also be injured following exposure to aquatic stress factors, resulting in sublethal cellular lesions and alterations in their physiology. Until recently, the question of whether or not the injured pathogens retain their ability to cause disease had not been addressed. The presently available data suggest that there is a temporary reduction or loss of virulence in pathogenic bacteria after injury and under suitable conditions virulence is fully restored following recovery from injury (Singh et al., 1985; Singh and McFeters, 1986; Singh et al., 1986; Singh and McFeters, 1987; Walsh and Bissonnette, 1987). The consequences of injury and repair processes in enteric pathogenic bacteria commonly associated with drinking water are not fully understood but it is the purpose of this chapter to discuss some of the known events following injury in aquatic environments resembling potable water. Furthermore, the revival from injury, their subsequent growth, and the virulence of these organisms will be discussed.

17.1   Occurrence of Injured Enteropathogens in Drinking Water

Enteropathogens are not well adapted to survive, grow, or multiply in drinking water because of limited nutrients and exposure to harmful physical and chemical agents present in the environment. However, some enteric pathogens, including *Salmonella typhimurium*, *Shigella sonnei*, and *Yersinia enterocolitica*, have been reported to persist within fresh water protozoa and escape inactivation or injury in water treated with high doses (2 to 10 mg/l) of chlorine (King et al., 1988). Some of the factors that can cause stress or injury to free-living microorganisms have been described by McFeters and Camper (1983) and LeChevallier and McFeters (1985). In spite of unfavorable conditions, pathogens can remain viable in water and cause various gastric illnesses after being ingested. The survival time of some waterborne enteropathogens in different aquatic environments and the infectious doses in humans are listed in Table 17.1. McFeters et al. (1974) reported longer survival of some enteric pathogens than the coliforms in well water and emphasized that under some circumstances the absence of indicator organisms may be insignificant when considering the risk of the waterborne acquisition of diseases. Further, Rhodes and Kator (1988) found that *E. coli* die-off and sublethal stress in filtered estuarine water were inversely related to temperature. However, low temperatures ($<10\,^{\circ}$C) favored the survival of *Salmonella* sp. which exhibited less stress and die-off than the *E. coli* cultures. Epidemiological studies by Cabelli et al. (1982) revealed the association of gastrointestinal disorders with recreational waters that were in compliance with the microbiological guidelines of the United States Environmental Protection Agency. Enteropathogens were also found in outbreaks of waterborne illnesses where coliforms were either undetected or detected only in low numbers (Seligmann and Reitler, 1965; Boring et al., 1971). Detection of such pathogens could be far less efficient when a portion of the cell population is injured and the infective dose is low, as seen in *Shigella* (Table 17.1). It is believed that the causative agent in a large number of waterborne outbreaks remains undetected. According to one estimate, the etiological agent was determined in only 56% of waterborne outbreaks from 1920 to 1980 (Craun,

Table 17.1 Waterborne enteropathogens

Pathogens	Survival time in aquatic environment	Reference	Infectious dose (number of viable cells)	Reference
Vibrio cholerae	7.2 h[a]	McFeters et al., 1974	10^6–10^{11}	Cash et al., 1974
Shigella dysenteriae	22.4 h[a]	McFeters et al., 1974	10^1–10^2	Levine et al., 1973
Enterotoxigenic *Escherichia coli*	13.0 h[b]	Grimes and Colwell, 1986	10^8–10^{10}	Levine et al., 1977
Salmonella typhimurium	16.0 h[a]	McFeters et al., 1974	$<10^3$	Blaser et al., 1982
S. typhi	16.0 h[a]	McFeters et al., 1974	10^5	Keswick, 1984
Yersinia enterocolitica	540 days	Highsmith et al., 1977	3.5×10^9	Morris and Feeley, 1976
Campylobacter jejuni	3 days[c]	Rollins and Colwell, 1986	5.0×10^2	Robinson, 1981
	10 days[d]	Rollins and Colwell, 1986	8×10^2–2×10^9	Black et al., 1988

[a] The time required for a 50% reduction in the initial population in diffusion chambers suspended in a stream.
[b] The time cells became nonculturable in a diffusion chamber suspended in a sea water microcosm.
[c] The time cells became nonculturable in an inoculated stream-water microcosm with constant shaking.
[d] The time cells became nonculturable in an inoculated stream-water microcosm without shaking.

1986). The increase in waterborne gastroenteritis of unknown etiology during recent years could be due to as yet unrecognized agents (Craun, 1986) or inadequate methodology employed for the detection and isolation of environmentally stressed pathogens.

There is a wide gap in the information on the effect of various physical and chemical agents on waterborne enteropathogens. Studies in our laboratory have shown that enteropathogens, when placed in water containing sublethal concentrations of either of two important stressors, chlorine or copper (McFeters and Camper, 1983; Domek et al., 1984), caused irreversible (lethal) lesions in some of the exposed cells and reversible lesions (injury) in a large proportion of the remaining cell population (Table 17.2). It is obvious from these data that the susceptibility of different organisms to chlorine and copper varies over a wide range. The sensitivity of an organism to a stressor is dependent on several factors, including the cell concentration, contact time, and temperature, but under similar experimental conditions *Y. enterocolitica*, *S. typhimurium*, and *Shigella* sp. were reported to be more resistant to the action of chlorine than the coliform bacteria (Table 17.2). Higher concentrations of chlorine (0.9 to 1.5

Table 17.2 Injury or lethality of pathogens and coliform bacteria exposed to sublethal concentrations of copper or chlorine

Organism	Number cells/ml	Chlorine (0.25–1.60 mg/l) Lethality (%)[a]	Injury (%)[b]	Copper (0.6–1.0 mg/l) Lethality (%)	Injury (%)
Enterotoxigenic *Escherichia coli* (4 strains)[c]	5×10^8	48–85	87–94	63–83	84–95
Y. enterocolitica (1 serotype 0:8)[d]	5×10^8	52	91	30	89
S. typhimurium (7 strains)[e]	1×10^6	30–50	>90	NK[f]	NK
S. dysenteriae (1 strain)[e]	1×10^6	<50	>90	NK	NK
Coliform bacteria (11 strains)[e]	1×10^6	<50	>90	NK	NK
C. jejuni (3 strains)[g]	7×10^4	98	96	NK	NK

[a] % lethality $= \dfrac{\text{CFU at time } T^0 - \text{CFU at time } T^1}{\text{CFU nonselective medium}} \times 100$

[b] % Injury $= \dfrac{\text{CFU nonselective medium} - \text{CFU selective medium}}{\text{CFU nonselective medium}} \times 100$

[c] Singh et al., 1986.
[d] Singh and McFeters, 1987.
[e] LeChevallier et al., 1985.
[f] Not known.
[g] Blaser et al., 1986; Chlorine concentration: 0.1 mg per liter.

mg of chlorine per liter) were required to induce >90% injury in these cultures as compared with the levels (0.25 to 0.5 mg of chlorine per liter) responsible for the same degree of injury in coliform bacteria. These data and the results reported by others (King et al., 1988) suggest that some pathogens may remain unaffected under conditions that are detrimental to coliform bacteria. Other enteropathogens, like *Campylobacter jejuni* (Blaser et al., 1986) and enterotoxigenic *E. coli* (ETEC) (Singh et al., 1986), were inactivated at lower doses of chlorine and caused substantial injury in the remaining cell population.

17.2 Effect of Injury on Enteropathogenic Bacteria

Cellular Damage   Stevenson (1978) showed that bacteria decrease in size and activity because of aquatic stresses. However, normal development ensues once a particular stress has been removed. Other investigators (McFeters and Camper, 1983; Domek et al., 1987) have shown that sublethal concentrations of chlorine and copper caused multiple cellular lesions in indicator bacteria, and the cytoplasmic membrane was suggested to be one of the common targets of these stressors.

Information on the damaged cellular components and altered physiological activities of stressed pathogens is limited. Using transmission electron and epifluorescence microscopy, Baker et al. (1983) reported a decrease in volume by 85% for *Vibrio cholerae* cells placed in a nutrient-free aquatic microcosm. These cells developed into coccoid forms with remnant cell walls but restored their normal size within 2 h after nutrient supplementation. *V. cholerae* placed in a low-nutrient microcosm showed a rapid decline in total lipids, carbohydrates, and depletion of poly-β-hydroxybutyrate in 7 days (Hood et al., 1986). However, an extended period of starvation (30 days) caused a slow but constant decline in cellular DNA and protein. Walsh and Bissonnette (1983) reported that sublethal concentrations of chlorine caused damage to the surface adhesins of ETEC cells and reduced their ability to attach to human leukocytes in an *in vitro* assay system. LeChevallier et al. (1985) suggested fimbrial damage of chlorine-exposed ETEC and *S. typhimurium*. It was also reported that chlorine altered surface structures of *Y. enterocolitica* cells that may be essential for imparting invasive properties to this organism. Alteration(s) in cell surface sugar components of *Y. enterocolitica* following exposure to copper has also been suggested by Singh et al. (1985). The exposed cells showed increased binding to a mannose-N-acetylglucosamine-specific lectin as compared with the unexposed population.

Altered Physiology   The injured coliforms showed an immediate reduction in intracellular concentration of ATP, aerobic respiration, glucose transport and utilization, and the accumulation of lactate, ethanol, and acetate, as well as alteration of other activities related to the cytoplasmic membrane (McFeters and Camper, 1983: Domek et al., 1987). Singh and McFeters (1986) examined the effect of copper injury on enterotoxin production of ETEC strains and reported that heat-stable enterotoxin production in the injured cultures was delayed. However, permanent impairment of this property was not observed. These results were substantiated by similar observations of Walsh and Bissonnette (1987), who reported reduced heat-labile enterotoxin (LT) production by chlorine-injured ETEC cultures immediately after injury, during resuscitation, and at least 1.5 h following repair of injury. However, LT levels were comparable in both uninjured and chlorine-injured cultures after 24 h of incubation, indicating a temporary loss of this activity. Other investigators (Palmer et al., 1984) reported both biochemical and genetic changes in an ETEC strain exposed to natural water containing sublethal concentrations of toxic chemicals. After 19 days, alterations in cellular membrane proteins were observed and the enterotoxin-related plasmids could not be detected. However, the cells remained viable and returned to normal when transferred to a nutrient-rich medium devoid of chemical toxins. Several other enteropathogens like *S. enteritidis* (Roszak et al., 1984), ETEC, *V. cholerae*, *Salmonella* and *Shigella* spp. (Colwell et al., 1985), and *C. jejuni* (Rollins and Colwell, 1986), became dormant after prolonged exposure to simulated natural aquatic environments, and did not grow when transferred to laboratory culture media. This condition, termed as "viable but nonculturable", has been described as a mechanism of bacterial survival in the natural environment (Roszak and Colwell, 1987). The relationship between nonculturable and injured bacteria is currently not known. However, in both cases organisms are undetectable by generally accepted cultural methodologies and thus may be hazardous to public health.

Effect on Virulence   Alteration in the virulence characteristics of pathogenic bacteria can occur in response to various physical and chemical factors in water. Although most of these agents have been reported to be injurious to bacteria, the possibility of enhanced pathogenicity cannot be ruled out because of chromosomal or plasmid-borne gene mutation by various physical and chemical agents. Such a phenomenon could be of enormous public health significance. However, the enhanced virulence of enteropathogens resulting from exposure to factors in drinking water has not been documented.

Results of a preliminary study in this laboratory in 1974 (Thomson and McFeters, unpublished results) revealed that a strain of ETEC, when suspended in chlorinated water for four days, lost its virulence in orally inoculated infant mice. Whether the loss was due to inability of the culture to produce toxin or its failure to colonize the mouse gut was not determined in that study. Colonization of the gut is regarded to be a prerequisite in the pathogenesis of ETEC and the role of adhesins on cell surfaces in the colonization of the human intestine has been elucidated by Evans et al. (1978). In chlorine-exposed ETEC cultures, damaged cellular adhesins and loss of ability of the cells to adhere to human leucocytes in an in vitro test were reported by Walsh and Bissonnette (1983). They did not determine the virulence of those cultures using in vivo systems. The lost ability of chlorine-exposed *Y. enterocolitica* to invade HeLa cells in an in vitro system (LeChevallier et al., 1985) and intestinal epithelial cells in an in vivo mouse model (LeChevallier et al., 1987) was related to the reduced virulence of this pathogen in intraperitoneally inoculated mice. Singh et al. (1985) further suggested that the liver plays an important role in the nonimmune rapid clearance of copper-injured cells of *Y. enterocolitica* in mice, and thus may be responsible for the reduced virulence in parenterally inoculated animals.

Under nutritional limitations and/or in the presence of stressors, bacteria do not express plasmids that encode for virulence (Palmer et al., 1984) and may exhibit a temporary loss of pathogenicity if stressed cultures are tested before the genetic and biochemical lesions are repaired. It is, therefore, important to exercise caution in interpreting and correlating the results of pathogenicity assessment made in different test systems.

17.3 Revival of Injured Pathogens and Their Virulence

The importance of injury and repair in bacteria causing food spoilage and foodborne disease is well recognized. A simple resuscitation step is usually incorporated into commonly recommended isolation and detection procedures for the repair of injured bacteria. In general, injured organisms can recover upon being held in a nonselective medium at a favorable temperature and pH. The time required for complete repair varies considerably with the environmental conditions and the type and intensity of the cellular damage (Mossel and Van Netten, 1984).

An increase in the number of colony-forming units (CFUs) on a selective medium without a substantial increase in CFUs on nonselective medium is regarded as an indication of the repair process. It was recognized by Hurst and Hughes (1981) that increased tolerance for a selective medium did not coincide with the revival of other physiological functions during resuscitation of an injured culture. Repair of different lesions proceeded independently and appeared to be dependent on the environment in which the organism was placed after injury. It was also observed that copper-injured ETEC strains regained their ability to tolerate deoxycholate and resumed active multiplication before their ability to produce heat-stable enterotoxin was fully restored (Singh and McFeters, 1986). It is, therefore, important to understand the repair process for different lesions in injured enteropathogens, in both in vitro and in vivo systems, because such information might be used advantageously to kill the bacteria or render them ineffective by interfering with their repair processes.

Revival of three copper-injured ETEC strains in brain heart infusion (BHI) broth and a defined amino acid salt (AAS) medium was studied at 35°C (Singh and McFeters, 1986). At timed intervals the samples were plated on a nonselective tryptic soy medium supplemented with 1% lactose and 0.3% yeast extract (TLY) agar and on selective TLYD (TLY agar supplemented with 0.1% sodium deoxycholate). The deoxycholate-tolerant population was determined during the revival process by dividing the number of CFUs on TLYD by the number of CFUs on TLY and multiplying by 100. When plotted against time, the percent increase represented the recovery of the deoxycholate tolerance in the injured cells (Figure 17.1). In these experiments, recovery was seen before the onset of active multiplication of the cells. This indicated that repair and not growth was involved in the recovery process, as suggested by others (Mossel and Van Netten, 1984). It was also shown that the copper-injured cells recovered their tolerance to sodium deoxycholate faster in nutrient-rich BHI broth than in the defined AAS medium. The process of recovery was slower in AAS medium but went on to completion after an extended incubation period.

The repair process in enteropathogens was not only studied in in vitro systems but was also observed in the small intestine of experimentally inoculated mice (Singh et al., 1986). Figure 17.2 shows the growth and repair of uninjured and copper-injured ETEC after intraluminal inoculation of anesthetized mice. An increase in the number of CFUs on TLY agar represents growth while the revival of injury is indicated by an increase in CFUs on TLYD agar without substantial cell multiplication. A substantial portion of the injured population recovered during the first two hours of intraluminal incubation. However, both multiplication and growth occurred later during the incubation

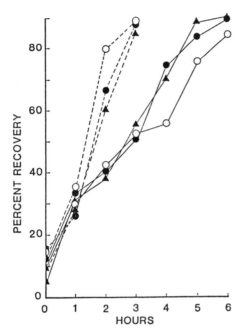

Figure 17.1. Recovery of deoxycholate tolerance in copper-injured *Escherichia coli* strains H10407 (●), E6 (○), and TX-432 (▲) when placed in brain-heart infusion broth (– – – –) and an amino acid salt medium (———).

(2 to 4 h). It was also found that injured cells were able to recover, grow, and cause pathological changes in the mammalian gut.

These observations indicate that ETEC strains retain genotypic determinants required for virulence after simulated environmental stress in water. The significance of these results was further substantiated by the finding that an additional stress in chlorine-injured *Y. enterocolitica*, induced in the stomach of orogastrically inoculated mice, did not alter the virulence of the culture (Singh and McFeters, 1987). Presumably, rapid recovery from injury in these cultures occurred in the small intestine, the site which was reported as favorable for the repair of injured pathogens (Singh et al., 1986). However, in vivo expression of virulence appears to be influenced by the nature of a stress factor causing the injury. Thus, copper exposure of ETEC strain H10407 elicited a reduced fluid accumulation response after intraluminal administration in rabbits (Singh et al., 1986). Also, oral administration of copper-exposed *Y. enterocolitica* resulted in slightly reduced virulence for mice (Singh and McFeters, 1987). This was partly due to the delayed recovery of copper-injured ETEC cells in the small intestine (Singh et al., 1986) and enhanced death of the injured *Y. enterocolitica* culture during passage through the stomachs of inoculated mice (Singh and McFeters, 1987). It is apparent that both reduced viability and delayed revival of injured bacteria occurred after additional stress.

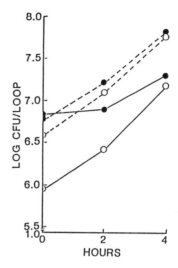

Figure 17.2. Comparison of growth of uninjured (---) and copper-injured (———) *Escherichia coli* strain H10407 cells after intraluminal inoculation in mice. Symbols: ● = CFU on TLY; ○ = CFU on TLYD

17.4   Significance of Injured Enteropathogens

The importance of enumerating injured index organisms in assessing the potability of water has been recognized in Section 920 of *Standard Methods for the Examination of Water and Waste Water* (APHA, 1985). The presently available information indicates that enteric pathogenic bacteria are also injured by aquatic stressors and that the injured pathogens, as well as indicator organisms, may still be present in drinking water but escape detection because they do not grow on routinely used selective media. Also, the potential health hazard of an injured pathogen remains undiminished since properties associated with virulence are not irreversibly lost. The limited information available indicates that injured pathogens can grow and express virulence phenotypes, such as production of enterotoxins, following recovery in both in vivo and in vitro systems. Further, the increased sensitivity of injured pathogens to selective agents may, in part, explain the failure of attempts to associate etiological agents with a drinking water source in several outbreaks (Craun, 1986). However, there is a need for more information on the effects of a wide range of physical and chemical aquatic stressors on different pathogenic bacteria and to study the process of recovery in different systems.

17.5   Conclusions

Exposure of enteropathogenic bacteria to stress factors in water induces sublethal lesions in a large proportion of the population. As a consequence, the cells become more sensitive to the selective agents and show either reduction or loss of virulence. Prolonged exposure can lead to altered surface proteins, loss of virulence plasmids, and the synthesis of enterotoxins or induction of a dormant stage. All these processes have been shown to be reversible when the organisms are placed under more favorable conditions. The fate of injured organisms is determined by the nature of the environment in which they are placed after injury. The injured organisms that are already debilitated die rapidly in water at room temperature, and in the presence of selective agents. However, these organisms can recover in a nonselective medium even at low nutrient concentrations and express undiminished virulence. Recovery from injury can also occur in the mammalian small intestine and the time required for complete repair in vitro or in vivo may depend on the severity of injury as well as the nature of the environment. Thus, the virulence potential of orally inoculated cultures is only slightly reduced or undiminished. The reduced virulence of parenterally inoculated or orally administered cultures may be due to the rapid nonimmune clearance of the injured organisms or their inability to colonize the small intestinal epithelium. There is a need to further investigate the disease potential of injured organisms in different environments and extend these studies to other enteropathogens. Efforts should also be made to improve the routine culture methodology used in aquatic studies for the detection of not only injured index bacteria but pathogenic organisms as well. Further advances in these areas will provide useful information in efforts to control waterborne diseases worldwide.

We thank Anne Camper, Helga Pac, and Lauren Boer for technical assistance, Barry Pyle for critically reading the manuscript, and Nancy Burns and Renée Cook for clerical assistance.
The study was supported by Public Health grant AI 19089 from the National Institute of Allergy and Infectious Disease and grant 14-08-0001-G1493 from the United States Geological survey.

References

Baker, R.M., Singleton, F.L., and Hood, M.A. 1983. Effects of nutrient deprivation on *Vibrio cholerae*. *Applied and Environmental Microbiology* 46:930–940.
Black, R.E., Levine, M.M., Clements, M.L., Hughes, T.P., and Blaser, M.J. 1988. Experimental *Campylobacter jejuni* infection in humans. *The Journal of Infectious Diseases* 157:472–479.
Blaser, M.J. and Newman, L.S. 1982. A review of human salmonellosis: I. Infective dose. *Reviews of Infectious Diseases* 4:1096—1106.
Blaser, M.J., Smith, P.F., Wang, W.-L.L., and Hoff, J.C. 1986. Inactivation of *Campylobacter jejuni* by chlorine and monochloramine. *Applied and Environmental Microbiology* 51:307–311.
Boring, J.R. III, Martin, W.T., and Elliott, L.M. 1971. Isolation of *Salmonella typhimurium* from municipal water, Riverside Calif., 1965. *American Journal of Epidemiology* 93:49–54.
Cabelli, V.J., Dufour, A.P., McCabe, L.J., and Levin, M.A. 1982. Swimming-associated gastroenteritis and water quality. *American Journal of Epidemiology* 115:606–616.
Cash, R.A., Music, S.I., Libonati, J.P., Snyder, M.J., Wenzel, R.D., and Hornick, R.B. 1974.

Response of man to infection with *Vibrio cholerae*. I. Clinical, serologic, bacteriologic responses to a known inoculum. *The Journal of Infectious Diseases* 129:45–52.

Colwell, R.R., Brayton, P.R., Grimes, D.J., Roszak, D.B., Huq, S.A., and Palmer, L.M. 1985. Viable but non-culturable *Vibrio cholerae* and related pathogens in the environment: implications for release of genetically engineered microorganisms *Bio/Technology*. 3:817–820.

Craun, G.F. 1984. Health aspects of ground water pollution. pp. 135–179 in Bitton, G. and Gerba, C.P. (editors), *Groundwater Pollution Microbiology*, John Wiley & Sons, New York.

Craun, G.F. 1986. Statistics of waterborne outbreaks in the U.S. (1920–1980). pp. 73–159. in Craun, G.F. (editor) *Waterborne Diseases in the United States*, CRC Press, Inc., Boca Raton, Florida.

Domek, M.J., LeChevallier, M.W., Cameron, S.C., and McFeters, G.A. 1984. Evidence for the role of copper in the injury process of coliforms in drinking water. *Applied and Environmental Microbiology* 48:289–293.

Domek, M.J., Robbins, J.E., Anderson, M.E., and McFeters, G.A. 1987. Metabolism of *Escherichia coli* injured by copper. *Canadian Journal of Microbiology* 33:57–62.

Evans, D.G., Satterwhite, T.K., Evans, Jr., D.J., and DuPont, H.L. 1978. Differences in serological responses and excretion patterns of volunteers challenged with enterotoxigenic *Escherichia coli* with and without the colonization factor antigen. *Infection and Immunity* 19:883–888.

Grimes, D.J. and Colwell, R.R. 1986. Viability and virulence of *Escherichia coli* suspended by membrane chamber in semitropical ocean water. *FEMS Microbiology Letters* 34:161–165.

Highsmith, A.K., Feeley, J.C., Skaliy, P., Wells, J.G., and Wood, B.T. 1977. Isolation of *Yersinia enterocolitica* from well water and growth in distilled water. *Applied and Environmental Microbiology* 34:745–750.

Hood, M.A., Guckert, J.B., White, D.C., and Deck, F. 1986. Effect of nutrient deprivation on lipid, carbohydrate, DNA, RNA, and protein levels in *Vibrio cholerae*. *Applied and Environmental Microbiology* 52:788–793.

Hurst, A. and Hughes, A. 1981. Repair of salt tolerance and recovery of lost D-alanine and magnesium following sublethal heating of *Staphylococcus aureus* are independent events. *Canadian Journal of Microbiology* 27:627–632.

Keswick, B.H. 1984. Sources of ground water pollution. pp. 39–64 in Bitton, G. and Gerba, C.P. (Editors) *Groundwater Pollution Microbiology*, John Wiley and Sons, New York.

King, C.H., Shotts, Jr., E.B., Wooley, R.E., and Porter, K.G. 1988. Survival of coliform and bacterial pathogens within protozoa during chlorination. *Applied and Environmental Microbiology* 54:3023–3033.

LeChevallier, M.W. and McFeters, G.A. 1984. Recent advances in coliform methodology for water analysis. *Journal of Environmental Health* 47:5–9.

LeChevallier, M.W. and McFeters, G.A. 1985. Interaction between heterotrophic plate count bacteria and coliform organisms. *Applied and Environmental Microbiology* 49:1338–1341.

LeChevallier, M.W., Schiemann, D.A., and McFeters, G.A. 1987. Factors contributing to the reduced invasiveness of chlorine-injured *Yersinia enterocolitica*. *Applied and Environmental Microbiology* 53:1358–1364.

LeChevallier, M.W., Singh, A., Schiemann, D.A., and McFeters, G.A. 1985. Changes in virulence of waterborne enteropathogens with chlorine injury. *Applied and Environmental Microbiology* 50:412–419.

Levine, M.M., Caplan, E.S., Waterman, D., Cash, R.A., Hornick, R.B., and Snyder, M.J. 1977. Diarrhea caused by *Escherichia coli* that produce only heat-stable enterotoxin. *Infection and Immunity* 17:78–82.

Levine, M.M., DuPont, H.L., Formal, S.B., Hornick, R.B., Takeuchi, A., Gangarosa, E.J.,

Snyder, M.J., and Libonati, J.P. 1973. Pathogenesis of *Shigella dysenteriae* 1 (Shiga) dysentery. *Journal of Infectious Diseases* 127:261–270.

McFeters, G.A., Bissonnette, G.K., Jezeski, J.J., Thomson, C.A., and Stuart, D.G. 1974. Comparative survival of indicator bacteria and enteric pathogens in well water. *Applied Microbiology* 27:823–829.

McFeters, G.A. and Camper, A.K. 1983. Enumeration of coliform bacteria exposed to chlorine. *Advances in Applied Microbiology* 29:177–193.

McFeters, G.A., LeChevallier, M.W., Singh, A., and Kippin, J.S. 1986. Health significance and occurrence of injured bacteria in drinking water. *Water Science Technology* 18: 227–231.

Morris, G.K., and Feeley, J.C. 1976. *Yersinia enterocolitica:* a review of its role in food hygiene. *Bulletin of World Health Organization* 54:79–85.

Mossel, D.A.A. and Van Netten, P. 1984. Harmful effects of selective media on stressed micro-organisms: nature and remedies. pp. 329–369 in Andrew, M.H.E., and Russell, A.D. (editors) Society for Applied Microbiology Symposium Series No 12. *The Revival of Injured Microbes.* Academic Press, London.

Palmer, L.M., Baya, A.M., Grimes, D.J., and Colwell, R.R. 1984. Molecular genetic and phenotypic alteration of *Escherichia coli* in natural water microcosms containing toxic chemicals. *FEMS Microbiology Letters* 21:169–173.

Rhodes, M.A. and Kator, H. 1988. Survival of *Escherichia coli* and *Salmonella* spp. in estuarine environments. *Applied and Environmental Microbiology* 54:2902–2907.

Robinson, D.A. 1981. Infective dose of *Campylobacter jejuni* in milk. *British Medical Journal* 282:1584.

Rollins, D.M. and Colwell, R.R. 1986. Viable but nonculturable stage of *Campylobacter jejuni* and its role in survival in the natural aquatic environment. *Applied and Environmental Microbiology* 52:531–538.

Roszak, D.B. and Colwell, R.R. 1987. Survival strategies of bacteria in the natural environment. *Microbiological Reviews* 51:365–379.

Roszak, D.B., Grimes, D.J., and Colwell, R.R. 1984. Viable but nonrecoverable stage of *Salmonella enteritidis* in aquatic systems. *Canadian Journal of Microbiology* 30:334–338.

Seligmann, R. and Reitler, R. 1965. Enteropathogens in water with low *Escherichia coli* titer. *Journal. American Water Works Association* 57:1572–1574.

Singh, A., LeChevallier, M.W., and McFeters, G.A. 1985. Reduced virulence of *Yersinia enterocolitica* by copper-induced injury. *Applied and Environmental Microbiology* 50: 406–411.

Singh, A. and McFeters, G.A. 1986. Recovery, growth, and production of heat-stable enterotoxin by *Escherichia coli* after copper-induced injury. *Applied and Environmental Microbiology* 51:738–742.

Singh, A. and McFeters, G.A. 1987. Survival and virulence of copper- and chlorine-stressed *Yersinia enterocolitica* in experimentally infected mice. *Applied and Environmental Microbiology* 53:1768–1774.

Singh, A., Yeager, R., and McFeters, G.A. 1986. Assessment of in vivo revival, growth, and pathogenicity of *Escherichia coli* strains after copper- and chlorine-induced injury. *Applied and Environmental Microbiology* 52:832–837.

Stevenson, L.H. 1978. A case for bacterial dormancy in aquatic system. *Microbial Ecology* 4:127–133.

Walsh, S.M. and Bissonnette, G.K. 1983. Chlorine-induced damage to surface adhesins during sublethal injury to enterotoxigenic *Escherichia coli*. *Applied and Environmental Microbiology* 45:1060–1065.

Walsh, S.M. and Bissonnette, G.K. 1987. Effect of chlorine injury on heat-labile enterotoxin production in enterotoxigenic *Escherichia coli*. *Canadian Journal of Microbiology* 33:1091–1096.

18

Viruses in Source and Drinking Water

Charles P. Gerba[1] and Joan B. Rose[2]

[1]Departments of Nutrition and Food Science and Microbiology and Immunology, University of Arizona, Tucson, Arizona and [2]Department of Environmental and Occupational Health, University of South Florida, Tampa, Florida

All warm blooded animals harbor enteric viruses which are excreted in fecal material and can find their way into the aquatic environment. In addition fish, other lower animals and even plants release viruses which can find their way into water. Fortunately, most viruses are highly host specific and only the enteric viruses of humans appear to offer the greatest health concern for waterborne transmission. Human enteric viruses are able to exist for extended periods in the environment and many may survive conventional water and wastewater treatment. Waterborne transmission has now been well documented for many of the enteric viruses. While a direct association with the presence of virus in a water supply and illness has not been demonstrated for all enteric viruses, the presence of any of these viruses in drinking water should be considered a major health concern.

18.1 Characteristics of Human Enteric Viruses

More than 110 types of human enteric viruses have been identified (Table 18.1) including the enteroviruses, hepatitis A virus (HAV), Norwalk virus, reovirus, rotavirus, and adenovirus. This list continues to grow as new human enteric viruses are discovered almost every year. Viruses are obligate intracellular parasites made up of a core of nucleic acid (RNA or DNA) surrounded by a protein coat. Although they cannot reproduce without the host cell, they can survive in the environment for extended periods of time. The enteric viruses are small, ranging in size from 20 nm to 85 nm. Because of their size and the requirement of a host system to grow, special techniques are used to detect and enumerate viruses.

Enteroviruses were first isolated from the environment more than 40 years ago (Paul and Trask, 1941) and have been the most extensively studied. En-

380

Table 18.1 Characteristics of enteric viruses

Virus group	Number of types	Size (nm)	Type of nucleic acid	Diseases
Enterovirus				
Poliovirus	3	20–30	RNA	Paralysis
				Aseptic meningitis
Coxsackievirus				
A	24	20–30	RNA	Herpangia
				Aseptic meningitis
				Respiratory illness
				Paralysis fever
B	6	20–30	RNA	Pleurodynia
				Aseptic meningitis
				Pericarditis
				Myocarditis
				Congenital heart anomalies
				Nephritis
Echovirus	34	20–30	RNA	Respiratory infection
				Aseptic meningitis
				Diarrhea Pericarditis
				Myocarditis Fever, rash
Reovirus	3	75–80	RNA	Respiratory disease
				Gastroenteritis
Adenovirus	41	68–85	DNA	Acute conjunctivitis
				Diarrhea Respiratory
				illness Eye infection
Hepatitis A Virus	1	27	RNA	Infectious hepatitis
Rotavirus	4	70	RNA	Infantile gastroenteritis
Norwalk agent (probably a calcivirus)	1?	27	RNA	Gastroenteritis
Astrovirus	4	28	RNA	Gastroenteritis
Calcivirus	1?		RNA	Gastroenteritis
Snow Mountain Agent (probably a calcivirus)	1?		RNA?	Gastroenteritis
Norwalk-like viruses	?		?	Gastroenteritis
Epidemic non A non B hepatitis	1?		RNA?	Hepatitis

teroviruses, including polioviruses, coxsackieviruses, and echoviruses, may cause diseases such as paralysis, meningitis, respiratory illness, and diarrhea. The adenoviruses can cause eye infections and respiratory disease. Three new types have been described which appear to largely cause gastroenteritis in children (Kurtz and Lee, 1987). The role of reoviruses in human disease has never been clearly defined, although the same serological types isolated from humans have been associated with serious illness in animals.

Since rotaviruses have a double-stranded RNA genome and a similar size and shape, they are believed to be closely related to the reoviruses. Human rotaviruses were first discovered in 1973 (Bishop et al., 1973) and were subsequently found to be a major cause of childhood gastroenteritis. This can be a particularly devastating disease in children under two years of age and causes 50 percent of the hospitalized cases of diarrheal illness in temperate climates (Blacklow and Cukor, 1981). Rotavirus may also infect the adult population and apparently is one of the primary agents of traveler's diarrhea (Bolivar et al., 1978). Electron microscopy, enzyme-linked immunosorbent assay (ELISA), and immunofluorescent techniques have been used for the detection of rotaviruses directly from environmental samples (Smith and Gerba, 1982). Growth in tissue culture has proven more difficult than with the enteroviruses and requires specialized techniques, such as the incorporation of trypsin into the culture media.

Hepatitis A virus (HAV) has been classified as enterovirus 72 (Melnick, 1982) and like the enteroviruses demonstrates acid stability and resistance to ether and possesses similar physical features. However, it appears to be the most thermally stable of the enteric viruses and has been shown to survive for extended periods of time in sea water and ground water (Sobsey, 1988). Infections with HAV cause liver degeneration and are characterized by jaundice; thus, through the centuries viral hepatitis has been easily recognized. Only since 1979 have techniques been developed for the propagation of HAV in laboratory and isolation from environmental samples (Provast and Hilleman, 1979; Hejkal et al., 1982). In 1980 a non-A, non-B hepatitis virus was discovered responsible for a massive waterborne outbreak in India (Wong, 1980). Although other similar epidemics of non-A, non-B hepatitis have been reported, there is still little available information concerning the characterization of this agent (Gust and Purcell, 1987).

The Norwalk virus was discovered in an outbreak of gastroenteritis in 1968 and is characterized by diarrhea and vomiting (Adler and Zickl, 1969). A major cause of waterborne disease in the United States (Kaplan et al., 1982), the Norwalk virus has yet to be isolated from environmental samples. The virus has not been propagated in the laboratory but appears to be related to the calcivirus group (Dolin et al., 1987). This agent appears to be a major cause of adult gastroenteritis with little association to diarrhea in infants (Dolin et al., 1987).

The name astrovirus was the name given in 1975 to describe a small round virus with a star-like appearance seen in the feces of children with gastroenteritis (Madeleley and Cosgrove, 1975). The virus causes a mild illness after incubation of 3 to 4 days. Five human serotypes have been described; antigenically distinct astroviruses infect a wide variety of animals. Outbreaks of illness have been associated with oysters and drinking water (Kutz and Lee, 1987).

The Snow Mountain Agent is a 27 to 32 nm virus which has been the etiologic agent of several outbreaks of water- and food-borne acute gastroenteritis (Dolin et al., 1987). The biophysical properties and polypeptide composition of this virus suggest a relationship to the calcivirus group and the Norwalk virus (Madore et al., 1986).

18.2   Waterborne Transmission of Enteric Viruses

The presence of any pathogenic microorganism in potable water has always been considered a potential public health concern. This is rightly so since there is a significant risk of contracting an infectious disease from the ingestion of sewage-contaminated water (Craun, 1986). The direct sewage contamination of drinking water supplies or obvious failures in treatment of sewage-contaminated source waters are well recognized as responsible factors for waterborne disease outbreaks in the United States (Lippy and Waltrip, 1984). The average annual number of documented waterborne disease outbreaks in the United States has increased almost four-fold since the 1960's.

Acute gastroenteritis of unknown etiology accounted for over 50% of documented waterborne disease outbreaks from 1946 to 1980 in the United States (Lippy and Waltrip, 1984). Viruses were identified as the cause of 12% of the total number of outbreaks during this period. The difficulty in the isolation of many enteric viruses from clinical and water samples probably accounts for the limited number so far identified as causes of waterborne disease. If we assume that all of the nonbacterial gastrointestinal illness being reported has a viral etiology, then as much as 64% could have been related to the presence of viruses in water. Undoubtedly, all of the nonbacterial gastroenteritis is not due to viruses, but as methods have improved for enteric viruses. The epidemiology of waterborne disease is complex, and it is doubtful that the true incidence of such disease will ever be elucidated.

18.3   Human Enteric Viruses in Surface Waters

Enteric viruses can be expected to occur in any surface water exposed to contamination by human fecal wastes. The actual number which may be encountered at any one time can be expected to vary greatly depending upon the incidence of enteric viral disease in the source population, time of year, source (raw vs. treated wastes), rainfall, etc. Also important are the methods used to concentrate and detect the viruses in the source. The vast majority of our data deals only with the occurrence of enteroviruses. Data are very limited on the occurrence of the other enteric viruses. The maximum number of viruses observed in surface fresh waters in several studies is summarized in Table 18.2. The lower numbers of enteric viruses seen in American rivers is probably due to the practice of wastewater disinfection, which is not commonly practiced in the rest of the world. Taking into consideration the limitations of current detection methodology, the true numbers of viruses present in surface waters is probably 10 to 100 times greater than that shown in Table 18.2.

18.4   Enteric Viruses in Ground Water

From 1971 to 1982 ground water was responsible for 51% of all waterborne outbreaks and 40% of all waterborne illness in the United States (Craun, 1986). Hepatitis A virus, Norwalk virus, and rotavirus have been implicated as causes of ground water disease outbreaks (Gerba et al., 1985). The most frequently reported source of contamination in outbreaks involving ground water is overflow or seepage of sewage from septic tanks. However, viruses in ground water may also originate from the land application of sewage and sewage sludges, domestic landfills, oxidation ponds, deep well injection of sewage, etc. (Gerba, 1987). Being smaller than enteric bacteria, viruses are capable of traveling long distances through soil and ground water, even hundreds of meters under the proper conditions (Gerba and Bitton, 1984). The degree of virus

Table 18.2 Maximum numbers of enteric viruses observed in surface waters

Water source	Country	Maximum number of viruses recovered (PFU or MPNICU/liter)*
Missouri River	United States	0.1
Seine River	France	191
Thames River	England	22
Lee River	England	12
Rhine River	France	283
Moselle River	France	0.2
Tanana River	United States	0.06
Rio Besos	Spain	55
Rio Lhobregat	Spain	55
Lake Ronkonkoma	United States	1.4
Wear River	England	6
Avon and Sowe Rivers	England	620
Oak Creek	United States	4.2
Serene Lakes, CA	United States	58
Creek, Humboldt Lake, CA	United States	414
River, MI	United States	0.8

Modified from Block (1983) and Bitton et al., 1985.
* PFU = Plaque forming units; MPNICU = Most probable number of infectious culture units.

movement is dependent on a number of factors, including the type of virus, the nature of the soil, and the climate.

The major factor limiting virus movement in ground water is its adsorption to the soil. Some viruses such as poliovirus adsorbs readily to most soils, while the coliphage MS-2 absorbs poorly (Goyal and Gerba, 1979). Viruses are also more mobile in sandy soils and in areas receiving intense periods of rainfall. Although septic tanks appear responsible for the majority of waterborne disease outbreaks caused by contaminated ground water, the amount of field work done on the occurrence of enteric viruses in ground water near septic tanks is very limited. This is because viruses would only be expected to occur in septic effluent if a household member was infected with an enteric virus. This is in contrast to domestic sewage effluents from communities where the population size is great enough so that at least some individuals will always be infected. For this reason, most of the work near septic tanks has only been done during or after a waterborne disease outbreak (Gerba, 1987).

However, viruses have frequently been isolated from ground water where the land application of sewage either for disposal or irrigation has bene practiced (Gerba, 1987).

Studies reporting the isolation of viruses from drinking water wells is shown in Table 18.3. Perhaps the most extensive studies on the occurrence of viruses in ground water were those of Marzouk et al. (1979). They found 20% of 99 samples collected from shallow ground water wells (3 m in depth) positive for enteric viruses. In a more recent study conducted in the southwest, viruses were also detected in almost 20% of all samples (Gerba et al., 1988). Water and well depth ranged from 1 to 30 meters of the wells sampled. Several of

Table 18.3 Isolation of viruses from wells used as sources of drinking water

Location	Virus
Florida	Echo 22/23
Germany	Echo 3, 6, 30; coxsackie B1, 4, 5; U*
India	U
Michigan	Polio 2
England	Polio 2
Israel	Polio 1, U
Ghana	Polio 1, coxsackie B3
Mexico	Rota; coxsackie B4, 6
Texas	Hepatitis A, coxsackie B3
Georgia	Hepatitis A
Maryland	Hepatitis A; polio 1; echo 27, 29
Bolivia	Rota
England	Polio 1, 2, 3; coxsackie B2
Norfolk Island	Rota, adeno 5, polio 1

Modified from Gerba (1987).
* U, unidentified.

the wells in both studies were providing non-disinfected water to communities or households. These studies have also demonstrated the presence of viruses in ground water in meeting current coliform standards of one per 100 ml (USEPA, 1975). This suggests that current coliform standards are not adequate to predict the occurrence of enteric viruses. This discrepancy arises from the larger volumes processed for viruses, greater transport, and survival of viruses in soils. The limited studies to date on the occurrence of viruses in ground water suggest that contamination of this resource is probably underestimated.

18.5   Reported Enteric Virus Isolation from Drinking Water

Coin (1966) was the first to isolate viruses from drinking water in the early 1960s in Paris, France. Other early reported isolations came from Romania (Nestor and Costin, 1976) and Russia (Rabyshko, 1974). Viral contamination of drinking water was subsequently reported from other locations around the world (Table 18.4).

The first report of enteric viruses in drinking water was attributed to inadequate treatment. In Coin's study, the treatment applied to the source water consisted of settling and marginal chlorination. Although other investigators did isolate viruses from water receiving complete treatment (coagulation-sedimentation, filtration, and disinfection), the initial reports from the United States were suspect due to possible laboratory contamination of samples (Akin, 1984). It is now apparent that viruses can be recovered from treated drinking water; approximately 53% of the reported isolations came from water with complete treatment while 26% came from water which was only disinfected and 15% from untreated water.

A few comprehensive surveys of virus contamination of drinking waters have been conducted in Canada, England, Israel, Netherlands, Romania and the United States (Table 18.5). In the Netherlands and United States studies, no viruses were detected in the finished water; however, in four other investigations viruses were recovered from ten to 71% of the sources or facilities tested. In most cases large volumes of water were filtered (20 to 1900 liters), and low numbers of viruses were recovered (0.0005 PFU to 0.87 PFU/l) with a small percentage of the samples positive (4 to 7%). Yet it is still disturbing that at some point many facilities are distributing drinking water containing viral pathogens.

The limitations of the methods used in environmental virology lend ambiguous significance to the intermittent isolation of low concentrations of viruses from drinking water. Currently, the adsorption/elution technique using electronegative or electropositive filters for concentration of viruses from water samples is the method of choice (APHA, 1985). This method depends on the use of an adsorbent material and subsequent deadsorption (or elution) of the viruses with a small volume of a proteinaceous liquid such as beef extract.

Table 18.4 Enteric virus isolation from treated drinking water[a]

Virus type recovered	Number samples positive/number samples collected	Water treatment	Location and date
Polio, other enterics	43/553	Disinfection	France 1960–62
Reo, entero	2/25	Complete[b]	France 1961–63
Coxsackie A	2/65	Complete	Romania 1962–71
Coxsackie A and B, echo, polio	9/64	Complete	Russia 1968–71
Nonpolio	NR[c]	Complete[e]	Italy (NR)
Reo, echo, polio	7/64	Complete	Massachusetts 1969–71
Polio	1/1	Untreated	Michigan 1970
Reo, entero	2/NR	Complete[d]	South Africa 1971
Polio	1/7	Complete	Massachusetts 1972–73
NR	5/100	Complete	South Africa, 1974
Polio	2/33	Complete	Costa Rica 1974–75
Polio, coxsackie A, untyped strains	8/220	Complete	Romania 1977–82
Reo?	1/464	Complete	South Africa (NR)
Polio	3/?	Complete	Southern State, 1972
Echo	1/10	Disinfection	Florida 1975
Polio	4/12	Complete	Virginia 1975
Polio	(6.7%)	Disinfection	Canada, 1975–76
Polio	1/42	Complete	Virginia 1976
NR	19/74	Disinfection	India 1978
Polio, echo	12/18	Disinfection	Israel 1978
Coxsackie B, rotavirus	13/31	Complete	Mexico 1978
Polio	31/31	Complete	Canada 1978–79
Reo	3/10	Untreated	Spain 1980
Adeno, rota and small round viruses	6/32	Untreated	Australia 1979–1980
Coxsackie B, adeno	7/7	NR	Peoples' Rep. China 1982
Polio, coxsackie B, echo	90/553	Mostly disinfection	United Kingdom 1979–82
Coxsackie	1/1	Filtered	California 1980
Hepatitis A, coxsackie B	3/6	Disinfection	Texas 1980
Polio	11/155	Complete	Canada 1983
Enterovirus rotavirus	19/23	Complete	Mexico 1982
NR	1/4	Disinfection	Germany 1982
NR	3/111	Untreated	Israel 1980–81
Polio	6/17	Disinfection	United Kingdom 1983–84
Hepatitis A	4/6	Untreated	Maryland 1981
Rota, polio, coxsackie	11/54	Complete	Mexico 1982–83
Rota	6/20	Complete	Colombia 1986
Rota	2/5	Complete	Bolivia 1986

[a] Adapted from: AWWA Committee on Viruses in Drinking Water, 1979; Melnick and Gerba, 1982; Berg, 1984; Akin, 1984; Toranzos et al., 1986, a, b; Rose et al., 1986.
[b] Complete treatment: coagulation-sedimentation, rapid filtration, and disinfection.
[c] NR, Not Reported.
[d] Further chlorination received before distribution; no viruses recovered after final treatment.
[e] An experimental plant that used complete treatment of a heavily polluted source water; finished product was not consumed.

Table 18.5 Major surveys of drinking water for viral contamination

Country	Number of facilities or communities surveyed	Number of facilities or communities where viruses were isolated	Reference
Canada	Seven treatment plants	Five	Payment et al., 1985
England	229 sources	89	Tyler, 1983
Israel	20 settlements	Three	Guttman-Bass and Fattal, 1985
Netherlands	Eight treatment plants	None	Van Olphen et al., 1984
United States	54 communities	None	Akin, 1984
Romania	Eight towns	Four	Nestor et al., 1978

This technique allows viruses to be concentrated from large volumes of water. Many types of adsorbent materials have been used (Gerba, 1983). Early attempts at virus isolation from drinking water involved the use of gauze pads as a virus adsorbent, although this technique was not quantitative or very reproducible. By the 1970's, cellulose nitrate membrane filters were being used to recover viruses from water; furthermore, it was found that the addition of divalent and trivalent cations and acidification of the water sample would enhance the virus recovery. In 1976 and 1978, respectively, the pleated cartridge system and electro-positive filters were introduced. Laboratory studies have shown recovery efficiencies of 50% using these filters, but it is unlikely that this is achieved under field conditions, as a variety of water constituents can decrease the recovery of viruses (Sobsey, 1986).

The development of field concentration methods was a major factor in the increased isolation of enteric viruses from drinking water, but laboratory innovations in the propagation and enumeration of viruses such as rotavirus and HAV have also contributed significantly more sensitivity to virus detection. Both have influenced the number of reports of virus isolation from drinking water as well, which has been increasing in the last 20 years (Rose and Gerba, 1986). From 1965 to 1970 only one report on the isolation of viruses from drinking water was published; but by the 1980s (1981 to 1985) 14 recoveries were documented, which included the first isolations of rotavirus and HAV from drinking water (Hejkal et al., 1982; Keswick et al., 1984). Although new and improved techniques may explain the increasing incidence of enteric viruses in drinking waters, an increased interest has also undoubtedly contributed to this rise. It may be surmised that as methods are developed for detection of the Norwalk virus and others and as more individuals become aware of these viruses, more enteric viral contamination of drinking water will be reported.

18.6 Techniques for Virus Detection

The first laboratory methods for the growth and isolation of animal viruses outside of animals were developed only within the last 40 years. The first group of animal viruses readily cultivated in tissue culture was the enteroviruses. Many of these viruses proved to be easily isolated from clinical and environment samples with primate and human kidney tissue culture. Most of the enteroviruses are readily isolated and quantitated by either observation for characteristic viral cytopathogenic effects in the tissue culture or by the observation of plaques, or areas of cell destruction in a monolayer of cells. By staining the cell monolayers with vital dyes, plaques appear as clear zones which can easily be counted. Many other enteric viruses, including reoviruses and adenoviruses, can be quantitated by similar methods.

Other enteric viruses are more difficult to grow in tissue culture and other methods such as immunofluorescence (IF), for rotavirus, or radioimmuno-focus-assay (RIFA), for hepatitis A are needed. For many of the enteric viruses (including the Norwalk agent), no method is currently available for their isolation in tissue culture.

Isolation of enteric viruses in tissue culture requires from 3 to 21 days. Identification can also require several additional weeks. Costs of collection and analysis of water samples for enteric viruses are significantly more expensive than bacteria. Typical costs currently run $300 to $400 for enterovirus analysis. More rapid techniques have been studied over the years to reduce the time and cost necessary for virus detection. Such techniques include immunofluorescence, radio-immunoassay (RIA), and enzyme linked-immunosorbent-assay (ELISA). All of these techniques require the production of a specific antibody against the virus one wishes to detect. While only requiring as little as 1 to 2 hours to complete, RIA and ELISA require the presence of large numbers of viruses (10,000 to 100,000 virus particles), making it difficult to use with environmental samples where concentrations of viruses would be expected to be many orders of magnitude less.

Immunofluorescence of infected cells can produce results within 24 to 48 hours and provides information on the infectivity of the virus. It has been successfully used to detect rotaviruses in drinking water (Rose et al., 1986). Unfortunately, for the application of this technique and others mentioned above, specific antibody is required for each virus which one wishes to detect. Thus, a large number of specific antibodies would be required. Application of this technique could have an advantage if a certain group of enteric viruses were used as indicators, and thus only a limited number of antibodies need be produced.

With the advent of recombinant DNA technology, it has become possible to use nucleic acid probes to detect enteric viruses in water (Jiang et al., 1986, 1987; Richardson et al., 1988). Gene probes are produced by inserting a specific fragment of viral complementary DNA (cDNA) into a plasmid, which multiplies in a bacterial host. The cloning procedure ensures that the sequences amplified in the bacterial host are absolutely free from host cellular sequences and thus ensures a high specificity of the probe.

A number of different techniques may be used to label the nucleic acid. Radioactive isotopes such as ^{32}P may be incorporated which can then be detected by x-ray sensitive film. Biotin-labeled probes can be detected with avidin in a colorimetric procedure (Kulski, 1985). A sample to be tested is treated first with phenol/chloroform and then chloroform alone to release the nucleic acid from the viral capsid. The nucleic acid is spotted (i.e., immobilized) onto a filter matrix of nitrocellulose or nylon membrane with the aid of a vacuum filter manifold. Viral nucleic acids are then hybridized to the labeled cDNA probe by placing the filter matrix in a hybridization solution containing the labeled cDNA probe.

Because related enteric viruses share some sequence homology within their genomes, group-specific probes can be developed which can be used to detect entire groups of related viruses (Rotbart et al., 1984). Thus, it has already been demonstrated that cDNA probes developed with the poliovirus type 1 genome are capable of detecting coxsackie, polio, and hepatitis A viruses (Rotbart et al., 1984). Gene probes have also been developed for other groups of enteric viruses such as rotavirus and adenovirus. Currently, most probes are capable of detecting as little as 0.5 to 2 pg of nucleic acid of 1,000 to 10,000 Tissue Culture Infectious Dose (TCID$_{50}$) of poliovirus (Rotbart et al., 1984).

Jiang et al. (1986) demonstrated the application of a cDNA to the detection of hepatitis A in estuarine water and sediment samples. The same group also used a single-stranded RNA probe labeled with ^{32}P and biotin. The biotin-labeled probe was found to be 10-fold less sensitive than the equivalent ^{32}P-labeled probe. The ^{32}P-labeled probe was capable of detecting as few as 180,000 physical viral particles as determined by electron microscopy.

Our laboratory has developed a ^{32}P-labeled cDNA probe which is capable of detecting 1 to 10 tissue culture infectious doses of poliovirus type 1 (approximately 100 to 1,000 physical particles) and hepatitis A (Margolin et al., 1986; Richardson et al., 1988). The high sensitivity was accomplished by using a highly labeled probe (2.0×10^9 cpm/μg DNA) and an amplification technique using the vector plasmid.

A comparison of poliovirus probe and cell culture isolation of enteroviruses from ground water and tap water was made. Of 43 samples tested to date, 8 were positive by cell culture and 11 by gene probe. Seven of the 11 positive gene probes were confirmed in cell culture. Failure to confirm one of the cell-culture-positive samples may indicate that the virus is another enterovirus. While our poliovirus probe will detect members of the coxsackie and echo virus group, it does so with a decreased sensitivity (Margolin et al., 1986). Three of the positive samples could not be confirmed in the Buffalo Green Monkey (BGM) cell line. Since many enteroviruses do not grow well in the BGM cell line, this does not necessarily mean they are false positives. Alternatively, the viruses detected in the samples may represent inactivated viruses or that only one or two viruses were present in the entire sample. The concentration of viruses in the samples, as determined by cell culture ranged from 0.03 to 26 tissue culture infectious doses per ml of concentrate, demonstrating the high sensitivity of the gene probe technology (Margolin et al., 1989).

A potential major disadvantage of gene probes is that they may not differentiate between infectious and noninfectious viruses. Unfortunately, no one has yet evaluated the ability of current gene probes to detect viruses inactivated by disinfection or after inactivation in natural waters. Although the significant rate of confirmation in cell culture in two studies suggests that this may not be a major problem for evaluating untreated waters, these studies can only be considered preliminary in nature. Gene probes could be used to more rapidly detect virus growth in cell culture, which would aid the speed of confirmation. It may also be possible to develop gene probes that could determine the fectivity of viruses in environmental samples.

While nonradioactive probes are generally 10-fold less sensitive than radioactively labeled probes, they may be useful in many situations and hold promise for the widespread use of virus detection to almost any water quality laboratory. Nonradioactive probe technology is generally more rapid, with results forthcoming within 2–3 hours.

It is clear that the development of gene probe technology will have a major impact on microbiology in the coming years which will greatly expand the horizon of the study of viruses in water in the future.

18.7 Significance of Enteric Viruses in Drinking Water

Only in a few cases have virus isolations from drinking water been linked to disease in the exposed population. This fact alone may seem to infer that viral contamination of drinking water supplies is not a significant problem and does not warrant major concern. This dangerous supposition could place a major burden of illness on the population.

Viruses were responsible for 12% of the waterborne outbreaks from 1946 to 1980 in the United States (Lippy and Waltrip, 1984), yet recovery of the viral agent from the contaminated water supply has not been routinely documented. This may be due to a lack of methods and facilities capable of doing such analyses, or because the investigating offices simply did not request virus analysis. Finally, it must be kept in mind that by conservative estimations only a small percentage of the waterborne outbreaks are ever detected or reported. Thus there is probably significantly more illness in the population due to viral contamination of drinking water than is recognized.

As mentioned previously, extensive studies of viruses in drinking water have reported no viruses or low levels of viruses. Thus, viruses may be present in numbers below the detection limits of the methods. The significance of low levels of contamination can be exemplified by the low infectious dose of viruses. Studies on minimum infectious dose (i.e. that dose which causes infection in 1% of the exposed population) indicates that as few as 1 to 2 tissue-culture plaque-forming-units of enteric viruses are capable of causing infection (Ward and Akin, 1984; Ward et al., 1986). The most recent studies on this topic were performed by Ward et al. (1986), who fed human rotavirus from a stool specimen of a hospitalized child to adult volunteers. The amount of virus required to cause infection (i.e. shedding of virus, serorconversion, or both) was equal to the minimum detectable in cell cultures. Over 50% of those who became infected also developed clinical illness.

Not everyone who may become infected with enteric viruses or parasites will become clinically ill. Asymptomatic infections are particularly common among some of the enteroviruses. The development of clinical illness depends on numerous factors, including the immune status of the host, age of the host, virulence of the microorganisms and type, strain of microorganism, and route of infection. The observed frequencies of symptomatic infections for various enteroviruses may range from 1% for poliovirus to more than 75% for some of the coxsackie B viruses (Cherry, 1981). The frequency of clinical hepatitis A viruses in adults is estimated at 75%. However, during waterborne outbreaks it has been observed as high as 97% (Lednar et al., 1985).

Mortality rates are also affected by many of the same factors that determine the likelihood of the development of clinical illness. The risk of mortality for hepatitis A virus is 0.6% in the United States (CDC, 1985). Mortality from other enteroviruses infections in North America and Europe has been reported to range from <0.1 to 1.8% (Assaad and Borecka, 1977). Utilizing this information in a probability risk assessment model, Gerba and Haas (1988) estimated that the risks of developing clinical illness from ingestion of one plaque-forming-unit of an enterovirus per 1000 liters over a lifetime were in the order of 10^{-1} and mortality as great as 10^{-3}.

The contamination of drinking water in many instances may have been due to defects or malfunctions in the treatment processes as was evident by the presence of coliform bacteria. During investigations of treatment systems after documented waterborne outbreaks it was found that 39% were attributed to inadequate or interrupted treatment of water sources (Craun, 1986). Therefore, it may be reasonable to assume that the detection of enteric viruses in treated drinking water is a result of treatment deficiencies. This indicates a problem with maintaining adequate treatment needed to meet drinking water standards rather than a problem with the standards themselves. However, as early as 1960 there was concern that coliform bacterial standards used to detect domestic pollution were not adequate in regard to viruses (Clarke and Kabler, 1964). It has been shown that viruses are more resistant to inactivation and removal by treatment processes than the bacteria and can survive for longer periods of time in the environment than the bacteria. In addition, larger volumes of water are usually examined for viruses (10 to 1,000 liters) than indicator bacteria. There have now been reported isolations of viruses from drinking water meeting U.S. bacteriological standards (1 coliform/100 ml), turbidity standards (1 Nephelometric Turbidity Unit) and containing recommended levels of free chlorine (0.5 mg/l) (Deetz et al., 1984; Payment et al., 1984; Rose et al., 1985; Tyler, 1984). In fact, Slade (1985) has reported the isolation of viruses from a chalk well with an excellent bacteriological and organoleptic quality where viruses survived one mg/l of free chlorine after a 15-minute contact time. These reports have initiated concerns in regards to the adequacy of the standards and guidelines governing the hygienic quality of potable waters. Water which was presumed to be safe based on currently used coliform standards should not be assumed to be free of viruses.

At present, a number of major issues need further investigation. Better and new methods with increased sensitivity for the detection of viruses from water should be developed. A new indicator system for the enteric viruses whereby contamination could be detected rapidly and inexpensively would be advantageous. A monitoring program may better define the incidence and significance of these viral pathogens in the aquatic environment. Treatment processes and deficiencies which influence virus removal and inactivation should be fully characterized.

Much progress has been made in the last decade concerning enteric viruses in water but with more answers have come more questions. Thus further research will be necessary to ensure that all drinking water will be safe and free from viral pathogens.

References

Adler, J. and Zickl, R. 1969. Winter vomiting disease. *J. Infect. Dis.* 119:668–673.

Akin, E. 1984. Occurrence of viruses in treated drinking water in the United States. *Wat. Sci. Tech.* 17:689–700.

American Water Works Association Committee. 1979. Committee report. Viruses in drinking water. *J. Amer. Water Works Assoc.* 71:441–444.

American Public Health Association. 1985. *Standard Methods for the Examination of Water and Wastewater.* 16th ed. Washington, D.C.

Assaad, F. and Borecka, I. 1977. Nine-year study of WHO virus reports on fatal virus infections. *Bull. Wld. Hlth. Org.* 55:445–453.

Berg, G. 1984. *Criteria Document for Waterborne Disease.* Office of Criteria and Standards, U.S. Environmental Protection Agency. Washington, D.C.

Bishop, R.F., Davidson, G.P., Holmes, I.H., and Ruck, B.J. 1973. Virus particles in epithelial cells of duodenal mucosa from children with acute nonbacterial gastroenteritis. *Lancet.* 2:1281–1283.

Bitton, G., Farrah, S.R., Montague, C., Binford, M.W., Scheverman, P.R., and Watson, A. 1985. *Survey of Virus Isolation Data from Environmental Samples.* EPA Contract No. 68-03-3196. Cincinnati, OH.

Blacklow, N.R. and Cukor, G. 1981. Viral gastroenteritis. *N. Eng. J. Med.* 304:397–406.

Block, J.C. 1983. Viruses in environmental waters, pp. 117–145 in Berg, G. (editor), *Viral Pollution of the Environment,* CRC Press, Boca Raton, FL.

Bolivar, R., Conklin, R.H., Vollet, J.J., Pickering, L.K., Dupont, H.L., Walters, D.L., and Kohl, S. 1978. Rotavirus in travelers' diarrhea: study of an adult student population in Mexico. *J. Infect. Dis.* 137:324–327.

CDC (Centers for Disease Control). 1985. Hepatitis surveillance. Report No. 49. CDC, Atlanta, GA.

Cherry, J.D. 1981. Nonpolio enteroviruses: coxsackieviruses, echoviruses, and enteroviruses, pp. 1316–1365 in Feigin, R.D. and Cherry, J.D. (editors), *Textbook of Pediatric Infectious Diseases*, Saunders, Philadelphia.

Clarke, N.A. and Kabler, P.W. 1964. Human enteric viruses in sewage. *Health Lab. Sci.* 1:44–50.

Coin, L. 1966. Modern microbiological and virological aspects of water pollution, pp. 1–10 in Jaag, O. (editor), *Advances in Water Pollution Research*, Pergamon Press, London.

Craun, G.F. 1986. *Waterborne Diseases in the United States*, CRC Press, Boca Raton, FL.

Dolin, R., Treanor, J.J., and Madore, P. 1987. Novel agents of viral gastroenteritis. *J. Infect. Dis.* 155:365–371.

Gerba, C.P. 1983. Methods for recovering viruses from the water environment, pp. 19–35 in Berg, G. (editor), *Viral Pollution of the Environment*, CRC Press, Boca Raton, FL.

Gerba, C.P. 1987. Transport and fate of viruses in soils: field studies, pp. 141–154 in Rao, V.C. and Melnick, J.L. (editors), *Human Viruses in Sediments, Sludges and Soils*, CRC Press, Boca Raton, FL.

Gerba, C.P. and Bitton, G. 1984. Microbiol pollutants, their survival and transport pattern to ground water, pp. 65–88 in Bitton, G. and Gerba, C.P. (editors), *Groundwater Pollution Microbiology*, John Wiley and Sons, N.Y.

Gerba, C.P. and Haas, C.N. 1988. Assessment of risks associated with enteric viruses in contaminated drinking water, pp. 489–494 in Lichtenberg, J.J., Winter, J.A., Weber, C.I., and Fradkin, L. (editors), *Chemical and Biological Characterization of Sludges, Sediments, Dredge Spoils, and Drilling Muds. ASTM STP 976*, American Society for Testing and Materials, Philadelphia.

Gerba, C.P., Rose, J.B., DeLeon, R., Toranzos, G., and Margolin, A. 1988. Viral contamination of ground water in Arizona. *J. Arizona-Nevada Acad. Sci.* 23:47–48.

Gerba, C.P., Singh, S.N., and Rose, J.B. 1985. Waterborne viral gastroenteritis and hepatitis. *CRC Crit. Rev. in Environmental Control* 15:213–236.

Goyal, S.M. and Gerba, C.P. 1979. Comparative adsorption of human enteroviruses, simian rotavirus, and selected bacteriophages to soils. *Appl. Environ. Microbiol.* 38:241–247.

Gust, I.D. and Purcell, R.H. 1987. Report of a workshop: waterborne non-A, non-B hepatitis. *J. Infect. Dis.* 156:630–635.

Guttman-Bass, N. and Fattal, B. 1984. Analysis of tap water for viruses: results of a survey. *Water Sci. and Technol.* 17:89–96.

Hejkal, T.W., Keswich, B., LaBelle, R.L., Gerba, C.P., Sanchez, Y., Dreesman, G., and Hafkin, B. 1982. Viruses in a community water supply associated with an outbreak of gastroenteritis and infectious hepatitis. *J. Amer. Water Works Assoc.* 74:318–321.

Jiang, S., Estes, M.K., and Metcalf, T.G. 1987. Detection of hepatitis A virus by hybridization with single stranded RNA probes. *Appl. Environ. Microbiol* 53:2487–2495.

Jiang, S., Estes, M.K., Metcalf, T.G., and Melnick, J.L. 1986. Detection of hepatitis A virus in seeded estuarine samples by hybridization with cDNA probes. *Appl. Environ. Microbiol.* 52:711–717.

Kaplan, J.E., Gary, G.W., Baron, R.C., Singh, N., Schonberger, L.B., Fieldman, R., and Greenberg, H.B. 1982. Epidemiology of Norwalk gastroenteritis and the role of Norwalk virus in outbreaks of acute nonbacterial gastroenteritis. *Ann. Internal Med.* 96:757–761.

Keswic, B.H., Gerba, C.P., DuPont, H.L., and Rose, J.B. 1984. Detection of enteric viruses in treated drinking water. *Appl. Environ. Microbiol.* 47:1290–1294.

Kulski, J.K. and Norval, M. 1985. Nucleic acid probes in diagnosis of viral diseases of man and brief review. *Arch. Virol.* 83:3–15.

Kurtz, J.B. and Lee, T.W. 1987. Astroviruses human and animal, pp. 92–101 in *Novel Diarrhea Viruses*, John Wiley and Sons, Chichester, OK.

Lednar, W.M., Lemon, S.M., Kirkpatrick, J.W., Redfield, R.R., Fields, M.L., and Kelley,

P.W. 1985. Frequency of illness associated with epidemic hepatitis A virus infection in adults. *Am. J. Epidemiol.* 122:226–233.

Lippy, E.C. and Waltrip, S.C. 1984. Waterborne disease outbreaks—1946–1980: a thirty-five-year perspective. *J. Amer. Water Works Assoc.* 76:60–67.

Madeleley, C.R. and Cosgrove, B.P. 1975. 28 nm particles in feces in infantile gastroenteritis. *Lancet* 2:451–452.

Madore, H.P., Treanor, J.J., and Dolin, R. 1986. Characterization of the Snow Mountain Agent of viral gastroenteritis. *J. Virol.* 58:487–492.

Marzouk, Y., Goyal, S.M., and Gerba, C.P. 1979. Prevalence of enteroviruses in ground water in Israel. *Ground Water* 17:487–491.

Melnick, J.L. 1982. Classification of hepatitis A virus as enterovirus type 72 and hepatitis B as Hepadnavirus type 1. *Intervirology* 18:105–106.

Margolin, A.B., Hewlett, M.J., and Gerba, C.P. 1986. Use of a cDNA dot-blot hybridization technique for detection of enteroviruses in water. pp. 187–95. Water Quality Technology Conference Proceedings, *Amer. Water Works Assoc.* Denver, CO.

Margolin, A.B., Richardson, K.J., DeLeon, R., and Gerba, C.P. 1989. Application of gene probes to the detection of enteroviruses in ground water. pp. 165–270, in Larson, R.A. (editor), *Biohazards in Drinking Water*, Lewis Publishers, Chelsea, MI.

Melnick, J.L. and Gerba, C.P. 1982. Viruses in surface and drinking waters.*Environ. Intl.* 7:3–7.

Nestor, I. and Costin, L. 1976. Presence of certain enteroviruses (coxsackie) in sewage effluents and river waters of Romania. *J. Hyg. Epidemiol. Microbiol. Immunol.* 20:137–149.

Nestor, I. Costin, L., Sourea, D., and Ionescu, N. 1978. Investigations on the presence of enteroviruses in drinking water. *Rev. Roum. Med.* 29:203–207.

Paul, J.R. and Trask, J.D. 1941. The virus of poliomyelitis in stools and sewage. *J. Am. Med. Assoc.* 116:493–497.

Payment, P., Trudel, M. and Plante, R. 1985. Elimination of viruses and indicator bacteria at each step of treatment during preparation of drinking water at seven water treatment plants. *Appl. Environ. Microbiol.* 49:1418–1428.

Provost, P.J. and Hilleman, M.R. 1979. Propagation of human hepatitis A virus in cell culture in vitro. *Proc. Soc. Exp. Biol. Med.* 160:213–221.

Rabyshko, E.V. 1974. Certain problems of the circulation of enteroviruses in the environmental objects. *Gig. Sanit.* 39:105–106.

Richardson, K.J., Margolin, A.B., and Gerba, C.P. 1988. A novel method for liberating viral nucleic acid for assay of water samples with cDNA probes. *J. Virological Methods* 22:13–21.

Rose, J.B., Gerba, C.P., Singh, S.N., Toranzos, G.A., and Keswick, B. 1986. Isolation of entero-and rotaviruses from a drinking water treatment facility. *J. Amer. Water Works Assoc.* 78:56–61.

Rose, J.B. and Gerba, C.P. 1986. A review of viruses in drinking water. *Current Practices in Environ. Engr.* 2:119–141.

Rotbart, H.A., Levin, M.J., and Villarreal, L.P. 1984. Use of subgenomic poliovirus DNA hybridization probes to detect major subgroups of enteroviruses. *J. Clin. Microbiol.* 20:1105–1108.

Slade, J.S. 1985. Viruses and bacteria in a chalk well. *Water Sci. Technol.* 17:111–125.

Smith, E.M. and Gerba, C.P. 1982. Development of a method for the detection of human rotavirus in water. *Appl. Environ. Microbiol.* 43:1440–1450.

Sobsey, M.D. 1976. Methods for detecting enteric viruses in water and wastewater, pp. 89–138 in Berg, G., Bodily, H.L., Lennette, E.H., Melnick, J.L., and Metcalf, T.G. (editors), *Viruses in Water*, Amer. Public Hlth. Assoc., Washington, D.C.

Toranzos, G.A., Gerba, C.P., Zapata, M., and Cardona, F. 1986a. Presence de virus enteriques dans des eaux de consommation a cochabamba (Bolivie). *Revue intern. des Sciences de l'eau* 2:91–93.

Toranzos, G.A., Hanssen, H., Gerba, C.P. 1986b. Occurrence of enteroviruses and rotaviruses in drinking water in Colombia. *Wat. Sci. Techn.* 18:109–114.

Tyler, J.M. 1982. Viruses in fresh and saline waters, pp. 42–63 in Bulter, M., Medlen, A.R., and Morris, R. (editors), *Viruses and Disinfection of Water and Wastewater,* University of Surrey, U.K.

United States Environmental Protection Agency (USEPA). 1975. National interim primary drinking water regulations. *Federal Register.* 73:59568–59588.

Van Olphen, M., Kapsenberg, J.G., Van de Baan, E., and Kroon, W.A. 1984. Removal of enteric viruses from surface water at eight waterworks in the Netherlands. *Appl. Environ. Microbiol.* 47:927–932.

Walter, R., Gerba, C.P., and Farrah, S.R. 1989. *Contamination of the Environment by Viruses and Methods of Control.* Lewis Publishers, Chelsea, MI., in press.

Ward, R.R. and Akin, E.W. 1984. Minimum infectious dose of animal viruses. *CRC Crit. Rev. Environ. Contr.* 14:297–310.

Ward, R.L., Bernstein, D.I., and Young, E.C. 1986. Human rotavirus studies in volunteers of infectious dose and serological response to infection. *J. Infect. Dis.* 154:871–877.

Wong, D.C., Purcell, R.H., Sreenivasan, M.A., Prasad, S.R., and Pavri, K.M. 1980. Epidemic and endemic hepatitis in India: evidence for a non-A, non-B hepatitis virus aetiology. *Lancet* ii:876–878.

Part 5

Methods and Monitoring in Drinking Water Microbiology

19

The Presence-Absence Test for Monitoring Drinking Water Quality

James A. Clark

Laboratory Services Branch, Ontario Ministry of the Environment, Rexdale, Ontario, Canada

Microbiological examinations of drinking water samples on a routine basis have been performed for many years to determine the quality of water in municipal distribution systems. Methods of analysis were developed chiefly to demonstrate the presence and numbers of total coliform bacteria in samples, so that a measure of the degree of pollution could be determined. Recently, the necessity for determining numbers of coliform bacteria in a sample was questioned based on studies which have shown that these organisms were irregularly distributed throughout municipal water systems (Pipes and Christian, 1984). Instead, the frequency of occurrence of samples positive for coliform bacteria was considered as being more representative of the overall microbiological water quality.

In earlier years, the standard test for total coliforms consisted of choosing selected volumes of a water sample and inoculating them into tubes of lactose broth with incubation at 37°C for 24–48 hours. Tubes suspected of being positive for total coliforms had acid and gas formation and inoculum from these tubes was streaked on eosin-methylene blue agar plates. Typical colonies were confirmed in lactose broth and brilliant green lactose bile 2% (Weiss and Hunter, 1939). A disadvantage of this test was that a confirmed positive numerical result could take three or four days, because incubation periods were at least 24 hours for each of the different stages of the test. For quantitative purposes, this test became known as the most probable number (MPN) procedure.

The membrane filter (MF) technique was introduced in the late 1950's as an alternative to the MPN procedure. The MF technique had the ability, when an appropriate volume of sample was chosen, to produce discrete colonies representative of individual bacterial cells. Total coliform bacteria formed green

399

metallic, sheen-like colonies when the membrane filter was placed on m-Endo medium. The colonies were initially transferred to confirmatory media, such as lactose broth and brilliant green lactose bile 2%. When sufficient confidence was gained in recognizing which colonies were likely those of total coliform bacteria, the confirmatory procedures were no longer performed on each and every colony. The MF technique was shown to give almost identical results to the MPN procedure, but had the added advantages of providing results in 24 hours with less work required to perform the analysis (Thomas, Woodward and Kabler, 1956).

19.1 Development of the Presence-Absence (P-A) Test

Drinking water samples submitted from properly operated water treatment plants and their associated distribution systems generally have good quality water and produce few positive tests for total coliform bacteria during the course of a year. This characteristic and the preponderance of negative membrane filter tests prompted initiatives to devise a simpler and less expensive method for examining drinking water samples. The idea of using a broth medium, similar to that used in the MPN tube test, but simplifying the test to a single tube or bottle capable of analyzing 100 ml of sample for total coliform bacteria was considered worth investigating.

The initial studies on a presence-absence (P-A) test used double strength MacConkey broth modified by the addition of 10 g/l of tryptone (Clark, 1968). The test was performed by adding 50 ml of the water sample to 50 ml of the double strength, modified broth in a glass, screw-capped bottle containing an inverted Durham tube to trap any gas produced. Before the sample was added, the double strength medium was removed from the inverted tube by tilting the bottle to introduce a pocket of air. After the sample was added and the medium and sample thoroughly mixed, the medium was reintroduced into the inverted tube by a second reverse tilting manipulation. P-A bottles were incubated at 35°C for five days and checked daily for growth, acid, or acid and gas production. Presumptive positive bottles had inoculum transferred to confirmatory media, such as brilliant green bile 2% lactose broth, EC broth or MacConkey broth. Inoculum was also streaked on a MacConkey agar plate.

Several years of comparative tests were conducted in which raw and drinking water samples submitted to the Ontario Ministry of the Environment (MOE) Laboratory for analysis were given parallel membrane filter and presence-absence tests. At that time, the MF test was performed using m-Endo broth in a saturated filter pad. Any sample producing coliform or sheen colony counts on the membrane filter had representative sheen colonies picked and subjected to confirmatory tests similar to that done for presumptive positive P-A bottles.

The results of the first study reported in 1968 revealed that the P-A test produced 478 confirmed positive results for coliforms, whereas the MF test produced only 317 confirmed positive results (Clark, 1968). The second and third studies, reported in 1969 and 1973, yielded similar results. The P-A test produced approximately double the number of confirmed positive results for members of the coliform group, when compared with the MF test, as shown in Table 19.1 (Clark, 1969, Clark and Vlassoff, 1973).

Confirmatory tests set up to differentiate fecal coliforms, total coliforms, and anaerogenic coliforms during this period were sometimes negative for these organisms, even when presumptive positive P-A bottles had an acid or acid and gas reaction. Further investigation demonstrated that other bacteria were also responsible for these types of reactions. These bacteria included fecal streptococci, aeromonads such as *Aeromonas hydrophila*, pseudomonads such as *Pseudomonas aeruginosa*, and clostridial organisms such as *Clostridium perfringens*. Later work also showed that many weak acid reactions in P-A bottles were caused by micrococcus or staphylococcus organisms such as *Staphylococcus aureus* (Clark, unpublished data). These other bacteria were considered as secondary indicators of poor water quality, particularly when coliforms were not isolated from either P-A bottles or MF tests. Although considered as secondary indicators to coliform bacteria, these other organisms have often been found in polluted waters and some have been reported as pathogens (Seidler, Allan, Lockman, Calwell, Joseph and Daily, 1980, Seyfried and Cook, 1984).

19.2   Indicator Bacteria Isolated from P-A Bottles

The demonstration that a variety of indicator organisms could be isolated from presumptive positive P-A bottles prompted an investigation to determine the frequency of occurrence of these organisms in raw water and drinking water samples. This information was reported in 1973 and covered the period from 1968 to 1971 (Clark and Vlassoff, 1973). A second study reported in 1982, covering the years 1979 and 1980, provided additional information (Clark,

Table 19.1 Comparison of membrane filter and presence-absence test results for three study periods

Type of result	Year reported		
	1968	1969	1973
MF +, P-A +	255	198	546
MF −, P-A +	223	260	780
MF +, P-A −	62	47	53
MF −, P-A −	8224	7816	13107
Total samples	8764	8321	14486

+ = Confirmed positive result for coliform bacteria
− = Coliform bacteria were not detected or organisms detected were not confirmed as coliforms.

Burger and Sabatinos, 1982). A tabulation of this information is summarized in Table 19.2. Frequency rates of recovering indicator bacteria from raw water samples were much higher in the 1973 study because surface water samples formed a larger portion of the samples analyzed than in the 1982 study, in which ground water samples were the predominant type of sample examined. The frequency recovery rates from drinking water samples were more nearly alike, although some variation did occur between the two study periods.

Fecal coliforms, total coliforms, and fecal streptococci had the highest recovery rates when raw surface water samples predominated among the samples analyzed in 1973. In the 1982 study, raw water samples, primarily from ground water supplies, had lower levels of fecal indicator bacteria with total coliforms and *Aeromonas* sp. being isolated more frequently. With drinking water samples, total coliforms had the highest recovery rate for both periods examined. The other bacterial indicators, which included fecal coliforms, fecal streptococci, *P. aeruginosa*, and *C. perfringens*, had higher recovery rates from drinking water samples in the 1973 study compared with the 1982 study. Similar high recovery frequencies of these indicators from the raw water samples examined in the 1973 study suggested a positive association between the indicator bacteria found in these two water types, but other less obvious factors may have influenced the recovery rates.

Another aspect of the study reported in 1973 provided further information on the bacterial indicators that could be isolated from presumptive positive P-A bottles. Sometimes, several indicators could be isolated in combination with each other and these multiple isolations appeared related to the degree of pollution in the samples. This relationship is shown in Table 19.3 by arranging the different combinations of indicators according to the coliform level at which

Table 19.2 Frequency of detection of indicator bacteria in raw and treated water samples

	Raw water[a]		Drinking water	
Indicator bacteria	1973	1982	1973	1982
Fecal coliforms	28.7%	0.6	1.6	0.3
Total coliforms	23.3	4.1	7.5	5.4
Aeromonas sp.	2.6	1.1	0.5	2.4
Fecal streptococci	20.4	0.4	1.2	0.2
Staphylococcus aureus	NA[b]	0.1	NA	0.1
Pseudomonas aeruginosa	5.4	0.1	0.4	0.2
Clostridium perfringens	6.6	—[c]	1.7	<0.1
Total samples	5438	11867	14486	41263

[a] The data give the percent of samples positive for the particular test. See text for explanation of variable recovery rates.
[b] NA = No analysis done at that time.
[c] — = Not detected.

they were isolated. As the number of total coliform colonies increased on membrane filter plates, the number of bacterial indicators found in combination with each other in parallel analyses with P-A tests was also shown to occur more frequently. When total coliform counts exceeded 100 per 100 ml, fecal coliforms were often isolated in combination with fecal streptococci, *P. aeruginosa*, and *C. perfringens*. Conversely, as total coliforms isolated by the MF method decreased to the next lower level, more of the indicators were present as double combinations, rather than triple combinations. Similarly, when total coliforms declined to the 1–10 per 100 ml level, fewer isolations of fecal coliforms occurred in P-A bottles in comparison with the number of isolations of only total coliforms. The shift was also evident at the <1 per 100 ml level, when fecal coliforms declined to 9% of their overall total. Total coliforms at this level formed 39% of the coliform isolations and with anaerogenic coliforms, the percentage rose to 62%.

Table 19.3 Effect of increasing coliform numbers on indicator bacteria combinations and on the response time to produce a presumptive positive P-A result

Indicator[a] bacteria found in P-A test	Total coliform MF counts				Time required to produce a presumptive positive P-A result				
	<1	1–10	11–100	>100	24h	48h	72h	96h	120h
FC/FS/PSA/CL			6	7	13				
FC/PSA/CL		2	5	9	15	1			
FC/FS/CL	2	3	34	39	74	4			
FC/FS/PSA		3	32	134	169				
FC/PSA		9	35	20	61	3			
FC/CL	12	14	57	19	92	8	2		
FC/FS	18	69	303	302	672	19	1		
FC	136	170	255	105	614	43	9		
TC/FS/PSA/CL		1	1		1	1			
TC/PSA/CL		3	3		5	1			
TC/FS/CL	2	3	8	5	18				
TC/FS/PSA		3	6	5	12	2			
TC/PSA	10	11	16	6	26	10	7		
TC/CL	23	30	39	19	80	28	3		
TC/FS	33	44	76	35	86	88	10	3	1
TC	644	401	302	91	522	532	200	114	70
AC/FS/CL		1			1				
AC/PSA	2	4			4	1	1		
AC/CL	8	5		1	3	10	1		
AC/FS	10	8	2	1	3	10	5	2	1
AC	315	116	51	15	52	165	133	92	55

[a] FC = Fecal coliforms   FS = Fecal streptococci
TC = Total coliforms   PSA = *Pseudomonas aeruginosa*
AC = Anaerogenic coliforms   CL = *Clostridium perfringens*

The amount of time required for a P-A bottle to produce a presumptive positive result after inoculation with a sample was shown to be partly related to the numbers of total coliforms in the sample. This relationship was indirectly demonstrated by comparing the two halves of Table 19.3. When the frequencies of the indicator bacteria, under the column headings for 24 to 120 hours to produce a presumptive positive P-A result, were compared with the corresponding frequencies under the column headings for coliform counts of <1 to >100 per 100 ml, the rapidity of the P-A reaction appeared reasonably well correlated with rising total coliform counts for the fecal coliform combinations. The relationship was not quite as apparent when the frequencies for other total coliform combinations were compared in the two halves of Table 19.3. An explanation for this difference may be that another factor, such as the degree of viability or injury of the bacterial cells, may be more responsible for reaction rates in P-A bottles. When the pollution in a water sample was reasonably fresh, the viability of bacterial cells would likely be less affected by environmental conditions and their lag phase might be shorter when introduced into a bacteriological growth medium.

The fecal coliform combinations produced presumptive positive reactions in the shortest period of time as 95% of the samples with these combinations showed this reaction within a 24 hour incubation period. Total coliform combinations had 77% of the P-A bottles with presumptive positives after an incubation period of 48 hours and 90% within 72 hours. Anaerogenic coliform combinations took more time to produce presumptive positive results with a 96 hour incubation period required to detect 90% of these organisms. This latter group might have originally functioned as fecal or total coliforms, but the stresses of the environment might have reduced or limited their physiological capabilities to respond when introduced into a bacteriological growth medium.

These results suggested that the standard time of 48 hours incubation for the presumptive portion of the MPN tube test for total coliforms might not be long enough. Additional incubation time might be required to allow injured or debilitated coliform cells to adjust or adapt to their new enriched environment in a tube or P-A bottle before growth and multiplication occurred. The extended incubation period for P-A bottles of four to five days might be partly responsible for the significantly greater number of positive P-A results, when compared with the results of parallel MF tests. On the other hand, the lower frequency of MF tests positive for coliform bacteria could be related to the composition of the recovery medium (McFeters, Cameron, and Le Chevallier, 1982).

19.3   Prediction of Coliform Presence by P-A Reactions

The analysis of drinking water samples by the P-A test resulted in various types of acid and gas reactions, when a presumptive positive test was produced. Although confirmatory tests were the best means of determining which indicator bacteria produced a presumptive positive result, the possibility of predicting the confirmed result by the type of reaction in the P-A bottle was considered useful in certain instances. Data from two studies reported in 1982 and 1985 were collected and tabulated in Table 19.4 (Clark, Burger and Sabatinos, 1982, Clark, 1985).

Acid reactions were separated into strong, medium, and weak reactions based on the yellowness of color development in the medium containing the bromocresol purple indicator. Gas production related to the amount of gas produced in the inverted tube in each P-A bottle. Later, in 1983, the use of inverted tubes in P-A bottles was discontinued as a laborsaving step, but not before parallel observations were made on P-A bottles relative to the volume of gas present in the inverted tubes and the degree of foaming which occurred when the medium in each P-A bottle was gently swirled to release dissolved gas. These observations were reported in the 1985 study as shown in Table 19.4 along with the comparable data from the 1982 report before the foaming reaction was put to use. Each of the acid, gas, and foam reaction totals was accompanied by the corresponding confirmation rate for total coliforms, where applicable.

Although gas and foam reactions were invariably accompanied by some degree of acidity in P-A bottles, the three types of reactions are displayed separately in Table 19.4. Acid reactions were more frequent than gas and foam reactions, as many of the indicator bacteria responsible for this type of reaction do not produce gas from lactose fermentation. The 1982 results for strong acid reactions show a confirmation rate no better than 54% for total coliforms. This rate changed to 76% in the 1985 report for coliform cultures producing strong

Table 19.4 Frequency of reactions in P-A bottles and their confirmation rate for total coliforms

Type of reaction	Reaction[a] symbol	Number of reactions in 1982	% confirmed	Number of reactions in 1985	% confirmed
Acid reactions	A	333	54	898	76
	(A)	122	41	728	42
	a	41	53	412	33
Gas reactions	G	78	94	258	99
	g	150	85	177	86
	(g)	50	62	88	67
Foam reactions	F	ND[b]		628	98
	f	ND		482	74

[a] A = bright yellow, strong acid
(A) = dark yellow, medium acid
a = slight yellow, weak acid

G = inverted tube with >10% gas
g = inverted tube with 10% gas
(g) = inverted tube with <10% gas

F = surface foam layer
f = slight foam at edge of medium

[b] ND = no data recorded during this period.

acid reactions. As other indicator bacteria, such as *Aeromonas* sp., were recovered from P-A bottles with acid reactions, this response was not considered suitable for predicting which indicator bacteria were present. In fact, in the 1982 study, acid reactions without gas resulted in *Aeromonas* sp., being isolated 28% of the time, coliforms 10%, and fecal streptococci 1%.

Gas and foam reactions, although produced less frequently, were more predictive for total coliform bacteria. In the 1982 study, production of >10% gas in the inverted tube resulted in a 94% confirmation rate for total coliforms. When the gas production volume in the inverted tube was 10% or less, the confirmation rates dropped to 85% and 62% respectively. Similar results were reported in 1985 (see Table 19.4). When no inverted tube was used in the P-A bottle, and the technician depended on the demonstration of foam production, the confirmation rate for total coliforms was 98%, if foaming was vigorous enough to cover the surface of the medium. When only slight foaming occurred, the confirmation rate fell to 74%. The high rate of total coliform confirmation in P-A bottles with abundant gas or foam production was considered sufficiently accurate for predictive purposes, that corrective action for water treatment plants could be recommended without waiting for the results of confirmation tests.

19.4 Identification of Isolates from P-A Tests

The isolation of indicator bacteria associated with presumptive positive P-A bottles was taken one step further by using Enterotubes (Hoffmann-La Roche Limited) to identify and group the isolates into bacterial genera (Clark and Pagel, 1977). The 1977 report showed that most of the cultures could be grouped into the genera *Enterobacter, Klebsiella, Citrobacter*, and *Escherichia* of the family *Enterobacteriaceae*. Cultures not belonging to this family frequently were members of the genus *Aeromonas*. When the identified cultures were separated into those isolated from MF tests and those from P-A tests, the frequencies of the genera from both tests were very similar except for *Klebsiella* and *Aeromonas*. The more frequent isolation of *Klebsiella* from MF tests was probably because of the large, moist, green sheen colonies formed by this organism on MF plates. As these colonies generally stood out from the others, they would tend to be selected over smaller, green sheen colonies for identification.

The more frequent isolation of *Aeromonas* from P-A tests can also be associated with the colony selection procedure. Technicians were requested to select mainly dark red or pink, lactose-fermenting colonies from MacConkey agar plates. However, inoculum streaked from presumptive positive P-A bottles onto MacConkey plates would often result in large, non-lactose-fermenting colonies, which were selected for identification if no lactose-fermenting colonies were present. The bacterial isolates from most of these colonies proved to belong to the genus *Aeromonas*.

The 1977 report also categorized the genera into isolates from raw water samples and those from drinking water samples. However, as this taxonomic identification of cultures encompassed several years, a more detailed listing of the cultures into species was reported in 1982 (Clark, Burger and Sabatinos, 1982). Table 19.5 summarizes the distribution of cultures found in raw water, drinking water and water from new distribution mains, if the isolation frequency for a species was at least 1% of the total cultures for one of the water types.

Escherichia coli was the most frequently isolated culture from raw water samples and was the only species to show a significant decline in frequency following water treatment. *Enterobacter cloacae* was the next most frequently isolated bacterium, but this organism showed a higher incidence in drinking water than in raw water. The next organism in frequency rank was *Aeromonas hydrophila*, which like *E. cloacae*, had a higher incidence in drinking water samples. Other species showing higher frequencies in drinking water included *E. agglomerans*, *Klebsiella oxytoca* and *Serratia* sp. Isolates of *K. pneumoniae*, *E. aerogenes*, and *Citrobacter freundii* had a similar incidence in both water types. With water samples from new distribution mains, *C. freundii* was most frequently isolated followed by *E. cloacae* and *A. hydrophila*. The higher incidence of *C. freundii* in water from new mains might be related to the amount of soil debris frequently associated with the construction of new distribution lines.

Table 19.5 Distribution of organisms isolated from raw, drinking, and new main water samples by P-A tests

Identification	Raw water	Drinking water	Water from new mains
Enterobacter cloacae	18%	26%	22%
E. agglomerans	3%	6%	3%
E. aerogenes	3%	3%	3%
E. hafniae	1%	<1%	<1%
Klebsiella pneumoniae	8%	8%	10%
K. oxytoca	3%	5%	6%
Escherichia coli	40%	19%	12%
Citrobacter freundii	6%	6%	23%
Serratia spp.	1%	2%	1%
Proteus spp.	1%	1%	<1%
Aeromonas hydrophila	9%	17%	17%
Others—oxidase positive	2%	1%	<1%
—oxidase negative	5%	6%	3%
Number of cultures	3036	7442	1036

19.5   Monitoring Drinking Water Quality by the P-A Test

The P-A test was developed primarily as an adjunct to current quantitative methods, i.e. the MF or MPN procedure, because of its qualitative nature. Whenever submissions of routine samples from a town or city had all or almost all samples regularly giving negative results by MF tests, the P-A test was substituted for the quantitative procedure. As an option, a number of samples, one in four or one in eight, were selected at random to receive both MF and P-A tests. This was done in case either total coliform or background colonies were sufficiently high on membrane filter plates to signal either unsafe or poor water quality. More samples could be analyzed within a given period of time by the P-A test, because its inoculation procedure was simpler and faster than performing an MF analysis.

A set of criteria definitions were developed to allow the use of the P-A test in place of either the MF or MPN procedures to monitor drinking water quality within Ontario (anonymous, 1984). An unsafe water quality result was represented by a P-A test, which produced a presumptive positive result within 48 hours. As all samples analyzed by the P-A test were kept for 24 hours before discarding, a sample giving a positive result within this time period could be located and reexamined by an MF test to provide a quantitative result within 48 hours. A telephone call would be made to the local MOE office, which in turn, would inform the appropriate staff responsible for submitting the positive sample(s) that a repeat sample(s) should be collected from the offending location, and on occasion, a sample from an adjacent site and another upstream on a feeder line. Repeat samples were analyzed by both MF and P-A tests. Personnel at the waterworks were also expected to check on chlorine residuals and take other corrective measures as necessary.

If a positive P-A result occurred after 48 hours, the water quality was categorized as poor by definition. This type of result was not expected to occur in more than 25% of the samples from a single submission, in successive submissions from the same sampling location, or in more than 10% of the samples submitted in any one month. If one or more aspects of this definition were exceeded, more extensive monitoring of the distribution system was recommended.

Increased interest in the P-A test, as a possible alternative for monitoring drinking water quality to current MF and MPN techniques, was shown by the introduction of a tentative presence-absence (P-A) coliform test in the 16th Edition of *Standard Methods* (APHA 1985) and by recent work on coliform densities in distribution systems (Pipes and Christian, 1984). Studies on patterns of dispersion of coliforms in a number of small water distribution systems demonstrated that coliforms were not randomly dispersed, but could occur sporadically in large numbers in a few samples, even though the arithmetic mean for a water system was quite low. The authors concluded that averaging the number of coliforms in samples collected for monitoring purposes provided a poor estimate of the mean coliform density within a distribution system.

Earlier, several statistical models were examined by using actual data obtained from confirmed coliform counts to determine if more accurate methods of defining the mean coliform density in a water distribution system could be found (Christian and Pipes, 1983). However, limitations in the analytical procedures for doing coliform analyses affected the manner in which statistical models could be successfully applied to accurately estimate coliform densities. Consequently, an alternative approach was proposed and was called frequency-of-occurrence monitoring (Pipes and Christian, 1984). This approach used a presence-absence concept in which the frequency of samples giving positive tests provided the criterion value to determine acceptable water quality within a distribution system. The term "frequency-of-occurrence monitoring" was chosen to avoid confusion with the P-A test. Although the latter test could be one of the procedures used to analyze samples, membrane filter tests and other versions of the fermentation tube method would also be acceptable. A more accurate estimate of coliform contamination within distribution systems was expected using the frequency-of-occurrence monitoring approach, as it would avoid the problem encountered by other statistical methods that were affected by high or low range data truncation.

19.6 Other Presence-Absence Studies

An early presence-absence test used triple strength nutrient broth to enhance coliform growth in samples before an inoculum was streaked onto eosin-methylene blue agar, followed by confirmation of typical coliform colonies (Weiss and Hunter, 1939). This method had several merits, but was not followed up by others until studies were done on a presence-absence test using MacConkey broth and later, a lactose and lauryl tryptose broth combination (Clark, 1968, Clark, Burger and Sabatinos, 1982). Subsequently, other workers compared the relative efficiency of the P-A test with the MF and MPN tests for detection of coliform bacteria and found the P-A method produced more positive tests (Jacobs, Zeigler, Reed, Stukel and Rice, 1986, Fujioka, Kungskulniti, and Nakasone, 1986). One study that compared MF and P-A tests found the MF test produced more positive tests than the P-A test (Pipes, Minnigh, Moyer and Troy, 1986). An examination of the data collected in their study over a three year period showed that the MF test performed better than the P-A test in only one of the three years.

Another study investigated the effect of delayed incubation on P-A test results as well as field inoculation of samples directly into bottles containing P-A medium. No statistically significant differences were found between field-inoculated and laboratory-inoculated tests and the delayed incubation tests showed no significant differences compared to immediate incubation of P-A tests (Pipes, Minnigh and Troy, 1986). The results of this study were of interest because sample bottles could be made available with additives to promote the growth of coliforms during transportation to the laboratory, if the frequency-of-occurrence monitoring system were put into practice. Enhanced coliform numbers in samples would likely shorten the response time to establish sample locations giving a positive test for coliforms.

19.7 Summary and Conclusions

The value of the presence-absence test has been assessed by many years of comparative analyses with the membrane filter technique. For the most part, the P-A test produced a higher number of positive tests, demonstrating greater sensitivity for detection of coliforms than the MF test. In addition, a variety of other indicator bacteria could be isolated from the initial inoculation of one P-A bottle, providing an opportunity to learn more about the nature of the pollution in the sample.

Identification of the indicator bacteria detected and tabulation of their frequencies in different combinations showed an association with pollution levels as determined by MF coliform counts and the rate at which a P-A test became positive. P-A bottles producing a strong acid reaction and vigorous production of gas or foam in the medium invariably proved to have coliforms present as shown by later confirmation tests. Identification of individual cultures from raw and drinking water samples showed *E. coli* predominated in raw water and *E. cloacae* in drinking water.

Monitoring drinking water quality by P-A tests was initially done in conjunction with MF tests to provide a quantification and comparison of coliforms in any sample found to show pollution. Recent work of others has suggested that if frequency-of-occurrence monitoring were given official status for determining drinking water quality, the P-A test or any variations thereof might be appropriate for accurately estimating coliform frequencies in a water supply. Implementation of frequency-of-occurrence monitoring would permit the continuance of MF or MPN procedures for water quality testing without the emphasis on coliform densities in individual samples. This new monitoring method would also foster the development of new analytical techniques which could take advantage of the presence-absence concept to automate or simplify microbiological methods for determining water quality within distribution systems.

References

American Public Health Association. 1985. Presence-absence (P-A) coliform test. pp. 882–884 in *Standard Methods for the Examination of Water and Wastewater*, 16th Edition. American Public Health Association Inc., Washington, D.C.

Anonymous, 1984. *Ontario Drinking Water Objectives*. Microbiological Characteristics—Health. pp. 9–12. A publication of the Ontario Ministry of the Environment.

Christian, R.R. and Pipes, W.O. 1983. Frequency distribution of coliforms in water distribution systems. *Appl. Environ. Microbiol.* 45:603–609.

Clark, J.A. 1968. A presence-absence (P-A) test providing sensitive and inexpensive detection of coliforms, fecal coliforms and fecal streptococci in municipal drinking water supplies. *Can. J. Microbiol.* 14:13–18.

Clark, J.A. 1969. The detection of various bacteria indicative of water pollution by a presence-absence (P-A) procedure. *Can. J. Microbiol.* 15:771–780.

Clark, J.A. and Vlassoff, L.T. 1973. Relationships among pollution indicator bacteria isolated from raw water and distribution systems by the presence-absence (P-A) test. *Health Lab. Sci.* 10:163–172.

Clark, J.A. and Pagel, J.E. 1977. Pollution indicator bacteria associated with municipal raw and drinking water supplies. *Can. J. Microbiol.* 23:465–470.

Clark, J.A., Burger, C.A. and Sabatinos, L.E. 1982. Characterization of indicator bacteria in municipal raw water, drinking water and new main water samples. *Can. J. Microbiol.* 28:1002–1013.

Clark, J.A. 1985. Performance characteristics of the presence-absence test. pp. 129–141 in *Proceedings of the AWWA Water Quality Technology Conference*, Houston, Texas.

Fujioka, R., Kungskulniti, N., and Nakasone, S. 1986. Evaluation of the presence-absence test for coliforms and the membrane filter method for heterotrophic bacteria. pp. 271–283 in *Proceedings of the AWWA Water Quality Technology Conference*, Portland, Oregon.

Jacobs, N.J., Zeigler, W.L., Reed, F.C., Stukel, T.A., and Rice, E.W. 1986. Comparison of membrane filter, multiple-fermentation-tube, and presence-absence techniques for detecting total coliforms in small community water systems. *Appl. Environ. Microbiol.* 51:1007–1012.

McFeters, G.A., Cameron, S.C., and Le Chevallier, M.W. 1982. Influence of diluents, media, and membrane filters on the detection of injured waterborne coliform bacteria. *Appl. Environ. Microbiol.* 43:97–103.

Pipes, W.O. and Christian, R.R. 1984. Estimating mean coliform densities of water distribution systems. *J. Am. Water Works Assoc.* 76:60–64.

Pipes, W.O., Minnigh, H.A., Moyer, B., and Troy, M.A. 1986. Comparison of Clark's presence-absence test and the membrane filter method for coliform detection in potable water samples. *Appl. Environ. Microbiol.* 52:439–443.

Pipes, W.O., Minnigh, H.A., and Troy, M. 1986. Field inoculation of Clark's P-A test for coliform detection. pp. 305–315 in *Proceedings of the AWWA Water Quality Technology Conference*, Portland, Oregon.

Seidler, R.J., Allan, D.A., Lockman, H., Colwell, R.R., Joseph, S.W., and Daily, O.P. 1980. Isolation, enumeration and characterization of *Aeromonas* from polluted water encountered in diving operations. *Appl. Environ. Microbiol.* 39:1010–1018.

Seyfried, P.L. and Cook, R.J. 1984. Otitis externa infections related to *Pseudomonas aeuginosa* levels in five Ontario Lakes. *Can. J. Public Health.* 75:83–91.

Thomas Jr., H.A., Woodward, R.L., and Kabler, P.W. 1956. Use of molecular filter membranes for water potability control. *J. Am. Water Works Assoc.* 48:1391–1402.

Weiss, J.E. and Hunter, C.A. 1939. Simplified bacteriological examination of water. *J. Am. Water Works Assoc.* 31:707–713.

20

Statistical Approaches to Monitoring

Charles N. Haas and Barbara Heller

Illinois Institute of Technology, Chicago, Illinois

20.1 Introduction

The assessment of microbiologically-associated risk from a given water supply requires knowledge of the dose-response function of potential consumers to contamination as well as the probability distribution of microorganisms in the water supply. Fuhs (1975) has presented this concept in the context of recreational water quality, however, it is equally valid in the context of drinking water quality.

Conventionally, the coliform group of organisms is used as a microbial water quality indicator for drinking water, although, to the author's knowledge, no explicit dose-response functions for this indicator group exist for finished drinking water. While there are dose-response relationships for bacterial and viral pathogens (Haas, 1983), little information on statistical distributions of pathogens in potable waters is available.

Therefore, this chapter is restricted to statistical distributions of indicator organisms, with the understanding that, in the context of risk assessment, substantial additional information is needed. In particular, it is desired to determine the distribution function of microorganisms in the water consumed. This point deserves emphasis since current microbial enumeration techniques are (with the possible exception of recent fluorescent microscopic methods) based on some type of growth process. Thus it becomes necessary to ascertain any contributions which the variability of the assay technique make to estimation of intrinsic environmental variability.

20.2 The Poisson Distribution

As a reference point for consideration of variability of an assay technique, the Poisson distribution is convenient. The Poisson distribution specifies that the probability of observing N organisms in a sample volume V from a water where the mean density of organisms per unit volume is u can be given by:

$$P(N) = (uV)^N e^{-uV}/N! \qquad\qquad 1$$

Eisenhart and Wilson (1943) have reviewed the early uses of this distribution, including application to the enumeration techniques for microorganisms.

The Poisson distribution can be formally derived from a set of assumptions. Let Λ be the (integral) number of bacteria that can occupy a certain volume of water. Let k be the fraction (between 0 and 1) of water from which the sample is taken which has bacteria present. Let N be the number of bacteria in the sample (determined by some type of assay). Then, by the binomial theory for sampling, the probability that N bacteria will be in the sample, providing that Λ is large with respect to N (so that sampling with replacement is a valid assumption), is given by:

$$P(N) = [\Lambda!/N!(\Lambda-N)!] \, k^N(1-k)^{\Lambda-N} \qquad\qquad 2$$

$$\text{for } N=0,1,\ldots,\Lambda$$

If Λ is large and both N/Λ and k are small, then application of the Stirling approximation to the factorial with expansion of logarithms to a one term Taylor series yields:

$$P(N) = (k\Lambda)^N e^{-k\Lambda}/N! \qquad\qquad 3$$

$$\text{for } N=0,1,\ldots,\Lambda$$

which, upon the substitution $uV=k\Lambda$ yields equation 1.

The Poisson distribution requires only one parameter to completely specify it—the mean density. Furthermore, the variance of microbial numbers which are found is identical to the mean. The Poisson distribution arises when the microorganisms have a random spatial distribution in the water being sampled. By this one means that:

1. The mean density per unit volume does not vary over the region being sampled.

2. There are no interactions among the microorganisms which affect their spatial distribution.

Condition 2 would be violated if, for example, the position of a given microorganism, at a given time, either encouraged or inhibited the presence of a neighboring organism (Diggle, 1983).

It is often found that the distribution of microorganisms in water samples does not follow the Poisson distribution, either in the sense that the precise distribution of equation 1 is not obeyed or that the variance among replicate samples is different (generally greater) from that given by the Poisson analysis.

There are many possible explanations for such features. First, some experimental evidence for failure of the Poisson assumption to describe experimental data will be presented. Second, mechanisms which could account for such effects will be discussed, then diagnostic tests for such behavior will be presented.

The statement by Eisenhart and Wilson (1943) on this point remains quite applicable:

> Whenever a very complex population . . . is studied . . . departure of the observed distributions . . . from the theoretical distribution is almost always noted . . . [T]he use of the data for drawing any profound conclusions is highly questionable. Instead steps should be taken to locate the origin of the abnormal variation, and, if possible, to eliminate it . . .

20.3   Experimental Evidence for Non-Poisson Behavior

In one of the first studies on membrane filtration quantitation of total coliforms, Thomas and Woodward (1955) noted that variances among replicate coliform analyses of the same sample were, in many cases, greater than those predicted from the Poisson distribution.

Pipes et al. (1977) reported that examination of replicates of a single sample of water as well as sets of samples of a water taken over a short time period generally formed a frequency distribution more consistent with the negative binomial or log-normal than with the Poisson. In a more extensive study of water distribution samples obtained over a period of time, similar conclusions were reached (Christian and Pipes, 1983). El-Shaarawi et al. (1981) demonstrated that the frequency distribution of total coliforms obtained during cruises of Lake Erie lasting several weeks were non-Poisson, but could be described by the negative binomial distribution.

The present authors have tested a number of data sets consisting of replicate coliform determinations on series of water samples taken over time (Haas and Heller, 1986). The majority of the data sets tested showed greater variability than predicted by the Poisson distribution, and of these, all were consistent with a negative binomial distribution. In particular, Poisson statistics were sufficient to characterize data in an extensive series of results obtained by McDaniels and Bordner (1983) for pure cultures of organisms. However, for indigenous coliforms there was a highly significant difference from Poisson behavior.

20.4   Mechanisms for Non-Poisson Behavior

There are two broad ways in which deviations from Poisson behavior could occur between repeated samples from a water. The first mechanism involves a systematic interaction between bacteria which renders the probability that a single bacterium being sampled (k in equation 2) is no longer independent of the number of bacteria in the sample. Stated differently, the number of bacteria is dependent on spatial or temporal location of the sample. In other words, conditions for homogeneous Poisson processes are not met. The second mechanism is that, while the Poisson distribution applies to the water body being sampled, there is a variable inefficiency in the enumeration procedure itself which masks the true environmental distribution.

Clearly, the two mechanisms have different consequences. In the first case, denoted as the "ecological" model, the distribution of microorganisms is, in fact, not random in the environment. Hence, any susceptible human or non-human exposed to this environment will experience a similar non-random distribution of risk. For the second mechanism, identified as the "analytical" model, however, the masking effect not only means that the distribution determined by the laboratory quantitation method is not precisely that experienced by a susceptible individual, but also very likely that the average microbial density is improperly estimated.

Statisticians have developed many alternatives to the Poisson model by modifying the formula for the distribution from various points of view. Some of these techniques are primarily mathematical, while others are motivated by idealized physical processes such as "clumping" or "contagion". But it is important to distinguish between a given statistical model which may appear to fit the data (empirical approach) and a physical mechanisms which may lead to that model. Any given "alternative" model may be explained by several possible physical mechanisms. The fact that a given model may be *derived* mathematically is not proof that such a mechanism underlies the physical process.

20.5 The Probability Generating Function

For several points below, it is useful to introduce the concept of the probability generating function (p.g.f.) $g(z)$ of the variable z. This function of the original random discrete variable x is defined as the expectation (average) of the quantity z^x. This function has the following property:

$$P(N) = g^{(N)}(0)/N! \qquad\qquad 4$$

In other words, the probability of a random discrete variable obtaining a value of N is obtained from the N'th term in a Taylor series expansion of the generating function evaluated at zero.

The p.g.f. has several additional useful properties (Douglas, 1980):

1. If $g(z)$ and $h(z)$ are the p.g.f.'s for two random variables, then the p.g.f. for the sum of the two random variables is given by $g(z)h(z)$.

2. If a random number of independent identically distributed random

variates are to be summed, and if $g_N(z)$ is the p.g.f. for the distribution of the *number* of summed terms and $g_X(z)$ is the p.g.f. for the distribution of the *values* of the summed terms, then the p.g.f. for the distribution of the *summed values* is given by $g_N(g_X(z))$—i.e., one of the p.g.f.s becomes the argument for the other. This type of process yields a distribution called stopped or convoluted.

3. If $g(z,a)$ is a p.g.f. with a parameter "a" (for example, the mean), and if "a" itself is a random variable with a cumulative distribution function $G(a)$ (which is the probability that "a" has a value less than or equal to "a"), then the p.g.f. for the overall process is given by:

$$\int g(z,a) \, dG(a) \qquad \qquad 5$$

where the integration is taken over the entire range of values which "a" can assume. The distribution resulting from this process is called a mixture distribution.

For the particular important case of the Poisson distribution, the p.g.f. is given by:

$$g(z) = \exp(-uV + uVz) \qquad \qquad 6$$

where u is the mean concentration of microorganisms and V is the sample volume.

20.6   Ecological Mechanisms for Non-Poisson Distributions

One possible mechanism which may account for non-Poisson distributions of counts amongst replicates is the existence of clumps of organisms (Pielou, 1977). If there are randomly distributed *clumps* of bacteria in the water, then the probability that a given water sample of volume V will contain J clumps may be given by the Poisson distribution as:

$$P(J) = (qV)^J e^{-qV}/J! \qquad \qquad 7$$

where q is the mean number of clumps per unit volume. If it is further assumed that the number of *enumerable* bacteria per clump is distributed according to the logarithmic distribution such that the probability of recovering K bacteria from a single clump (where K is a nonzero positive integer) is given by:

$$P(K) = -[1/1n(1-a)]a^K/K! \qquad \qquad 8$$

where a is the parameter of the distribution, and the mean number of bacteria in a clump is given as $-a/[(1-a)1n(1-a)]$ (Engen, 1978). This problem results in a stopped distribution. Since the p.g.f. of the logarithmic distribution is $1n(1-az)/1n(1-a)$, by property 2 of section 20.5, the p.g.f. for the overall number of bacteria in a sample is given by:

$$g(z) = \exp\{-\Lambda[1-1n(1-az)/1n(1-a)]\} \qquad 9$$

with suitable substitutions, this may be simplified to (Pielou, 1969):

$$(Q-Pz)^{-k} \qquad 10$$

with $\Lambda = k\ 1n(Q)$
$a = P/Q$
$P = Q-1$

Equation 10 is the p.g.f. for the negative binomial distribution with a mean (uV) equal to kP and variance of kPQ. From use of the definition of the p.g.f., the probability of obtaining N bacteria in a sample volume of V can be written as:

$$P(N) = [\Gamma(k+N)/\Gamma(k)N!]\ (uV/k)^N\ [k/(k+uV)]^{k+N} \qquad 11$$

There are biologically plausible reasons for acceptance of this type of distribution. It has been established that coliforms can associate with particulates in water to form aggregates of bacteria adsorbed to solid matter, and that there are differences amongst coliforms species in this ability, with *E. coli* being less likely to adsorb (Schillinger and Gannon, 1982). Providing only that the coliform-containing particles were randomly distributed, and that the number of free colony-forming units per particle was logarithmically distributed, the negative binomial would result. Clearly, the negative binomial distribution is only one type of compound distribution that could result. Distributions of individuals per clump other than the logarithmic would result in another compound distribution.

An interesting and important result is given by Douglas (1980) for all Poisson-stopped distributions, where the distribution of clusters is Poisson, but with another distribution characterizing the cluster size. If the cluster size distribution disallows empty clusters and contains some clusters with more than one bacterium, then the variance of the compound distribution will be greater than Poisson.

20.7 Analytical Mechanisms for Non-Poisson Behavior

It is entirely possible that methods intrinsic to the analytical or sampling methodology may impose a non-Poisson distribution on replicated counts. For example, suppose that individual microorganisms are, intrinsically, Poisson distributed with identical means in replicate samples. However, suppose further that they have nonidentical probabilities of being recovered by the enumeration methodology. In the simplest case, if some fraction Θ of the samples are recoverable, then the probabilities of recovering N organisms in a sample are give by:

$$P(0) = 1-\Theta[1-\exp(-\Lambda)]$$
$$P(N,\text{ for } N>0) = \Theta\Lambda^N\exp(-\Lambda)/N! \qquad 12$$

This is the Poisson with added zeros distribution, which has a mean of $\Theta\Lambda$ and a variance of $\Theta\Lambda[1+\Lambda(1-\Theta)]$. Clearly, the variance/mean ratio is greater than (the Poisson value of) unity (for $\Theta < 1$) (Pielou, 1969). Also, clearly, this results in underestimating the true mean density of bacteria in the water being sampled.

A second possibility is that the bacteria being sampled are enumerated with a variable efficiency. This may be regarded as a mixture of the Poisson distribution by a frequency distribution of enumerability. In the particular case where the enumerability is given by a Pearson Type III distribution (equation 13), application of point 3 in section 20.5 pertaining to a mixture distribution leads to the p.g.f. for a negative binomial distribution with mean of pk (Pielou, 1969).

$$\text{Prob } (\Lambda) = [1/p^k\Gamma(k)]\Lambda^{k-1}\exp(-\Lambda/p) \qquad 13$$

Clearly, since the negative binomial distribution can be generated from either a "clumping" mechanisms or by variation in the enumeration technique, the simple observation of a negative binomial distribution is insufficient to allow one to determine whether such variations are real (i.e., environmental) or artificial (i.e., analytical) and what mechanisms might be responsible for such observations. Furthermore, since it has been shown (Feller, 1943) that any distribution that is produced as a mixture may also be produced as a stopped distribution, the mere observation of a distribution other than the Poisson is insufficient to conclusively decide between an environmental mechanism and an analytical mechanism for its generation.

A variety of mechanisms are available which might produce analytical distributions other than Poisson. Perhaps the most likely of these is the variability of plates or culture vessels in which the sampled microorganisms are grown prior to detection. This rationale was employed by Margolin et al. (1981) in their application of the negative binomial distribution to the description of variability between replicates in the Ames mutagenicity assay. Such variability could be intrinsic to the formulation of the medium itself, due to interaction of the medium with competitors or commensal organisms in the sample, or due to the variability in recovery of sublethally injured organisms. It is known that other indigenous bacteria (noncoliforms) may hinder the enumerability of coliforms (LeChevallier and McFeters, 1985a). It is also known that bacteria in waters, particularly those which have been disinfected, may have been sublethally injured, and thus, perhaps susceptible to the pressures of ordinary differential media (Bissonnette et al., 1975; McFeters et al., 1982, 1986; LeChevallier and McFeters, 1984, 1985b).

In the particular case in which either tissue culture assays or animals are used to enumerate microorganisms, it is well understood that "host variability" is a potential problem and can lead to deviations from Poisson behavior. The quantitative effects on the distribution of microorganisms resulting from such variability have been reviewed elsewhere (Haas, 1983; Worcester, 1954; Moran, 1954).

A negative binomial distribution between samples at various dilutions can also result if there is imprecision in the preparation of serial dilutions routinely used in analytical microbiology (Chase and Hoel, 1975).

20.8   Other "Contagious Distributions"

The negative binomial and the Poisson with added zero's model are only two of many examples of distributions for discrete variables which have greater than Poisson variability. The more common two-parameter distributions, in addition to the negative binomial, are the Neyman Type A, the Polya-Aeppli and the Thomas distributions. These are tabulated in Table 20.1, along with the formulas for the mean, variance/mean ratio, and probability of zero organisms.

In addition to those noted in Table 20.1, there is a wide number of other distributions with over or under dispersion relative to the Poisson (with two or more parameters). Some of these may be grouped into families. For example, this is demonstrated in the work of Gurland (1958) where two families of mixture distributions from the Poisson are derived in terms of the confluent and ordinary hypergeometric functions. From a practical point of view, however, when overdispersion is present, many of the various distributions may be found to fit the data equally well (Bliss, 1953; Neyman and Scott, 1957). This is not surprising since the overdisperse distributions tend to differ only in the higher-order moments, and only with respect to the fine details of the dependency of the variance/mean ratio on the mean.

20.9   Effects of Distribution on Quantal Assays

In the presence of overdispersion, such as in the case of the negative binomial distribution, the frequency of both low and high colony counts will be increased. As an example, Figure 20.1 plots the expected frequency of various colony counts in water containing a mean of 2 organisms per unit volume sampled for the Poisson distribution and for negative binomial distributions at various "k" values.

From the previous discussion, however, if a distribution other than the Poisson is observed in replicate samples and if an analytical explanation cannot be refuted, it is possible that a systematic underestimate of the mean concentration of organisms can be made. One therefore desires to test whether a given data set is or is not consistent with the Poisson distribution. Of course, for a large number of replicates of a single sample it is possible for any distribution to estimate its parameters (e.g. by maximum likelihood) and ascertain goodness-of-fit between the experimental and computed frequency distributions using standard techniques (for example, Kolmogorov-Smirnov test). However,

Table 20.1 Two-parameter contagious distributions

Name	Mean	Variance/Mean	P(0)	P(x)
Negative Binomial	μ	$1+\mu/k$	$(1+\mu/k)^{-k}$	$[\Gamma(k+N)/\Gamma(k)N!]\,(\mu/k)^x\,[k/(k+\mu)]^{k+x}$
Neyman Type A	ab	$1+b$	$\exp(-a+ae^{-b})$	$(e^{-ab}/x!)\sum_{k=0}^{\infty}(ae^{-b})^k k^x/k!$
Polya-Aeppli	ab	$1+2b$	$\exp[-ab/(1+b)]$	$P(0)\,[b/(1+b)]^x \times \sum_{s=1}^{x}(x-1)![a/(1+b)]^s/[s!(x-s)!(s-1)!]$
Thomas	$a(1+b)$	$1+b+b/(1+b)$	$\exp(-a)$	$P(0)b^x/x!\sum_{s=1}^{x}\{x!/[s!(x-s)!]\}(ae^{-b}/b)^s s^{x-s}$

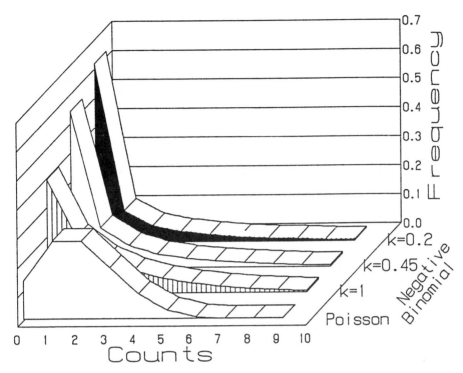

Figure 20.1 Histogram of colony counts for Poisson and various negative binomial distributions at a mean density of 2 organisms/volume.

water quality data of this type are not always available. Hence, a variety of alternative tests have been developed.

The authors have reviewed methods for analyzing this problem (Haas and Heller, 1986). For the most part, two types of data sets may be distinguished:

1. A large number of replicates of a single sample such as 20 replicates of a single water. In this case, the data may also consist of the mean and standard deviation of the population of replicates.

2. A series of samples taken over the course of time, with each sample replicated an identical number of times. For example, triplicate MF determinations on each of 40 samples. In this case the accessible data consist of numbers of colonies on each of the replicate plates for each of the samples.

For the first type of data set, where there are a number of replicates of a single sample (i.e. from water with a constant mean density), Fisher et al. (1922) proposed an index of dispersion as a test for the Poisson distribution (actually as a test of whether the variance is as would be expected for a Poisson distribution with the observed mean) computed as:

$$D^2 = \sum(x_i - \bar{x})^2/\bar{x} = (N-1)\text{variance}/\text{mean} \qquad 14$$

where the summation is taken over all observed colony counts (x_i) and N is the number of replicates. The null hypothesis that the variance is equal to that predicted from the Poisson distribution results in a D^2 which is distributed according to the Chi-squared distribution with $N-1$ degrees of freedom. This test can be incorporated into a control chart for routine analytical use (Eisenhart and Wilson, 1943; Cowell and Morisetti, 1969).

For the second type of data set, there are at least two approaches to testing agreement with the Poisson distribution for plate count data. One approach is to use equation 14 to compute a D^2 value for each set of replicates (Churchill and Eisenhart; Wilson and Kullman, 1931). The distribution of D^2 values could then be compared to the chi-squared distribution at $(N-1)$ degrees of freedom. However, recent workers have proposed the following alternative test as a more powerful one than the use of the D^2 (Collings and Margolin, 1985):

$$T_c = \sum_i \sum_j (x_{ij} - \bar{x}_{i.})^2/\bar{x}_{..} \qquad 15$$

where x_{ij} is the count for replicate j of the experiment from water sample i (i.e. with mean of i), $\bar{x}_{i.}$ is the mean count of experiment j, and $\bar{x}_{..}$ is the overall mean count. If the computed value of the statistic T_c is greater than the critical chi-square distribution at degrees of freedom determined from equation 16, then there is overdispersion present.

$$df = \sum_i (N_i - 1)\bar{x}_{i.}/\bar{x}_{..} \qquad 16$$

In equation 16, N_i is the number of replicates of the ith sample.

For the purpose of ascertaining goodness-of-fit to a negative binomial distribution, a test has also been developed for type 2 data in which the mean (and possibly the "k" value) differ between samples (Heller, 1986). Haas and Heller (1986) applied this test to data when the Poisson distribution could be rejected by the D^2 test and found conformance to the negative binomial.

20.10   Distribution Effects on All-or-None Assays

As shown in Figure 20.1, for overdisperse distributions such as the negative binomial, the expected frequency of negative tubes is greater than predicted from the Poisson distribution. When the objective is to infer the mean density from the percent positives, there may be a substantial bias which arises if the Poisson distribution is assumed but is, in fact, not true.

The simplest all-or-none assay is one in which a single dilution of a sample is inoculated into a number (T) of replicate culture vessels (nutrient medium, tissue cultures or even animals) and the number of infected cultures (P) after some period of time determined. From the theory of maximum likelihood, assuming the Poisson distribution (cf. Cochran, 1950), if a volume (V) is analyzed, the mean density may be given by:

$$\mu = -(1/V) \ln[(T-P)/T] \qquad\qquad 17$$

This can be used either in the context of dilution tubes or animals, or by counting replicate areas on a culture medium or surface and scoring only for presence or absence of organisms.

The method can readily be extended to situations when multiple dilutions are present and MPN tables may be computed (American Public Health Association, 1985) or direct maximization of the likelihood function may be carried out.

Furthermore, if the Poisson distribution is assumed, confidence limits about the estimated mean may be constructed by a variety of procedures (DeMan, 1975; Loyer and Hamilton, 1984).

However, even in the case of a single dilution experiment, if an all-or-none assay is used as a basis for inference of the mean density by application of equation 16, a strong bias in the downward direction will exist if the negative binomial (or another overdisperse distribution) characterizes sample variability (Wadley, 1950). Furthermore, if an overdisperse distribution arises via one of the analytical mechanisms outlined above, there is also a methodological (as well as a statistical) bias that could lead to an underestimate of mean microbial density.

Hence, it is strongly desirable to have a means to ascertain that the distribution of replicates in a dilution count (or all-or-none) experiment conforms to the Poisson distribution.

If a sufficiently large number of replicates are inoculated at each dilution, and the average concentration estimated from maximum likelihood equations for multiple dilutions, the number of positive replicates at each dilution can be predicted directly from equation 1 (taking $1 - P(0)$ as the probability of positive replicates). The expected and observed number of positive replicates can be compared using the χ^2 goodness of fit test (Haldane, 1939; Chang et al., 1958).

However, in drinking water microbiology, particularly in the case of the MPN coliform procedure, the use of a small number of replicates (rather than a large number) is the common practice. Hence, the above approach is not applicable. Several authors have studied the problem of testing conformance with the Poisson assumption (usually referred to as a test of homogeneity) in such systems.

Savage and Halvorson (1941) monitored the growth of *Staphylococcus aureus* on inhibitory media using a decimal dilution tube technique with ten replicates per dilution. Although the details of their calculation are unclear, they classified the dilution scores (i.e. 10–5–2–0, where the numbers represent the number of positives found in consecutively increasing dilutions of sample) that occurred into "common" and "uncommon" codes (i.e., likely and unlikely). They further characterized the uncommon codes into gradual (with several intervening dilutions with some but not all positive) and abrupt (with no intervening intermediate dilution) subclasses and found that an unfavorable medium, particularly at short incubation times, produced greater proportions of uncommon codes especially of the gradual type.

Woodward (1957) prepared tables of common and uncommon codes in the most frequently used three decimal dilution fivefold replicate protocol. The details of his preparation of these tables remain unclear, despite discussion by Loyer and Hamilton (1984). Nevertheless, Woodward applied his calculations to a data set of total coliform measurements and found good agreement between the predicted frequency of uncommon codes and their observed frequencies, implicitly validating the Poisson assumption.

A major advance in homogeneity testing of dilution experiments was made by Moran (1954b) who proposed the statistic T, computed from:

$$T = \sum_i P_i(N-P_i) \qquad\qquad 18$$

where the summation is over all dilutions from a single sample, P_i is the number of positives at that dilution, and N is the total number of replicates at each dilution. Providing that the number of dilutions is sufficiently large so that at low dilution all replicates are positive and at high dilution all replicates are negative, the theoretical mean and variance of the T statistic (assuming Poisson replication errors) can be calculated. However, particularly for tenfold dilutions, it is somewhat dependent on the mean density in a periodic manner. Then the quantity $[T-\text{mean}(T)]/[\text{var}(T)]^{1/2}$ is approximately normally distributed. Moran, (1954b) tabulated the mean(T) and var(T) for various dilution ratios and replications.

Armitage and Spicer (1956) analyzed the sensitivity of Moran's test of heterogeneity and found that, in general, unless the number of replicates per dilution was high (in excess of 10) and unless the mean enumerability is low with positive skewness, the sensitivity of the Moran test to departures from Poisson behavior is relatively poor.

Stevens (1958) developed an alternative test, referred to as the "range of transition" and defined as "the number of [dilutions], counted inclusively, from the first where not all the tubes are fertile to the last where not all the tubes are sterile." In other words, the series N c_1 c_2 0 has a range of transition equal to 2 providing that neither c_1 nor c_2 are equal to either zero or N (the total number of tubes inoculated per dilution). For dilution factors of 2, 4, and 10, with various numbers of tubes per dilution, for infinitely long dilution series (so that the range is always bracketed by N and 0) Stevens calculated the values of his statistic under the assumption of the Poisson distribution. As with the Moran statistic, in the case of decimal dilutions, a substantial periodicity was observed. Armitage and Bartsch (1960) compared the Stevens range of transition to other tests of homogeneity (but not the Moran statistic) and found it to be relatively sensitive. However, they suggested the use of a modification defined as the "number of positive results at dilutions beyond that at which the first negative occurs" particularly when a single replicate per dilution is used.

Both the Stevens tests and the Moran tests of heterogeneity rely on an assumption of an infinitely long sequence of dilutions so that there is a bracketing of dilutions between completely infected, e.g. 5 tubes of 5 positive, and completely sterile, e.g. 0 tubes of 5 positive. In drinking water practice this is rarely the case, and most often the least diluted sample (greatest volume) yields less than 100 percent infected tubes. Haas and Heller (1988) used a modification of the Stevens test for the assessment of homogeneity in the MPN total coliform procedure and constructed tables for the 5 tube 3 decimal dilution protocol most common in coliform analysis. The test was applied to a large data set from routine raw water coliform monitoring, and heterogeneity was found. In numerical studies, it was determined that, at the large sample size used (about 1000 samples), departures in the form of the negative binomial distribution with k values as great as 5 (i.e., only 20 percent excess coefficient of variation) could be detected.

20.11   Conclusions

Little attention has been paid, historically, to the frequency distribution of microorganisms in the environment. In particular, many studies which have been purported to study such distributions have failed to distinguish between real (i.e. environmental) distribution characteristics and those which might be contributed by analytical methods. There are, as discussed, a number of statistical techniques for isolating these analytical contributions. Workers who investigate microbial density distributions should carefully consider the effect of analytical sampling errors both in their experimental designs and in their data analysis techniques. Those responsible for the development of environmental standards should take cognizance of such analytical variability.

References

American Public Health Association. 1985. *Standard Methods for the Examination of Water and Wastewater*, APHA, Washington, D.C.

Armitage, P. and Bartsch, G.E. 1960. The detection of host variability in a dilution series with single observations. *Biometrics* 16:582–592.

Armitage, P. and Spicer, C.C. 1956. The detection of variation in host susceptibility in dilution counting experiments. *Journal of Hygiene* 54:401–14.

Bissonnette, G.K., Jezeski, J.J., McFeters, G.A., and Stuart, D.S. 1975. Influence of environmental stress on enumeration of indicator bacteria from natural waters. *Applied Microbiology* 29:186–194.

Bliss, C.I. 1953. Fitting the negative binomial distribution to biological data. *Biometrics* 9:176–200.

Chang, S.L., Berg, G., Busch, K.A., Stevenson, R.E., Clarke, N.A. and Kabler, P.W. 1958.

Application of the Most Probable Number Method for estimating concentrations of animal viruses by the tissue culture technique. *Virology* 6:27–42.

Chase, G.R. and Hoel, D.G. 1975. Serial dilutions: error effects and optimal designs. *Biometrika*, 62–329–34.

Christian, R.R. and Pipes, W.O. 1983. Frequency distribution of coliforms in water distribution systems. *Applied and Environmental Microbiology* 45:603–9.

Collings, B.J. and Margolin, B.H. 1985. Testing goodness of fit for the Poisson distribution when observations are not identically distributed. *Journal of the American Statistical Association* 80:411–418.

Cochoran, W.G. 1950. Estimation of bacterial densities by means of the Most Probable Number. *Biometrics* 6:105–17.

Cowell, N.D. and Morisetti, M.D. 1969. Microbiological techniques—some statistical aspects. *Journal of Science of Food and Agriculture* 20:573–579.

DeMan, J.C. 1975. The probability of Most Probable Numbers. *Eur. J. Appl. Micro.* 1:67–78.

Diggle, P.J. 1983. *Statistical Analysis of Spatial Point Patterns*. Academic Press, London.

Douglas, J.B. 1980. *Analysis With Standard Contagious Distributions*. International Co-operative Publishing House, Fairland, MD.

Eisenhart, C. and Wilson, P.W. 1943. Statistical methods and control in bacteriology. *Bacteriological Reviews* 7:57–137.

El-Shaarawi, A.H., Esterby, S.R., and Dutka, B.J. 1981. Bacterial density in water determined by Poisson or negative binomial distributions. *Applied and Environmental Microbiology* 41:107–116.

Engen, S. 1978. *Stochastic Abundance Models*. Chapman and Hall, London.

Feller, W. 1943. On a general class of contagious distributions. *Annals of Mathematical Statistics* 14:389–400.

Fisher, R.A., Thornton, H.G., and Mackenzie, W.A. 1922. The accuracy of the plating method of estimating the density of bacterial populations. With particular reference to the use of Thornton's agar medium with soil samples. *Annals of Applied Biology* 9:325–339.

Fuhs, G.W. 1975. A probabilistic model of bathing beach safety. *The Science of the Total Environment* 4:165–175.

Gurland, J. 1958. A generalized class of contagious distributions. *Biometrics* 14:229–249.

Haas, C.N. 1983. Estimation of risk due to low doses of microorganisms: a comparison of alternative methodologies. *American Journal of Epidemiology* 118:573–582.

Haas, C.N. and Heller, B. 1986. Statistics of enumerating total coliforms in water samples by membrane filter procedures. *Water Research* 20:525–30.

Haas, C.N. and Heller, B. 1988. Test of the validity of the Poisson assumption for the analysis of MPN results. *Applied and Environmental Microbiology* 54:2996–3002.

Haldane, J.B.S. 1939. Sampling errors in the determination of bacterial or virus density by the dilution method. *Journal of Hygiene* 39:389–93.

Heller, B. 1986. A goodness-of-fit test for the negative binomial distribution applicable to large sets of small samples. pp. 215–220 in *Statistical Aspects of Water Quality Monitoring*, A.H. El-Shaarawi and R.E. Kwiatkowski (eds.). Elsevier, Amsterdam.

LeChevallier, M.W. and McFeters, G.A. 1984. Recent advances in coliform methodology for water analysis. *Journal of Environmental Health* 47:5–9.

LeChevallier, M.W. and McFeters, G.A. 1985a. Interactions between heterotrophic plate count bacteria and coliform organisms. *Applied and Environmental Microbiology* 49:1338–1341.

LeChevallier, M.W. and McFeters, G.A. 1985b. Enumerating injured coliforms in drinking water. *Journal of the American Water Works Association* 77:81–87.

Loyer, M.W. and Hamilton, M.A. 1984. Interval estimation of the density of organisms using a serial dilution experiment. *Biometrics* 40:907–16.

Margolin, B.H., Kaplan, N., and Zeigler, E. 1981. Statistical analysis of the Ames *Salmonella*/microsome test. *Proceedings of the National Academy of Sciences USA* 78:3779–83.

McFeters, G.A., Cameron, S.C., and LeChevallier, M.W. 1982. Influence of diluents, media and membrane filters on the detection of injured waterborne coliform bacteria. *Applied and Environmental Microbiology* 43:97–103.

McFeters, G.A., Kippin, J.S., and LeChevallier, M.W. 1986. Injured coliforms in drinking water. *Applied and Environmental Microbiology* 51:1–5.

McDaniels, A.E. and Bordner, R.H. 1983. Effects of holding time and temperature on coliform numbers in drinking water. *Journal of the American Water Works Association* 75:458–63.

Moran, P.A.P. 1954a. The dilution assay of viruses. *Journal of Hygiene* 52:189–93.

Moran, P.A.P. 1954b. The dilution assay of viruses. II. *Journal of Hygiene* 52:444–446.

Neyman, J. and Scott, E.L. 1957. On a mathematical theory of populations conceived as conglomerations of clusters. *Cold Spring Harbor Symposium on Quantitative Biology* 22:109–120.

Pielou, E.C. 1969. *An Introduction to Mathematical Ecology*. Wiley-Interscience, New York.

Pielou, E.C. 1977. *Mathematical Ecology*, 2nd edition, Wiley-Interscience, New York.

Pipes, W.O., Ward, P., and Ahn, S.H. 1977. Frequency distribution for coliforms in water. *Journal of the American Water Works Association* 69:664–668.

Savage, G.M. and Halvorson, H.O. 1941. The effect of culture environments on results obtained with the dilution method of determining bacterial population. *Journal of Bacteriology* 41:355–362.

Schillinger, J.E. and Gannon, J.J. 1982. Coliform attachment to suspended particles in stormwater. Technical Completion Report on Annual Cooperative Project A-111-MICH, Institute of Water Research, Michigan State University, East Lansing.

Stevens, W.L. 1958. Dilution series: a statistical test of technique. *Journal of the Royal Statistical Society, Part B* 20:205–214.

Thomas, H.A. and Woodward, R.L. 1955. Estimation of coliform density by the membrane filter and the fermentation tube techniques. *American Journal of Public Health* 45:1431–7.

Wadley, F.M. 1950. Limitations of the "Zero Method" of population counts. *Science* 119:689–90.

Wilson, P.W. and Kullman, E.D. 1931. A statistical inquiry into methods for estimating numbers of *Rhizobia*. *Journal of Bacteriology* 22:71–90.

Woodward, R.L. 1957. How probable is the Most Probable Number? *Journal of the American Water Works Association* 49:1060–1068.

Worcester, J. 1954. How many organisms? *Biometrics* 10:227–34.

21

Microbiological Methods and Monitoring of Drinking Water

Wesley O. Pipes

Department of Civil Engineering and The Environmental Studies Institute,
Drexel University, Philadelphia, Pennsylvania

21.1 Introduction

Microbiological monitoring of potable water in distribution systems has been practiced in the United States and other countries since early in the 20th century. It is an attempt to determine the safety of the water in relation to the possibility of transmission of waterborne disease. Water samples are collected and tested for the presence of indicator bacteria. One source of the indicator bacteria is fecal material. If the indicators are found, the water may be contaminated with fecal material and, therefore, may not be safe to drink. If the indicator bacteria are not found, it may be inferred that the water is not contaminated with fecal material and will not transmit waterborne disease. Microbiological monitoring of drinking water in practice is far from the ideal expressed in the two previous sentences.

There are a number of issues about microbiological monitoring which have never been fully resolved but continue to be debated. These issues can be categorized into microbiological issues, water quality issues, and sampling issues (Table 21.1). This chapter is concerned primarily with the sampling issues but certain of the water quality and microbiological issues are also addressed in order to place the sampling issues in proper context (also see Chapter 10).

The first water quality issue of Table 21.1 is a key to understanding the continuing lack of resolution of the other issues. The coliform group of bacteria has been used much more than any other indicator group for monitoring potable water. This group consists of *Escherichia coli* which usually comes from fecal material plus other bacteria which are derived from nonfecal sources. Under certain circumstances, some coliforms from nonfecal sources can grow in water distribution systems (Allen et al., 1980) (see also Chapters 10–12). Thus, finding coliform bacteria in a water distribution system does not necessarily (or even usually) indicate a threat of transmission of waterborne disease. Eliminating all the coliform bacteria from a water distribution system

Table 21.1 Issues of microbiological monitoring of water distribution systems

Microbiological issues
1. What group of indicator organisms to use
2. How to preserve the samples during transit to the laboratory
3. What laboratory examination procedure to use
4. How to measure the density of indicators in the sample
5. What confirmatory procedures (or additional tests) to use

Water quality issues
1. What is the significance of the indicators when found
2. What parameter of indicator presence to use
3. What indicator level (standard) is protective of public health
4. How should the results of additional (repeat or check) samples be used
5. What measures should be taken when the standard is exceeded

Sampling issues
1. What time period is appropriate for water quality evaluation
2. How many samples to collect
3. Where to collect the samples
4. When (during the week and the day) to collect the samples
5. What additional samples to collect following an unsatisfactory sample

once they become established can be very expensive but, if that is not done, continued monitoring for coliform bacteria in an attempt to identify new contamination is futile.

There are, in general, three ways in which a water distribution system can become contaminated with coliform bacteria. 1) Contamination can be the result of a cross connection between a water main and a sewer. 2) There is a possibility that a few coliforms and even pathogens might slip past the treatment processes which are intended to remove or inactivate them. 3) Coliforms may gain access to the water from any free air/water interface in the distribution system, such as in an elevated storage tank, where dust and aerosols may enter.

Cross connections present the most serious threat of waterborne disease. Fortunately, most of the time the flow in a cross connection is from the water main to the sewer and only if the pressure in the water main drops to less than atmospheric does the cross connection result in contamination of drinking water. The contamination may be confined to the building where the cross connection occurs or it may spread in a portion of the distribution system.

The second most serious threat of waterborne disease comes from inadequate disinfection of the water before it is pumped into the distribution system. It is very likely that almost any body of surface water will contain some of the pathogens of waterborne disease and even some ground water supplies are contaminated. The pathogens and indicators of fecal contamination can be removed by treatment processes such as sedimentation and filtration and/or inactivated by chemical disinfection. Disinfection of water supplies is still not universal and even when it is practiced it is not always adequate.

Bacterial contamination of stored water from the air or by rodents or birds has not been shown to be a source of the pathogens of waterborne disease. However, coliforms do originate in storage tanks or other reservoirs and this must be taken into consideration in monitoring programs.

The objectives of the water utility in keeping the customers safe from waterborne disease are clear. These are: 1) the water should be completely disinfected before it is pumped into the distribution system; 2) the water distribution system should be intact and cross connections between water mains and sewers should be located and eliminated; and 3) stored water should be protected from contamination and, if necessary, disinfected again before being returned to the distribution system. In addition, many water utilities maintain a disinfectant residual in the water to provide some immediate protection against cross connection contamination and to suppress the growth of coliforms and other bacteria in the distribution system.

In contrast to the above, the objectives of microbiological monitoring are not at all clear. Possible objectives include the following: 1) to demonstrate that the water is properly disinfected during treatment; 2) to search for cross connections so that they may be eliminated; and 3) to evaluate some overall measure of the presence of indicators and, from that, the possible degree of risk of contracting waterborne disease. All three objectives are attained if the water is monitored adequately and indicators are not found. However, if some coliforms are found in some samples, the significance of the positive results must be interpreted and a decision must be made about what corrective action to take. It is also possible that no direct relationship between microbiological monitoring for coliforms and the risk of waterborne disease is assumed and the monitoring is intended to be a reminder of the need to disinfect the water and otherwise be concerned about the microbiological content of the water.

Microbiological monitoring of water distribution systems is a legal requirement in the United States and in most other countries. The standards and regulations governing monitoring have been formulated by expert committees and the objectives which the committees had in mind are, in some way, embodied in the standards and regulations. In the next section an attempt is made to infer the objectives of monitoring from some of the details of the standards and regulations. This is followed by an examination of two models used in research studies of microbiological monitoring of water distribution systems and a discussion of how the research results can be applied for future monitoring and regulatory purposes.

21.2   Drinking Water Standards

Bacteriological quality standard for potable water were first adopted in 1914. The 1914 standards (Table 21.2) are known as the "Treasury Department Standards" because the U.S. Department of the Treasury had the legal responsibility

Table 21.2 U.S. Treasury Department standards—1914

Maximum limits of permissible bacteriological impurity:
1. The total number of bacteria developing on standard agar plates, incubated 24 hours at 37°C, shall not exceed 100 per cubic centimeter. The estimate shall be made from not less than two plates and should be reliable and accurate.
2. Not more than one out of five 10 c.c. portions of any sample examined shall show the presence of *B. coli*. [The testing procedure demonstrates the presence of facultative organisms which produce gas from lactose fermentation.]

for interstate commerce at that time. The next four sets of standards (Tables 21.3–21.5) were adopted by the U.S. Public Health Service (PHS). The Treasury Department and PHS standards were not generally enforceable except when applied to water provided for common carriers in interstate commerce. These standards were advisory to state and local health departments. The 1977 regulations (Table 21.6) established the maximum microbiological contaminant levels (MCL's) of the National Interim Primary Drinking Water Regulations adopted under the 1974 Federal Safe Drinking Water Act (PL 93–523) by the U.S. Environmental Protection Agency (EPA) and are legal requirements. The tables summarizing the microbiological drinking water standards follow the form of presentation of Olivieri et al. (1983).

Microbiological Issues   The most important microbiological issue is the selection of the indicator organisms. The indicator used for bacteriological drinking water standards has been the coliform group. The indicator was originally specified as *Bacterium coli* which was defined in the 1914 and 1925 standards as the organism which ferments lactose with the production of acid and gas. Later, when it was realized that there are several species belonging to four (or more) different genera which produce gas from the fermentation of lactose, the term "coliform group" came into use. In the late 1940's a subgroup of coliforms which are able to ferment lactose at 44.5°C was split off as "the fecal coliform group."

The coliform group continues to be used as the indicator in spite of the fact that it has been well demonstrated in recent years that coliform colonies

Table 21.3 U.S. Public Health Service standards—1925

1. Not more than 10% of all the 10 c.c. standard portions examined shall show the presence of *B. coli*.
2. Occasionally, three or more of the five 10 c.c. standard portions of a standard sample may show the presence of *B. coli*. This shall not be allowed if it occurs in more than:
   a. 5% of the samples when 20 or more samples are examined;
   b. One sample when less than 20 samples are examined.

Table 21.4 U.S. Public Health Service standards—1943 and 1946

These standards allow for use of either 10 ml or 100 ml portions with the multiple tube test.

If 10 ml portions are used:
1. No more than 10% of all the 10 ml portions examined per month shall show the presence of the coliform group.
2. Occasionally 3 or more of the five 10 ml portions of a sample may show the presence of the coliform group. This shall not be allowable if it occurs in consecutive samples or in more than:
   a. 5% of the samples when 20 or more samples are examined per month.
   b. One sample when less than 20 samples are examined per month.

If 100 ml portions used:
3. No more than 60% of all the 100 ml portions examined per month shall show the presence of the coliform group.
4. Occasionally all of the five 100 ml portions of a sample may show the presence of the coliform group. This shall not be allowed if it occurs in consecutive samples or in more than:
   a. 20% of the samples when 5 or more samples are examined per month.
   b. One sample when less than 5 samples are examined per month.
5. Additional samples following an unsatisfactory sample: When three or more of the five 10 ml portions or all 5 of the 100 ml portions, constituting a standard sample, show the presence of the coliform group, daily samples shall be collected promptly and examined until the results be of satisfactory quality.

Added for 1946 Standards:
a. Samples collected following an unsatisfactory sample as defined in (2) or (4) shall not be included in the determination of the number of samples examined each month.
b. Subsequent unsatisfactory samples in the daily series required by (5) shall not be used as a basis for prohibiting the supply provided that (i) immediate, active efforts are made to locate the cause of contamination, (ii) immediate action is taken to eliminate such cause and (iii) samples taken following such remedial action are satisfactory.

growing on the interior surfaces of water mains are extremely widespread (Allen et al., 1980). Such attached colonies shed bacteria into water at times of changes in the velocity of flow (Goshko et al., 1983) and are quite difficult to eliminate from the system (Hudson et al., 1983). Attachment of the coliforms to particles and surfaces (Herson et al., 1987, LeChevallier et al., 1988) provides some protection from the bactericidal effects of chlorine. The attached coliforms have not been found to be related to any health effects but their presence in a water distribution system results in violations of the standards. Water utilities are reluctant to spend a great deal of time and money eliminating what they consider to be a "nonproblem" and thus do not support the continued use of the coliform standard.

Table 21.5 U.S. Public Health Service standards—1962

These standards allow for use of either the fermentation tube technique or the membrane filter method.

If the fermentation tube technique is used with either 10 ml or 100 ml portions, the standards are identical to the 1946 USPHS standards except for a few minor changes in wording.

If membrane filter method is used:
1. The arithmetic mean coliform count of all standard samples examined per month shall not exceed one per 100 ml
2. Coliform colonies per standard sample shall not exceed 3/50 ml, 4/100 ml, 7/200 ml or 13/500 ml in more than 5% of the samples examined each month or in more than one sample if fewer than 20 samples are examined.
3. When coliform colonies in a standard sample exceed the limits for a single sample, additional samples shall be collected promptly and examined until the results show the water to be of satisfactory quality. These unsatisfactory samples are regarded in the same manner as those used for the fermentation tube test.

The microbiological issues numbers 3 and 5 (Table 21.1) are closely related to the issue of the identity of the indicator to be used because the indicator group is operationally defined by the techniques used to separate and identify them. It is futile to define an indicator group without a simple, reliable test to determine if it is present. Two methods of testing for coliform bacteria in drinking water are in current use (American Public Health Association, 1985), the fermentation tube technique (FT) and the membrane filter method (MF). Either type of test may be used to give a qualitative indication of the presence

Table 21.6 U.S. Environmental Protection Agency—1977 National Interim Primary Drinking Water Regulations

Maximum Contaminant Level: These standards also allow for the use of either the fermentation tube technique or the membrane filter method.

If membrane filter method is used:
1. The arithmetic mean coliform count of all standard samples examined per month shall not exceed 1 per 100 ml.
2. The number of coliform bacteria shall not exceed 4/100 ml in:
   a. More than one sample when less than 20 are examined per month; or
   b. More than 5% of the samples when more than 20 are examined per month.
3. When the coliform count in a single sample exceeds 4/100 ml, at least two consecutive daily check samples shall be collected and examined until the results from two consecutive samples show less than one coliform bacterium per 100 ml

If the fermentation tube technique is used with either 10 ml or 100 ml portions the standards are essentially the same as the 1946 USPHS standards.

or absence of the coliform group in a water sample and, under the proper conditions, to give an estimate of the coliform density of the sample.

The FT technique involves inoculating a measured volume of water, such as 10 ml or 100 ml, in a test tube with a broth containing lactose. If fermentation occurs and gas is produced, this is taken as a positive presumptive test for coliforms. Various media for the FT technique are used, differing mostly in the inhibitors added to prevent noncoliforms from growing and in the dyes used to indicate pH change. If one fermentation tube per sample is used, the test is called a presence absence (P-A) test (see Clark 1969 and Chapter 20). If several fermentation tubes are inoculated with portions of the same water sample, the test is called the multiple tube test. If there are both positive and negative portions in the multiple tube test, the results may be used for calculation of an estimate of the coliform density of the sample which is called the most probably number (MPN). The multiple tube test is frequently called the MPN test.

The MF method requires that some volume of water be filtered through a membrane filter which retains the bacteria on its surface. The filter is then placed on a nutrient medium which contains chemical compounds which react with metabolic products produced by coliforms. After incubation the bacterial colonies which have caused the characteristic indicator reaction are counted as presumptive coliforms. It is assumed that each colony represents one coliform organism and the coliform density is calculated as the colony count per volume filtered. If no coliform colonies are present or if so many colonies are present that they merge and overlap (about 80 colonies per filter), the coliform density of the water sample is indeterminate.

The revision of the microbiological MCL adopted in June 1989 to go into effect in January 1991 presents a slight departure from the exclusive use of the total coliform group as the indicator. The requirement is that water samples will be initially examined for total coliforms by a P-A test. If coliforms are found in any sample, the test culture is examined for the presence of fecal coliforms (or *E. coli*). If fecal coliforms (or *E. coli*) are found, the consumers have to be notified immediately to boil drinking water. If more than 5% of the samples collected over a period of a month show the presence of any coliforms, the water system is in violation of the microbiological MCL and the consumers must be notified by mail. This is a compromise between the use of the total coliform group and the fecal coliform group (or *E. coli*). It may prove effective but that is yet to be determined.

Water Quality Issues   Two different parameters have been used for the bacteriological drinking water standards. The first is the frequency of occurrence of "unsatisfactory samples" which are those with coliform densities above some preselected density (single sample limit). The second is some measure of the total number of coliforms in the distribution system (such as the average). The standards have been stated in terms of the sample results rather than in terms of what would be considered to be an acceptable level of coliform contamination of the water distribution system itself. Thus, the level of coliform bacteria in the water in the distribution system which is acceptable as part of any of the standards is not immediately apparent but must be inferred from the way the standards were written.

A correlation between coliform densities in water and the incidence of waterborne disease was postulated by Kehr and Butterfield (1943). However, attempts to find data supporting this hypothetical correlation have not been successful (Batik, Craun, and Pipes, 1984). When a waterborne disease outbreak occurs, there usually is a very high level of coliform contamination but there are many documented cases of high coliform densities in potable water samples which were not related to disease outbreaks. Thus, there is no direct evidence that either parameter of the level of coliform contamination correlates better with waterborne disease outbreaks than the other.

The 1914 standards (Table 21.2) specified that no sample should have more than 1 out of five 10 ml portions with coliforms present. This is now known to be equivalent to an MPN of 2.2 per 100 ml; however, the method of calculating an MPN estimate of bacterial density from a multiple tube test was not published until 1915 (McCrady, 1915). The inference is that the committee which recommended this standard was not thinking about a statistical parameter but believed that any portion of the water would be safe if it had no more than 1 "*E. coli*" per 50 ml. It seems that they established a density limit which they thought was safe and specified that none of the water should exceed that density. If they had been thinking about estimating a statistical parameter, they would have specified some number of samples needed to make the water quality evaluation.

In the 1925 PHS Standards (Table 21.3) the single sample limit was increased from 1 to 2 out of 5 ten ml portions positive (MPN = 5.1 per 100 ml) and the acceptable frequency of occurrence of samples above the single sample limit was increased from 0 to 5%. This is a much less stringent standard than the one proposed in 1914. The only apparent explanation for the relaxation of the standard in 1925 is that there must have been a number of water systems which were unable to meet the 1914 limit and which were not experiencing waterborne disease outbreaks. If most water systems had been able to meet the 1914 limit, there would have been no need to relax the standard and, if waterborne disease outbreaks had been occurring in those systems which did not meet the limit, it would have been prudent to retain the limit as it had been.

To compensate for the relaxation of the single sample limit and the frequency of occurrence of 0, another rule which limits coliform occurrence to no more than 10% of all of the 10 ml portions was added. This second rule may be interpreted in two different ways: 1) If 10% of all 10 ml portions from all samples were positive and the bacteriological water quality was homogeneous throughout the system, this would be equivalent to an MPN of 1.1 coliform bacteria per 100 ml for the entire distributions system. The first interpretation of the second rule is that the committee assumed that water quality would be homogeneous in water distribution systems and was attempting to limit the average coliform density to about 1 per 100 ml. This is the interpretation which is recorded in Table 21.7 but it is actually a rather silly interpretation. The assumption that water quality is homogeneous throughout a water distribution system is tantamount to assuming that the water in the system is well mixed; i.e., if coliform bacteria were introduced into the system at any point they would immediately be dispersed randomly throughout the entire system. This does not happen and it is difficult to believe that the expert committee would make that bad a mistake. 2) The other interpretation is that the second rule was intended to be another type of frequency-of-occurrence rule. The first rule allows all of the samples examined to have two positive tubes but limits the samples with three or more positive tubes to no more than 5%. The second rule means that at least half of the samples have to have no positive tubes and, if some of the positive samples have more than one positive tube, the fraction of the samples with no positive tubes must increase. This explanation of the second interpretation is very complex but the major point that it greatly reduces the allowable frequency of occurrence of samples with coliforms present beyond the restrictions of the first rule is clear.

The 1943 revision (Table 21.4) carried forward the same rules for the use of the FT technique with five 10 ml portions of each sample. Rules were also added for the use of five 100 ml. portions per sample. The rules for 100 ml

Table 21.7 U.S. Environmental Protection Agency—1989 National Primary Drinking Water Regulations: Total Coliforms

Maximum Contaminant Level goal: zero

Maximum Contaminant Levels: based on presence/absence of total coliforms in 100 ml sample, rather than on estimate of coliform density.

a. If system analyses fewer than 40 samples/month, no more than one sample/month can be coliform-positive.

b. If system analyses at least 20 samples/month, no more than 5% of samples can be coliform-positive.

Repeat Samples:	If coliforms are detected in any sample, the system must collect 3 repeat samples from the same location as the original sample and near-by locations. Systems collecting fewer than 5 samples per month must collect enough repeat samples to give at least 5 samples in the month of the positive samples and increase sampling the next month to 5 samples.
Additional Tests:	Any positive sample must be tested for the presence of fecal coliforms or *E. coli*. If fecal coliforms or *E. coli* are found, the public must be notified.
Sanitary Survey:	Any system examining fewer than 5 samples per month must have a sanitary survey every 5 years (10 years for non-community systems).

portions are theoretically much easier to meet than the rules for 10 ml portions; that is, a distribution system may have a much higher level of coliform contamination and still not violate them but there is no record that they have ever been used for routine monitoring of a water distribution system. Also, a new provision requiring additional samples following "unsatisfactory samples" was added. However, "unsatisfactory samples" were not defined.

The two rules of the microbiological standards were not changed in 1946, but "unsatisfactory samples" were defined as those in which coliform bacteria were present at levels above the single sample limit. Also, it was made clear that the additional samples following an unsatisfactory sample were to be collected every day from the same tap until a sample with coliform bacteria less than the single sample limit was obtained. These additional samples were not to be used in calculating compliance with the standard nor were they to be counted as part of the number of samples per month required by the standards.

In 1962 the use of the total coliform group as the primary bacterial indicator for the drinking water standards was continued and the MF method of analysis was adopted as an alternative to the FT method. The single sample limit for the MF method was given as 3/50 ml, 4/100 ml, 7/200 ml, or 13/500 ml (Table 21.5). There is a clear inference that the utility could choose the sample volume for the MF method somewhere between 50 ml and 500 ml. The frequency of occurrence of samples above the single sample limit was kept at 5% of the total number of samples examined each month or no more than 1 sample if fewer than 20 samples per month were examined. There is a clear parallel between the single sample limit and frequency of occurrence for the MF method and the FT technique. This infers that there is also an intended parallel between the MF and FT versions of the other rule. The average MF coliform colony count of all samples examined in a month's time was set at no more than 1 per 100 ml. This suggests that the committee which recommended the MF average rule in 1962 interpreted the 1925 FT rule of no more than 10% of all 10 ml tubes positive as an attempt to limit the average coliform density of the system. As noted previously, this interpretation does not make sense because it would have to be based on the assumption that all of the water in the distribution system was well mixed. The intent of the MF average rule to set a limit of the average coliform density of the water in the distribution seems clear. The MF average rule of 1 per 100 ml is much more stringent than the MF frequency of occurrence rule of no more than 5% of the samples with MF counts of more than 4 per 100 ml (Dempsey and Pipes, 1986); that is, there are levels of coliform contamination which would cause a violation of the average rule but not the frequency of occurrence rule.

The National Interim Primary Drinking Water Regulations are the responsibility of the U.S. Environmental Protection Agency. The maximum microbiological contaminant level (MCL) of the Interim Regulations (Table 21.6) is much the same as the 1962 Public Health Service standards. The only difference is that the additional samples following an unsatisfactory sample are now called "check samples" and must be collected daily until a sample with no coliform bacteria (not just coliforms at a density below the single sample limit) are collected on two (not one) consecutive days. The Interim Primary Drinking Water Regulations did not make the rules about coliforms in the water distribution system more difficult to meet but they did make the requirements for check samples much more stringent.

The revised microbiological MCL (adopted in June 1989) introduced a number of changes. The limit on the average MF coliform count of the samples examined is eliminated along with the (supposedly) parallel FT rule that no more than 10% of all the 10 ml portions of the samples examined shall have coliforms present. The single sample limit is reduced to no coliform bacteria in a 100 ml sample and the frequency of occurrence of unsatisfactory samples remains at 5% or one sample if fewer than 40 samples per month are collected and examined. If a sample with coliforms is found, additional samples must be collected and the results of the examinations of the additional samples are used in determining compliance with the MCL.

Sampling Issues   In the previous water quality standards and regulations, the sampling issues have not been addressed as clearly as the water quality issues. In many cases, the resolution has been left to the discretion of the state regulatory agency or even the local utility. The probability of finding coliform bacteria in samples from a water distribution system depends upon where the samples are collected as well as upon the number of samples examined. Thus, the state enforcement agencies or local utilities may have control of whether or not the MCL is violated in any given month.

Time for Evaluation   The period of time for water quality evaluation has always been one month. This is for administrative convenience and has not been related to changes in the microbiological water quality in distribution systems. The available evidence indicates that there can be abrupt changes from week to week (and probably from day to day) but there are times when no changes occur over periods longer than one month (Pipes et al., 1987). Additional research is needed in order to come to a full understanding of the time factor.

Number of Samples   The 1914 Treasury Department standards did not include any requirements for a minimum number of samples to be used for evaluation of microbiological water quality. Apparently, that committee was thinking in terms of an absolute level of safety and judged that, if there were no more than one *"Bacterium coli"* in 50 ml, the water would be safe to drink.

Subsequent PHS and EPA standards have included sampling frequency requirements with the number of samples required for making the water quality evaluation increasing as the population served increases. The sampling frequency requirements of the 1946 drinking water standards are presented as a graph in Figure 21.1 which was published in the tenth edition of Standard

Methods (American Public Health Association, 1955). There are minor differences in the frequency tables accompanying other standards but all require at least 1 sample per month for the smallest water systems with increasing numbers of samples for larger systems up to 500 per month for the largest systems. In the 1977 Interim Primary Drinking Water Regulations, the systems required to take only one sample per month were those serving fewer than 1000 people. Systems serving between 1000 and 2500 people are required to take 2 samples per month and so forth. The rationale for requiring more samples for systems serving more people has never been explained. A common assumption is that population served is a surrogate for size and the larger systems require more samples just because they are larger. That assumption is not consistent with statistical sampling theory and a different justification is needed.

Locations for Sample Collection   Decisions about when during the month and where to collect the bacteriological samples have been left to the judgment of the state agencies responsible for enforcement or to the local water utilities. There is a great deal of variation in the practices associated with microbiological monitoring of water systems in the United States. It is common for fixed sampling locations to be used from month to month because it is believed that the results from sampling one location can represent a relatively large area of a distribution system (Pipes et al., 1985).

Check Sampling   It seems clear that the requirements for additional sampling after an unsatisfactory sample in the 1943, 1946, 1962 and 1977 standards were intended to check for a cross connection at the point of sampling and

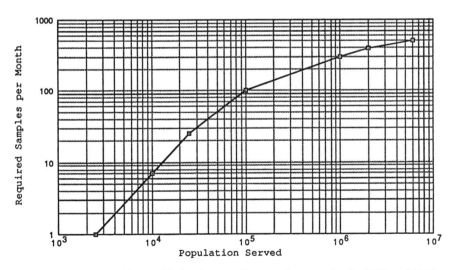

Figure 21.1   U.S. Public Health Service sampling requirements for the 1943 and 1946 drinking water standards.

were not directed toward determining the level of coliform contamination in the system as a whole (Pipes and Minnigh, 1987). The 1989 revision of the microbiological MCL requires three additional samples following an unsatisfactory sample with one at the location of the original unsatisfactory sample and the two others at nearby locations. For systems collecting only one sample per month, four additional samples are required following a positive sample. This regulation is intended to provide more samples for evaluation of the overall level of coliform contamination in the system but it also includes a check for continuing contamination at the location of the original positive sample.

21.3   Research on Microbiological Monitoring

Since the adoption of the National Interim Primary Drinking Water Regulation in 1977, several investigations have been directed toward providing better resolutions of the issues listed in Table 21.1. These studies were intended to provide a rational basis for the revision of the microbiological MCL.

Two conceptual models of water distribution systems, the mechanistic model and the empirical model, have been formulated. The starting point for the mechanistic model is information about the direction and velocities of the flow of water in the distribution system. If the mechanistic model were correct, it would be possible to predict the spread of contamination throughout the system from the point of entry of the contaminant. In contrast to this, the empirical model does not make use of any information about the flow of water but treats the distribution system as if it were a "black box." All sampling locations are considered to be equal and overall parameters of the level of contamination are inferred from the results of examination of the samples.

One aspect of microbiological monitoring which needs to be addressed before proceeding to a discussion of the models of sampling is the accuracy of determination of the coliforms in a sample. Not all of the coliforms which occur in drinking water samples will grow and produce the characteristic reactions needed for recognition as coliforms in the media used for testing (Bissonette et al., 1977). Even when all the coliforms grow and produce the characteristic reaction, two or more coliforms may end up in the same fermentation tube or may fall close enough together on a membrane filter that only one colony is produced. Thus, there are errors in demonstrating the presence of coliforms in some samples and in the determination of their numbers when they are found. Monitoring and the models of monitoring deal with "detectable coliforms" and it must be kept in mind that there are coliforms which are not detected.

Mechanistic Model   The mechanistic model is used as a working hypothesis by many water utilities for establishing sampling locations. It is assumed that the water is usually not contaminated when it is pumped into the distribution system and that, when a contamination event occurs, the bacteria remain in the water and are detectable for some relatively long time. This assumption has a number of corollaries; e.g., if contamination is found at one location, it will be found at all downstream locations and, if it is not found at a particular location it will not be found at upstream locations. Thus, if the system is sampled near the point at which the water is pumped into the system and also at the most distant point from the input and contamination is not found, then there is no contamination in between. If the mechanistic model were correct, microbiological monitoring could be reduced to a finite number of sampling locations which could be used to examine the entire system.

The mechanistic model was introduced in a paper presented at the 1978 annual meeting of the American Water Works Association (Pipes and Christian, 1978) as an attempt to explain some existing sampling practices. Subsequently, the model was tested during a three year project which involved intensive sampling of several small water distribution systems.

The results from the study indicated that the assumptions underlying the mechanistic model were incorrect (Pipes and Christian, 1982). If contamination is found at one location, it does not necessarily occur at all downstream locations. If contamination is not found at a location, the probability of finding contamination is not any less upstream of that location than anywhere else in the system. A water sample from one location represents only the microbiological water quality at that location and at that time and it is not possible to track contamination from one sampling location to downstream locations.

The time factor associated with contamination in a water distribution system has only been explored to a small degree. The assumption of the mechanistic model is that contamination or lack of contamination will persist at a sampling location for a relatively long period of time. The research results have failed to turn up evidence of persistence of contamination at any location and locations which have yielded negative samples do not seem to have any lower probability of yielding positive samples when they are sampled again (Pipes and Christian, 1982).

One result of testing the mechanistic model is that it has been shown that water systems are sometimes divided into areas isolated in such a manner that the microbiological water quality may differ between the areas. Isolated areas are selected so that the flow of water from one isolated area to another is unidirectional away the water source toward the periphery of the system but each isolated area includes locations where the direction of flow in the water main may change in response to changes in water usage. Contamination can exist in one isolated area without being transported into any of the other areas by the normal flow of water. The level of contamination may differ from one isolated area to another during a period of sampling (Pipes and Christian, 1982). Thus, adequate sampling of the entire water distribution system requires sampling of all isolated areas of the system. The concept of isolated areas with different levels of coliform contamination can provide a justification of the need for larger numbers of samples for larger systems which is consistent with sampling theory.

The mechanistic model of sampling continues to be used as a basis for selecting sampling locations and interpreting sampling results by many utilities (Pipes et al., 1985) although the assumptions on which it is based have been shown to be incorrect. One deterrent to discarding the mechanistic model is that without it, sampling a water distribution system becomes "a needle in the haystack" type of problem. There are many places and many different times where contamination may occur. Collection of enough samples to demonstrate that coliform contamination of the system is a rare event becomes an enormous problem.

Empirical Model   The empirical model of sampling is not based on any assumptions about a sampling location representing the water quality over an area of the system or over any period of time. The entire distribution system is conceived as being made up of a finite (but large) number of sampling locations. Contamination (or the absence of contamination) found at a particular location will not necessarily be found at any nearby location or at the same location at any other time (even a few seconds later). Enough samples should be taken so that the probability of detecting coliforms, if they are present, is high enough to give evidence of adequate protection of public health.

Since the time for water quality evaluation is one month, all the water passing through the system in a month's time is considered to represent the "population" of potential unit volume samples which could be examined. The unit volume for U.S. practice is 100 ml and this same unit volume is used in many other countries. This discussion will be in terms of 100 ml samples but it should be remembered that there is no particular scientific basis for the 100 ml sample and other unit volumes could be used just as readily.

Development of the empirical model depends upon selection of a probability density function to represent the frequency of occurrence of various counts of indicator bacteria in unit volume samples (see Chapter 20). When data from nine water distribution systems were analyzed, it was found that the coliform counts showed a right skewed distribution and could be described by either the negative binomial and lognormal distribution (Christian and Pipes, 1983). Other distributions such as the gamma distribution and the Pearson type II could have also been used but they would not have offered any advantage over the two which were used. A multimodal Poisson distribution could have been used but the multimodal Poisson is mathematically identical to the negative binomial (see Chapter 20).

In subsequent studies (Pipes and Christian, 1984; Dempsey and Pipes, 1986; Pipes and Minnigh, 1987) the lognormal distribution has been used to represent the densities of coliforms in samples from water distribution systems because it is familiar to some water works personnel, the computations are somewhat easier than those for the negative binomial and it is the distribution of a continuous variable. The negative binomial distribution deals with the frequency of counts. It may appear that coliform densities in 100 ml samples are integral counts but the underlying variable is the coliform density, which is a continuous variable. Coliform colony counts in individual samples are whole numbers and the volume removed from the sample for examination is 100 ml. For example, coliform densities of one per 153 ml (6.5 per liter) or 1 per 22.5 ml (4.4 per 100 ml) or other similar densities probably exist in water distribution systems; it is the selection of a unit volume for examination which forces the data to be determined as an integral count. Finding no coliforms in a 100 ml sample does not mean that there are no coliforms in the water, it just means that the density is probably less than one per 100 ml at that location. The use of the lognormal distribution to represent coliform counts in a water distribution system is consistent with the earlier studies of Velz (1949) and Thomas (1952).

21.4   Applications of Research Results

The demonstration that the mechanistic model is incorrect and that the occurrence of coliform counts in samples from a distribution system or from an isolated area of a distribution system can be represented by a lognormal distribution forms a conceptual basis for resolving some of the issues of microbiological monitoring. Additional research results continue to become available each year and a basis is being built for future improvements in microbiological monitoring of water systems.

Choice of an Indicator Organism   There is a great deal of dissatisfaction with the total coliform group as the indicator for monitoring drinking water; yet, there is a great reluctance to give it up because there is so much experience with it and the incidence of waterborne disease has decreased since 1914. The trend is to use a second indicator in addition to the total coliform group. Clark's presence-absence test (see Chapter 19) was developed to allow testing for a number of indicators besides total coliforms using a single fermentation tube (Clark 1969) and it allows the detection of coliforms in the presence of large numbers of other bacteria which may interfere with other coliform tests (Clark 1980). There are several procedures available to test for *Escherichia coli* simultaneously with a total coliform test and one of these has recently received a national test as an indicator for monitoring drinking water (Edberg et al., 1988).

There is a provision in the 1989 revision of the microbiological MCL which would require testing for either fecal coliforms or *E. coli*. This may be an opening which will develop into a change in the indicator used.

The Choice of a Water Quality Parameter   The parameters of coliform densities which have been used in the standards and regulations are the frequency of occurrence of samples above some single sample limit and the arithmetic mean of all the samples for the month. Apparently these parameters were selected intuitively by committees who developed the standards in 1914, 1925, and 1962.

It requires two parameters to represent fully the level of coliform contamination in a water distribution system. When the lognormal distribution is used, these two parameters can be the geometric mean (GM) and the geometric standard deviation (GSD). If coliform density is represented by the random variable X and the logarithm of X ($Y = \ln X$) is normally distributed with mean μ_y and variance σ_y^2, then X has a lognormal distribution with mean μ_x and variance σ_x^2. The GM of x is $\ln \mu_y$ and the GSD is $\ln \sigma_y$. The level of coliform contamination in a water distribution system can be represented by a line plotted on logarithmic probability paper, or as a single point on an arithmetic plot of the GM versus the GSD (Dempsey and Pipes, 1986). This approach makes it possible to compare graphically the relative stringency of various rules which might be used in regulations and visualize different levels of contamination.

It requires a relatively large number of samples (50 or more) for estimation of two parameters of the level of coliform contamination in a small water distribution system or in an isolated area of a large system. It is not practical to require water utilities to collect and examine enough samples in a month's time to estimate the GM and GSD for the various isolated areas of the distribution systems. Thus, the rule for determination of the acceptability of the bacteriological quality of the water must be based on a single parameter which cannot fully represent the level of coliform contamination. However, it is possible to consider which single parameter of coliform contamination would be most likely to be related to any possible health effects.

Arithmetic Mean as the Parameter   The arithmetic mean has the desirable property that it represents the total number of coliform bacteria passing through the system in a month's time because the average density multiplied by the volume of water gives the total number of coliforms. Clearly, the total number of coliforms passing through the system in a month's time represents the level of contamination quantitatively. However, the mean coliform density is very difficult to estimate accurately and there is no particular reason to expect that the risk of waterborne disease might be related to it.

One difficulty of estimating the mean density is data truncation. When 100 ml is used as the unit volume for examination, densities between 0 and 1 are recorded as less than 1 per 100 ml and counts of more than 80 are recorded as TNTC (too numerous to count). Thus, densities between 1 and 80 are the only ones which can be determined precisely. Truncation of FT data is even more severe; the MPN for 1 of 5 ten ml tubes positive is 2.2 per 100 ml and for 4 of 5 ten ml tubes positive is 16.1 per 100 ml.

Pipes and Christian (1983) found that the average MF coliform colony count, calculated by using 0 for low-range indeterminate and >80 for high-range indeterminate, always underestimated the mean density of coliform bacteria because of data truncation and the large variance of the coliform counts.

Estimating the parameters of a lognormal distribution representing the coliform data and then calculating the arithmetic mean from the lognormal parameters always gave a higher arithmetic mean than that calculated from the sample counts. Stukel (1986) later showed on a theoretical basis that estimating an arithmetic mean of samples from a lognormal distribution always gives an underestimate of the actual value. Haas and Heller (1988) have proposed a maximum likelihood procedure for estimating the arithmetic mean from truncated data (see Chapter 20) but there has not been time for evaluation of their method for regulations or for epidemiological investigations.

Geometric Mean as the Parameter   The geometric mean of indicator bacteria counts has been used as a parameter for representing the level of contamination of receiving waters and bathing beaches (El Shaarawi and Pipes, 1983) probably because it has been known for many years that those data can be fitted to a lognormal distribution (Velz, 1949). If some of the counts are zero (as always happens in the monitoring of water systems), the estimation of the geometric mean requires sophisticated mathematical manipulation. As compared with the arithmetic mean, the geometric mean discounts the samples with high coliform densities. Since the primary concern is the possibility of waterborne disease, it would seem that samples with high coliform densities should be emphasized rather than discounted. The geometric mean is not directly related to the total number of bacteria in the water as the arithmetic mean is. Also, as in the case of the arithmetic mean there is no a priori reason to expect that the risk of waterborne disease will be related to the geometric mean. Thus, this parameter does not appear to be desirable for use in regulations.

Frequency of Occurrence as the Parameter   The frequency of occurrence of indicator densities above some preselected level can be determined rather precisely by a reasonable number of samples. If coliform densities are related to the risk of waterborne disease, it may be assumed that the water with the highest densities will represent the greatest risk. When waterborne disease outbreaks occur, the water is usually found to contain densities much greater than 1 per 100 ml of coliforms, fecal coliforms, and *Escherichia coli*. The frequency of occurrence parameter for indicator organisms has not been used for epidemiological studies of waterborne disease, so it is possible that a relationship could be found if it were used.

Numerical Value for Frequency of Occurrence   If the single sample limit is set at none per unit volume, the numerical values for the frequency of occurrence parameter and its confidence limits can be estimated from the presence or absence of coliforms in a series of samples by using binomial probabilities. Also, if the GM and GSD for the densities of coliforms in a water system are known, the probability of occurrence of coliform densities above some value such as none (less than 1) per 100 ml or more than 4 per 100 ml is easy to calculate from the standardized normal distribution. Thus, the frequency of occurrence can be calculated for a simulation of water system sampling, if a level of contamination is assumed.

It was noted in a previous section that the MF average rule of 1 per 100 ml of the 1977 regulations is much more stringent than the frequency of occurrence of no more than 5% of the samples with coliform counts greater than 4 per 100 ml. Since the average rule was eliminated for the 1989 MCL revision, the single sample limit or the allowable fraction with coliforms present needs to be adjusted to avoid making the MCL much less stringent. In a theoretical study, it was found that lowering the single sample limit to 1 per 100 ml while keeping the 5% frequency of occurrence would make it approximately equivalent to the mean density of 1 per 100 ml for values of the GM and GSD which have been found to occur in water distribution systems (Dempsey and Pipes, 1986). A simulation study and analysis of monitoring data from four states confirmed this finding (Pipes et al., 1986).

In the proposed revision of the microbiological MCL the frequency of occurrence has been kept at 5% while the single sample limit will be reduced to 0 per 100 ml. This would make the revised MCl more stringent than the existing MCL if enough samples were collected so that the frequency of occurrence could be estimated with a 95% confidence limit of ±5%. However, the number of samples required for monitoring is not being increased but has been decreased in some cases, so that the net effect is a reduced stringency of the MCL. For any level of coliform contamination it will be easier to meet the 1989 revised MCL than the 1977 MCL.

Numbers of Samples Required   The original proposal for the revision of the MCL would have increased the number of samples required for the smallest systems (those serving fewer than 1000 consumers) from 1 to 5 per month. This was intended to provide 60 samples over a period of 12 months and the MCL would have been no more than 5% (3/60) of the samples in any 12 month period with coliforms present (Pipes et al., 1983). The 95% confidence interval for 3 positive samples out of 60 total samples is approximately ±5% which gives some appearance of statistical reliability. Since there would now be no need to estimate the coliform density of samples but just to determine their presence or absence, the examination cost per sample will be reduced and this partially compensates for cost of collection of extra samples. The extension of the time for evaluation of compliance with the MCL is not necessarily desirable but it does give enough samples for some degree of statistical reliability without increasing the number of samples required per month to an impossibly expensive level. The recommended approach of examining 5 samples per month and using the P-A test was field-tested successfully in ten small systems over a one year period (see Chapter 19 and Jacobs et al., 1986).

The proposal to increase the number of samples per month for the smallest systems has drawn a great deal of opposition from water utilities and the state agencies which have the responsibility of enforcing the MCL's. The opposition has been so strong that the provision requiring more samples has been dropped from the proposed revision and the table of samples required versus population served has been changed so that no systems will be required to take more samples than they do under the present regulation; some are required to take fewer.

The 5% frequency of occurrence in the MCL can be misunderstood. It applies only to the very large systems serving more than 50,000 people which collect more than 40 samples per month. Over 95% of the community water systems in the U.S. are required to take fewer than 40 samples per month (about 80% of them are required to take only 1 sample per month) and are allowed one sample per month with coliforms present without violating the MCL. It is the second sample with coliforms present which causes a violation of the MCL. The probability of finding two samples with coliforms present in a month's time is directly related to the number of samples collected each month (Dempsey and Pipes, 1986). Thus, the very small systems, which are the group which experience the most waterborne disease outbreaks, can have very high levels of coliform contamination without violating the MCL.

A really determined effort to demonstrate that the frequency of occurrence of coliform bacteria in the water distribution system was less than 5% would require more samples than most water utilities are willing to collect and examine in a month. Since that is the case, it might be a better use of resources to drop the idea of routine monitoring and make periodic intensive studies of each of the small systems on a rotating time schedule.

Selection of Sampling Locations   The best approach for selection of sampling locations for microbiological monitoring of a water distribution system is stratified random sampling. The entire water distribution system should be divided into isolated areas; i.e., parts of the system where contamination could exist without being detected in samples from locations in other parts of the system. After the stratification, the samples for the entire system are apportioned among the isolated areas and individual sampling locations in each area are selected by a random process. One way of accomplishing the random selection of sampling locations is to number each service connection in an isolated area and then select the locations to be sampled each month from a table of random numbers. In this way, each sampling location has an equal chance of being selected and there are no locations in the system where contamination can exist indefinitely without any chance of being detected.

Although the stratified random sampling approach has been used for research (Pipes and Christian, 1982), it is not used by water utilities nor recommended by state agencies for monitoring water distribution systems. Permission from the customer is needed to sample any service connection and some locations selected for sampling may not be accessible on the day of sampling. This requires a procedure for selection of alternative sampling locations by the sampler in the field. Samplers may want to avoid certain areas of the community for security reasons. The practical problems associated with a stratified random sampling program prevent it from being used.

The concept of stratified random sampling does provide a rationale for collecting more samples from larger water distribution systems. The larger number of isolated areas of the distribution system require a larger number of samples in order to sample all areas equally. However, utilities do not identify isolated areas of their distribution system and in the usual monitoring programs some isolated areas are not sampled.

Check Sampling   Requirements for collection of additional samples following an unsatisfactory sample were introduced in the 1943 PHS standards and they have been part of the monitoring process ever since. It is clear that this requirement was originally intended as a check for a cross connection at the location of the original unsatisfactory sample. In the 1977 regulations the name of these samples was changed from *additional* samples to *check* samples, inferring that they were intended as a check for continuing contamination. There is no record of any cross connection being located during a monitoring program as a result of this provision. It has been pointed out that the nature of the contamination from a cross connection is such that repeated sampling on successive days probably would not continue to find coliforms (Pipes and Minnigh, 1987).

Another purpose for the additional sampling requirements can be to increase the number of samples for estimation of the frequency of occurrence. If only a few samples are taken and coliforms are found in one, the frequency of occurrence for that sampling period can be high; but, if additional samples are collected and no more coliforms are found, the frequency of occurrence is reduced. For instance, if 1 of 5 original samples has coliforms, the frequency of occurrence is 20%. If 5 additional samples are taken and none have coliforms, the frequency of occurrence is reduced to 10% for that sampling period. On the other hand, if additional samples with coliforms are found, the high frequency of occurrence is confirmed and it is clear that remedial action is needed. One appropriate response to finding coliform in samples from a distribution system is to collect more samples to check the accuracy of the original finding. If the additional sampling is used in this manner, it does not have to be sampling at the location of the original sample; indeed, it is better to sample throughout the system.

21.5 Summary

Monitoring of water distribution systems for the presence of bacteria which may indicate contamination with fecal material has been practiced in the U.S. and other countries for many years. This monitoring was originally intended to determine if the water is free of the pathogens which cause waterborne infectious diseases. However, the coliform group, which is the indicator used, is not well correlated with fecal contamination or with the presence of pathogens. Also, the selection of the numbers of samples and the locations for sampling is such that fecal contamination of parts of most distribution systems could persist indefinitely without being detected by routine monitoring. The rationale for continuing to require monitoring under the present circumstances is not at all clear.

The standards and regulations which govern the microbiological monitoring of drinking water in the U.S. have evolved slowly from the first standards in 1914 to the new regulations adopted in 1989. There has been research on drinking water microbiology, the water quality issues related to monitoring, and the methodology of sampling, but very few of the research findings have had much effect on the regulations. There is a continuing need both for using the available research findings for development of more rational regulations and standards and for research studies which are designed to resolve the issues of microbiological monitoring of water quality. Some of the most significant research needs are as follows:

1.   The most pressing need is for a bacterial indicator which is directly related to fecal contamination and/or to the presence of pathogens which can cause waterborne disease. Since there are now methods to identify *E. coli* separately from the other coliforms, it is possible to investigate its use as such an indicator, which was the intent of the 1914 standards.

2.   A relationship between some level of coliform contamination in water systems and the probability of waterborne disease outbreaks has never been demonstrated. The development of an indicator to replace the coliform group will require adequate epidemiological studies.

3.   It is possible that a relationship between the occurrence of the coliform group and waterborne disease exists. The frequency of occurrence parameter as a measure of coliform or *E. coli* contamination has not been tested versus the incidence of waterborne disease outbreaks.

4.   With the tools which are now available, it is possible to test for adequacy of disinfection and for post-treatment contamination throughout a water distribution system. It would take many months to conduct such a testing program on a large water distribution program but the program would have more value than the present routine monitoring programs.

5.   The time persistence of microbiological water quality in water systems has not been adequately described. The causes of abrupt changes in microbiological water quality need to be determined from case studies.

6.   The future studies should be designed to provide concepts and information which is directly applicable to development of better regulations and standards.

In the 1925 revision of the microbiological drinking water standards, the single sample limit of 1 of 5 ten ml tubes positive was relaxed to 2 of 5 ten ml tubes positive and the frequency of occurrence limit was relaxed from no samples to 5% of the samples. In the 1989 revision, the single sample limit was decreased from more than 4 per 100 ml to less than 1 per 100 ml and the frequency of occurrence was left at 5%. This may appear to be a tightening of the MCL rule but since the average rule of coliform per 100 ml was dropped and the number of samples not increased (but decreased for some systems), it will be easier for all water distribution systems to meet the revised MCL. The only justification for such a relaxation of the microbiological MCL rule is a recognition that coliform bacteria occur in large numbers in many water distribution systems where there is no problem with waterborne diseases.

References

Allen, M., Taylor, R.M. and Geldreich, E.E. 1980. The occurrence of microorganisms in water main encrustation. *J. Am. Water Works Assoc.* 72:616–625.

APHA. 1955. *Standard Methods for the Examination of Water and Wastewater*, Tenth Edition. APHA, WPCF, and AWWA, Washington, DC. 522 pp.

American Public Health Association. 1985. *Standard Methods for the Examination of Water and Wastewater*, Sixteenth Edition. Washington, DC. 1284 pp.

Batik, O., Craun, G.F., and Pipes, W.O. 1983. Routine monitoring and waterborne disease outbreaks. *Jour. Envir. Health* 45:227–230.

Christian, R.R. and Pipes, W.O. 1983. Frequency distributions of coliforms in water distribution systems. *Appl. Environ. Microbiol.* 45:603–609.

Bissonette, G.K., Jezeski, J.J., McFetters, G.A. and Stuart, D.G. 1977. Evaluation of recovery methods to detect coliforms in water. *Appl. Environ. Microbiol.* 45:603–609.

Clark, J.A. 1969. The detection of various bacteria indicative of water pollution by a presence-absence (P-A) procedure. *Can. Jour. Microbiol.* 15:771–780.

Clark, J.A. 1980. The influence of increasing numbers of nonindicator organisms on the detection of coliforms by the membrane filter and presence-absence tests. *Can. Jour. Microbiol.* 26:827–832.

Dempsey, K.W. and Pipes, W.O. 1986. Evaluating relative stringencies of existing and proposed microbiological MCL's. *Jour. Am. Water Works Assocn.* 78(11):47–54.

Edberg, S.C., Allen, M.J., Smith, D.B., and The National Collaborative Study. 1988. National field evaluation of a defined substrate method for the simultaneous enumeration of total coliforms and *Escherichia coli* from drinking water: Comparison with the standard multiple tube fermentation method. *Appl. Environ. Microbiol.* 54:1595–1601.

El Shaarawi, A.H. and Pipes, W.O. 1982. Enumeration and statistical inference, pp. 43–66 in Pipes, W.O. (ed), *Bacterial Indicators of Pollution*, CRC Press, Inc., Boca Raton, FL.

Goshko, M.A., Pipes, W.O., and Christian, R.R. 1983. Coliform occurrence and chlorine residual in small water distribution systems. *Jour. Am. Water Works Assocn.* 75(7):372–378.

Haas, C.N. and Heller, B. 1988. Averaging of TNTC counts, *Appl. Environ. Microbiol.* 54:2069–2072.

Herson, D.S., McGonigle, B., Payer, M.A., and Baker, K.H. 1987. Attachment as a factor in the protection of *Enterobacter cloacae* from chlorination. *Appl. Environ. Microbiol.* 53:1178–1180.

Hudson, L.D., Hankins, J.W., and Battaglia, M. 1983. Coliforms in a water distribution system: A remedial approach., *Jour. Am. Water Works Assocn.* 75:564–568.

Jacobs, N.J., Zigler, W.L., Reed, F.C., Stukel, T.A., and Rice, E.W. 1986. Comparison of membrane filter, multiple-fermentation-tube, and presence-absence techniques for detecting total coliforms in small community water systems. *Appl. Environ. Microbiol.* 51:1007–1012.

Kehr, R.W. and Butterfield, C.T. 1943. Notes on the relationship between coliforms and enteric pathogens. *Pub. Health. Repts.* 58:589–596.

LeChevallier, M.W., Cawthon, C.D., and Lee, R.G. 1988. Factors promoting survival in chlorinated water supplies. *Appl. Environ. Microbiol.* 54:649–654.

McCrady, M.H. 1915. The numerical interpretation of fermentation tube results. *Jour. Infect. Dis.* 17;183–206.

Olivieri, V.P., Ballestero, J., Cabelli, V.J., Chamberlin, C., Cliver, D., DuFour, A., Ginsberg, W., Healy, G., Highsmith, A., Read, R., Reasoner, D., and Tobin, R. 1983. Measurement of microbial quality. pp. II-1–II-42 in Berger, P.S. and Argaman, Y. (editors), *Assessment of Microbiology and Turbidity Standards for Drinking Water*, EPA 570/9-83-001. Office of Drinking Water, U.S. Environmental Protection Agency, Washington, DC.

Pipes, W.O. and Christian, R.R. 1978. A sampling model for coliforms in water distribution systems. *Proc. 1978 Annual Conf.*, paper 33-5, American Water Works Association, Denver, CO.

Pipes, W.O. and Christian, R.R. 1982. Sampling Frequency-Microbiological drinking water Regulations, EPA-570/9-82-001. Office of Drinking Water, U.S. Environmental Protection Agency, Washington, DC.

Pipes, W.O. and Christian, R.R. 1984. Estimating mean coliform densities of water distribution systems. *Jour. Am. Water Works Assocn.* 76(11):60–64.

Pipes, W.O. and Minnigh, H.A. 1987. Significance and interpretation of repeat sampling results, *Tech. Conf. Proc. Advances in Water Analysis and Treatment*, paper ST-11, Am. Water Works Assocn., Denver, CO.

Pipes, W.O., Troy, M.A. and Minnigh, H.A. 1986. Empirical evaluation of present and possible future microbiological MCL rules, pp. 1561–1567, *Proc. 1986 Annual Conf.*, Am. Water Works Assocn., Denver, CO.

Pipes, W.O., Minnigh, H.A. and Troy, M.A. 1986. Field inoculation of Clark's presence-absence test for coliform detection, *Tech. Conf. Proc. Advances in Water Analysis and Treatment*, paper 3B-5, pp. 305–315, Am. Water Works Assocn., Denver, CO.

Pipes, W.O., Bordner, R., Christian, R.R., El Shaarawi, A.H., Fuhs, G.W., Kennedy, H., Means, E. Moser, R., and Victoreen, H. 1983. Monitoring of microbiological quality. pp. III-1–III-45 in Berger, P.S. and Argaman, Y. (editors), *Assessment of Microbiology and Turbidity Standards for Drinking Water*, EPA 570/9-83-001. Office of Drinking Water, U.S. Environmental Protection Agency, Washington, D.C.

Pipes, W.O., Burlingame, G.A., Becker, R.J., Christian, R.R., Geldreich, E.E., Ginsberg, W., Means, E.G., Spitzer, E.F., Standridge, J., Victoreen, H.T., and Wentworth, N. 1985. Committee report: Current practice in bacteriological sampling, *Jour. Am. Water Works Assocn.* 77(9):75–81.

Pipes, W.O., Mueller, K., Troy, M.A., and Minnigh, H.A. 1987. Frequency-of-occurrence monitoring for coliform bacteria in small water systems. *Jour. Am. Water Works Assocn.* 79(11):59–63.

Public Health Service. 1962. *Drinking Water Standards*. U.S. Department of Health, Education and Welfare, Washington, DC.

Thomas, H.A. 1952. On averaging the results of coliform tests. *Jour. Bost. Soc. Civil Engrs.* 39:253–261.

Velz, C.J. 1951. Graphical approach to statistics, Part 4: Evaluation of bacterial density. *Water and Sewage Works* 98:67–74.

22

Monitoring Heterotrophic Bacteria in Potable Water

Donald J. Reasoner

Drinking Water Research Division, Risk Reduction Engineering Laboratory, U.S. Environmental Protection Agency, Cincinnati, Ohio

22.1 Introduction

Measurements of bacterial populations in water have been used since the beginning of sanitary bacteriology (see Chapter 21), and interest in the interpretation of the results has occupied many researchers over the intervening years. In addition to the coliform count, a more generalized bacterial counting procedure has also been used. In the United States, standardization of the methodology for general bacterial counts using a plate-counting procedure began with the activities of the Committee of Bacteriologists of the American Water Works Association from 1895 to 1898 (Prescott et al., 1950). Continued work on standardization of methods for detection of bacteria led to the inclusion of those methods in the first edition (1905) of what is now known as *Standard Methods for the Examination of Water and Wastewater* (*Standard Methods*, hereafter).

The early plate count procedures utilized nutrient gelatin and a 20°C incubation temperature to keep the gelatin in a solidified state. Only after the introduction of agar as a solidifying agent for bacterial growth media was it possible to use incubation temperatures greater than 28°C. The second edition (1912) of *Standard Methods* introduced the 37°C, 24 h plate count using nutrient agar, but also included the 20°C, 48 h plate count using nutrient gelatin. Inclusion of a 20°C plate count was continued through the 12th edition (1965) of *Standard Methods*, but was dropped from the 13th (1971) through the 15th (1980) editions. The 16th edition (1985) of *Standard Methods* reintroduced an option of a 20°C plate count. In fact, the 16th edition of *Standard Methods* introduced several changes in bacterial plate count procedures by including the membrane filter (MF) and spread plate (SP) methods, the use of a low nutrient medium (R2A medium, Reasoner and Geldreich, 1979; 1985), and the use of different incubation temperatures (20°C, 28°C, and 35°C) and different incubation times for the determination of heterotrophic plate count (HPC) bacteria levels in water.

During the past decade, interest in assessing the bacterial quality of drinking water, particularly during distribution, has renewed interest in bacterial plate count methods. These activities resulted in the development of new media and the addition of the spread plate and membrane filter methods to the 16th edition (1985) of *Standard Methods*. Many research studies in recent years have used media and methods different from the traditional pour plate (PP) procedure. Results from such studies caused reevaluation of the methods and resulted in a name change for this section of *Standard Methods* from the "Standard Plate Count" to the "Heterotrophic Plate Count" (abbreviated HPC). The heterotrophic plate count enumerates aerobic and facultative aerobic bacteria found in water that are capable of growth on simple organic compounds (primarily carbohydrates, amino acids and peptides) found in the culture medium, and under incubation time and temperature conditions specified. The net result of the more recent research and changes in methods for enumerating the general heterotrophic bacterial populations in drinking water is that more options are now available and water utilities are being encouraged to make use of these methods as tools to gain better information on the effects of water treatment processes and distribution on the bacteriological quality of drinking water.

This paper will deal primarily with the HPC methods that have been used, and those that are currently in use, in the United States for the enumeration of bacteria in potable water treatment and distribution. However, other studies that are germane to the topic will also be used to illustrate the uses of the HPC measurement. The statistical aspects of plate count procedures will not be dealt with due to space constraints, and the reader is referred to other sources such as Jennison and Wadsworth (1940), Eisenhart and Wilson (1943), Snyder (1947), and Cowell and Morisetti (1969) (see also Chapter 20).

22.2 Uses of Heterotrophic Plate Count Measurements

The purpose of water treatment is to provide a safe water supply for the consumer by using a series of unit processes to reduce levels of turbidity, and of chemical and microbial pollutants. However, beyond treatment, the quality of the treated water must be maintained during storage and distribution to the consumer.

The HPC is a useful tool for:

1. Monitoring the efficiency of water treatment processes, including disinfection.

2. Obtaining supplemental information on HPC levels that may interfere with coliform detection in water samples collected for coliform compliance monitoring (U.S.E.P.A., 1975).

3. Assessing changes in finished water quality during distribution and storage, and distribution system cleanliness.

4. Assessing microbial growth on materials used in construction of potable water treatment and distribution systems.

5. Measuring bacterial regrowth or aftergrowth potential in treated drinking water.

6. Monitoring microbial population changes following treatment modifications such as a change in the type of disinfectant used.

22.3   Methodology Considerations

The pour plate (PP) technique was first described by Koch in 1883 (Koch, 1912; McNew, 1938) as a method for obtaining pure cultures of bacteria from single cells. In the United States, the PP method with incubation at 35°C for 48 h has been used extensively for bacterial enumeration in drinking water. The spread plate method has not been used much, primarily because it was not in *Standard Methods* until the 16 edition (1985). The MF method (Taylor and Geldreich, 1979) was developed to provide greater flexibility in enumerating bacteria in treated drinking water. It permits the analysis of sample volumes larger than one ml, which is the primary limitation of the PP and SP methods. However, it was also not in *Standard Methods* until the 16th edition (1985). The advantages and disadvantages of each method are considered below.

Pour Plate   The PP method's greatest asset is its simplicity and ease of use. However, it has two drawbacks that may make it the least desirable method for enumerating bacteria in water. Bacteria in low-nutrient aquatic environments are generally considered to be physiologically stressed. As such, they have been shown to be susceptible to secondary stress caused by warming due to the melted agar tempered to 43°C to 46°C (Stapert et al., 1962; Klein and Wu, 1974). Klein and Wu (1974) showed that warming either by melted, tempered agar or by pre-warming agar spread plates to 45°C caused significantly decreased recoveries of bacteria compared to counts obtained by the SP method. They concluded that the PP method is not suitable for enumeration of heat-sensitive and possibly nutritionally stressed heterotrophic bacteria in water. As a result of this, at least in part, the PP procedure almost invariably yields lower bacterial counts than does the SP method (Buck and Cleverdon, 1960; van Soestbergen and Lee, 1969; Means et al., 1981; Taylor et al., 1983; Reasoner and Geldreich, 1985). Bacteria are also submerged in the agar in the PP method and the colonies tend to be smaller and do not show colony morphology characteristics that may be useful in identification. The other drawback is that the maximum sample volume that can be examined using the PP method is 1.0 ml, thus limiting the usefulness of the procedure when analysis of a larger volume is needed.

Spread Plate   The SP procedure is no more difficult than the PP, but does require preparation of the plates ahead of time and the use of a sterile bent glass spreading rod to distribute the sample over the surface of the agar. Because of these requirements, the procedure is viewed by some as being more tedious than the PP procedure. Also, because the plates must be prepared ahead, they must be checked for contamination before the sample is planted; and they must not be used while too moist, or allowed to become too dehydrated. Clark (1971) showed that pre-drying plates to a weight loss of 3–4 g/15 ml plate will allow the use of an inoculum volume up to 1.0 ml without affecting bacterial recovery. However, the recommended volumes for planting are 0.1 to 0.5 ml. The SP method for the HPC provides conditions that are more conducive to the growth of bacteria in water than the PP method. Consequently, as indicated above, the SP almost invariably yields higher colony counts than does the PP method. Since all bacteria in the sample are planted on the surface of the agar medium, their growth is not constrained by being surrounded by the agar. Thus, it is easy to observe differences in colony shape, texture, pigmentation, size, and other characteristics.

The precision of the SP method has been compared to that of the PP method in several studies. Comparisons of the PP and SP methods have provided mixed results regarding which method provides the best precision, but on balance, it would appear that the SP method is as precise as the PP method. Snyder (1947), studying relative errors of bacteriological plate counting methods, concluded that the SP method was as precise as any of the methods studied. Van Soestbergen and Lee (1969) concluded that the PP method was more precise than the SP method but yielded lower counts, whereas Taylor et al. (1983) found the SP method to be more precise than the PP method. A study of the accuracy and precision of the SP method for enumerating aerobic heterotrophic bacteria in environmental water samples was conducted by Kaper et al. (1978). Their results are also applicable to the SP method for HPC bacteria in potable water.

A disadvantage that the SP method shares with the PP method is that the maximum sample volume that can be analyzed is 1.0 ml using conventional culture dishes.

Membrane Filter   The advent of the MF method for the HPC analysis brought greater flexibility to the plate count procedure for drinking water by allowing the analysis of sample volumes larger than 1.0 ml. This permits the enumeration of very low concentrations of bacteria in finished drinking water and is limited only by the volume of water than can be drawn through the membrane without plugging the pores, or the deposition of so much material on the surface of the membrane that interference with growth of the bacteria occurs.

Comparisons of bacterial counts obtained by the MF method with those obtained by the PP method indicate that the MF method gives results equal to or greater than the PP method. Table 22.1 provides a summary of PP, SP, and MF results in various studies where the same medium was used with different HPC methods. The ratios of the SP and MF results to the PP results, or of the MF to the SP results show the relative performance of the methods.

In comparisons of the MF to the SP method, however, there have been mixed results because the choice of medium strongly affects the bacterial counts along with incubation temperature and even the water source. LeChevallier et al. (1980) showed that the MF method using m-SPC medium (now referred to as m-HPC medium in *Standard Methods*; hereafter, m-HPC will be used) was superior to the standard PP method for recovering bacteria from chlorinated drinking water, but there was no significant difference between the two techniques for raw water In a comparison of plate count media and methods Reasoner and Geldreich (1985) found that at incubation temperatures of 28°C or 20°C the MF method using either m-HPC medium or R2A medium yielded significantly higher bacterial counts than the standard PP method in the analysis of drinking water. At 35°C, however, m-HPC results were equal to or only slightly higher than the standard PP results. Fujioka et al. (1986) compared the standard pour plate with the MF method using m-HPC agar and R2A agar for the enumeration of bacteria in drinking water. Their results showed that the PP method recovered bacteria from only 80.5% of the water samples, but 100% of the samples yielded bacteria by the MF method using either m-HPC or R2A medium.

Other studies provided similar results, but the role of the type of medium used for the HPC determination must be recognized and evaluated. Table 22.2 provides a comparison of HPC results obtained in several studies in which different media/methods were used for enumeration of HPC bacteria in drinking water.

Evaluation of a modification of the MF method designed to permit simplified monitoring of HPC bacteria in water was reported by Green et al. (1982). The SPC Sampler made by Millipore Corporation utilizes m-HPC medium and was compared to the standard PP method and with the regular m-HPC MF method. The results indicated that there was excellent correlation between values obtained with the SPC Sampler and the regular m-HPC MF test and between the SPC Sampler results and the standard PP, but there was somewhat greater scatter of the data points with the PP procedure. On the other hand, Silley (1985) found that the Millipore SPC Sampler was not as good as the PP method with incubation at 20°C or 37°C for the enumeration of viable bacteria in a process water system. Pour plate counts were significantly higher than the SPC Sampler counts at both temperatures. Thus, the application of the HPC analysis dictates that comparisons of HPC methods are necessary to determine which method provides the highest HPC numbers consistent with the information needs of the user.

Table 22.1 Comparison of HPC results by pour plate, spread plate, and membrane filter methods on the same medium within each referenced study

Temperature (°C)	Method (CFU/ml)			Ratio			Reference
	PP	SP	MF	PP/MF	PP/SP	MF/SP	
16	3,100	6,300	—	—	0.49	—	Buck and Cleverdon, 1960
RT	42	—	113	0.37	—	—	Stapert et al., 1962
37	108	115	—	—	0.94	—	van Soestbergen and Lee, 1969
35	23	110	—	—	0.21	—	Klein and Wu, 1974
	62	200	—	—	0.31	—	
20	170	230	—	—	0.74	—	Means et al., 1981
35	210	240	—	—	0.87	—	
20	740	3,100	—	—	0.24	—	Taylor et al., 1983
20	1,020	—	986	1.03	—	—	Maul et al., 1985
35	430	510	270	1.59	0.84	0.53	Reasoner et al., 1985
28	4,000	7,000	4,600	0.86	0.57	0.65	
20	2,700	6,100	3,900	0.69	0.44	0.64	
35	3.3	4.9	—	—	0.67	—	Lombardo et al., 1986
28	4.0	5.2	—	—	0.76	—	
28	—	5.2	4.8	—	—	0.92	Gibbs and Hayes, 1988
22	100	710*	—	—	0.14	—	

RT = room temperature. CFU, colony-forming units. PP, pour plate. SP, spread plate. MF, membrane filter.
* Based on ratio given in the reference and arbitrarily assigning the value of 100 CFU/ml as the pour plate mean.

Table 22.2 Comparison of HPC results using different media

Tempera-	Method (CFU/ml)			
ture (°C)	PP	SP	MF	Reference
35	3,137 (SPC)*	—	4,273 (m-HPC)	Taylor and Geldreich, 1979
20	170 (SPC)	440 (R2A)	510 (m-HPC)	Means et al., 1981
20	—	4,000 (R2A)	12 (m-HPC)	Fiksdal et al., 1982
	—	1,000 (R2A)	110 (m-HPC)	
35	—	20 (R2A)	6 (m-HPC)	
	—	4 (R2A)	<1 (m-HPC)	
35	277 (SPC)	—	283 (m-HPC)	Green et al., 1983
20	1,123 (NA)	—	1,217 (m-HPC)	Maul et al., 1985
	1,192 (NA)	—	1,192 (R2A)	
35	22 (SPC)	200 (R2A)	32 (m-HPC)	Reasoner and Geldreich, 1985
28	80 (SPC)	360 (R2A)	140 (m-HPC)	
20	22 (SPC)	90 (R2A)	47 (m-HPC)	
35	53 (SPC)	—	66.7 (m-HPC)	Fujioka et al., 1986
	53 (SPC)	—	57.1 (R2A)	
26	590 (SPC)	1,550 (R2A)	—	Stetzenbach et al., 1986
22	100 (YEA)	710 (YEA)	—	Gibbs and Hayes, 1988
	440 (R2A)	—	—	
	100 (YEA)	3,900 (R2A)	—	

CFU, colony-forming units.
* Media: SPC = standard plate count agar; NA = nutrient agar; YEA = yeast extract agar, R2A medium (Reasoner and Geldreich, 1979), m-HPC medium (Taylor and Geldreich, 1979; m-HPC was published originally as m-SPC medium)

Medium and Incubation Temperature and Time Considerations

The early editions of *Standard Methods* provided for plate counts at two different temperatures, 20°C (room temperature count) and 37°C (body temperature count). The 37°C plate count was believed to give an indication of the presence of rapid-growing bacteria more likely to be related to pathogenic types that might be present in sewage pollution. The 20°C plate count was used for enumeration of characteristic water bacteria that tend to develop more slowly. The use of 35°C incubation in the United States for the plate count is founded in the early use of the body temperature count, but practical compromise resulted in lowering the incubation temperature by 2°C. This was apparently done because of lack of incubator space in laboratories that performed both milk and water analyses. The interested reader desiring a more comprehensive historical perspective on the room temperature count versus the body temperature count should consult Prescott et al. (1950).

During the late 1970s, interest in enumerating and identifying the noncoliform bacteria in water supplies resulted in the development of new media for use in the plate count procedure, re-evaluation of the incubation temperature requirements, and changing the name from the Standard Plate Count to the Heterotrophic Plate count (HPC) in the 16th edition of *Standard Methods*. Directly linked to, and in some cases integrated with, these changes were a variety of studies on the bacteriology of drinking water treatment and distributions systems.

Recognition of the need for a MF method for enumeration of bacteria in treated distribution water led to the development of m-HPC (m-SPC) agar by Taylor and Geldreich (1979). This medium was developed to provide results equivalent to the standard plate count (SPC) agar or tryptone glucose yeast extract (TGY) agar; the two media specified in *Standard Methods* for the 35°C plate count. The rationale for this was that the plate count data base developed by utilities that used the standard PP method would not be substantially changed if the results by the MF plate count were equivalent to those of the standard PP method. Haas et al. (1982) evaluated the m-HPC method as a substitute for the standard PP. They concluded that the m-HPC method was equivalent to the standard PP method when applied to raw, partially treated, or finished drinking water. However, they also cautioned that since their sampling represented only two systems in one geographical region, the method would have to be tested further for general applicability. The m-HPC medium in combination with the MF provided great flexibility in the volume of sample that could be examined and therefore increased the limit of detection of the method.

Shortly thereafter, interest in trying to maximize bacterial recoveries in order to develop a profile of typical bacteria that could be expected to be found in treated drinking water resulted in the development of R2A agar (Reasoner and Geldreich, 1979, 1985). In contrast to the standard plate count media and to m-HPC medium, R2A agar is a low-nutrient medium with respect to the concentrations of individual ingredients and the total composition, but has a greater variety of nutrients. Using the standard 35°C incubation temperature in a comparison of the HPC results on R2A spread plates and R2A MF plates versus the standard PP method and the m-HPC MF method, R2A medium yielded higher counts than the other media. The length of incubation is important with R2A medium; because of its lower nutrient concentration, bacteria found in drinking water tend to develop more slowly and incubation for 5 to 7 days at 20°C or 28°C provides the highest bacterial counts on this medium. Figure 22.1 shows relative differences in HPC results obtained on SP and MF plates with R2A, m-HPC and *Standard Methods* plate count agar (SMA) at temperatures of 20°C, 28°C, and 35°C and incubation from 2 days to 14 days. As can be seen, 35°C incubation is far inferior to either 28°C or 20°C, regardless of medium used.

R2A medium has been used for enumeration of HPC bacteria in various types of potable water including well-water (Stetzenbach et al., 1986), product water from point-of-use granular activated carbon (GAC) treatment devices

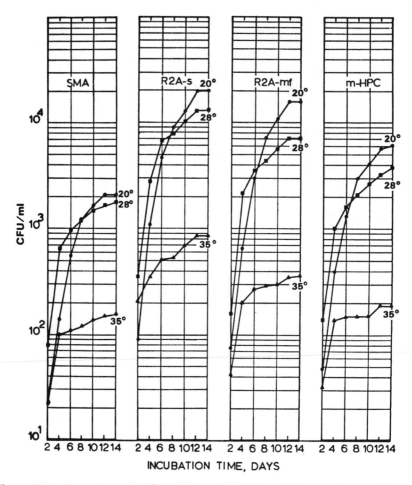

Figure 22.1. Comparison of 20°C, 28°C, and 35°C bacterial counts of treated distribution water. CFU, colony-forming units. SMA, Standard Methods plate count agar. R2A-s, R2A medium with spread plates. R2A-mf, R2A medium with membrane filters. m-HPC, HPC medium with membrane filters. Bacterial counts obtained using media and methods options in *Standard Methods* (16th ed.). Data adapted from Reasoner and Geldreich (1985).

(Reasoner et al., 1987), water treated by aeration to remove volatile organic chemicals (Fronk-Leist and Love, 1984), for enumeration of bacteria associated with GAC from drinking water (Camper et al., 1986), and for examination of fouling of reverse osmosis membranes (Ridgway et al., 1983, 1984).

The influence of incubation temperature was also shown to be very important in maximizing the bacterial count obtained with any of the HPC media. Table 22.3 shows, for several studies, the effect of the length of incubation, as well as differences in media and incubation temperatures on the HPC results. In general, the longer the incubation, the higher the HPC becomes.

Table 22.3 Effect of length of incubation on HPC results

Temperature (°C)	Medium and method	Bacterial count (CFU/ml)							Reference
		Incubation time, days							
		1	2	3	4	5	6	7	
35	SPC-PP	—	2,300	—	5,000	—	—	6,200	Klein and Wu, 1974
35	R2A-SP	—	10	—	—	—	—	20	Fiksdal et al., 1982
20	R2A-SP	—	0	—	—	—	—	4,000	
35	R2A-MF	—	5	—	—	—	—	10	
20	R2A-MF	—	0	—	—	—	—	40	
20	m-HPC-MF	—	190	945	1,217	—	—	—	Maul et al., 1985
20	R2A-MF	—	287	904	1,192	—	—	—	
20	NA-PP	—	389	855	1,123	—	—	—	
35	SPC-PP	—	22	—	100	—	110	115	Reasoner and Geldreich, 1985
28	SPC-PP	—	90	—	640	—	950	1,000	
20	SPC-PP	—	22	—	130	—	570	900	
35	R2A-SP	—	200	—	340	—	500	510	
28	R2A-SP	—	360	—	2,800	—	6,700	7,200	
20	R2A-SP	—	90	—	1,100	—	4,700	6,100	
35	R2A-MF	—	41	—	200	—	270	280	
28	R2A-MF	—	160	—	2,200	—	3,500	4,000	
20	R2A-MF	—	75	—	650	—	3,000	4,900	
35	m-HPC-MF	—	32	—	140	—	150	150	
28	m-HPC-MF	—	140	—	1,000	—	1,700	1,900	
20	m-HPC-MF	—	48	—	400	—	1,600	2,000	
35	PCA-PP	1.3	16.8	—	30.8	34.2	34.8	34.8	Silley, 1985
20	PCA-PP	1.7	21.2	—	101	114.3	121.3	125.8	
22	YEA-PP	—	—	100*	—	—	—	1,800	Gibbs and Hayes, 1988

CFU, colony-forming units.
* Based on plating ratio given in the reference and arbitrarily assigning the value of 100 CFU/ml as the pour plate mean.

In the study by Reasoner and Geldreich (1985), all media yielded the lowest bacterial counts when 35°C incubation was used, regardless of the length of incubation or plating method (Figure 22.1 and Table 22.3). Incubation at 20°C yielded the highest counts on all media when incubation was extended to 12 to 14 days. The 28°C counts were higher than the 20°C counts from day 2 through day 6 of incubation, and for most purposes 28°C appeared to be the best incubation temperature.

Lombardo et al. (1986), using the MF method, compared 18°C, 28°C, and 35°C incubation temperatures and various media (TGY, R2A, and 0.5 strength R2A) for the HPC analysis on drinking water in Southern Nevada. From their results they concluded that incubation at 28°C provided the optimal recovery of HPC bacteria and that R2A and 0.5 strength R2A media provided better recoveries than TGY and m-HPC media. Incubation at 35°C yielded the least recovery of bacteria regardless of the medium used. Incubation at 18°C gave lower counts than 28°C even though the plates were incubated for a week longer. They recommended that R2A medium and 28°C incubation for 7 days be used for HPC analyses of drinking water.

Other studies have also compared the use of different incubation temperatures, times, and culture media for enumeration of heterotrophic bacteria in drinking water. Means et al. (1981) analyzed water samples from two distribution systems using *Standard Methods* plate count agar pour plates and spread plates incubated at 35°C and 20°C for 48 h, R2A spread plates incubated for 72 h at 20°C, and m-HPC MF plates incubated 96 h at 35°C. Their results indicated that m-HPC MF plates incubated 96 h recovered significantly higher total and pigmented bacteria than the other media and plating methods. The standard PP method underestimated the bacterial populations in the two distribution systems by 79% and 66%, respectively, compared to the m-HPC results. However, it is likely that the results were biased to the extent that m-HPC plates were incubated twice as long as recommended (96 h versus 48 h) either in the original publication (Taylor and Geldreich, 1979) or in *Standard Methods*, and the R2A plates were incubated at a lower temperature and for less time (20°C for 72 h) than recommended (28°C for 5–7 days). Fiksdal et al. (1982) compared bacterial recoveries from drinking water using spread and MF plates of R2A medium, *Standard Methods* plate count agar, and casein-peptone-starch (CPS) agar, and MF plates of m-HPC agar. Incubation temperatures of 20°C and 35°C with incubation up to 7 days were used; plates were counted after 2 days and 7 days. For all media used as spread plates and for the MF m-HPC plates, incubation for 7 days at 20°C gave the highest counts. For chlorinated water samples, 35°C incubation for 2 days was optimum for high counts on all media, whereas for nonchlorinated water, 20°C was optimum.

In a study comparing the standard PP with the MF using R2A and m-HPC media, Maul et al. (1985) concluded that the MF technique is a preferable alternative to the standard PP method. They found that the plate counts obtained with the various media could differ appreciably when the incubation time was less than 72 h, but from 72 h onward, the results from TGY agar, R2A agar and m-HPC were similar for the drinking water samples they analyzed. They found that incubation beyond 72 h was of limited usefulness because of the relatively small increase in both bacterial recovery and statistically consistent results.

In the United Kingdom, the standard technique for the HPC is the pour plate using yeast extract agar (YEA) incubated 3 days at 20°C to 22°C. Gibbs and Hayes (1988) compared YEA pour and spread plates with R2A pour and spread plates, and 7 day counts with 3 day counts for enumerating bacteria in drinking water samples. By comparing the 7 day:3 day count ratios, they found that R2A medium gave 4-fold greater counts at 3 days, and 8-fold greater counts at 7 days than did YEA. Spread plates gave 7-fold greater average counts than the pour plates at both 3 days and 7 days, and R2A spread plate counts were, on average, 39-fold and 27-fold greater than the average counts on YEA pour and spread plates, respectively. R2A 7 day spread plate counts were an average of 520-fold greater than the 3 day YEA pour plate average count. The authors concluded that the gross underestimation of the bacterial numbers by the standard U.K. method could interfere with assessing public health effects that might be associated with a heterotrophic bacteria in water supplies and that a review of the recommended procedures was in order.

Thus, there is considerable evidence from a variety of studies that the PP method, a rich medium, 35°C incubation temperature, and a short incubation period are all factors that result in reduced enumeration of bacteria that may be present in potable water. This combination of conditions is very selective for that segment of the bacterial flora that can grow very rapidly, some of which may, indeed, originate from a pollution event such as sewage contamination. However, most often there is a much larger segment of the bacterial flora for which the most suitable growth conditions are met by the use of either the SP or the MF method using a medium that has a variety of nutrients at lower concentrations, a lower incubation temperature, and incubation for longer than 72 h. In some cases, a nutrient rich medium also works very well with the SP or MF method, lower incubation temperature and increased time of incubation.

There is no single colony formation method and set of conditions that will allow enumeration of all bacteria that may be found in water (Buck, 1979); therefore, given the choice of methods and media available, the user should evaluate their effectiveness for the types of water that are most often analyzed. Goulder (1976) stated that "Although plate counts represent, at best, only a small proportion of total bacteria it is nevertheless worth using a medium which gives highest attainable counts because any increase in the proportion counted increases the sensitivity of the method as an indicator of gross change in bacteriological quality." This statement seems appropriate for drinking water bacteriological quality as well.

22.4   Applications of HPC Monitoring

Treatment Processes   One of the simplest applications of the HPC is that of monitoring the reductions that occur in the bacterial load of the water as it passes through the various unit processes of drinking water treatment. While there are considerable data available on coliform and pathogen removal or inactivation through the treatment processes, there is little published information available on HPC reductions except, perhaps, as related to the disinfected finished water as it enters the distribution system. It is probable that HPC reductions generally parallel coliform reductions through treatment processes. Geldreich (1988) reported the following cumulative coliform reductions from raw water through treatment: raw water, 0%; raw water storage, 50%; coagulation and sedimentation, 62%; rapid sand filtration, 96%; and disinfection, 99% or greater.

In studies describing the effects of application of disinfection at alternative points during water treatment, Cummins and Nash (1979) used the standard PP technique with incubation at 35°C for 48 h. When chlorine was applied to the raw water before clarification, the HPC in the finished water was reduced by 87.1%. When chlorine was applied to stored raw water, HPC reductions of >98.7% and >99% were recorded. Finally, when chlorine was applied between clarification and settling, or between coagulation and settling and filtration, HPC reductions of >99.6% and >99.98%, respectively, were found.

Information gathered by Hutchinson and Ridgway (1977) indicated that pre-treatment of water using coarse sand and rapid filtration rates (roughing filters) removed 50–80% of plate count bacteria (37°C plate count), while simple sedimentation removed about 50%.

The treatment process was shown by Armstrong et al. (1982) to select for antibiotic-resistant bacteria. HPC bacteria were enumerated using the m-HPC MF method with incubation at 35°C for 48 h. Raw water HPC levels ranged from 5.4×10^3 CFU/ml to 1.8×10^4 CFU/ml. Flash mixing of raw water with chlorine resulted in reductions in HPC levels that ranged from 640-fold to 3,800-fold. Clearwell samples showed a 600,000-fold decrease in HPC compared to the raw water levels. Although the flash mixing treatment reduced the overall HPC numbers there was an increase in the percentage of the HPC organisms that were multiply-antibiotic resistant (MAR); 15.8% and 18.2% MAR in the raw water, compared to 57.1% and 43.5% MAR in the clear well. MAR bacteria were also shown to be selected during water distribution.

Brazos et al. (1988) studied seasonal effects on total bacterial removal in a rapid sand filtration plant. Using the standard PP incubated at 35°C for 96 h for the HPC analysis, the HPC decreased by 2 logs (98.89%) from the raw water through pretreatment (sedimentation, lime addition, ferrous sulfate coagulation, chlorination, and primary sedimentation). Secondary treatment (primary recarbonation, additional lime, coagulant and activated carbon if needed, secondary sedimentation, recarbonation, chlorination, rapid sand filtration, and ammonium sulfate) resulted in a cumulative HPC reduction of >99.9%.

Data reported by Ridgway (1978) indicated that the source water played a role in the regrowth of bacteria in the distribution system once the chlorine residual was lost. Bacteria were enumerated using the PP method, YEA agar, and incubation at either 37°C for 48 h, or at 22°C for 7 days. Reductions of HPC levels from the raw source water to the finished water after treatment (but before entering the distribution system) ranged from <1 log to 2 logs for upland catchment water, and from 1 log to 4 logs for river derived water. Treatment of the water ranged from impoundment + slow sand filtration + chlorination, to impoundment + prechlorination + coagulation + rapid sand filtration + chlorination (or chlorination + ammoniation), and other combinations in between. Reductions in the HPC between treatment steps were not given. Regrowth of bacteria during distribution was related to the loss of free-chlorine residual, the type of water, and the level of organic matter in the finished water.

Data from Lombardo et al. (1986) showed a >3 log reduction (99.98%) in the HPC from the raw lake water to the finished treatment effluent. The treatment chain consisted of prechlorination + aluminum sulfate coagulation + flocculation + mixed media filtration + post-chlorination.

Very little information is available on disinfection of the general heterotrophic bacterial population in water; most disinfection studies have focused on specific microbial pathogens or on the coliform group of bacteria. Ward et al. (1983) used the m-HPC MF method with incubation for 7 days at 22°C to evaluate the inactivation of HPC bacteria in water from a finished water reservoir that received fully treated chlorinated water. They found a 1.5 log reduction of HPC bacteria after 10 min contact with monochloramines (1.7 mg/l, pH 8, Cl_2-N weight ratio of 3:1), followed by a gradual, additional 2 log decrease after 60 min of contact. A 2 log decrease occurred after 1 min contact with free chlorine (0.45 mg/l), followed by a >3 log decrease after 10 min contact. There was then a very gradual decrease until the HPC was only 2 CFU/ml after 60 min contact time. The HPC bacteria inactivation curves showed a multi-component die-off indicating the inactivation of bacteria with varying sensitivity to disinfection.

As part of a broader study involving total trihalomethane levels and total coliform bacteria, MacLeod and Zimmerman (1987) examined the effect on the HPC bacteria of the changeover from the use of a free-chlorine residual to the use of a combined-chlorine residual as the secondary disinfectant. The HPC bacteria were enumerated by the 35°C, 48 h incubation, PP method. Their results indicated that bacteriological quality decreased following the conversion to chloramines since there was only a small but significant increase in the HPC level.

Storage and Distribution   Monitoring the bacteriological quality of the storage and distribution system is probably the most common and familiar use of the HPC analysis. Changes that occur in the bacteriological quality of the treated water from the time it enters the distribution system until it reaches the customer can be determined using HPC methods and an appropriately designed sampling program. Bacteriological quality changes may cause aesthetic problems involving taste and odor development, discolored water, slime growths, and economic problems including corrosion of pipes and biodeterioration of materials (Water Research Center, 1976).

Bacterial numbers tend to increase during distribution (see Chapters 10–12) and the density reached is influenced by a number of factors including the bacterial quality of the finished water entering the system, temperature, residence time, presence or absence of a disinfectant residual, construction materials, and availability of nutrients for growth (Geldreich, 1973; Hutchinson and Ridgway, 1977; van der Kooij, 1981; Dott, 1983; Olson and Nagy, 1984). A variety of studies during recent years have focused on the distribution system and factors that affect bacterial quality as measured by HPC methods.

The occurrence of pigmented bacteria in raw water, water at various stages of treatment, and in distribution water was investigated by Reasoner and Geldreich (1980) and Reasoner et al., (1989). HPC media and methods used were the standard PP, m-HPC MF and R2A MF, all incubated at 35°C. Counts of total colonies and differential counts of the various pigmented colonies were made after 2, 4, and 7 days of incubation. Results from this study showed that m-HPC and R2A media consistently recovered more HPC bacteria, including pigmented forms, than the standard PP method. The 7 day incubation was found to provide the best pigmented counts on both m-HPC and R2A media. Pigmented bacteria could be found in water from all sample locations and comprised a significant percentage of the HPC population of the treated distribution water. Variations in the percentage occurrences of yellow, orange, and pink bacteria appeared to be seasonally related at the distribution site, and yellow and orange pigmented forms consistently present. HPC numbers showed a stepwise reduction at successive treatment steps, resulting in a >4 log reduction in HPC bacteria in the chlorinated filter bed effluent compared to the raw water.

LeChevallier, et al. (1980) found that the number of HPC bacteria (m-HPC method) in drinking water ranged from <0.02 to $>1.0 \times 10^4$ CFU/ml. Bacterial levels in dead-end areas without a free chlorine residual were 23 times higher than the levels in water where there was a free chlorine residual. Turbidity was significantly correlated with HPC numbers in the distribution water and 1 to 3 log increases in HPC were common when turbidities exceeded 5 nephelometric turbidity units (NTU). Highest HPC densities coincided with water temperatures approaching 20°C in late summer. Identification of HPC bacteria in the distribution water indicated that $>30\%$ of the bacteria were opportunistic pathogens, and about 35% were coliform antagonists. LeChevallier et al. (1980) concluded that enumeration of HPC bacteria was useful to indicate the presence of opportunistic pathogens, the potential for coliform suppression, and deterioration in drinking water quality.

The relationship of HPC numbers to turbidity and free chlorine in two distribution systems was studied by Reilly and Kippin (1983). HPC numbers were determined using the PP method with incubation at 35 °C for 48 h. Their results indicated that the type and levels of turbidity found in the water supplies studied did not affect the frequency of occurrence of HPC bacteria. They also found that there was a definite reduction of HPC numbers as the free chlorine residual increased.

An intensive study of a water distribution system was conducted in which spatial and temporal HPC variations were examined (Maul et al., 1985a). HPC data were obtained using the MF method and R2A medium with incubation at 20 °C for 72 h. The sampling program was designed to study the dispersion patterns of HPC in the system and the data were subjected to intensive statistical analyses. An increase in HPC numbers in the last 2 of 6 surveys corresponded to an increase in the mean water temperature. Mean water temperatures during each of the 6 surveys were 7.8, 8.0, 7.3, 8.8, 12.3 and 16.7 °C. Based on the statistical analyses, the distribution system could be divided into different zones of bacterial density that reflected a structured pattern of spatial HPC heterogeneity. High densities of HPC bacteria were more likely to occur in peripheral locations far from the treatment plant rather than close to it. The incidence of HPC bacteria appeared to be inversely related to the chlorine residual present. Free and total chlorine residuals decreased rapidly with distance from the treatment plant and could not be detected in the peripheral parts of the system where the HPC levels were highest. Based on the results from this study, a sampling design for monitoring water quality was developed (Maul et al., 1985b).

The relationship of bacteriological quality to turbidity and particle counts in distribution water was studied by McCoy and Olson (1986) using HPC results from R2A MF plates incubated at 23 °C for 7 days. Water samples from locations at the beginning (upstream) and within the distribution system (downstream) in each of three distribution systems (two surface water supplies and a ground water supply) were sampled twice per month over a one year period. Particle counts, turbidity, and HPC numbers were higher in surface water samples than in groundwater samples. Comparison of HPC numbers with total bacterial counts by epifluorescence direct counts showed a positive linear relationship between these two parameters. Mean HPC levels at the upstream site in the groundwater supplied system ranged from 0.62 CFU/ml to 1,420 CFU/ml and from 1.03 CFU/ml to 547 CFU/ml at the downstream site. For the surface-supplied systems, mean HPC levels at the upstream sites ranged from 1,490 CFU/ml to 8,200 CFU/ml and 0.41 CFU/ml to 790 CFU/ml, and at the respective downstream sites from 110 CFU/ml to 2,593 CFU/ml and from 0.96 CFU/ml to 840 CFU/ml. HPC counts were found to underestimate the actual number of bacterial cells by a factor of about 500. Turbidity was found to be a linear function of total particle concentration, but not of the number of bacterial cells. Degradation of bacterial water quality was shown to be the result of unpredictable intermittent events that occurred within the system, at which time there was a correlation between turbidity and the concentration of HPC bacteria.

Olson and Hanami (1981) used plate count agar spread plates incubated at 35°C for 48 h to evaluate seasonal variations of HPC bacteria in two water distribution systems, one chlorinated and the other unchlorinated. Their results indicated seasonal fluctuations in the HPC and showed that the predominant bacterial genera in the chlorinated system changed between the source and the distribution sampling site. This suggested that chemical or physical factors in the distribution line influenced the establishment and growth of certain bacterial groups through the water main. The predominant bacterial genera were *Acinetobacter, Pseudomonas/Alcaligenes,* and *Flavobacterium.* In the unchlorinated system, seasonal variation also occurred and the predominant genera were *Pseudomonas/Alcaligenes, Acinetobacter, Flavobacterium* and *Klebsiella.*

In a study of the microbiological quality of water from dead-end areas of a distribution system, Rae (1982) compared water quality of samples collected from inside taps of homes nearest the dead-end hydrants with the quality of samples collected from the hydrants themselves, and with the influence of treatment (filtered vs. unfiltered) on the quality of the water supplying the dead-end. HPC numbers were determined using the standard PP method and 35°C incubation for 96 h; counts were determined at 24 and 48 h. After 96 h at 35°C, the plates were placed at 22°C for an additional 72 h. Other microbiological parameters investigated were total coliforms, fecal coliforms, fecal streptococci, fungi, and algae. From the results, Rae (1985) concluded that distance from the treatment plant had the greatest effect on the density of microbial populations present in 17 dead-ends within the study distribution system. Water flowing more than 5 miles from the treatment plant before being dead-ended had >2-fold higher HPC levels than water from dead-ends within one mile of the treatment plant. Comparison of chlorine residual with the HPC/ml showed that the HPC ranged from 410 CFU/ml at 0.45 mg chlorine/l to as high as 16,000 CFU/ml at 0.2 mg chlorine/l and 8,500 CFU/ml in the presence of 0.65 mg chlorine/l. Water supplying the dead-end areas and the water from the houses tapped nearest the dead-end hydrants contained from 6 to about 740 CFU/ml and 2 to 340 CFU/ml, respectively.

In a study of chlorine depletion in household plumbing systems, Brazos et al. (1986) used the standard PP method (35°C incubation for 48 h) for determining HPC numbers and a modified acridine orange direct count (AODC) for enumeration of total bacterial cells. They found that both HPC and total bacterial numbers by direct count increased dramatically in water in household plumbing systems where chlorine levels were completely dissipated during storage. The mean HPC increased by >2 logs to 40,000 CFU/ml and the mean AODC increased by >1 log to 90,000 CFU/ml compared to the levels found in water from the distribution mains. The ratio of HPC/AODC bacteria increased from 5% in the distribution system water to 41% in the household plumbing system, indicating that a greater proportion of the AODC organisms were recoverable from the plumbing system by the HPC method.

The effect on water quality of distribution line cleaning using polyurethane foam cubes was reported by Glicker (1987). Among parameters monitored was the HPC using the *Standard Methods* procedure. The results of monitoring at user taps during this study indicated that there was no statistically significant difference between pre- and post-cleaning HPC levels in the water provided to the users.

Bacterial regrowth versus aftergrowth in treated distribution water was examined in a 15 month study by Brazos and O'Connor (1988). Regrowth organisms were defined as those bacteria that derive from viable cells of disinfectant injured bacteria and those that are in a state of exogenous dormancy; i.e., are neither active nor dead. These are organisms that originated in the source water. Aftergrowth organisms were defined as those organisms that are contributed to the distribution water by detachment from the surfaces of the distribution system pipes or other sources external to the source water. HPC organisms were enumerated by the PP method with incubation at 35°C for 96 h, and total bacterial cells were enumerated using a modified acridine orange direct count (AODC). The average HPC levels in the water leaving the clear well during the study period was 30 CFU/ml, and varied over a 2 log range during distribution. Mean HPC levels decreased by about one-half at the two distribution sites nearest the treatment plant and increased 3 to 4 logs at each of the more distant distribution sites. The average HPC at the most remote site was 140 CFU/ml. Lowest HPC values were found during the warm water temperature periods, possibly indicating increased disinfectant effectiveness. Results from the AODC analyses indicated that the bacterial numbers found throughout the distribution system were very close to the numbers discharged from the clear well. Results from experiments during the study indicated that bacterial regrowth contributed most significantly to increases in bacterial numbers in the water, and increases occurred rapidly in dechlorinated distribution water. During low water temperature (0–7°C) periods total bacteria averaged 1 to 2 logs higher ($>10^6$ CFU/ml) than during the warm water (20–29°C) periods (10^4–10^5 CFU/ml). Because of the higher direct counts during the low temperature periods, the authors concluded that there was impaired bacterial removal during the treatment process.

While investigating a biofilm problem in a distribution system, LeChevallier et al. (1987) found that HPC densities (R2A spread plate, 7 days at 20–24°C) increased as the water flowed from the treatment plant through the study area; HPC levels ranged from 3.2×10^2 to 1.3×10^7 CFU/ml (geometric mean, 3.65×10^4 CFU/ml) and often exceeded 10^6 CFU/ml at the ends of branch lines. The standard PP medium recovered an average of only 10 CFU/ml (range, 0.7 to 160 CFU/ml) and underestimated HPC levels in the dead-end areas by as much as 2.5×10^5 fold. Floc material from zinc orthophosphate used for corrosion control contained 2.4×10^6 CFU/ml, the cement-lined pipe surface had 2.0×10^4 CFU/20 cm², and tubercle material contained 1.3×10^7/g.

Along with analyses for free and total chlorine and chloramines, free and total ammonia-nitrogen, and coliforms, HPC analyses were useful to Wolfe et al. (1988) in understanding the development of nitrification problems in finished water reservoirs containing chloraminated water. The PP method with TGE agar, or R2A MF plates were used for the HPC determinations. R2A plates were incubated 7 days at 28°C. In the first case, difficulty was encountered in maintaining the total chlorine residual in the water leaving the reservoir. HPC analyses indicated that the PP bacterial counts were increasing while the chloramine and ammonia-nitrogen concentrations were dropping; HPC levels by the PP method reached as high as 8.5×10^4 CFU/ml. When breakpoint chlorination of the influent water was started, PP HPCs were reduced to $<$ 100 CFU/ml within a week. Analysis of HPC data collected previous to the nitrification problem suggested that there was some indication of increased biological activity in the reservoir preceding the episode. About one month before the episode, the PP HPC levels in the reservoir began increasing, reaching 1.0×10^3 CFU/ml, but the HPC data from R2A plates indicated that the trend toward increasing bacterial numbers began about 2 months before the episode.

In the second case involving another covered finished water reservoir, the first indication of nitrification was difficulty in maintaining a constant chloramine residual in the effluent from the reservoir. The R2A HPC was about 10–30 fold greater in the reservoir than in the influent water. Within a 1 to 2 day period, the R2A HPC level increased to $>1.2 \times 10^4$ CFU/ml. The reservoir was taken out of service and breakpoint chlorination initiated. Several days later, the R2A HPC was <50 CFU/ml, and the concentrations of nitrite and ammonia-oxidizing bacteria were below detectable levels. Conditions that led to the growth of ammonia-oxidizing bacteria included optimal pH and temperature, and availability of free ammonia. Increased detention times and possibly the dark environment because of the reservoir covers may have favored the development of these bacteria.

An unusual HPC procedure used for HPC monitoring in the distribution system was reported by Lewis (1988). Plate count agar pour plates with incubation at 35°C for 48 h were used for daily analyses of the finished water at the treatment plant. For distribution system monitoring, a modified m-HPC MF method was used employing incubation of the MF on m-HPC medium for only 20 to 22 h at 35°C, followed by staining the MF to improve colony contrast. HPC numbers reported ranged from about 30 CFU/ml to >100 CFU/ ml. No comparisons with other HPC methods were given.

Biofilms and Bacterial Growth on Construction Materials   Microbial growth associated with the surfaces of basins and other structures of water supply treatment processes and the pipes of the distribution and household plumbing systems has been assessed using HPC enumeration methods. Additionally, the HPC has been used in studies of bacteriologically mediated corrosion.

Interest in biofilm formation within water distribution systems has heightened in the past several years as more incidents of persistent coliform problems throughout the year as well as during warm water periods have been reported (see Chapter 12). Nagy and Olson (1986) used spread plates of plate count agar, m-HPC, and R2A media (7 day incubation at 20–22°C) for colony counts of distribution pipe samples. Bacterial numbers ranged from 11 CFU/100 cm² of pipe surface to 5.75 $\times$ 10⁶ CFU/100 cm². Regression analysis of the HPC against pipe years in service indicated that there was approximately a one log increase in HPC bacteria for every 10 years the pipe had been in service. Bacterial population densities on drinking water pipe surfaces are relatively low and unevenly distributed, possibly due to water velocity, low nutrient levels, low inoculum levels as a result of treatment, and the presence of a disinfectant residual. The types of bacteria present on pipe surfaces were similar to those reported in drinking water.

Donlan and Pipes (1987) studied the growth of microorganisms on cast iron pipe surfaces exposed to drinking water. Small cast iron cylinders were exposed to drinking water during warm (15–20°C) and cold (4–15°C) water periods. Some cast iron cylinders were exposed directly to the water by insertion into the distribution main through a corporation sampling device, while others were exposed to drinking water inside 100 liter plastic tanks. Exposure times were 28–41 days (short term) or 67–115 days (long term). HPC bacteria were quantified using R2A medium. HPC levels in the distribution water at sites selected for the study ranged from 4 to 219 CFU/ml during cold water periods, and from 10 to 8,300 CFU/ml during warm water periods. Bacterial densities on the cast iron cylinders exposed to the water at the same sites ranged from 2.1 $\times$ 10⁴ CFU/cm² to 7.59 $\times$ 10⁷ CFU/cm² during warm water periods, and from 56 CFU/cm² to 1.58 $\times$ 10⁵ CFU/cm² during cold water periods. HPC levels on cast iron cylinders exposed during warm water periods to water inside 100 liter tanks had HPC levels that ranged from 1.44 $\times$ 10⁵ CFU/cm² to 1.74 $\times$ 10⁷ CFU/cm². A strong positive relationship was found between the HPC of the bulk water and the HPC of the cylinder surfaces. There was a strong negative relationship between the chlorine concentration and the cylinder HPC under warm water conditions, but this correlation may have been site dependent. There was also a strong correlation between water temperature and the HPC of the cast iron cylinders. Water velocity was found to limit HPC growth on the cylinders either by increasing the exposure to monochloramine or by removing attached cells.

Inactivation of biofilm bacteria was studied by LeChevallier et al. (1988) using R2A medium to enumerate surviving bacteria, both in biofilms of HPC bacteria grown on metal coupons and on GAC, and in biofilms of *Klebsiella pneumoniae* grown on clean glass slides. HPC biofilm densities on the metal coupons were generally 10⁶ CFU/side of the coupon. Results indicated that the attached bacteria (biofilm organisms) were more difficult to inactivate than those that were freely suspended in the water, and monochloramine inactivated biofilm bacteria better than free chlorine or chlorine dioxide.

In a study of biologically mediated corrosion, Lee et al. (1980) used the standard PP method with plate count agar and incubation at 25°C for 7 days to enumerate heterotrophic bacteria in water. A simulated pipe system was used consisting of four pipe loops constructed of PVC and low carbon steel pipes and fed with tap water. HPC levels of up to 5×10^5 CFU/ml were found in the PVC systems, compared with 10^3 to 10^4 CFU/ml in the steel pipe systems. Corrosion of cast iron coupons occurred upon exposure to water containing a mixture of bacteria isolated from the simulated distribution systems. Analysis of test specimen corrosion products showed that organisms present in the water were also present in the corrosion material.

The growth of bacteria on construction materials and coatings used in water treatment, storage and distribution systems constitutes another area where HPC methods are used extensively. There is an extensive literature on the growth of microorganisms on materials in contact with drinking water, consisting largely of work done in Europe. The interested reader should consult Colbourne (1985) and Schoenen (1986) for access to the literature on this topic.

21.5 Conclusions and Research Needs

The use of HPC methods has provided significant information about bacterial quality changes in water during treatment, storage and distribution. The methods have proven their usefulness and they will continue to be used for monitoring water quality. In fact, utilities may find the HPC data increasingly useful as they are pushed to control the levels of potentially harmful disinfection by-products in their waters. The use of alternative disinfectants as part of the strategy to reduce disinfection by-product concentrations, and possibly the reduction of applied disinfectant dosages, may mean that greater reliance will be placed on HPC analyses to determine or predict significant water quality changes that will require some sort of treatment response.

Although it is known that high concentrations of HPC bacteria can develop in favorable locations in a distribution system, there is a paucity of data on human health effects resulting from exposure to these organisms following ingestion or through the respiratory tract (aerosols). To be sure, there is information about aerosol exposure to a waterborne pathogen such as *Legionella* sp. (see Chapter 16), and about skin infections due to *Pseudomonas aeruginosa* or some mycobacteria acquired by exposure to contaminated bathing water, but documented reports of adverse health effects due to high numbers of HPC organisms in drinking water are scarce to nonexistent. The increasing numbers of immunologically compromised people who are at risk to infections by organisms that are generally thought to be nonpathogenic suggests that research in this area is sorely needed.

The existence of a survival stage for many bacteria in natural environments raises the question of how important this phenomenon is to the sanitary quality of potable water and consumer health. A variety of bacteria have been shown to enter a state in which they are viable but nonculturable (Roszak and Colwell, 1987). Since these organisms cannot be enumerated by the cultural methods presently in use, other methods for their quantitation must be used. The application of methods to detect viable but nonculturable organisms in drinking water and the assessment of the importance of these organisms to human health are areas which need to be investigated.

This paper has been reviewed in accordance with the U.S. Environmental Protection Agency's peer and administrative review policies and approved for publication.

References

American Public Health Association. 1985. *Standard Methods for the Examination of Water and Wastewater*, 16th Ed., American Public Health Association, 1015 Fifteenth Street NW, Washington, D.C. 10005. 1267 pp.

Armstrong, J.L., Calomiris, J.J., Shigeno, D.S. and Seidler, R.J. 1982. Drug resistant bacteria in drinking water. pp. 263–276 in Technology Conference Proceedings, WQTC–9, 1981, *Advances in Laboratory Techniques for Quality Control*, American Water Works Association, Denver, Colorado.

Brazos, B.J., O'Connor, J.T., and Abcouwer, S. 1986. Kinetics of chlorine depletion and microbial growth in household plumbing systems. pp. 239–274 in Technology Conference Proceedings, WQTC–13, 1985, *Advances in Water Analysis and Treatment*, American Water Works Association, Denver, Colorado.

Brazos, B.J. and O'Connor, J.T. 1988. Relative contributions of regrowth and aftergrowth to the number of bacteria in a drinking water distribution system. pp. 433–467 in Technology Conference Proceedings, WQTC-15, *Issues and Answers for Today's Water Quality Professional*, American Water Works Association, Denver, Colorado.

Brazos, B.J., O'Connor, J.T., and Lenau, C.W. 1988. Seasonal effects on total bacterial removals in a rapid sand filtration plant. pp. 25–54 in: *Water Quality: A Realistic Perspective*. Seminar/Workshop, Ann Arbor, Michigan, February, 1988. Sponsored by the Univ. of Michigan College of Engineering, Michigan Section of the American Water Works Association, Michigan Water Pollution Control Association, and Michigan Department of Public Health.

Buck, J.D. 1979. The plate count in aquatic microbiology. pp. 19–28 in Costerton, J. W. and Colwell, R. R. (editors), *Native Aquatic Bacteria: Enumeration, Activity, and Ecology*, ASTM STP 695, American Society for Testing and Materials, Philadelphia, PA.

Buck, J.D. and Cleverdon, R.C. 1960. The spread plate as a method for the enumeration of marine bacteria. *Limnology and Oceanography* 5:78–80.

Camper, A.K., LeChevallier, M.W., Broadaway, S.C., and McFeters, G.A. 1986. Bacteria associated with granular activated carbon particles in drinking water. *Applied and Environmental Microbiology* 52:434–438.

Clark, D.S. 1971. Studies on the surface plate method of counting bacteria. *Canadian Journal of Microbiology* 17:943–946.

Colbourne, J.S. 1985. Materials usage and their effects on the microbiological quality of water supplies. *Journal of Applied Microbiology* 59:47S–59S. *Symposium Supplement*

Cowell, N.D. and Morisetti, M.D. 1969. Microbiological techniques—some statistical aspects. *Jour. of Science, Food and Agriculture* 20:573–579.

Cummins, B.B. and Nash, H.D. 1979. Microbiological implications of alternative treatment. Paper no. 2B-1, in Technology Conference Proceedings, WQTC-6, 1978, *New Laboratory Tools for Quality Control in Water Treatment and Distribution*, American Water Works Association, Denver, Colorado.

Donlan, R.M. and Pipes, W.O. 1987. Pipewall biofilm in drinking water mains. pp. 637–

660 in Technology Conference Proceedings, WQTC-14, 1986, *Advances in Water Analysis and Treatment.*, American Water Works Association, Denver, Colorado.

Dott, W. 1983. Qualitative and quantitative examination of bacteria found in aquatic habitats. 6. Bacterial regrowth in drinking water. *Zentralb. Bakteriol. Mikrobiol. Hygiene Abt l.Orig. B* 178:263–279.

Eisenhart, C. and Wilson, P.W. 1943. Statistical methods and control in bacteriology. *Bacteriological Reviews* 7:57–137.

Fiksdal, L., Vik, E.A., Mills, A., and Staley, J.T. 1982. Nonstandard methods for enumerating bacteria in drinking water. *Jour. American Water Works Association* 74:313–318.

Fronk-Leist, C.A. and Love, O.T., Jr. 1984. Evaluating the quality of water treated by redwood slat tower aerators. pp. 253–264 in Technology Conference Proceedings, WQTC-11, 1983, *Water Quality and Treatment: Advances in Laboratory Technology*, American Water Works Association, Denver, Colorado.

Fujioka, R., Kungskulniti, N., and Nakasone, S. 1986. Evaluation of the presence-absence test for coliforms and the membrane filtration method for heterotrophic bacteria. pp. 271–283, in Technology conference Proceedings, WQTC-13, 1985, *Advances in Water Analysis and Treatment*, American Water Works Association, Denver, Colorado.

Geldreich, E.E., Nash, H.D., Reasoner, D.J., and Taylor, R.H. 1972. The necessity of controlling bacterial populations in potable water: community water supply. *Jour. American Water Works Association* 64:596–602.

Geldreich, E.E. 1974. Is the total count necessary? pp. VII–1–VII–8 in Technology Conference Proceedings, WQTC-1973, *Water Quality*, American Water Works Association, Denver, Colorado.

Geldreich, E.E. 1988. Coliform non-compliance nightmares in water supply distribution systems. pp. 55–74, in *Water Quality: A Realistic Perspective*. Seminar/Workshop, Ann Arbor, Michigan, February, 1988. Sponsored by The Univ. of Michigan College of Engineering, Michigan Section of the American Water Works Association, Michigan Water Pollution Control Association and Michigan Department of Public Health.

Gibbs, R.A. and Hayes, C.R. 1988. The use of R2A medium and the spread plate method for the enumeration of heterotrophic bacteria in drinking water. *Letters in Applied Microbiology* 6:19–21.

Glicker, J. 1987. Evaluating the effectiveness of improving distribution system water quality. pp. 747–763 in Technology Conference Proceedings, WQTC-14, 1986, *Advances in Water Analysis and Treatment*, American Water Works Association, Denver, Colorado.

Goulder, R. 1976. Evaluation of media for counting viable bacteria in estuaries. *Journal of Applied Bacteriology* 41:351–355.

Green, B.L., Taylor R.H., and Geldreich, E.E. 1982. The SPC Sampler: a simple procedure for monitoring the bacteriologic quality of water. pp. 125–133 in Technology Conference Proceedings, WQTC-9, 1981, *Advances in Laboratory Techniques for Quality Control*, American Water Works Association, Denver, Colorado.

Haas, C.N., Meyer, M.A., and Paller, M.S. 1982. Analytical note: evaluation of the m-SPC method as a substitute for the standard plate count in water microbiology. *Jour. American Water Works Association* 74:322.

Hutchinson, M. and Ridgway, J. W. 1977. Microbiological aspects of drinking water. pp. 179–218 in Skinner, F. A. and Shewan, J. M. (editors), *Aquatic Microbiology*, Society for Applied Bacteriology Symposium Series No. 6, Academic Press, London, New York, San Francisco.

Jennison, M.W. and Wadsworth, G.P. 1940. Evaluation of the errors involved in estimating bacterial numbers by the plating method. *Journal of Bacteriology* 39:389–397.

Kaper, J.B., Mills, A.L., and Colwell, R.R. 1978. Evaluation of the accuracy and precision of enumerating aerobic heterotrophs in water samples by the spread plate method. *Applied and Environmental Microbiology* 35:756–761.

Klein, D.A. and Wu, S. 1974. Stress: a factor to be considered in heterotrophic micro-organism enumeration from aquatic environments. *Applied Microbiology* 27:427–431.

Koch, R. 1912. Ueber die neuen Untersuchungsmethoden zum Nachweis der Micro-kosmen in Boden, Luft und Wasser. Reprinted in *Ges. Werke von R. Koch* 1:274–284. Berlin.

LeChevallier, M.W., Seidler, R.J., and Evans, T.M. 1980. Enumeration and characteri-zation of standard plate count bacteria in chlorinated and raw water supplies. *Applied and Environmental Microbiology* 40:922–930.

LeChevallier, M.W., Babcock, T.M., and Lee, R.G. 1987. Examination and characteri-zation of distribution system biofilms. *Applied and Environmental Microbiology* 53:2714–2724.

LeChevallier, M.W., Cawthon, C.D., and Lee, R.G. 1988. Inactivation of biofilm bacteria. *Applied and Environmental Microbiology* 54:2492–2499.

Lee, S.H., O'Connor, J.T., and Banerji, S.K. 1980. Biologically mediated corrosion and its effects on water quality in distribution systems. *Jour. American Water Works As-sociation* 72:636–645.

Lewis, C.M. 1988. The water utility laboratory: Preparing for and dealing with micro-biological water quality emergencies. pp. 613–630 in Technology Conference Pro-ceedings, WQTC-15, 1987, *Issues and Answers for Today's Water Quality Professional*, American Water Works Association, Denver, Colorado.

Lombardo, L.R., West, P.R., and Holbrook, J.L. 1986. A comparison of various media and incubation temperatures used in the heterotrophic plate count analysis. pp. 251–270 in Technology Conference Proceedings, WQTC-13, 1985, *Advances in Water Analysis and Treatment*, American Water Works Association, Denver, Colorado.

MacLeod, B.W. and Zimmerman, J.A. 1987. Selected effects on distribution water quality as a result of conversion to chloramines. pp. 619–636 in Technology Conference Proceedings, WQTC-14, 1986, *Advances in Water Analysis and Treatment*, American Water Works Association, Denver, Colorado.

Maul, A., Block, J.C., and El-Shaarawi, A.H. 1985. Statistical approach for comparison between methods of bacterial enumeration in drinking water. *Journal of Microbiolog-ical Methods* 4:67–77.

Maul, A., El-Shaarawi, A.H., and Block, J.C. 1985a. Heterotrophic bacteria in water distribution systems. I. Spatial and temporal variation. *The Science of the Total En-vironment* 44:201–214.

Maul, A., El-Shaarawi, A.H., and Block, J.C. 1985b. Heterotrophic bacteria in water distribution systems. II. Sampling design for monitoring. *The Science of the Total Environment* 44:215–224.

McCoy, W.F. and Olson, B.H. 1986. Relationship among turbidity particle counts and bacteriological quality within water distribution lines. *Water Research* 20:1023–1029.

McNew, G.L. 1938. Dispersion and growth of bacterial cells suspended in agar. *Phy-topathology* 28:387–401.

Means, E.G., Hanami, L., Ridgway, H.F., and Olson, B.H. 1981. Evaluating mediums and plating techniques for enumerating bacteria in water distribution systems. *Jour. American Water Works Association* 73:585–590.

Nagy, L.A., Kelly, A.J., Thun, M.A., and Olson, B.H. 1983. Biofilm composition, formation and control in the Los Angeles aqueduct system, pp. 141–160 in Technology Con-ference Proceedings, WQTC-10, 1982, *The Laboratory's Role in Water Quality*, Amer-ican Water Works Association, Denver, Colorado.

Nagy, L.A. and Olson, B.H. 1986. Occurrence and significance of bacteria, fungi and yeasts associated with distribution pipe surfaces. pp. 213–238 in Technology Con-ference Proceedings, WQTC-14, 1985, *Advances in Water Analysis and Treatment*, American Water Works Association, Denver, Colorado.

Olson, B.H. and Hanami, L. 1981. Seasonal variation of bacterial populations in water distribution systems, pp. 137–150 in Technology Conference Proceedings, WQTC-

8, 1980, *Advances in Laboratory Techniques for Quality Control*, American Water Works Association, Denver, Colorado.

Olson, B.H. and Nagy L.A. 1984. Microbiology of potable water. *Advances in Applied Microbiology* 30:73–132.

Prescott, S.C., Winslow, C-E. A., and McCrady, M. 1950. *Water Bacteriology*, 6th edition. John Wiley and Sons, Inc., New York. 368 pp.

Rae, J.F. 1982. Algae and bacteria: dead-end hazard. pp. 233–244 in Technology Conference Proceedings, WQTC-9, 1981, *Advances in Laboratory Techniques for Quality Control*, American Water Works Association, Denver, Colorado.

Reasoner, D.J. and Geldreich, E.E. 1980. Significance of pigmented bacteria in water supplies. pp. 187–196 in Technology Conference Proceedings, WQTC-7, 1979, *Advances in Laboratory Techniques for Quality Control*, American Water Works Association, Denver, Colorado.

Reasoner, D.J. and Geldreich, E.E. 1979. A new medium for the enumeration and subculture of bacteria from potable water. Abstract N 7, p. 180, Abstracts of the Annual Meeting of the American Society for Microbiology, American Society for Microbiology, Washington, D.C.

Reasoner, D.J. and Geldreich, E.E. 1985. A new medium for the enumeration and subculture of bacteria from potable water. *Applied and Environmental Microbiology* 49:1–7.

Reasoner, D.J., Blannon, J.C., and Geldreich, E.E. 1987. Microbiological characteristics of third faucet point-of-use devices. *Jour. American Water Works Association* 79:59–66.

Reasoner, D.J., Blannon, J.C., Geldreich, E.E., and Barnick, J. 1989. Non-photosynthetic pigmented bacteria in a potable water treatment and distribution system. *Applied and Environmental Microbiology* 55:912–921.

Reilly, J.K. and Kippin, J.S. 1983. Relationship of bacterial counts with turbidity and free chlorine in two distribution systems. *Jour. American Water Works Association* 75:309–312.

Ridgway, H.F., Kelly, A., Justice, C., and Olson, B.H. 1983. Microbial fouling of reverse-osmosis membranes used in advanced wastewater treatment technology: chemical, bacteriological, and ultrastructural analyses. *Applied and Environmental Microbiology* 45:1066–1084.

Ridgway, H.F., Justice, C.A., Whittaker, C., Argo, D.G., and Olson, B. H. 1984. Biofilm fouling of RO membranes—its nature and effect on treatment of water for reuse. *Jour. American Water Works Association* 76:94–102.

Ridgway, J.W. 1978. The role of plate counts in the surveillance of potable water. Paper No. 3 in *Bacteriological Problems in Water Supply, A Water Research Seminar*, May 18, 1978, Medmenham Laboratory, Henley Road, Medmenham, P. O. Box 16, Marlow, Bucks, SL7 2HD.

Roszak, D.B. and Colwell, R.R. 1987. Survival strategies of bacteria in the natural environment. *Microbiological Reviews* 51:365–379.

Schoenen, D. 1986. Microbial growth due to materials used in drinking water systems. pp. 627–647 in Rehm, H. -J. and Reed, G. (editors), *Biotechnology* (Volume 8), VCH Verlagsgesellschaft, Weinheim, West Germany.

Silley, P. 1985. Evaluation of total-count samples against the traditional pour plate method for enumeration of total viable counts of bacteria in a process water system. *Letters in Applied Microbiology* 1:41–43.

Snyder, T.L. 1947. The relative errors of bacteriological plate counting methods. *Journal of Bacteriology* 54:641–654.

Stapert, E.M., Sokolski, W.T., and Northam, J.I. 1962. The factor of temperature in the better recovery of bacteria from water by filtration. *Canadian Journal of Microbiology* 8:809–810.

Stetzenbach, L.D., Kelley, L.M., and Sinclair, N.A. 1986. Isolation, identification and growth of well-water bacteria. *Groundwater* 24:6–10.

Taylor, R. H. and Geldreich, E. E. 1979. A new membrane filter procedure for bacterial counts in potable water and swimming pool samples. *Jour. American Water Works Association* 71:402–405.

Taylor, R.H., Allen, M.J., and Geldreich, E.E. 1983. Standard plate count: a comparison of pour plate and spread plate methods. *Jour. American Water Works Association* 75:35–37.

United States Environmental Protection Agency. 1975. Interim Primary Drinking Water Standards. *Federal Register* 40:11990–11996. Washington, D.C.

Van der Kooij, D. 1981. Multiplication of bacteria in drinking water. *Antonie von Leeuwenhoek Journal of Microbiology* 47:281–282.

Van Soestbergen, A.A. and Lee, C.H. 1969. Pour plates or streak plates. *Applied Microbiology* 18:1092–1093.

Ward, N.R., Wolfe, R.L., Means, E.G., and Olson, B.H. 1983. The inactivation of total count and selected gram-negative bacteria by inorganic monochloramines and dichloramines. pp. 81–97 in Technology Conference Proceedings, WQTC-10, 1982, *The Laboratory's Role in Water Quality*, American Water Works Association, Denver, Colorado.

Water Research Centre. 1976. Deterioration of bacteriological quality of water during distribution. *Notes on Water Research* No. 6, October.

Wolfe, R. L., Means, E. G. III, Davis, M. J., and Barrett, S. E. 1988. Biological nitrification in covered reservoirs containing chloraminated water. *Jour. American Water Works Association* 80:109–114.

23

Enumeration, Occurrence, and Significance of Injured Indicator Bacteria in Drinking Water

Gordon A. McFeters

Department of Microbiology, Montana State University, Bozeman, Montana

Indicator bacteria of sanitary significance in drinking water are considered allochthonous since they are usually transient and not long-term occupants of that kind of ecosystem. This concept accurately describes the status of most indicator bacteria including coliforms, as well as the enteropathogenic bacteria in aquatic systems, since they reproduce within the guts of animals and are not well adapted to the chemical and physical conditions in water. As a result, it is not surprising that these bacteria can become physiologically damaged with aquatic exposure.

The most prominent characteristic of injured waterborne indicator bacteria is an elevated sensitivity to selective media containing surface-active ingredients. For that reason, injury is defined as the sub-lethal physiological and structural consequence(s) resulting from exposure to injurious factors within aquatic environments. This is manifested by the inability of injured cells to reproduce under selective or restrictive conditions that are tolerated by uninjured cells. Consequently, water quality assessments can be inaccurate, since currently accepted media contain selective ingredients. Therefore, a medium that is nonselective or complete must be used as a reference in determining the extent of injury. This is illustrated in Figure 23.1, where a suspension of *Escherichia coli* was exposed to surface water using a membrane diffusion chamber (McFeters and Stuart, 1972). Similar results have also been seen for fecal streptococci (Bissonnette, 1974). Injury is seen as a progressive difference in enumeration efficiency using TSY (nonselective) and the selective media (Bissonnette, 1974; Bissonnette et al., 1975). It is usually assumed that both injured and noninjured cells are capable of growth on the nonselective reference medium, while only the noninjured bacteria can form colonies on se-

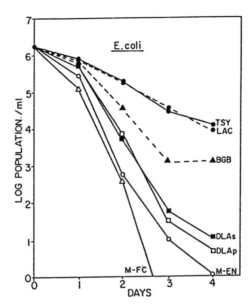

Figure 23.1   Death of *Escherichia coli* strain C320MP25 in a membrane filter chamber immersed in river water as shown by recovery in various media and counting procedures. Lactose broth–most probable number (LAC MPN) (●), Brilliant green lactose bile broth–most probable number (BGB MPN) (▲), deoxycholate lactose agar (DLA) surface-overlay plate (■), DLA pour plate (□), M-Endo MF membrane filtration (○), and M-FC membrane filtration (Δ) procedures during a 4-day exposure period. TSY counts (●) represent average counts on TSY medium when using MPN, pour plate, surface-overlay plate, and membrane filtration procedures.

lective media. The following equation has been used to quantify the degree of injury in specified populations (CFU, colony-forming units):

$$\% \text{ Injury} = \frac{\text{CFU reference} - \text{CFU selective}}{\text{CFU reference}} \times 100$$

Calculations to determine percentages of cells that are injured must be viewed as relative, since different selective media recover injured bacteria with varying efficiencies, as seen in Figure 23.1. However, this kind of calculation can provide useful information concerning the relative level of injury within a waterborne population since bacteria in natural systems are thought to respond to stress individually, as pointed out by Mossel and vanNetten (1984). That is, following stress the bacteria are present in at least three states: irreversibly inactivated, not injured, and those that are injured to varying degrees.

The definition and concept of injury includes the capability of debilitated cells to repair the cellular damage and regain tolerance for growth under restrictive or selective test conditions. This is illustrated in Figure 23.2. Suspensions of *E. coli* that were uninjured (0-time) and injured (2-days) were incubated in a nonselective medium and the recovery from injury followed as a convergence of the plate counts on TSY and DLA media, as well as the resumption of growth (Bissonnette, 1974). Therefore, injured coliforms are capable of resuscitation and can be enumerated if suitable media are used.

The purpose of this chapter is to discuss injured indicator bacteria in drinking water. It is not intended that the review of the literature will be all-inclusive, but much of the existing knowledge describing this phenomenon in drinking water will be presented. The development of methods to detect injured coliform bacteria in water will be presented with some information on the occurrence of stressed coliforms in potable water. Finally, the significance of injured bacteria will be discussed within the context of studies describing injury and its influence on properties relating to the virulence of some waterborne enteric pathogenic bacteria. Other reviews of this subject have also been published (LeChevallier and McFeters, 1984 and 1985; McFeters et al., 1987).

23.1　Causes of Injury in Drinking Water

A variety of chemical, physical, and biological factors associated with water have the potential to cause injury in enteric bacteria. This is particularly true in the case of potable water that has been treated with disinfectants. However,

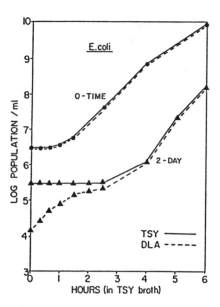

Figure 23.2.　Recovery from injury in TSY broth for *Escherichia coli* C320MP25 cells having been exposed to a stream environment for 2 days. Control of 0-h cells (Δ) and 2-day-exposed cells (▲) were enumerated over a 6 h growth period in TSY broth using TSY (———) and DLA (– – –) surface-overlay plates.

natural aquatic systems which may serve as source water can also be stressful environments for enteric bacteria due to the presence of subtle and often multiple antimicrobial factors. These environmental stressors may have an additive effect, as pointed out in the case of foods by Mossel and vanNetten (1984).

Disinfectants   Disinfectants and other biocides used to treat water probably represent the major cause of bacterial injury in aquatic systems (Hoadley and Cheng, 1974; Camper and McFeters, 1979; McFeters and Camper, 1983). While bacterial lethality is the intended consequence of water disinfection, the application of disinfectants in ways that lead to sub-optimal biocidal activity promotes injury. For example, the level of injured coliforms in treated drinking water can increase significantly when the disinfectant contact time is reduced for the purpose of minimizing trihalomethane formation (J. Kippin, personal communication). Variations in other discretionary operational variables in water treatment can, likewise, promote injury. In addition, the concentration of disinfectant usually varies widely in drinking water, from >1 mg/l at the point of application to below the level of detection at the extremities of the distribution system, and will fluctuate temporarily (O'Connor et al., 1975; Goshko et al., 1983). Thus, sub-optimal levels of disinfectant may occur during and after water treatment, leading to injury.

Chlorine is used widely as a disinfectant for both potable water and wastewater. As early as 1935, there was evidence that chlorine caused a reversible form of bacterial inactivation (Mudge and Smith, 1935). Schusner et al. (1971) convincingly described the reversible inactivation of *E. coli* by chlorine in foods, followed a few years later by two reports of the same process associated with tap water and wastewater (Braswell and Hoadley, 1974; Hoadley and Cheng, 1974).

Studies were initiated in our laboratories to characterize the process of chlorine injury under conditions similar to those in drinking water and study the recovery process from the physiological standpoint (Camper and McFeters, 1979; McFeters and Camper, 1983). The results of these studies revealed that reproducible injury could be attained at low chlorine concentrations (ca. 0.5 mg/l) and the process was a function of both concentration and time of exposure. We further demonstrated that there was an extended lag phase, similar to that seen in different examples of injury by others (Bissonnette et al., 1975; Schusner et al., 1971) when chlorine-injured bacteria were placed in a nonrestrictive medium to allow recovery and growth.

Alternative antimicrobial agents, such as ozone, chloramines, chlorine dioxide, and ultraviolet radiation, are gaining more attention for use in the disinfection of potable water because of growing concerns regarding toxic substances that are produced by chlorination (USEPA, 1975). At the present time, there is little specific information concerning the tendency of these treatments to injure bacteria.

Other stressors   The results of field studies performed in our laboratory suggested that metals may act as stressors in natural aquatic systems. The results of these studies revealed copper concentrations found in many municipal drinking water systems within the U.S. caused significant levels of injury (Domek et al., 1984). These findings support the addition of ethylenediametetraacetic acid (EDTA) to water samples to chelate metals prior to analysis, as suggested in *Standard Methods for the Examination of Water and Wastewater* (APHA, 1985) and by Domek et al. (1984).

A variety of physical factors are also of potential significance as stressors in potable and source water, although only a few of these have been investigated in detail. The work of Chamberlin and Mitchell (1979), Kapuscinski and Mitchell (1981), and Fujioka and Marikawa (1982) documented the injurious effects of ultraviolet irradiation on waterborne indicator bacteria. Acidic aquatic environments have also been implicated in causing injury in coliform bacteria (Hackney and Bissonnette, 1978; Double and Bissonnette, 1980).

Biological interactions provide another potential mechanism to explain the injury of enteric bacteria within aquatic systems. It is generally appreciated that excessive populations of noncoliform heterotrophic bacteria can suppress the enumeration of coliforms in water (Geldreich et al., 1972), although the precise mechanism is unclear. The possibility of biological interactions causing injury was investigated in our laboratory by examining the influence of pseudomonad populations on the enumeration efficiency and injury of a coliform. Significant levels of injury (ca. 55%) were seen when pseudomonads exceeded the coliforms by a ratio of greater than 100,000:1 (LeChevallier and McFeters, 1985).

23.2   Consequences of Injury

The cytoplasmic membrane is frequently described as the site of cellular damage resulting from sublethal injury caused by different stressors in aquatic environments (Mossel and Corry, 1977). This is because the physiological consequences of such stresses that have been reported have been the loss of functions mediated within the cytoplasmic membrane (Beauchat, 1978). These include compromised nutrient transport, oxidative phosphorylation, and loss of intracellular constituents. However, the overall cellular response is thought to be the cumulative effect of a number of inter-related events, as pointed out by Russell (1984). This explanation is reasonable, since the cytoplasmic membrane is the outermost cellular structure of major physiological consequence and would be the first target for reactive chemicals in the environment.

Studies within aquatic systems comparable to source and potable waters have revealed that the loss of a number of membrane-related functions was associated with injury. For instance, the physiological effects of the range of chlorine concentrations seen in drinking water were studied. These effects include greatly reduced nutrient transport and intracellular ATP concentrations (Camper and McFeters, 1979). The findings of other studies have provided further support for the hypothesis that the primary site of chlorine action is the inner membrane (Venkobacchar et al., 1977; Haas and Engelbrecht, 1980).

The physiological effects of copper-induced injury have also been studied because of the potential role of this metal for causing injury in drinking water. The results of these experiments revealed that injury from copper resulted in major metabolic alterations and reduced (>75%) aerobic respiration (Domek et al., 1984; Domek et al., 1987). These reports again suggest compromised physiological functions related to membrane damage following injury. Earlier studies by Zaske (1980) also examined the cellular consequences of metal-induced injury in natural surface water. This study demonstrated that progressive injury was accompanied by increased sensitivity to lysozyme.

Injury caused by other factors has also been investigated from the physiological perspective. Kapuschinski and Mitchell (1981), and Fujioka and Narikawa (1982) documented the injurious effects of sunlight on enteric bacteria and noted damage to the catalase system. Martin et al. (1976) reported evidence suggesting that low pH caused the same kind of damage in *E. coli*.

The manifestation of injury that is of the most practical importance is the elevated sensitivity of stressed bacteria to selective media (LeChevallier and McFeters, 1984 and 1985). This results in injured bacteria becoming incapable of growth on accepted media used in the microbiological determination of water potability (Bissonnette et al., 1977; LeChevallier et al., 1982; McFeters et al., 1982; LeChevallier and McFeters, 1984 and 1985). Therefore, the use of presently accepted media and methods to enumerate indicator bacteria in water can lead to an under-estimation of the actual number of viable bacteria present. Such data give an overly optimistic evaluation of water quality (McFeters et al., 1986), which could lead to a significant underestimation of the potential health risk. The significance of this hypothesis will be discussed in a subsequent section of this chapter.

23.3   Detection and Occurrence of Injured Indicator Bacteria in Drinking Water

Sublethally injured indicator bacteria are capable of recovering the ability to grow on selective media following incubation under suitable conditions (Bissonnette et al., 1977; LeChevallier et al., 1982; LeChevallier and McFeters, 1984 and 1985). This observation suggested that media and techniques to enumerate injured indicator bacteria could be improved.

Highly variable coliform recovery efficiencies from natural samples have been observed for some time (Hoadley and Cheng, 1974; Dutka and Tobin, 1976; Bissonnette et al., 1977). This finding may be explained by the data in Figure 23.1, which show a progressive discrepancy in efficiencies of enumeration for a variety of media as bacteria were exposed in a stressful aquatic environment. In addition, the results of another study examining the performance of a variety of selective media showed that the media most commonly used in the microbiological analysis of water (m-Endo and m-FC) were in the lower half of those tested for the recovery of coliforms injured in flowing drinking water (McFeters et al., 1982). Surface-active selective ingredients were the specific constituents of media responsible for this phenomenon, as reported by Schusner et al. (1971). This was confirmed in a later study (McFeters et al., 1982) that also demonstrated that coliforms injured in drinking water became uniquely sensitive to desoxycholate, while control cells were unaffected.

Most of the earliest efforts to devise media and methods for the enumeration of injured waterborne indicator bacteria concentrated on fecal coliforms. Section 920 in *Standard Methods for the Examination of Water and Wastewater* (APHA, 1985), entitled "Stressed Organisms," describes these analytical approaches. Most of these procedures involve an "enrichment" or "resuscitation" step in a nonselective medium and at a nonrestrictive temperature (Rose et al., 1975; Green et al., 1977; Lin, 1976; Stuart et al., 1976). A new medium recovered over three-fold more fecal coliforms from wastewater than m-FC when a resuscitation step was included (LeChevallier et al., 1984).

Media have also been developed to optimize the enumeration of total coliforms from drinking water. The earlier efforts resulted in a procedure that involved a two-hour preincubation of membrane filters on a pad saturated with lauryl tryptose broth (LTB), followed by 22 hours on m-Endo medium (McCarthy et al., 1961). A more recently developed medium, designated m-T7, was formulated specifically for the improved recovery of injured total coliforms in drinking water (LeChevallier et al., 1983) because of a number of problems with existing media. This medium will be included in Section 920 of the 17th edition of *Standard Methods for the Examination of Water and Wastewater* (APHA). The formulation of m-T7 is given in Table 23.1, and it is commercially available. It should be stressed that the addition of penicillin, after autoclaving, at a concentration of 1.0 mg/l, is important to provide sufficient selectivity. This medium was initially evaluated for the recovery of total coliforms from drinking water, since that was its intended use. m-T7 performed better than m-Endo, either with or without LTB preincubation, recovering 43% more coliforms in the initial testing of 67 surface and treated potable water samples. Subsequent testing of this medium with drinking water samples has revealed similar or better performance in most instances than was seen in the initial comparisons (J. Kippin and M. LeChevallier, personal communication). A number of additional new media are being proposed for the enumeration of total coliforms from drinking water but at the time of this writing, a study comparing the new media had not been completed. However, in one report (Rice, 1987), the efficiency of m-T7 and m-Endo to enumerate total coliforms in chlorinated sewage was comparable. Such studies using dilute sewage, rather than drinking water samples, are likely to yield results that do not accurately reflect the performance of these media in their intended role, since it has been documented that coliforms respond differently when isolated from treated drinking water when compared with other waters to which chlorine was added (LeChevallier et al., 1983). However, recent findings (Watters and McFeters,

Table 23.1 Formulation of m-T7 medium

Ingredient[a]	Quantity per liter of distilled water	
Proteose peptone 3	5	g
Yeast extract	3	g
Lactose	20	g
Tergitol 7 (25-percent solution)	0.4	ml
Polyoxyethylene ether W1	5	g
Brom thymol blue	0.1	g
Brom cresol purple	0.1	g
Agar	15	g
Penicillin G[b]	1.0	mg

The medium was autoclaved at 121°C for 15 min., and the final pH was adjusted aseptically to 7.4 with 0.1 N NaOH.

[a] All ingredients were manufactured by Difco Laboratories, Detroit, MI, except polyoxyethylene ether W1, penicillin G, and brom cresol purple, which was obtained from Baker Chemical Co., Phillipsburg, NJ.

[b] Penicillin G (1650 units) should be added to the medium after autoclaving (when tempered to 46–50°C). Media prepared with penicillin G should be used within one week when stored at 4°C.

unpublished results) showing that chloramine-injured coliforms recover better on media containing sodium sulfite help interpret the observations made by Rice (1987) since chlorine is converted largely to chloramines in sewage and other waters with elevated levels of organic constituents. In another study where over 600 drinking water samples from small communities were analyzed, m-T7 performed better than any of the other new media tested, while recovering coliforms twice as often as m-Endo (B. Caldwell, personal communication). The results of another recent study to evaluate a new medium indicated that m-T7 recovered higher numbers of chlorine injured *E. coli* than any of the other selective media tested (Freier and Hartman, 1987).

Laboratory practices can also influence the outcome of efforts to enumerate injured indicator bacteria. For example, the addition of sodium thiosulfate to neutralize residual chlorine and EDTA to chelate metals is important to prevent further cellular damage prior to analysis (APHA, 1985). In addition, it has been demonstrated that exposure in cold buffer containing low levels of organic materials (i.e., 0.1% tryptone) optimizes recovery when dilution is required (McFeters et al., 1982).

23.4   Occurrence of Injured Indicator Bacteria in Drinking Water

Data available on the incidence of injured indicator bacteria in drinking water is somewhat limited, although more information of this kind is becoming available. The results of surveys to detect injured coliforms within drinking water systems with m-T7 medium have been published (LeChevallier et al., 1982; LeChevallier and McFeters, 1985; McFeters et al., 1986). These data are summarized in Table 23.2, along with some unpublished findings. Although some site specific differences have been seen, the percentages of injured coliforms observed in the water of participating municipal drinking water systems ranged from 43% to 99%. Samples analyzed included water taken immediately after conventional treatment, during and after the backwash cycle, at various points in the distribution system and following the break, and subsequent repair and disinfection of a distribution main. The average counts of indicator bacteria detected using m-Endo LES medium in one study was <1.0 confirmed coliforms per 100 ml, while m-T7 yielded 5.7 to 67.5 confirmed coliforms per 100 ml. In some cases, such as in backwash water and following the repair and disinfection of a broken main, the majority of the samples giving positive results on m-T7 produced no detectable coliforms on m-Endo medium. The relatively high incidence of these false negative coliform analyses, shown in Table 23.2, offers a potential mechanism by which viable indicator bacteria can penetrate treatment barriers and gain access to the distribution system without being detected by conventional methods. Predictably, injury values in most of the chlorinated systems were high, while source and cistern waters appear to contain fewer injured coliforms. On the other hand, the use of m-T7 has not revealed injured coliforms in some disinfected systems (D. Smith and M. LeChevallier, personal communication). These findings indicate that injured bacteria may comprise the majority of coliforms in some operating drinking water systems and that these bacteria are largely undetected when accepted analytical methods are used.

With a growing awareness of the injury concept and the increasing availability of new methods and media that are capable of enumerating injured bacteria, more such data will become available. As discussed below, this information may be used by treatment plant operators for the early detection and control of microbiological problems.

23.5 Significance of Injured Bacteria in Water

It is difficult to reach a firm conclusion regarding the significance of injured enteric bacteria in water because the importance of uninjured coliforms and other indicator bacteria is often debated. However, it is possible to discuss this consideration from the perspective of the information presented here and common practice within the drinking water industry in the U.S.

Some dissatisfaction with the current indicator bacteria has been based upon instances where coliforms were undetected or found in very low numbers associated with outbreaks of waterborne diseases (Muller, 1964; Seligman and Reitler, 1965; Boring et al., 1971; The National Research Council, 1977; Craun, 1981). While it is clear that some such instances might be explained by disease caused by agents such as viruses that are more persistent than the indicator

Table 23.2 Injured coliforms in drinking water

Location/Date	Type of water	Number of samples	Mean Coliforms/100 ml		% Injury	% False negatives[a]
			m-Endo	m-T7		
Midwest US/1986	After filtration	9	1.1	3.9	71	55
Midwest US/1986	Distribution	13	2.4	7.9	69	23
Northeast US/1985–86	Distribution	86	1.9	4.8	64	54
Northeast US/1985–86	Raw	86	14.5	16.8	14	—
Northeast US/1985	During treatment	320	1.4	1.9	26	24
Caribbean/1985	Cisterns	13	15.2	20.5	26	—
East US/1986	After treatment	4	2.0[b]	209.0	99	100
Northwest US/1984–86	Small systems	552	62.0[c]	139.0[c]	55[d]	—
Northwest US/1984–86	Small systems	45	3.1	6.3	51	—
Northeast US/1986	After chlorinated backwash	7	0.1	1.5	92	14
West US/1986–87	After unchlorinated backwash	133	2.4	8.1	70	40
West US/1986–87	After chlorinated backwash	37	0.03	13.4	99	97
Southeast/1986	Raw (surface)	24	598.0	2828.0	79	—
Southeast/1986	Well (mineralized)	51	254.0	505.0	49	—
Southeast/1986	Well	122	2.0	3.0	33	—
Southeast/1986	Distribution	280	1.0	4.0	75	—

a "% False Negatives" represents the percentage of coliforms that failed to produce colonies of m-Endo medium but were enumerated on m-T7

b MPN values

c Values are % positive for coliforms

d Estimate

bacteria, injury might also be a factor obscuring the usefulness of indicator bacteria to signal health hazards in drinking water.

There are a number of justifications for the enumeration of injured coliforms and recognition of their significance. Most of these relate to treated drinking water where their presence is of regulatory importance. Hence, detection of coliforms is unwelcome to those responsible for the operation of the drinking water systems. Understandably, such individuals might view the use of methods that detect both injured and uninjured cells as a threat, since the practice could lead to coliform estimates that are as much as ten-fold greater than when accepted media and methods are used. However, the detection of the entire viable population of coliforms, including injured cells, provides an increased margin of safety by increasing analytical sensitivity and providing a greater capability to detect developing problems within the system. This could allow the initiation of remedial action earlier and before the difficulty becomes a crisis. Under these circumstances, such an analytical capability would provide the operator of the system a distinct advantage in the early detection of developing problems. This increased sensitivity could also be useful in epidemiological investigations of waterborne disease outbreaks. For example, it is possible that instances of waterborne morbidity, where coliforms are either undetected or detected in low numbers, might be explained in part by injured indicator bacteria. In such a case, the use of media that enumerate injured cells, such as m-T7, might show that coliforms are, in fact, present in those waters and provide a more sensitive and earlier warning of public health threats from pathogens.

There are also significant regulatory implications relating to injured coliforms. As pointed out previously, some within the drinking water industry regard the enumeration of injured coliforms in an adversarial context, since it could result in detection of higher numbers of coliforms that might lead to the possibility that a system would violate existing regulations. That fear, however, is largely without substance, since systems with high-quality drinking water are likely to see little change in coliform recoveries (Geldreich et al., 1978; Seidler et al., 1981). Another important regulatory concern relates to the adoption of the presence-absence (PA) concept (see Chapter 19 and Clark, 1985) into the Federal drinking water regulations. This could result in water quality test findings that would be erroneous in some circumstances. Sufficient latitude will be provided in the selection of the PA medium/method to allow the analyst to use a combination that would not detect injured cells. Such a circumstance could result in an absence (A) test result between 50% and >90% of the time when viable coliforms are actually present. This estimate, which is based upon the results presented earlier, indicates that such applications of the PA concept could lead to a significant level of false negative coliform tests, as shown in Table 23.2. On the other hand, if appropriate media/methods are used, more sensitive and accurate results would be obtained.

Published information indicates that pathogens, upon aquatic exposure and injury, retain virulence properties that may be expressed and pose a potential health hazard following recovery, as discussed in Chapter 17. This realization provides the rationale for giving injured but viable enteropathogens the same significance as freshly cultured cells since they can recover and remain a health threat. In addition, it must be remembered that other enteropathogenic bacteria are less sensitive to aquatic injury than coliforms. This finding underscores the importance of detecting injured coliforms as a valid indication of potential health hazards from enteropathogenic bacteria in water. Hence, the bacteria that are currently afforded public health significance in water, including coliforms and pathogens, are also meaningful when injured.

23.6   Conclusions

The findings discussed here support the following conclusions.

1. Injury of indicator and pathogenic bacteria can occur within drinking water systems.
2. Methods are available that effectively detect at least a proportion of injured coliforms.
3. The vast majority of coliforms within these systems may be undetected because sublethal stress leads to decreased detection on conventional media.
4. Pathogenic bacteria also become injured but at higher levels of stressing agents.

As a result, injured coliforms are of public health significance and their detection affords an added measure of sensitivity to assist in the early detection of treatment deficiencies or contamination within domestic potable water systems.

The author is particularly grateful to the following colleagues and former students: Gary K. Bissonnette, Susan C. Broadaway, Karen Bucklin, Bruce Caldwell, Anne K. Camper, Wiliam S. Dockins, Matthew J. Domek, Paulette Jakanoski, James J. Jezeski, Joyce S. Kippin, Mark W. LeChevallier, Helga Pac, Barry H. Pyle, Donald A. Schiemann, John E. Schillinger, Ajaib Singh, David G. Stuart, Tom Trock, and Susan K. Zaske. In addition, the encouragement of Edwin E. Geldreich, Robert H. Bordner, and Donald J. Reasoner is acknowledged. The technical and clerical assistance of Jerrie Beyrodt, Nancy Burns, Marie Martin, Pamela Blevins, Renée Cook and Debbie Powell is also appreciated.
Studies done in this laboratory were supported by grants from the U.S. Environmental Protection Agency (R807092, R805230, CR810015), U.S. Department of the Interior under the Water Resources Research Act of 1964 (Public Law 88–379) and administered through the Montana Joint Water Resources Research Center (OWRR A-092, OWRR A-099, OWRR B-035, B-040), the National Institute of Allergy and Infectious Disease of the Public Health Service (AI19089), the National Institute of Arthritis, Diabetes and Digestive and Kidney Disease of the Public Health Service (AM33510), the U.S. Geological Survey (14 08 0001 61493), and a grant from the American Water Works Association Research Foundation.

References

APHA, AWWA, and WPCF. 1985. *Standard Methods for the Examination of Water and Wastewater* (16th edition). Washington, D.C. 1268 pp.

Beauchat, L.R. 1978. Injury and the repair of gram-negative bacteria with special consideration of the involvement of the cytoplasmic membrane. *Advances in Applied Microbiology* 23:219–243.

Bissonnette, G.K. 1974. *Recovery Characteristics of Bacteria Injured in the Natural Aquatic Environment* (Ph.D. thesis). Montana State University, Bozeman, Montana. 161 pp.

Bissonnette, G.K., Jezeski, J.J., McFeters, G.A., and Stuart, D.G. 1977. Evaluation of recovery methods to detect coliforms in water. *Applied and Environmental Microbiology* 33:590–595.

Bissonnette, G.K., Jezeski, J.J., McFeters, G.A., and Stuart, D.G. 1975. Influence of environmental stress on enumeration of indicator bacteria from natural waters. *Applied Microbiology* 29:186–194.

Boring, J.R. (III), Martin, W.T., and Elliott, L.M. 1971. Isolation of *Salmonella typhimurium* from municipal water, Riverside, California, 1965. *American Journal of Epidemiology* 93:49–54.

Braswell, J.R. and Hoadley, A.W. 1974. Recovery of *Escherichia coli* from chlorinated secondary sewage. *Applied Microbiology* 28:328–329.

Camper, A.K. and McFeters, G.A. 1979. Chlorine injury and the enumeration of waterborne coliform bacteria. *Applied and Environmental Microbiology* 37:633–641.

Chamberlin, C.E. and Mitchell, R. 1978. A decay model for enteric bacteria in natural waters. pp. 325–348 in Mitchell, R. (editor), *Water Pollution Microbiology*, John Wiley and Sons, New York.

Clark, J.A. 1985. Performance characteristics of the presence-absence test. pp. 129–134 in AWWA, *Proceedings of AWWA Water Quality Technology Conference*, Denver, CO.

Craun, G.F. 1981. Disease outbreaks caused by drinking water. *Journal Water Pollution Control Federation* 53:1134–1138.

Domek, M.J., LeChevallier, M.W., Cameron, S.C., and McFeters, G.A. 1984. Evidence for the role of copper in the injury process of coliforms in drinking water. *Applied and Environmental Microbiology* 48:289–293.

Domek, M.J., Robbins, J.E., Anderson, M.E., and McFeters, G.A. 1987. Metabolism of *Escherichia coli* injured by copper. *Canadian Journal of Microbiology* 33:57–62.

Dutka, B.J. and Tobin, R.E. 1976. Study on the efficiency of four procedures of enumerating coliforms in water. *Canadian Journal of Microbiology* 22:630–632.

Freier, T.A. and Hartman, D.A. 1987. Improved membrane filtration media for enumeration of total coliforms and *Escherichia coli* from sewage and surface water. *Applied and Environmental Microbiology* 53:1246–1250.

Fujioka, R.S., and Narikawa, O.T. 1982. Effect of sunlight on enumeration of indicator bacteria under field conditions. *Applied and Environmental Microbiology* 44:395–401.

Geldreich, E.E., Nash, H.D., Reasoner, D.J., and Taylor, R.H. 1972. The necessity of controlling bacterial populations in potable waters: community water supply. *Journal of American Water Works Association* 64:596–602.

Geldreich, E.E., Allen, M.J., and Taylor, R.H. 1978. Interferences to coliform detection in potable water supplies, pp. 13–20 in Hendricks, C.W. (editor), *Evaluation of the Microbiology Standards for Drinking Water*, EPA-570/9-78-00C, U.S. Environmental Protection Agency, Washington, D.C.

Goshko, M.A., Pipes, W.O., and Christian, R.R. 1983. Coliform occurrence and chlorine residual in small water distribution systems. *Journal of American Water Works Association* 75:371–374.

Green, B.L., Clausen, E.M., and Litsky, W. 1977. Two-temperature membrane filter method for enumerating fecal coliform bacteria from chlorinated effluents. *Applied and Environmental Microbiology* 33:1259–1264.

Haas, C.W. and Engelbrecht, R.S. 1980. Physiological alterations of vegetative micro-organisms resulting from chlorination. *Journal Water Pollution Control Federation* 52:1976–1989.

Hackney, C.R. and Bissonnette, G.K. 1978. Recovery of indicator bacteria in acid mine streams. *Journal Water Pollution Control Federation* 52:1947–1952.

Hoadley, A.W. and Cheng, C.M. 1974. Recovery of indicator bacteria on selective media. *Journal of Applied Bacteriology* 37:45–57.

Kapuscinski, R.B. and Mitchell, R. 1981. Solar radiation induces sublethal injury in *Escherichia coli* in seawater. *Applied and Environmental Microbiology* 41:670–674.

LeChevallier, M.W. and McFeters, G.A. 1985. Interactions between heterotrophic plate count bacteria and coliform organisms. *Applied and Environmental Microbiology* 49:1338–1341.

LeChevallier, M.W. and McFeters, G.A. 1985. Enumerating injured coliforms in drinking water. *Journal of American Water Works Association* 77:81–87.

LeChevallier, M.W., Cameron, S.C., and McFeters, G.A. 1983. New medium for the recovery of coliform bacteria from drinking water. *Applied and Environmental Microbiology* 45:484–492.

LeChevallier, M.W. and McFeters, G.A. 1984. Recent advances in coliform methodology. *Journal of Environmental Health* 47:5–9.

LeChevallier, M.W., Jakanoski, P.E., and McFeters, G.A. 1984. Evaluation of m-T7 agar as a fecal coliform medium. *Applied and Environmental Microbiology* 48:371–375.

Lin, S.D. 1976. Membrane filter method for the recovery of fecal coliforms in chlorinated sewage. *Applied and Environmental Microbiology* 32:547–552.

Lisle, T. 1986. Comparison of standard media with respect to recovery of stressed bacteria. pp. 71–86 in American Water Works Association, Proceedings of AWWA Water Quality Technology Conference, Portland, OR.

Martin, S.E., Flowers, R.S., and Ordal, Z.J. 1976. Catalase: its effect on microbial enumeration. *Applied and Environmental Microbiology* 32:731–734.

McCarthy, J.A., Delaney, J.E., and Grasso, R.J. 1961. Measuring coliforms in water. *Water and Sewage Works* 108:238–243.

McFeters, G.A., LeChevallier, M.W., Singh, A., and Kippin, J.S. 1986. Health significance and occurrence of injured bacteria in drinking water. *Water Science and Technology* 18:227–231.

McFeters, G.A. and Stuart, D.G. 1972. Survival of coliform bacteria in natural waters: Field and laboratory studies with membrane-filter chambers. *Applied Microbiology* 24:805–811.

McFeters, G.A., Kippin, J.S., and LeChevallier, M.W. 1986. Injured coliforms in drinking water. *Applied and Environmental Microbiology* 51:1–5.

McFeters, G.A. and Camper, A.K. 1983. Enumeration of indicator bacteria exposed to chlorine. *Advances in Applied Microbiology* 20:177–193.

McFeters, G.A., Cameron, S.C., and LeChevallier, M.W. 1982. Influence of diluents, media, and membrane filters on the detection of injured waterborne coliform bacteria. *Applied and Environmental Microbiology* 43:97–103.

Mossel, D.A.A. and Corry, J.E.L. 1977. Detection and enumeration of sublethally injured pathogenic and index bacteria in foods and water processed for safety. *Alimenta* (Special Issue on Microbiology) 16:19–34.

Mossel, D.A.A. and Van Netten, P. 1984. Harmful effects of selective media on stressed microorganisms: Nature and remedies. pp. 329–371 in Andrew, M.H.E. and Russell, A.D. (editors), *The Revival of Injured Microbes*, Academic Press, London.

Mudge, C.S. and Smith, F.R. 1935. Relation of action on chlorine to bacterial death. *American Journal of Public Health* 25:442–447.

Muller, G. 1964. What are the lessons of the Hamburg flood for drinking water hygiene standards? *Archiv für Hygiene und Bakteriologie* 148:321–326.

National Research Council, The. 1977. *Drinking Water and Health*, Vol. 1. National Academy of Science, Washington, D.C. 939 pp.

O'Connor, J.T., Hash, L., and Edwards, A.B. 1975. Deterioration of water quality in distribution systems. *Journal of American Water Works Association* 67:113–116.

Rice, E.W., Fox, K.R., Nash, H.D., Read, E.J., and Smith, A.B. 1987. Comparison of media for recovery of total coliform bacteria from chemically treated water. *Applied and Environmental Microbiology* 53:1571–1573.

Rose, R.E., Geldreich, E.E., and Litsky, W. 1975. Improved membrane filter method for fecal coliform analysis. *Applied Microbiology* 29:532–536.

Russell, A.D. 1984. Potential sites of damage in microorganisms exposed to chemical or physical agents. pp. 1–18 in Andrew, M.H.E. and Russell, A.D. (editors), *The Revival of Injured Microbes*, Academic Press, London.

Schusner, D.L., Busta, F.F., and Speck, M.L. 1971. Inhibition of injured *Escherichia coli* by several selective agents. *Applied Microbiology* 21:41–45.

Seidler, R.J., Evans, T.M., Kaufman, J.R., Waarvick, C.E., and LeChevallier, M.W. 1981. Limitations of standard coliform enumeration techniques. *Journal of American Water Works Association* 73:538–542.

Seligman, R. and Reitler, R. 1965. Enteropathogens in water with low *Escherichia coli* titers. *Journal of American Water Works Association* 57:1572–1574.

Stevens, A.P., Grasso, R.J., and Delaney, J.E. 1974. Measurement of fecal coliforms in estuarine water. pp. 132–136 in Wilt, D.S. (editor), *Proceeding of the 8th National Shellfish Sanitation Workshop*, U.S. Department of Health, Education, and Welfare, Washington, D.C.

Stuart, D.G., McFeters, G.A., and Schillinger, J.E. 1977. Membrane filter technique for the quantification of stressed fecal coliforms in the aquatic environment. *Applied and Environmental Microbiology* 34:42–46.

U.S. Environmental Protection Agency. 1975. Water programs: secondary treatment information. *Federal Register* 40:34522.

Venkobacchar, C., Iyengar, L., and Rao, A.V.S.P. 1977. Mechanism of disinfection: effect of chlorine on cell membrane function. *Water Research* 11:727–729.

Zaske, S.K., Dockins, W.S., and McFeters, G.A. 1980. Cell envelope damage in *Escherichia coli* caused by short-term stress in water. *Applied and Environmental Microbiology* 40:386–390.

Appendix

The EPA Total Coliform and Surface Water Treatment Rules

The Environmental Protection Agency (EPA) has issued final rules governing the total number of coliform bacteria in drinking water and the treatment of surface and ground water sources used for drinking water. These rules take effect on December 31, 1990. Copies of the complete rules can be obtained by calling the Safe Drinking Water Hotline at 800-426-4791 and requesting a copy of the Federal Register for June 29, 1989 (40 CFR Parts 141 and 142).

The EPA Total Coliform Rule

Under this rule, the EPA requires a maximum containment level goal of *zero* for total coliforms, including both fecal coliforms and *Escherichia coli*. The total coliform rule applies to all public water supplies and describes the maximum levels for various contaminants, as well as monitoring and analytical requirements. The most important components of this rule are summarized below:

- Compliance is based on presence/absence of total coliform density, rather than an estimate of coliform density.
- Maximum containment level (MCL) for systems analyzing at least 40 samples/month: no more than 5% of the monthly samples may be coliform-positive.
- MCL for systems analyzing fewer than 40 samples/month: no more than one sample/month may be total coliform-positive.
- A public water system must demonstrate compliance with the MCL for total coliforms each month it is required to monitor.
- MCL violations must be reported to the state no later than the end of the next business day after the system learns of the violation.
- Each public water system must sample according to a written sample siting plan. Plans are subject to state review and revision.

- Monthly monitoring requirements are based on population served. Approximately one sample per 1000 population is required.
- A system must collect a set of repeat samples for each total coliform-positive routine sample and have it analyzed for total coliforms. At least one sample must be from the same tap as the original sample; other repeat samples must be collected from within five service connections of the original sample. At least one must be upstream and one must be downstream. All repeat samples must be collected within 24 hours of being notified of the original result, except where the state waives this requirement on a case-by-case basis.
- Repeat sampling requirements now also include additional samples collected the month following the original contaminated sample and check samples. The table below specifies sample numbers for this requirement.

Routine and repeat sampling requirements

No. of routine samples/ month	No. of repeat samples	No. of routine samples next month
1 or less	4	5
2	3	5
3	3	5
4	3	5
5 or more	3	As per normal

- If total coliforms are detected in any repeat sample, the system must collect another set of repeat samples, as before, unless the state has been notified and has reduced or eliminated the need for the remaining repeat samples.
- If any sample is total coliform-positive, the system must analyze that total coliform-positive culture to determine if fecal coliforms are present. If any repeat sample is fecal coliform positive, or if a fecal coliform-positive original sample is followed by a total coliform-positive repeat sample, the system is in acute violation of the MCL for total coliforms and the public must be notified by electronic media.
- If samples have heavy heterotrophic plate count (HPC) growth (non-coliform bacteria) but no coliforms are detected, the water system must be resampled for total coliforms.

The EPA Surface Water Treatment Rule

This rule promulgates national primary drinking water regulations for public water systems that use surface water sources or ground water sources under the direct influence of surface water. Ground water sources under the influence of surface water include those supplies which have significant occurrence of insects or other macroorganisms, algae, or large-diameter pathogens such as *Giardia*. Direct influence will be determined on a case-by-case basis and must be completed by June 1993 for community supplies.

The basis of the surface water treatment rule is to ensure supplies achieve at least 99.9 percent removal and/or inactivation of *Giardia* cysts, and at least 99.99 percent removal and/or inactivation of all viruses.

Public water systems which use filtration must report to the state, on a monthly basis, information regarding filtered water turbidity, disinfectant residual concentration in the water entering the distribution system, and disinfectant residual concentrations and/or HPC measurements in the distribution system. Systems must also report waterborne disease outbreaks, turbidity measurements over 5 NTU, and failure to maintain a disinfectant residual of 0.2 mg/l at the point of entry to the distribution system for more than 4 hours.

For both filtered and unfiltered surface water systems, turbidity measurements must be performed and recorded every 4 hours that the system serves water to the public. Conventional and direct filtration facilities must maintain filtered water turbidity levels less than or equal to 0.5 NTU at least 95 percent of the time. Unfiltered supplies and technologies such as slow sand filtration or diatomaceous earth filtration are allowed less than or equal to 1 NTU at least 95 percent of the time.

Criteria to avoid filtration are essentially as previously proposed. Systems which do not install filtration must meet strict raw water quality and monitoring requirements, watershed control agreements, and treatment by disinfection with an array of back-up systems and/or automatic system shut-down.

The Surface Water Treatment Rule also defines disinfectant concentration and contact time (CT) requirements.

Index